FUNDAMENTALS OF CIRCUIT ANALYSIS

FUNDAMENTALS OF CIRCUIT ANALYSIS

Kendall L. Su
Georgia Institute of Technology

WAVELAND PRESS, INC.
Prospect Heights, Illinois

For information about this book, write or call:

Waveland Press, Inc.
P.O. Box 400
Prospect Heights, Illinois 60070
(708) 634-0081

Copyright © 1993 by Waveland Press, Inc.

ISBN 0-88133-701-3

All rights reserved. No part of this book may be reproduced, stored in a retrieval system, or transmitted in any form or by any means without permission in writing from the publisher.

Printed in the United States of America

7 6 5 4 3

Contents

Preface   xiii

Chapter 1   Introduction   1
- 1.1   Basic symbols and notations   2
- 1.2   Physical quantities and units   4
  - 1.2.1   Current   4
  - 1.2.2   Potential difference or voltage   6
  - 1.2.3   Power   7
  - 1.2.4   Energy   9
- 1.3   Independent sources   9
- 1.4   Controlled sources   11
- 1.5   Ideal operational amplifier   13
- 1.6   Summary   15
-       Problems   15

Chapter 2   Simple Memoryless Circuits   19
- 2.1   The resistor and Ohm' law   19
- 2.2   Kirchhoff's laws   22
  - 2.2.1   Kirchhoff's current law (KCL)   24
  - 2.2.2   Kirchhoff's voltage law (KVL)   26
- 2.3   Number of independent voltages and currents   28
- 2.4   Analysis of simple resistive circuits   32
  - 2.4.1   Resistances in series   32
  - 2.4.2   Voltage-division rule   32
  - 2.4.3   Resistances in parallel   35
  - 2.4.4   Current-division rule   36
  - 2.4.5   Resistive circuits with both series and parallel combinations   38
- 2.5   Ladder and ladder-like analyses   40
- 2.6   Delta-wye and wye-delta transformations   46

2.7 Circuits with ideal op amps 50
    2.7.1 The noninverting voltage amplifier 50
    2.7.2 The voltage follower 51
    2.7.3 The inverting voltage amplifier 51
    2.7.4 The current-to-voltage converter 52
    2.7.5 The voltage-to-current converter 53
2.8 Duality 54
2.9 Summary 56
    Problems 57

Chapter 3 General Analysis Methods 75

3.1 General considerations 75
    3.1.1 Cutset analysis 76
    3.1.2 Loop analysis 77
    3.1.3 Alternative methods of choosing voltage variables 79
    3.1.4 Alternative methods of choosing current variables 81
3.2 Node analysis 83
    3.2.1 Networks with resistors and independent current sources 83
    3.2.2 Networks containing controlled current sources 90
    3.2.3 Networks containing voltage sources 93
3.3 Mesh analysis 99
    3.3.1 Networks with resistors and independent voltage sources 99
    3.3.2 Networks containing controlled voltage sources 102
    3.3.3 Networks containing current sources 105
3.4 Duality 111
3.5 Some concluding remarks 113
    Problems 114

Chapter 4 Network Properties and Theorems 129

4.1 Linear elements and networks 129
    4.1.1 Proportionality property 130
    4.1.2 Additivity property 131

CONTENTS

	4.1.3	Linearity property 132
4.2	Superposition theorem 133	
4.3	Equivalence of a linear two-terminal network 139	
	4.3.1	Determination of parameters of the equivalent circuits of a two-terminal network 147
	4.3.2	Norton's and Thévenin's equivalent circuits 152
	4.3.3	Thévenin's and Norton's theorems 157
	4.3.4	Determination of Norton's and Thévenin's equivalent circuits by external means 159
4.4	Source transformation 162	
4.5	Source shifting 169	
4.6	Tellegen's theorem 176	
	4.6.1	Conservation of power 178
	4.6.2	Tellegen's corollary 179
	4.6.3	Reciprocity theorem 182
4.7	Maximum power transfer 184	
4.8	Conclusions 187	
	Problems 187	

Chapter 5 Memory Elements 207

- 5.1 The capacitor 207
- 5.2 Energy stored in a capacitor 213
- 5.3 Series and parallel combinations of capacitors, voltage- and current-division rules 215
- 5.4 The inductor 220
- 5.5 Energy stored in an inductance 223
- 5.6 Series and parallel combinations of inductors, voltage- and current-division rules 225
- 5.7 Simple circuits containing inductors and capacitors 228
- 5.8 Singularity functions 232
  - 5.8.1 The unit step function 232
  - 5.8.2 The unit ramp function 236
  - 5.8.3 The unit impulse function 240
  - 5.8.4 Other singularity functions 245
- 5.9 Concluding remarks 249
  - Problems 250

Chapter 6 Simple Circuits with Memory Elements 261

- 6.1 Series RL circuit 261
  - 6.1.1 Step response 261
  - 6.1.2 Impulse response 267
  - 6.1.3 Source-free response 268
  - 6.1.4 Responses to other excitations 276
  - 6.1.5 Ramp response 278
- 6.2 Series RC circuit 279
  - 6.2.1 Step response 280
  - 6.2.2 Impulse response 282
  - 6.2.3 Source-free response 284
  - 6.2.4 Response to other excitations 286
  - 6.2.5 Ramp response 287
- 6.3 Thévenin's and Norton's equivalent of memory elements with initial energies 288
- 6.4 RL and RC parallel circuits 291
- 6.5 First-order op amp circuits 295
  - 6.5.1 The integrator 296
  - 6.5.2 The differentiator 296
  - 6.5.3 The lossy integrator 297
- 6.6 Responses of RLC circuits 299
  - 6.6.1 Step response of an RLC series circuit 299
  - 6.6.2 Impulse response of RLC series circuit 305
  - 6.6.3 Responses of an RLC parallel circuit 306
  - 6.6.4 Responses to other excitations 307
- 6.7 Other second-order circuits 309
- 6.8 Concluding remarks 313
  - Problems 314

Chapter 7 Steady-State Analysis of Circuits with Exponential and Sinusoidal Excitations 327

- 7.1 Complex numbers and arithmetic 328
  - 7.1.1 Rectangular form 328
  - 7.1.2 Exponential form 330
  - 7.1.3 Polar form 332
  - 7.1.4 Other terminology 333

CONTENTS

- 7.2 Analysis of circuits with exponential excitation 336
- 7.3 Steady-state analysis using complex amplitudes 340
    - 7.3.1 Relationship of complex amplitudes in a resistance 340
    - 7.3.2 Relationship of complex amplitudes in an inductance 341
    - 7.3.3 Relationship of complex amplitudes in a capacitance 342
    - 7.3.4 The impedance and the admittance 342
- 7.4 Interpretations and applications 347
- 7.5 Steady-state solution of circuits with sinusoidal excitations 353
- 7.6 Network relationships as functions of s 357
- 7.7 An alternative approach to relating networks with sinusoidal and exponential excitations 361
- 7.8 Summary 363
    - Problems 363

Chapter 8 Analysis of AC Circuits   369

- 8.1 The sinusoids 370
    - 8.1.1 Terminology of sinusoids 370
    - 8.1.2 Conversion of sinusoids 371
    - 8.1.3 Combination of sinusoids 374
- 8.2 The effective value of a periodic function 376
    - 8.2.1 The root-mean-square (rms) value of a periodic function 376
    - 8.2.2 The effective value of sinusoids 379
- 8.3 Phasors 381
- 8.4 Circuit analysis in the phasor domain 383
    - 8.4.1 The impedance and the admittance 383
    - 8.4.2 Relationship among phasors, phasor diagrams 385
    - 8.4.3 Series-parallel combinations 389
    - 8.4.4 Voltage-division and current-division rules 391
    - 8.4.5 Delta-wye and wye-delta equivalence 395
    - 8.4.6 General methods of analysis 397
- 8.5 Network theorems in the phasor domain 402
    - 8.5.1 Superposition 402
    - 8.5.2 Equivalent circuits of two-terminal networks

　　　　　　—Thévenin's and Norton's theorems　404
　　　8.5.3　Other theorems and techniques　406
8.6　Special aspects of ac circuits　406
8.7　The phasor circle diagrams　414
　　　8.7.1　General theory　414
　　　8.7.2　Impedance and admittance loci of simple branches
　　　　　　417
8.8　Resonance　426
　　　8.8.1　Series resonance circuit　427
　　　8.8.2　Parallel resonance circuit　431
8.9　Summary　432
　　　Problems　433

Chapter 9　Alternating-Current Powers　453

9.1　The power in a resistance　454
9.2　The power in an inductance　457
9.3　The power in a capacitance　459
9.4　The inductive and capacitive reactive powers　462
9.5　General ac powers　464
9.6　The wattmeter　467
9.7　The apparent power and the power factor　469
9.8　The complex power　471
9.9　Examples of various methods of ac power calculation
　　　472
9.10　Conservation of real and reactive powers　475
9.11　Power factor correction　479
9.12　Maximum power transfer　482
9.13　Summary　486
　　　Problems　487

Chapter 10　Two-Port Networks　497

10.1　One-port and two-port　498
10.2　Two-port parameters and matrices　501
10.3　Relationships among two-port parameters
　　　516
10.4　Circuit models of two-ports with known parameters
　　　524
10.5　Relationships in a terminated two-port　529

CONTENTS

10.6 Special two-ports 533
    10.6.1 Reciprocal two-port 533
    10.6.2 Symmetric two-port 534
    10.6.3 Tee and pi two-ports 535
    10.6.4 Padding 536
    10.6.5 Balanced two-port 539
    10.6.6 Balanced and unbalanced ladders 539
    10.6.7 Symmetric lattice 540
    10.6.8 Ideal transformer 543
10.7 Interconnection of two-ports 549
    10.7.1 Series-series combination 549
    10.7.2 Parallel-parallel combination 552
    10.7.3 Series-parallel combination 554
    10.7.4 Parallel-series combination 555
    10.7.5 Series and parallel combinations without ideal transformers 556
    10.7.6 Cascade combination 564
10.8 Concluding remarks 566
    Problems 567

Chapter 11 The Mutual Inductance 587

11.1 The magnetic coupling between two coils 587
    11.1.1 The equality of M_{12} and M_{21} 590
    11.1.2 The signs of the mutual terms 591
    11.1.3 Magnetic coupling among several coils 594
    11.1.4 Energy delivered to a pair of coupled coils 597
    11.1.5 Coefficient of coupling 599
    11.1.6 Mutual inductance in ac circuits 600
11.2 Illustrative examples 600
11.3 Equivalent circuits of coupled coils 607
11.4 Limiting cases of coupled coils 612
    11.4.1 The voltage transformer 612
    11.4.2 The current transformer 613
    11.4.3 The ideal transformer 614
11.5 Summary 614
    Problems 615

Chapter 12 Multiterminal Networks   623

12.1   The indefinite admittance matrix   623
12.2   Properties of the IAM   629
    12.2.1   The zero-sum property   629
    12.2.2   Current invariance   632
    12.2.3   The IAM of a network with no internal node   633
12.3   Modification and applications of the IAM   637
    12.3.1   Grounding of a terminal   637
    12.3.2   Reordering of terminals   640
    12.3.3   Terminal combination   641
    12.3.4   Parallel combination   643
    12.3.5   Interconnecting multiterminal networks   646
    12.3.6   Internalizing a number of terminals   648
12.4   Networks with multiterminal subnetworks   654
12.5   Relationship between a four-terminal network and a two-port   658
12.6   Summary   660
    Problems   660

Appendix A Matrix Algebra   677

A.1   Definitions   677
A.2   Algebraic rules of matrices   679
A.3   Special matrices   680
A.4   Some useful theorems   684
A.5   Matrix notation in a set of linear simultaneous equations   685
A.6   Partitioning of matrices   686

Appendix B Answers to Problems   689

Index   707

Preface

This text is primarily intended to be used in an introductory course for electrical and computer engineering majors. It is to serve as a text for a preparatory course for other, higher-level courses in several areas in electrical engineering. It is an expanded treatment of the first part of an earlier book, *Fundamentals of Circuits, Electronics, and Signal Analysis*. This text concentrates on the basic circuit-analysis part of its predecessor.

In the last two decades, there has been increased diversity in the organization of electrical engineering curriculums among colleges. In most schools, however, the basic circuits course is still retained, although the emphases placed and time devoted to this subject vary widely from school to school. It is my belief that this apparent variation in emphasis is more in form than in substance. Basic circuit theory remains the most fundamental body of knowledge and form of training for almost all electrical engineers.

This text is designed to be flexible enough that it can be used in a variety of curriculum structures. With some judicious choices of topics, it can be used in a course as short as one-quarter three-hour in length. If all topics are to be treated in detail, the material may require a two-quarter four-hour or a two-semester three-hour sequence.

This text assumes that the student has had a basic college physics course in electromagnetism as well as a standard college mathematics course in calculus. A course in differential equations in not a prerequisite.

With the wide availability of sophisticated calculators and personal computers, the "number crunching" part of circuit analysis is not emphasized in this book. For example, in analyzing a three-node circuit, the major effort is to write the three node equations. Once this is done, the solution of these equations is regarded as a routine matter in this book (although examples of standard methods of accomplishing this are given in early part of the text). We encourage students to employ reasonable means to bypass the tedium of manual computation, which adds little to the understanding of basic theory and principle, so they can concentrate on the broader view of the subject matters.

There are several widely-used softwares that can be adapted for circuit analysis. Many contemporary texts are tied to certain specific softwares. We purposely choose not to do this, not because there is anything wrong with learning to use standard softwares, but because we believe students will benefit more when this is done elsewhere in the curriculum. Learning to use a software that can analyze circuits does not help students understand circuit theory. In fact, it can be distracting.

On the other hand, prepared softwares have their places in an engineering curriculum. While familiarizing the student with some of these softwares, the focus should be on the interpretation of the results and the solution of some problems not easily handled manually. An electronic-circuit analysis program should be used as an adjunct to electronics courses. At Georgia Tech, softwares are used in laboratory courses, as well as other more advanced courses.

This text limits its scope to linear networks. The treatment of several topics differ substantially from most contemporary textbooks. Thévenin's and Norton's theorems are treated as two special cases of the general terminal characteristics of a two-terminal network. Tellegen's theorem is shown as an extension of Kirchhoff's laws first. Then it is used to show the conservation of power, including complex power. Also, the application of Tellegen's theorem to a certain class of networks leads to the reciprocity theorem. A separate chapter is devoted to the multiterminal networks. Several other classical topics—such as duality, source shifting, circle diagrams, two-port manipulation—are included in this volume as optional topics of instruction.

K. L. S.

Chapter 1

Introduction

Circuit analysis remains one of the major basic subjects in electrical engineering. Although, in recent years, much of the task of analyzing electric circuits has been relegated to computers, understanding the basic principles of circuit analysis remains the cornerstone of a career in electrical engineering. In this world of rapidly changing technology, the fundamentals have actually become more important in the education of an engineer. Hot topics and technologies-in-vogue may come and go, but the fundamentals remain the same. As new technology emerges, engineers frequently find themselves relying more and more on the basics.

Basic circuit theory is important not only as a tool used to analyze electric circuits, but also as an excellent educational vehicle to instill analytical discipline and foster logical thinking. In addition, it helps a student to develop good study habits. A mastery of this subject will lay a sound foundation on which to build a career in electrical engineering.

The tools developed and concepts expounded in this study are, to a great extent, common to a variety of systems, such as mechanical, acoustic, hydraulic, and communication systems. Hence, the techniques and principles covered here are often applicable to many other engineering problems.

We shall begin our study with a brief review of several physical quantities and then define the components used in electric circuits.

1.1 Basic symbols and notations

A circuit *element* is usually a mathematical model of a physical device. It represents the external electrical behavior of the device in mathematical terms. In representing a physical device by a circuit element, we almost always employ some approximation. Hence, it is extremely important to keep in mind the limited ability of the circuit element to represent its real-world counterpart accurately. Within these limitations, however, we shall regard these models as the *exact* representation of the corresponding device and apply all facilities and finesses at our disposal to attack the problem at hand.

We shall assume that the student has some familiarity, from physics courses, with the basic electromagnetic quantities listed in Table 1.1. We will further elaborate on some of these quantities as occasions arise. Certain frequently used prefixes that indicate multiples or submultiples are given in Table 1.2 for convenience.

Table 1.1 Some basic physical quantities

Quantity	Unit	Abbreviation
Time	second	s
Electric charge	coulomb	C
Electric current	ampere	A
Voltage (potential difference)	volt	V
Magnetic flux	weber	Wb
Energy	joule	J
Power	watt	W

Table 1.2 Prefixes for multiples and submultiples

Multiple or submultiple	Prefix	Abbreviation
10^{12}	tera	T
10^9	giga	G
10^6	mega	M
10^3	kilo	k
10^{-3}	milli	m
10^{-6}	micro	μ
10^{-9}	nano	n
10^{-12}	pico	p

INTRODUCTION

A *terminal* is simply a connecting point or junction in a circuit. The physical counterpart of a terminal may be either a terminal post or a soldered joint. It is represented by a small dot—solid or hollow—as shown in Fig. 1.1(a).

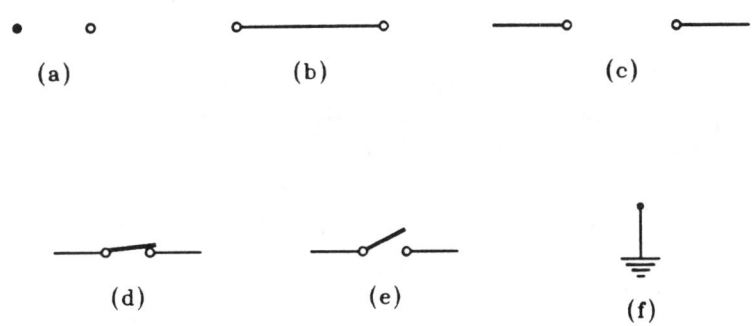

Figure 1.1: (a) Terminals, (b) short-circuited terminals, (c) open, (d) switch closed, (e) switch open, and (f) ground.

A *short circuit* (or simply *short*), is a path along which an electric current is free to flow. A short circuit may represent a highly conducting wire. It is also frequently used to connect points in a circuit that have the same potential. It is represented symbolically by a solid line, as shown in Fig. 1.1(b).

An *open circuit* is a condition in which no electric current can flow between two points. This situation is represented by the lack of a path, as illustrated in Fig. 1.1(c).

A *switch* connected between two terminals places a short circuit between the two terminals when it is *closed* as shown in Fig. 1.1(d), and an open circuit when it is *open* as shown in Fig. 1.1(e).

A *grounded terminal* or *ground* is a terminal or point whose absolute potential is assumed to be zero. A grounded terminal may be merely one whose potential is used for reference purposes. Or, it may be the representation of an actual grounding, achieved by physically connecting that point to earth. The symbol for a ground is shown in Fig. 1.1(f).

1.2 Physical quantities and units

The three basic physical quantities in the physical world are *mass* (usually in kilograms), *distance* (usually in meters), and *time* (usually in seconds). The units cited here are known as the KMS system. When a physical system includes electricity, another basic quantity, the *electric charge*, is added. Electric charge shall be denoted by q.

1.2.1 Current

A major purpose of a circuit is to serve as a conduit for charges to move along specific paths. The rate of motion of charge is an electric *current*. An electric current is denoted by i or I and is represented by the arrows as shown in Fig. 1.2. We shall use the symbol shown in Fig. 1.2(a). Many textbooks use that in Fig. 1.2(b). The student should realize that there is no difference between these two representations. It is strictly a matter of personal preference and habit.

(a)   (b)

Figure 1.2: Symbols for the current.

Mathematically, a current is the time rate of change or flow of electric charge, or

$$i = \frac{dq}{dt} \tag{1.1}$$

Hence, the current has the unit of coulombs per second. A coulomb per second is known as an ampere, abbreviated A.

Ordinarily, a current is carried through an electric path by the motion of electrons. Each electron carries a negative charge of 1.602×10^{-19} C. The direction of a current flow is the direction of the positive charge movement. Hence, physically, a current flow is in the opposite direction as that of the electron flow.

EXAMPLE 1. The electric charge $q(t)$ passing a certain junction is given in Fig. 1.3. Find the current $i(t)$ flowing through this junction.

INTRODUCTION

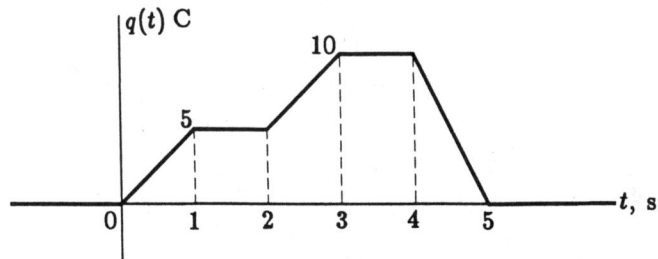

Figure 1.3: Charge $q(t)$ of Example 1.

SOLUTION Differentiating $q(t)$ of Fig. 1.3, we obtain the $i(t)$. The variation of $i(t)$ is shown below.

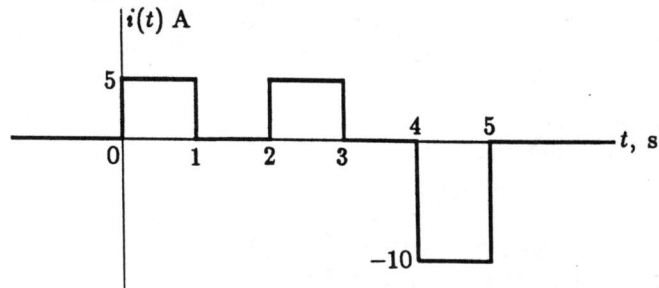

EXERCISE

1.2.1 Obtain the current corresponding to the charge $q(t)$ given.

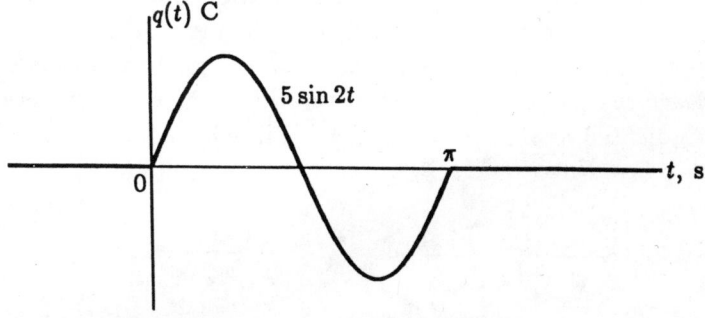

Ans. $10\cos 2t$ A, $0 < t < \pi$; 0, otherwise.

1.2.2 Potential difference or voltage

In an electric field, electric charges are subjected to a force. Thus, some energy must be expended to move a charge from one point to another. The energy or work required to move one coulomb of charge from point A to point B is known as the potential difference between points A and B. A potential difference is denoted by e, v, E, or V. The potential difference has the unit of joules per coulomb. This unit is known as a *volt*, abbreviated V. The absolute potential at a point A is the potential of point A with the ground as the reference and shall be denoted by E_A. The potential difference between two points, say A and B, is $E_A - E_B$ and is the work required to move a charge of one coulomb from B to A.

In electrical engineering, the potential difference is more commonly known as the *voltage*. A *voltage rise* from point B to point A is the amount of absolute potential by which point A exceeds point B. A *voltage drop* from point C to point D is the amount of absolute potential by which point C exceeds point D.

If point 1 is at 10 V and point 2 is at 6 V, we may describe the potential difference between these two points in any of the following ways.

The voltage rise from 2 to 1 is 4 V.
The voltage rise from 1 to 2 is −4 V.
The voltage drop from 1 to 2 is 4 V.
The voltage drop from 2 to1 is −4 V.

These four statements are all equivalent.

In a circuit, the voltage between two points is indicated by a pair of plus and minus signs as shown in Fig. 1.4(a). The voltage E denotes the voltage rise from the minus sign to the plus sign—the voltage rise from b to a, in this situation. The same status may also be indicated as shown in Fig. 1.4(b), in which the voltage rise from a to b is $-E$.

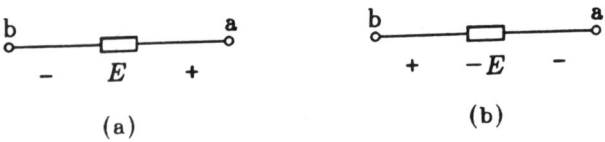

Figure 1.4: Symbols for voltages.

INTRODUCTION

Sometimes a double-subscript notation is used to denote the voltage between two points. In Fig. 1.4, the voltage rise from b to a may be denoted by E_{ab}. Thus, we may interpret the double-subscripted notation as $E_{ab} = E_a - E_b = E$. Accordingly, $E_{ba} = E_b - E_a = -E$. This convention is used a great deal in electronics. Another interpretation of the double-subscripted voltage E_{ab} is "the voltage of point a with point b as reference." As one can see, E_{ab} is also the voltage drop from a to b.

EXERCISE

1.2.2 We have $E_a = 5$ V, $E_b = 10$ V, $E_c = -2$ V, $E_d = 10$ V, and $E_e = -4$ V. Obtain E_1, E_2, E_3, E_{ba}, E_{ae}, E_{bc}, E_{bd}, and E_{cd}.

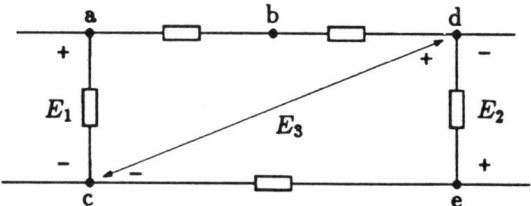

Ans. 7 V, -14 V, 12 V, 5 V, 9 V, 12 V, 0, -12 V.

1.2.3 Power

Fig. 1.5 shows a circuit element with its current and voltage labeled. The power *delivered to* that element at any instant is

$$p(t) = e(t)i(t) \qquad (1.2)$$

If e is in volts and i is in amperes, then p is in *watts*, abbreviated W.

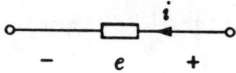

Figure 1.5: A general element with a current and a voltage.

When we apply Eq. (1.2), it is essential that we observe the relative directions of the voltage and current. The power is *delivered to* or *absorbed by* the element only when the current is flowing from the plus terminal to the minus terminal *through the element.*

If one of the quantities is reversed in direction while the other remains unchanged, then Eq. (1.2) gives the power *delivered by* or *supplied by* the element.

EXAMPLE 2. Calculate the power absorbed by the element in each of the situations in Fig. 1.6.

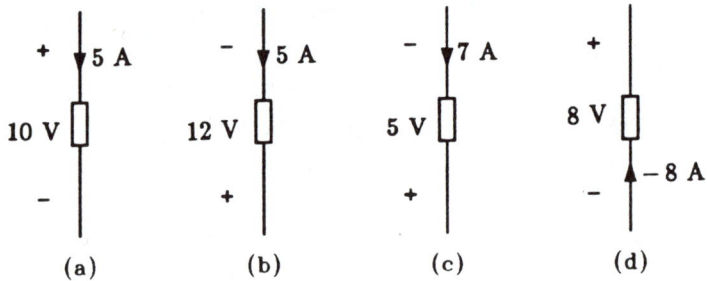

Figure 1.6: Arrangements used in Example 2.

SOLUTION (a) $p = 10 \times 5 = 50$ W
(b) $p = -12 \times 5 = -60$ W
(c) $p = -5 \times 7 = -35$ W
(d) $p = -8 \times (-8) = 64$ W

EXERCISE

1.2.3 Calculate the power supplied by each of the following elements.

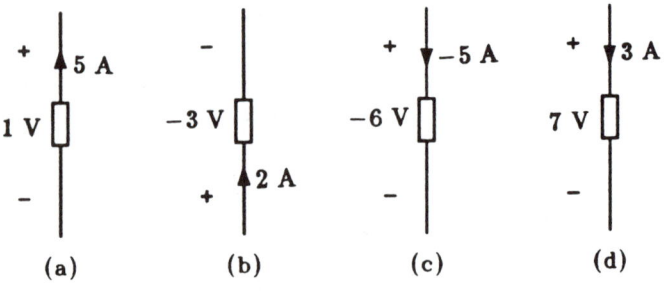

Ans. 5 W, 6 W, -30 W, -21 W.

INTRODUCTION

1.2.4 Energy

The power is the time rate of change of energy. Referring to Eq. (1.2), we have

$$p(t) = \frac{dw}{dt} = e(t)i(t) \qquad (1.3)$$

where $w(t)$ denotes the total energy delivered to the element, or

$$w(t) = \int_{-\infty}^{t} p(\lambda)\,d\lambda = \int_{-\infty}^{t} e(\lambda)i(\lambda)\,d\lambda \qquad (1.4)$$

It is understood that $w(-\infty) = 0$. Also, the total energy delivered to an element from $t = t_1$ to $t = t_2$ is

$$w(t_2) - w(t_1) = \int_{t_1}^{t_2} p(\lambda)\,d\lambda = \int_{t_1}^{t_2} e(\lambda)i(\lambda)\,d\lambda \qquad (1.5)$$

If the power is in watts and time is in seconds, then the energy is in *joules*, abbreviated J. A watt is one joules per second.

1.3 Independent sources

There are two types of independent sources—the independent voltage source and the independent current source.

An *independent voltage source* is an idealized circuit element that constrains the potential difference between the two points to which it is connected *under any circumstance*. The symbol for an independent voltage source is shown in Fig. 1.7(a), where $e_{ba}(t) = e_s(t)$ must always be satisfied. Note that the plus and minus signs are placed inside the circle for neatness.

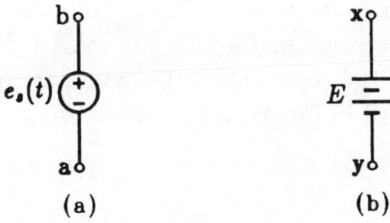

Figure 1.7: Symbols for the independent voltage source.

When $e_s(t)$ is a constant, say, E, it is a dc (direct-current) source and is sometimes called an (*ideal*) *battery* with the symbol shown in Fig. 1.7(b). The longer bar goes with the side of the source that is normally given the plus sign. Hence, in Fig. 1.7(b), $e_{xy}(t) = E$.

A *signal generator* or an *alternator* has characteristics that approach those of an ideal independent voltage source whose voltage varies sinusoidally with respect to time. A real battery has characteristics that are very close to those of a dc voltage source.

There is no restriction on the direction or amount of current that may flow through a voltage source. The direction and amount of the current flowing through a voltage source are determined entirely by the circuit to which the source is connected, not by the source itself.

A voltage source whose voltage is equal to zero is equivalent to a short circuit. Such a voltage source is said to be *idle* or *inactive*.

An *independent current source* in an idealized circuit element that constrains the current flowing through the branch in which it is inserted. The symbol for a current source is shown in Fig. 1.8, where $i_{\text{in}} = i_{\text{out}} = i_s(t)$ must always be satisfied.

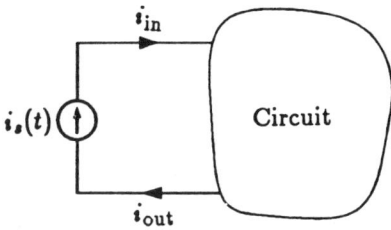

Figure 1.8: Symbol for the independent current source.

A current source whose current is equal to zero is equivalent to an open circuit. Such a current source is said to be *idle* or *inactive*. The voltage across a current source can be of either polarity and of any value and are determined, not by the source, but by the external circuit to which is it connected. Constant-current independent sources are often used in integrated circuits for biasing.

INTRODUCTION

EXERCISE

1.3.1 Calculate the powers *supplied by* the sources.

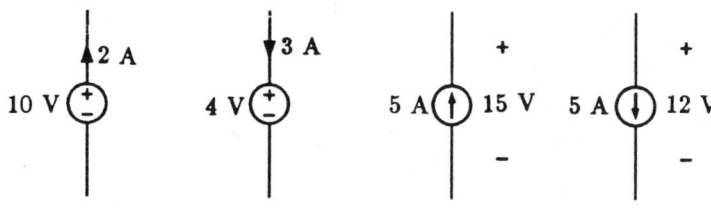

Ans. 20 W, −12 W, 75 W, −60 W.

1.4 Controlled sources

A *controlled* (or *dependent*) *source* is a voltage or current source whose strength is a function of another electrical quantity. The symbols for controlled sources are diamond-shaped boxes, as shown in Fig. 1.9. In some textbooks, the same symbol used for an independent source is also used for the controlled source. Philosophically, two distinct symbols are not necessary. A controlled source is indistinguishable from an independent one except that the strength of the controlled source is related to some other electrical quantity, while that of the independent one is not. However, for practical purposes, a special symbol—such as the diamond-shaped boxes—helps to call our attention to the fact that the source is not independent.

There are four possible types of controlled sources:

1. *Voltage-controlled voltage source* [Fig. 1.9(a)]. The ratio between the controlled voltage and the controlling voltage, μ, is known as the voltage amplification factor, the voltage gain, or simply the voltage ratio. This controlled source may be identified with the *ideal voltage amplifier*.

2. *Current-controlled current source* [Fig. 1.9(b)]. The ratio between the controlled current and the controlling current, α, is known as the current amplification factor, the current gain, or simply the current ratio. This controlled source may be identified with the *ideal current amplifier*.

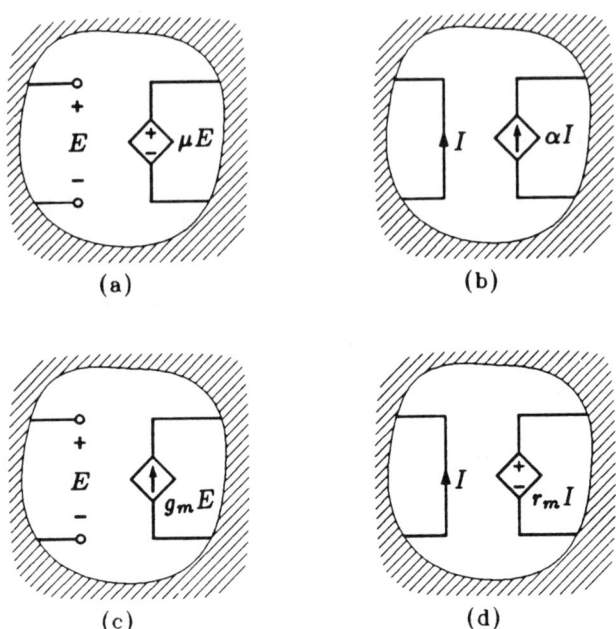

Figure 1.9: Symbols for controlled sources.

3. *Voltage-controlled current source* [Fig. 1.9(c)]. The ratio between the controlled current and the controlling voltage, g_m, has the dimension of the conductance[1] and is known as the transconductance or the mutual conductance. This controlled source is also known as a *voltage-to-current transducer or converter*.

4. *Current-controlled voltage source* [Fig. 1.9(d)]. The ratio between the controlled voltage and the controlling current, r_m, has the dimension of a resistance[2] and is known as the mutual resistance. This controlled source is also known as the *current-to-voltage transducer or converter*.

In these descriptions, all the proportionality constants—μ, α, g_m, and r_m—are tacitly assumed to be real constants (either positive or negative). Later, in ac circuits, these constants will be complex. For the majority of our circuit applications, this assumption is both re-

---
[1] See Section 2.1.
[2] See Section 2.1.

alistic and convenient. These controlled sources are both linear and memoryless. However, there are occasions in which these quantities may be either nonlinear or time-varying. Also, in other situations, some of these devices may become memory devices. For example, a controlled quantity may be proportional to the derivative or integral of the controlling quantity. We will limit ourselves to the linear type of controlled sources in this volume.

In Fig. 1.9(a) and (b), the unit of the controlling and the controlled quantities are the same. Hence, μ and α are dimensionless scalars. In Fig. 1.9(c) and (d), the controlling and the controlled quantities are of the opposite type—one is a voltage and the other a current; thus, the units of the controlled quantities should be specified. In this volume, in order to avoid cluttering up the circuit diagrams, unless specifically indicated, *the unit of every controlled voltage shall be understood to be volts, and that of every controlled current shall be understood to be amperes.* In other words, the *default* units of controlled sources are volts and amperes.

EXERCISE

1.4.1 Calculate the power supplied by the controlled sources if $E = 3$ V and $I = 6$ A.

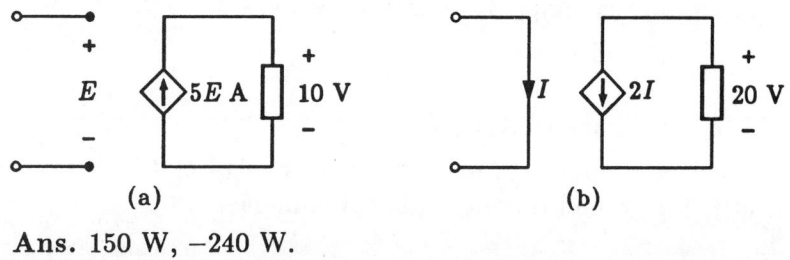

(a)   (b)

Ans. 150 W, −240 W.

1.5 Ideal operational amplifier (op amp)

One of the most widely used devices in electronic circuits is the *operational amplifier*, (also known as the *op amp*). Originally developed to perform certain operations, such as differentiation and integration, in analog computers, op amps find numerous applications in many areas of electrical engineering.

It is not our purpose to deal in any depth on the subject of op amps. We will accept it as a basic circuit building block in its idealized form, much like we accept voltage and current sources as idealized elements. Many students will be dealing with this device in greater detail in later courses. The study of op amps and their applications is a specialized area in itself.

As far as we are concerned here, an ideal op amp is a limiting case of a voltage-controlled voltage source. In the arrangement shown in Fig. 1.10, we have two input terminals. The plus terminal is known as the *noninverting* terminal. The minus terminal is known as the *inverting* terminal. The output voltage is $A_0(e_+ - e_-)$ in which all voltages are referred to ground.

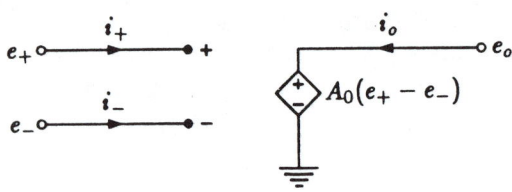

Figure 1.10: Arrangement of a controlled source that leads to the ideal op amp.

In an ideal op amp, $A_0 \to \infty$ and we have:
1. $i_+ = 0$ and $i_- = 0$.
2. $e_+ = e_-$.
3. e_o and i_o are arbitrary.

An ideal op amp is represented by the symbol of Fig. 1.11. Frequently, the ground is omitted, but it is implied.

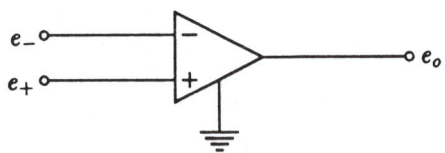

Figure 1.11: Symbol for an ideal op amp.

INTRODUCTION

1.6 Summary

We have introduced the basic physical and electrical quantities, their units, and how some of them are related to one another. We introduced the idea of an element in a circuit. The various types of sources, which are capable of supplying power to the rest of the circuit, are defined. These elements are idealized models of some practical power or energy sources. The op amp is also an idealized model of actual op amps that are in wide use in integrated circuits.

After the introduction of one more element in the next chapter, we are ready to interconnect some of these elements to form circuits and develop some methods for analyzing them.

Problems

1.1 The charge $q(t)$ passing a certain junction in a circuit is given in the graph. Give a dimensioned sketch of the associated current $i(t)$.

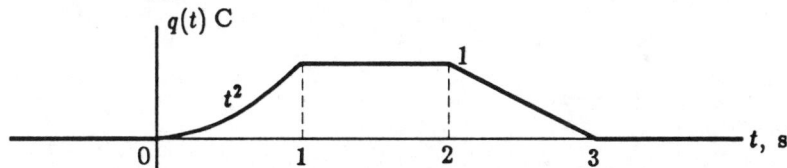

1.2 Give a dimensioned sketch for both the power delivered to the element and the total energy delivered to the element.

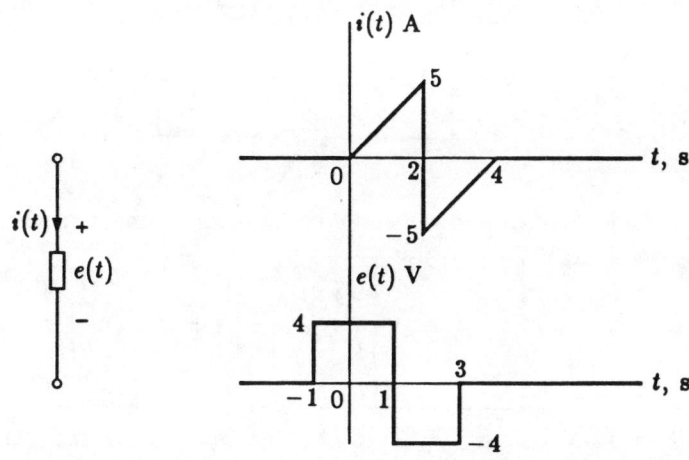

△ **1.3** The current at a certain junction in a circuit is $i(t) = \sin 10t$ A. What is the net charge passing through this junction between $t = 1$ and $t = 2$ s?

1.4 A charge, $q(t)$, is entering an element whose voltage is $e(t)$. The variations of both are given. Further, the direction of charge flow is such that it enters the element at the plus terminal of $e(t)$. Calculate the energy delivered to the element from $t = 0$ to $t = 6$ s.

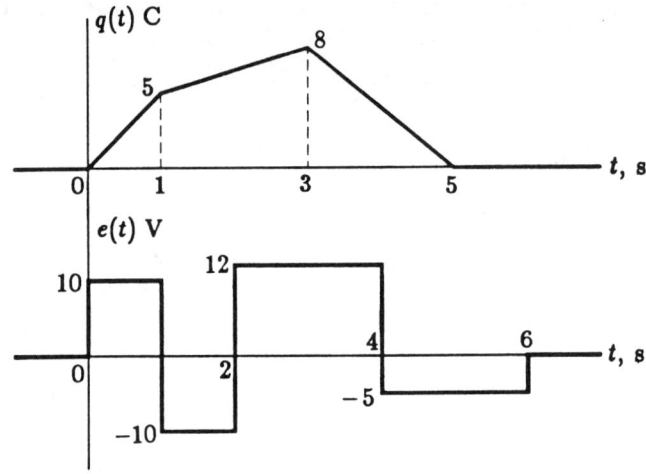

△△ **1.5** Calculate the power *delivered to* each element in the circuit.

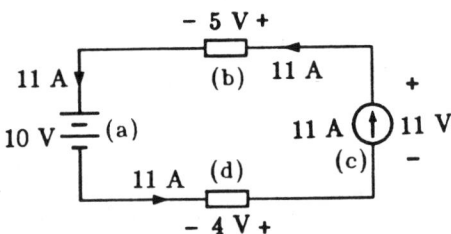

1.6 Calculate the power *supplied by* each element in the circuit. $E = 10$ V.

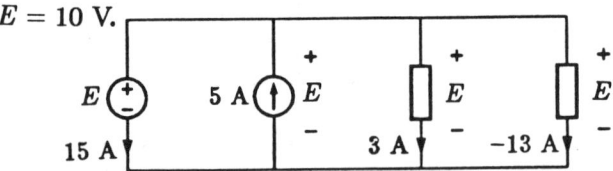

1.7 A 12-V battery is being charged by the current $i(t) = 100\times$

$\epsilon^{-5\times 10^{-4}t}$ A, starting at $t = 0$. Current $i(t)$ flows from the plus terminal to the minus terminal of the battery. Calculate the total charge delivered by the charger after one hour. Also, calculate the total energy gained by the battery after one hour.

▲▲ **1.8** The amount of charge stored in the battery is shown. Calculate the power supplied by the battery at $t = 100$, 300, 500 s.

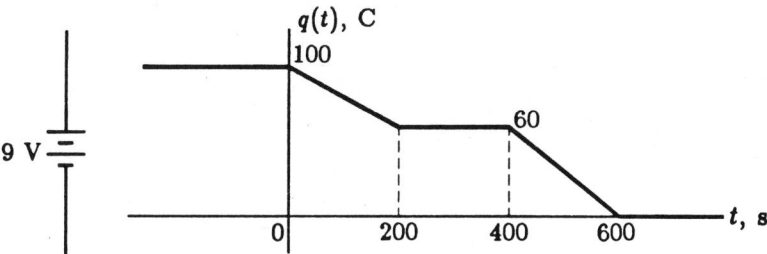

1.9 Calculate the power absorbed by each element.

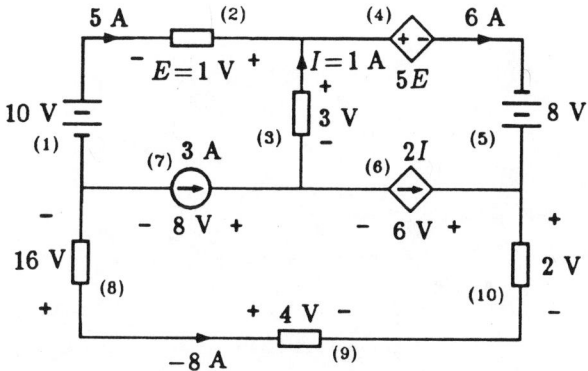

1.10 Discuss any inconsistency that may exist in each of the three arrangements shown.

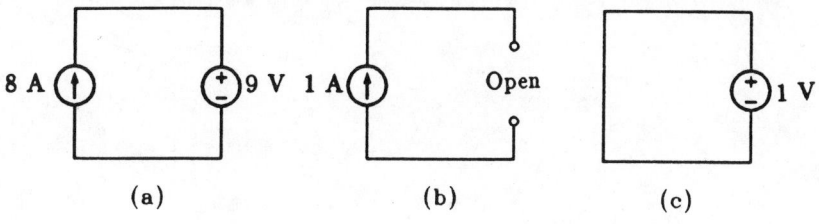

Chapter 2

Simple Memoryless Circuits

In addition to the various sources, including the op amp, introduced in the previous chapter, we now introduce one additional element—the resistor. We shall deal with certain simple circuits containing only sources and resistors in this chapter. These circuits, although simple, are both important and useful. The analysis of these relatively simple circuits provides us with a good starting point from which to begin the subject of circuit analysis.

2.1 The resistor and Ohm's law

A resistor is represented by the symbol shown in Fig. 2.1 and it must obey the relationship

$$e(t) = R\,i(t) \qquad (2.1)$$

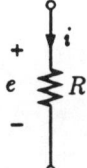

Figure 2.1: The symbol of a resistor.

which is known as *Ohm's law*. The proportionality constant, R, is known as the *resistance* of the resistor. When e is in volts and i is in amperes, R has the unit of ohms (frequently represented by the capital Greek letter omega, Ω.)

Eq. (2.1) is sometimes written as

$$i(t) = \frac{1}{R} e(t) = G\, e(t) \qquad (2.2)$$

where $G = 1/R$ is known as the *conductance* of the resistor. The unit for the conductance is the reciprocal of the ohm and is known as a mho frequently represented by the inverted capital Greek letter omega, $\mho$.[1]

A resistor that obeys Ohm's law is known as a linear time-invarying (LTI) resistor. It is linear in that its e-i curve is a straight line passing through the origin. It is time invarying when its resistance does not vary with time. We shall limit our scope to this type of resistor.

In applying Ohm's law, it is essential that we pay close attention to the relative directions of the voltage and current. In Fig. 2.1, the current flows *through the resistor* from the plus end to the minus end of the voltage polarity. We shall refer to this relative direction convention as the *standard Ohm's law* convention. In more complex circuits and in some analysis procedures, some of the assumed currents could very well happen to flow from the minus end to the plus end of some resistors. In that case, a negative sign must be placed on one side of Eq. (2.1) and Eq. (2.2).

Fig. 2.2 shows four possible relative directions of the voltage and the current for a given resistor. In (a) and (d), the voltages and currents follow the standard Ohm's law convention. Therefore

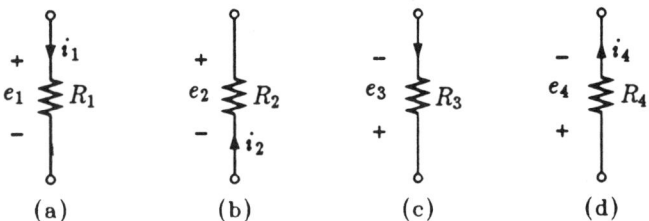

Figure 2.2: Relative voltage and current directions in a resistor.

---

[1] The international unit for conductance is the *siemen*, represented by the capital letter S. However, in the U.S., the mho is still widely used. We choose to retain the mho. The student should realize that a siemen and a mho are synonymous.

SIMPLE MEMORYLESS CIRCUITS

$$e_1 = R_1 i_1 \qquad e_4 = R_4 i_4 \tag{2.3}$$

However, in (b) and (c), the voltages and currents are contrary to the standard Ohm's law convention. Hence, it is necessary to write

$$e_2 = -R_2 i_2 \qquad e_3 = -R_3 i_3 \tag{2.4}$$

EXAMPLE 1. Obtain the unknown quantity in each resistor shown in Fig. 2.3.

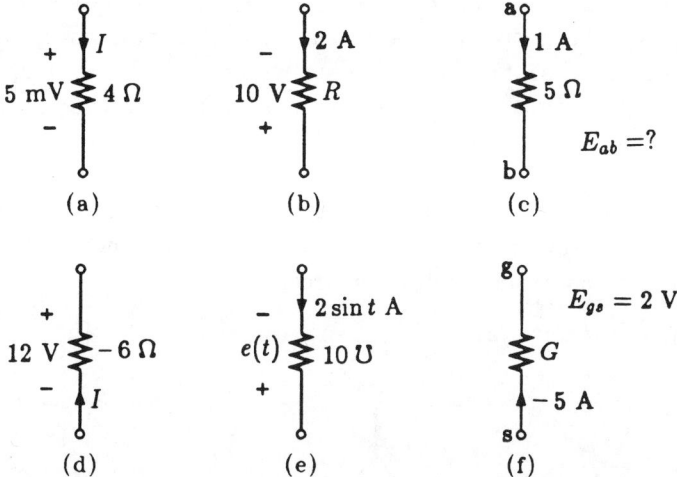

Figure 2.3: Arrangements used in Example 1.

SOLUTION

(a) $I = \dfrac{5 \text{ mV}}{4 \, \Omega} = 1.25 \text{ mA}$

(b) $R = -\dfrac{10 \text{ V}}{2 \text{ A}} = -5 \, \Omega$

(c) $E_{ab} = 5 \times 1 = 5 \text{ V}$

(d) $I = -\dfrac{12 \text{ V}}{-6 \, \Omega} = 2 \text{ A}$

(e) $e(t) = -\dfrac{2 \sin t \text{ A}}{10 \text{ U}} = -0.2 \sin t \text{ V}$

(f) $G = \dfrac{-5 \text{ A}}{-2 \text{ V}} = 2.5 \text{ }\mho$

The power delivered to a resistor with a resistance R is

$$p(t) = e(t)i(t) = R\,i^2(t) = G\,e^2(t) \tag{2.5}$$

Therefore, the power delivered to a resistor with a positive resistance is nonnegative. The power delivered to a resistor is consumed and dissipated as thermal energy or heat.

In Example 2, we show two circuits that simulate negative resistances. A negative resistance can also be simulated by other electronic circuits. Ordinarily, when we mention a resistor, we imply that it has a positive resistance unless it is specifically stated otherwise.

EXAMPLE 2. (a) Simulate a resistance of 5 Ω using a voltage-controlled current source. (b) Simulate a resistance of -10 Ω using a current-controlled voltage source.

SOLUTION (a) (b)

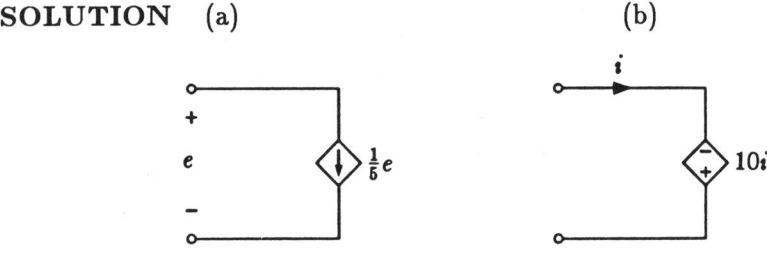

In Eq. (2.1), if the resistance is negative, then the resistor receives negative power. In other words, the resistor actually supplies positive power to the rest of the circuit. This should not be surprising if one recalls the discussion of the power delivered to an element in Subsection 1.2.3. A resistor with a negative resistance is a power-supplying device.

2.2 Kirchhoff's laws

Before we discuss the basic circuit laws, we should define a few terms.

SIMPLE MEMORYLESS CIRCUITS

A *node* is a terminal to which more than one element is connected.

A *branch* is any subnetwork that is connected to the rest of the circuit at only two terminals. Usually, a branch consists of only a single element, but sometime a combination of several elements as a whole may be considered a branch. In general, the internal complexity of a branch is arbitrary.

A *loop* is a closed path traced along a sequence of branches.

A *mesh* is a loop whose interior is empty.

A *cutset* is a minimum set of branches, when cut, divides the network into two subnetworks. When a *closed surface* (CS) is drawn to intersect a number of branches, those branches form a cutset. A closed surface becomes a closed boundary in a two-dimensional drawing.

In Fig. 2.4, terminals 1, 2, 3, and 4 are nodes. Whether or not

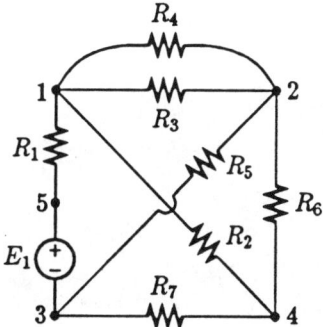

Figure 2.4: A simple circuit.

terminal 5 is considered a node is optional. If terminal 5 is to be a node, then source E_1 and resistor R_1 both become branches. If not, the combination of E_1 and R_1 will form a branch. Similarly, R_3 and R_4 may each be considered a branch. Or, the combination of them may be considered a single branch.

The circuit of Fig. 2.5(a) appears to have eight nodes. One can indeed consider the circuit to have eight nodes. Electrically, however, there are only two nontrivial nodes. This becomes more evident if we redraw the network as shown in Fig. 2.5(b). Alternatively, we may shade all points that must have the same potential, as we do in Fig. 2.5(c). Each shaded area may be thought of as an elongated node.

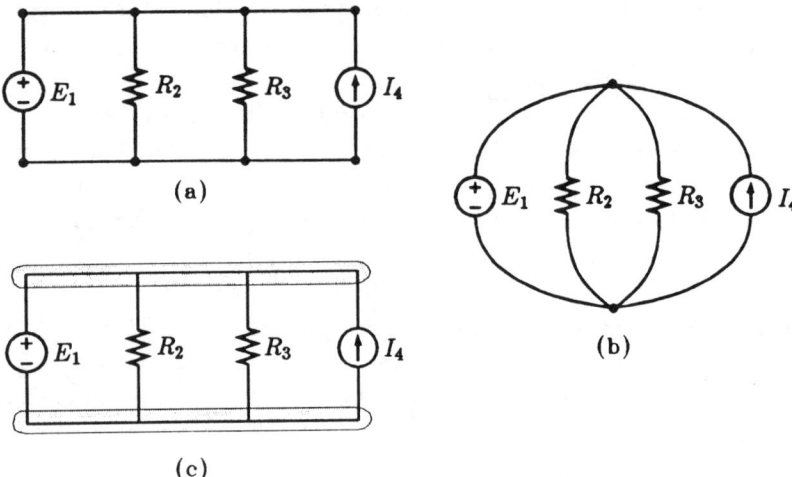

Figure 2.5: Different ways of drawing a two-node circuit.

Several loops are indicated for the circuit of Fig. 2.6. Loops 3 and 4 are also meshes.

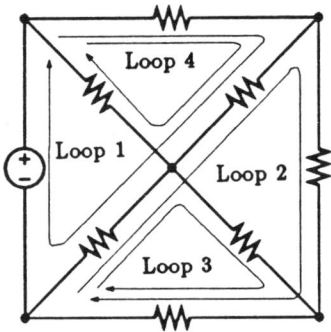

Figure 2.6: Examples of loops in a circuit.

2.2.1 Kirchhoff's current law (KCL)

In a circuit, charges cannot accumulate, disappear, or be created. This is known as the conservation of electric charge. KCL states that

> The algebraic sum of all currents entering any closed surface at any instant must be zero.

SIMPLE MEMORYLESS CIRCUITS

As an example of this law, consider the closed surface shown in Fig. 2.7. KCL requires

$$i_1 - i_2 + i_3 - i_4 = 0 \tag{2.6}$$

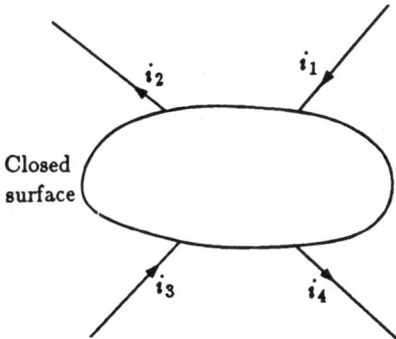

Figure 2.7: Example of the application of KCL.

Eq. (2.6) may also be written as

$$i_1 + i_3 = i_2 + i_4 \tag{2.7}$$

Eq. (2.7), in effect, states that the total current entering a closed surface must be equal to the total current leaving the closed surface.

EXAMPLE 3. Find all unknown currents in Fig. 2.8

SOLUTION   For CS 1,

$i_1 = 5 - 2 = 3$ A

For CS 2,

$i_2 = 5 - 2 - 4 = -1$ A

For CS 3

$i_3 = 5 - 8 - 4 = -7$ A

For CS 4,

$i_4 = -8 - 4 = -12$ A

Note CS 1 and CS 4 enclose only one node each. KCL certainly must be satisfied at every node. For example, at node 3 of Fig. 2.8, $-1 + 2 = -7 + 8$.

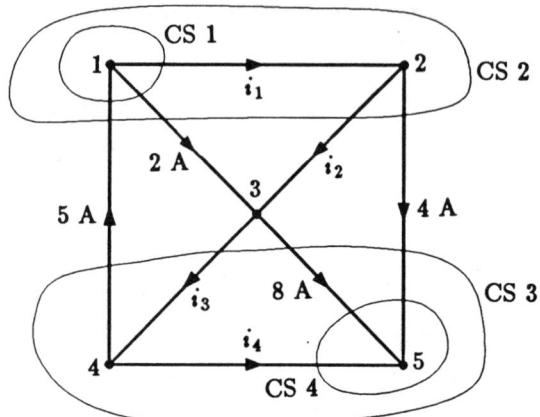

Figure 2.8: Arrangement used in Example 3.

2.2.2 Kirchhoff's voltage law (KVL)

Since the potential difference between a point and itself must be zero (this is known as the conservation of electric potential), we have Kirchhoff's voltage law, which states:

> The algebraic sum of all voltages around any loop at any instant must be zero.

As an example of this law, consider the loop in that part of a circuit shown in Fig. 2.9. KVL requires

$$e_1 - e_2 - e_3 + e_4 = 0 \tag{2.8}$$

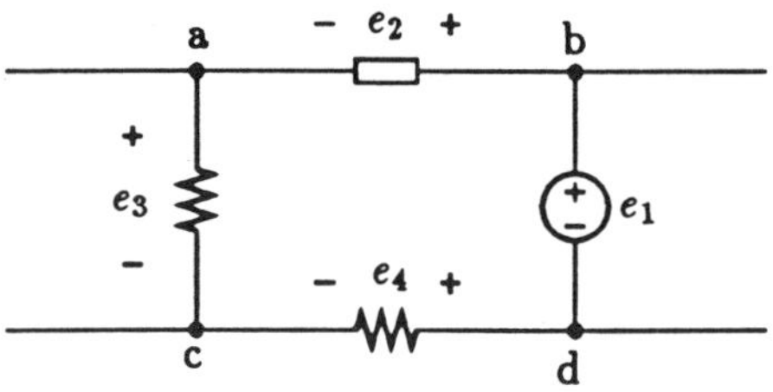

Figure 2.9: An example of the application of KVL in a loop.

KVL is sometimes stated alternatively as follows:

> The total voltage rise (or drop) from one node to another is independent of the path.

SIMPLE MEMORYLESS CIRCUITS

For the loop shown in Fig. 2.9, if we wish to find the voltage rise from d to b, we have

$$e_{bd} = e_1 = -e_4 + e_3 + e_2 \tag{2.9}$$

which is the same as Eq. (2.8).

EXAMPLE 4. Find all unknown voltages in the circuit of Fig. 2.10.

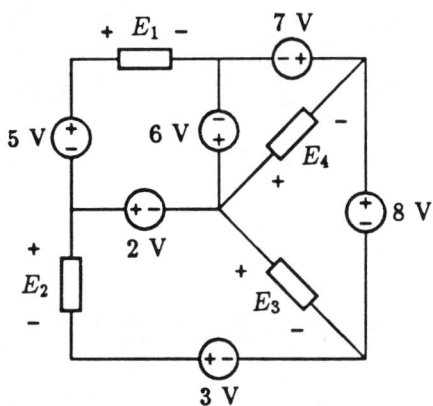

Figure 2.10: Circuit used for Example 4.

SOLUTION Applying KVL, we get

$$E_1 = 6 + 2 + 5 = 13 \text{ V}$$

$$E_2 = -3 + 8 - 7 + 6 + 2 = 6 \text{ V}$$

$$E_3 = 3 + E_2 - 2 = 7 \text{ V}$$

$$E_4 = -7 + 6 = -1 \text{ V}$$

EXERCISES

2.2.1 Find all unknown currents in the circuit.

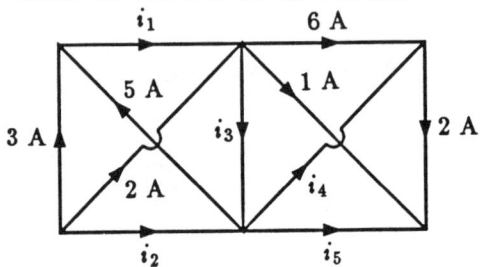

Ans. $i_1 = 8$ A, $i_2 = -5$ A, $i_3 = 3$ A, $i_4 = -4$ A, $i_5 = -3$ A.

2.2.2 Find all unknown voltages in the circuit.

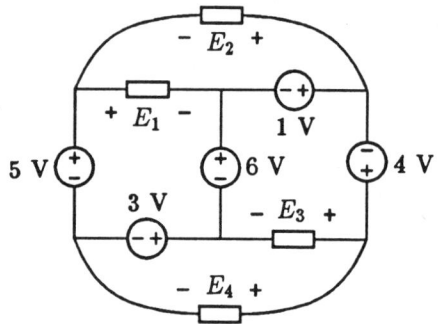

Ans. $E_1 = -4$ V, $E_2 = 5$ V, $E_3 = 11$ V, $E_4 = 14$ V.

2.3 Numbers of independent voltages and currents

Given a network, KVL and KCL must be satisfied by all loops and all closed surfaces, respectively. Usually, there are a large number of KVL and KCL equations we can write for a given network, and they must all be satisfied. It is quite plausible to say that many of these equations will be redundant if we simply approach the task of invoking KVL and KCL in a random manner. In a network with B nontrivial branches, there are B branch voltages and B branch currents. Therefore, there are $2B$ electrical quantities. Of these $2B$ quantities, the voltage or current in a branch may be fixed (if the branch happens to be a source) or the voltage and current are related to each other very simply (e.g., if the branch is made up of a linear resistor).

Among the B branch voltages and the B branch currents, they are not all independent. For example, in Fig. 2.11, it is obvious

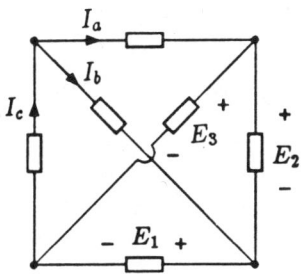

Figure 2.11: Examples of obvious interdependence of electrical quantities in a circuit.

that $E_3 = E_1 + E_2$ and $I_c = I_a + I_b$. Hence, these quantities are not all independent of one another. The question naturally arises as to how many branch voltages and branch currents are independent. This question is addressed in the branch of circuit theory known as *network topology*, which is a highly developed area. Here, we introduce some of the basic concepts and simple rules of this area so that we can apply KVL and KCL in a reasonably organized fashion.

A *network tree* is that part of a network that includes just enough branches to connect all nodes without forming any closed path. Each branch of a tree is known as a *twig*. For example, the branches shown in heavy lines on Fig. 2.12 form one possible tree of the network. Obviously, we can construct many network trees for a given network.

The following are some of the properties of a network tree:
1. A tree contains no closed loop.
2. The voltage difference between any two nodes in a network is completely defined by the twig voltages in a tree.
3. If no voltage outside a tree is known and if one or more of the twig voltages is not known, then some of the relative node voltages will be undefined.

FUNDAMENTALS OF CIRCUIT ANALYSIS

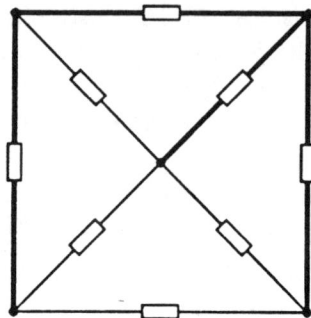

Figure 2.12: Example of a tree (heavy lines) for a network.

4. All voltages in a network are determinable if all twig voltages in a tree are known, because of item 2 above.

5. No voltage in a tree is expressible in terms of other voltages in that same tree since no closed path exists in a tree.

6. To form a tree, we start with one twig that connects two nodes. Each additional twig connects one additional node to the tree. All nodes will be connected when $N - 1$ twigs have been included in a tree, where N is the number of nontrivial nodes.

In the network of Fig. 2.12, there are five nodes. The four twigs that are shown in heavy lines form one tree.

A *network cotree* is that part of a network that complements a tree. Branches of a cotree are called *links* or *chords*. Those branches in Fig. 2.12 that are shown in light lines form a cotree.

A cotree associated with a tree has the following properties:

1. The currents in the links are independent. This can be seen by starting with a tree and adding one link at a time. Each addition of a link completes a loop in the network. Thus, any link current has a complete path to flow along without using any other link as its path.

2. All twig currents are expressible in terms of link currents, and link currents alone. This can be seen from the fact that a closed surface can always be drawn that intersects one twig, no other twig, and some links.

3. If one or more of the link currents are not known, then some of the twig currents will be undefined.

4. No closed surface can be drawn that intersect only links.

SIMPLE MEMORYLESS CIRCUITS

Hence, link currents are not expressible in terms of one another.

5. The number of links in a cotree is $B - (N - 1)$.

EXAMPLE 5. The network of Fig. 2.13 has six branches and four nodes. Express all link voltages in terms of twig voltages. Also express all twig currents in terms of link currents.

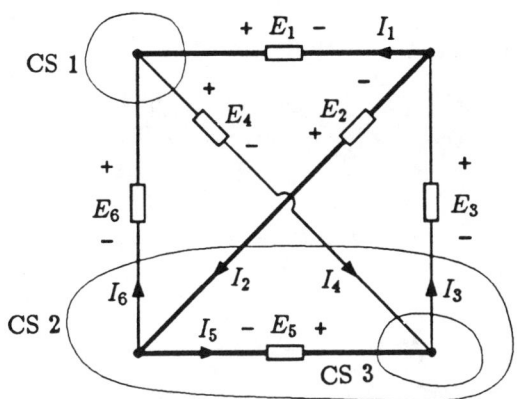

Figure 2.13: Circuit used in Example 5.

SOLUTION We obtain the link voltages by adding up the twig voltages along the tree from the minus terminal to the plus terminal of each link. We get

$$E_3 = -E_5 - E_2$$

$$E_4 = -E_5 - E_2 + E_1$$

$$E_6 = -E_2 + E_1$$

To obtain the twig currents, we draw one closed surface that intersects only one twig and then apply KCL to that closed surface. We get

$$I_1 = I_4 - I_6$$

$$I_2 = I_6 + I_3 - I_4$$

$$I_5 = I_3 - I_4$$

2.4 Analysis of simple resistive circuits

Having established the voltage-current relationship in a resistor (Ohm's law), defined the independent and controlled sources, and described Kirchhoff's laws, we are now in a position to examine some circuit relationships in some simple circuits.

2.4.1 Resistances in series

A number of resistors are said to be *in series* if the same current must flow through all of them. This connection is illustrated in Fig. 2.14. KVL gives

$$e = e_1 + e_2 + \cdots + e_n = i(R_1 + R_2 + \cdots + R_n) \tag{2.10}$$

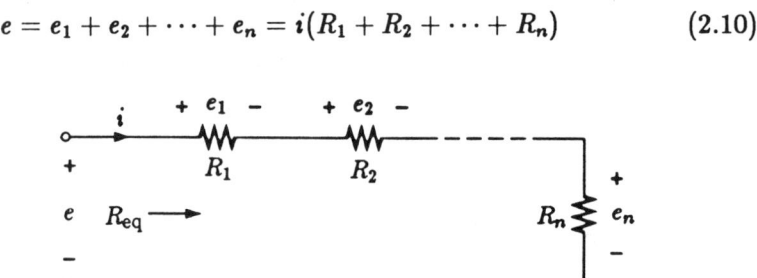

Figure 2.14: A number of resistors in series.

Hence, the equivalent resistance is

$$R_{eq} = \frac{e}{i} = R_1 + R_2 + \cdots + R_n \tag{2.11}$$

Eq. (2.11) states that the equivalent resistance of a number of resistors connected in series is the sum of all resistances.

2.4.2 Voltage-division rule

In Fig. 2.14, it is easily seen that

$$e_k = i\, R_k = \frac{R_k}{R_1 + R_2 + \cdots + R_n} \times e, \quad k = 1, 2, \ldots, n \tag{2.12}$$

Eq. (2.12) states that when a number of resistors are connected in series, the voltage across each resistor compared to the total voltage,

SIMPLE MEMORYLESS CIRCUITS

is the same as the resistance of that resistor compared to the total resistance.[2]

EXAMPLE 6. Calculate i, e_1, e_2, and e_3 in Fig. 2.15.

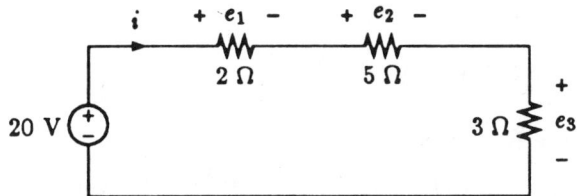

Figure 2.15: Circuit used in Example 6.

SOLUTION

$$i = \frac{20}{2+5+3} = 2 \text{ A}$$

$$e_1 = 2 \times 2 = \frac{2}{2+5+3} \times 20 = 4 \text{ V}$$

$$e_2 = 2 \times 5 = \frac{5}{2+5+3} \times 20 = 10 \text{ V}$$

$$e_3 = 2 \times 3 = \frac{3}{2+5+3} \times 20 = 6 \text{ V}$$

EXAMPLE 7. Calculate voltage e_{ab} in Fig. 2.16.

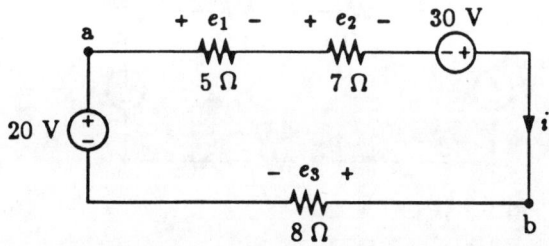

Figure 2.16: Circuit used in Example 7.

SOLUTION Using KVL, we have

---

[2]The voltage-division rule are also known as the *potentiometer rule*.

$(5+7+8)i = 20+30$

$i = \dfrac{50}{20} = 2.5$ A

$e_1 = 2.5 \times 5 = 12.5$ V

$e_2 = 2.5 \times 7 = 17.5$ V

$e_3 = 2.5 \times 8 = 20$ V

$e_{ab} = -30 + e_2 + e_1 = -e_3 + 20 = 0$

EXERCISES

2.4.1 Find e_1, e_2, and e_3.

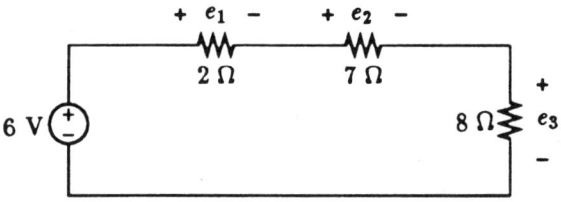

Ans. $e_1 = \dfrac{12}{17}$ V, $= \dfrac{42}{17}$ V, $e_3 = \dfrac{48}{17}$ V.

2.4.2 Determine e_{ad} and e_{bc}.

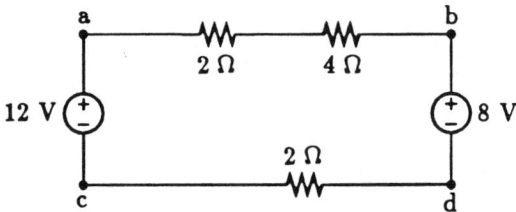

Ans. 11 V, 9 V.

2.4.3 Resistances in parallel

A number of resistors are connected in parallel if their terminal voltages are always the same. This combination is illustrated in Fig. 2.17. KCL requires

$$i = i_1 + i_2 + \cdots + i_n = \frac{e}{R_1} + \frac{e}{R_2} + \cdots + \frac{e}{R_n} \tag{2.13}$$

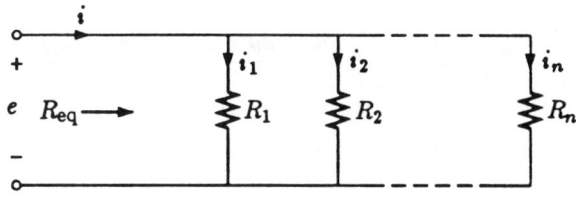

Figure 2.17: Resistors connected in parallel.

Hence

$$\frac{i}{e} = \frac{1}{R_{eq}} = \frac{1}{R_1} + \frac{1}{R_2} + \cdots + \frac{1}{R_n} \tag{2.14}$$

In terms of conductances[3]

$$G_{eq} = G_1 + G_2 + \cdots + G_n \tag{2.15}$$

Thus, the equivalent conductance of a number of resistors connected in parallel is the sum of all conductances.

If $n = 2$, we may prefer to write Eq. (2.14) as

$$R_{eq} = \frac{1}{\frac{1}{R_1} + \frac{1}{R_2}} = \frac{R_1 R_2}{R_1 + R_2} \tag{2.16}$$

If $n = 3$, Eq. (2.14) may be rearranged to read

$$R_{eq} = \frac{R_1 R_2 R_3}{R_1 R_2 + R_2 R_3 + R_3 R_1} \tag{2.17}$$

---

[3] Henceforth, it will be automatically understood that $G_k = 1/R_k$, for any k.

2.4.4 Current-division rule

In the arrangement of Fig. 2.17, we have

$$i = e(G_1 + G_2 + \cdots + G_n) \tag{2.18}$$

Hence

$$i_k = eG_k = \frac{G_k}{G_1 + G_2 + \cdots + G_n} \times i, \quad k = 1, 2, \ldots, n \tag{2.19}$$

Eq. (2.19) states that the individual current in a parallel combination compared to the total current is the same as the individual conductance compared to the total conductance.

If $n = 2$, the current-division rule becomes

$$i_1 = \frac{G_1}{G_1 + G_2} \times i = \frac{R_2}{R_1 + R_2} \times i \tag{2.20}$$

$$i_2 = \frac{G_2}{G_1 + G_2} \times i = \frac{R_1}{R_1 + R_2} \times i \tag{2.21}$$

EXAMPLE 8. Calculate e, i_1, i_2, and i_3 in the circuit of Fig. 2.18.

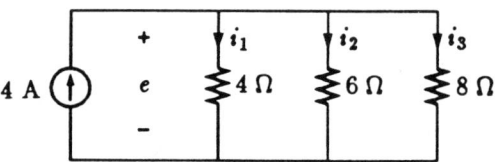

Figure 2.18: Circuit used in Example 8.

SOLUTION

$$R_{eq} = \frac{1}{\frac{1}{4} + \frac{1}{6} + \frac{1}{8}} = \frac{24}{13} \, \Omega$$

$$e = 4 \times R_{eq} = \frac{96}{13} \, V$$

$$i_1 = \frac{\frac{1}{4}}{\frac{1}{4}+\frac{1}{6}+\frac{1}{8}} \times 4 = \frac{24}{13} \text{ A}$$

$$i_2 = \frac{\frac{1}{6}}{\frac{1}{4}+\frac{1}{6}+\frac{1}{8}} \times 4 = \frac{16}{13} \text{ A}$$

$$i_3 = \frac{\frac{1}{8}}{\frac{1}{4}+\frac{1}{6}+\frac{1}{8}} \times 4 = \frac{12}{13} \text{ A}$$

Unlike the series connections, the parallel connections of circuit elements can sometimes be difficult to recognize. This is especially true when elements that are electrically in parallel but are not drawn graphically in parallel. For example, in Fig. 2.19, R_1, R_2, and R_3 are connected in parallel. So are R_4 and R_5, as well as R_6 and R_7.

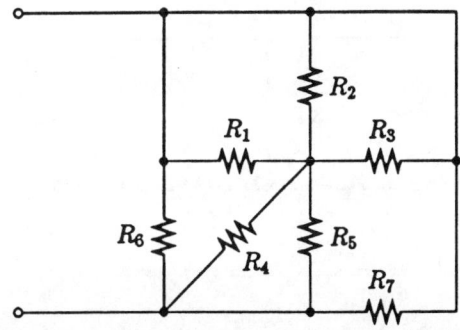

Figure 2.19: A circuit with several groups of parallel resistors.

It is helpful to adopt a notation to indicate the parallel combination of resistances. *The notation $R_1 \| R_2$ will be used to denote the parallel combination of R_1 and R_2.*

EXAMPLE 9. Find the equivalent resistance between terminals A and B of Fig. 2.20.

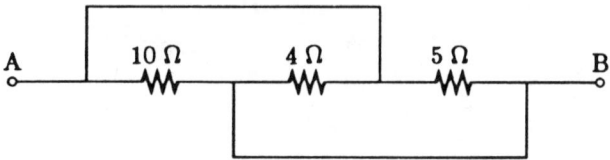

Figure 2.20: Arrangement used in Example 9.

SOLUTION The three resistors are connected in parallel. This can be seen as each resistor is connected electrically between terminals A and B. Hence

$$R_{eq} = 4\|5\|10 = \frac{1}{\frac{1}{4}+\frac{1}{5}+\frac{1}{10}} = \frac{20}{11}\ \Omega$$

EXERCISE

2.4.3 Calculate R_{eq}.

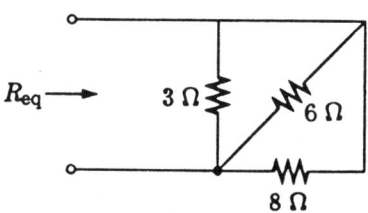

Ans. 1.6 Ω.

2.4.5 Resistive circuits with both series and parallel combinations

Many practical and useful circuits consist of both parallel and series combinations of elements, sometimes these connections are repeated many times. The following example should illustrate these situations.

EXAMPLE 10. Find all currents in the circuit of Fig. 2.21(a).

SIMPLE MEMORYLESS CIRCUITS

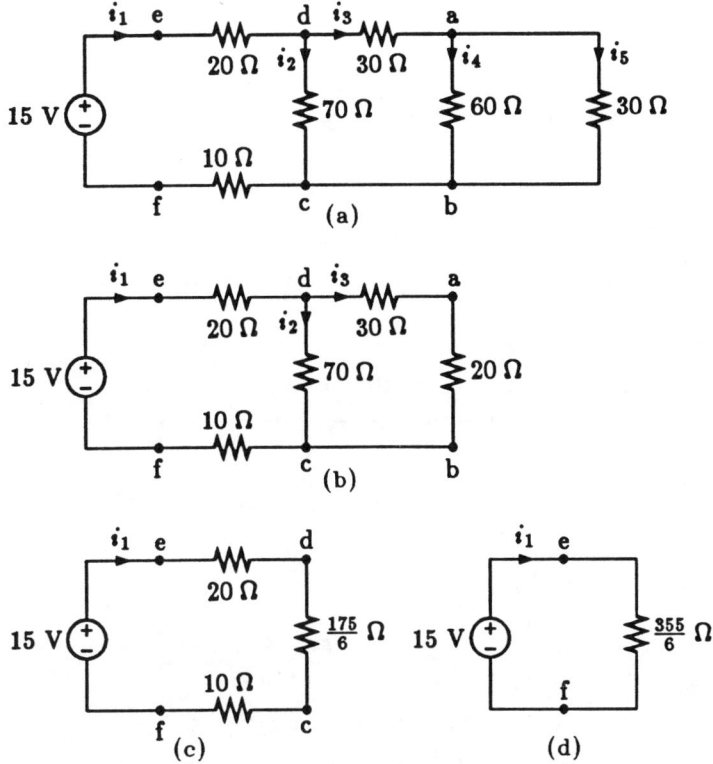

Figure 2.21: Circuits used in Example 10.

SOLUTION By applying the formulas for the series and parallel combinations of the resistances, we obtain the equivalent resistances in several steps as shown in parts (b) through (d). Hence, from Fig. 2.21(d), we have

$$i_1 = \frac{15}{\frac{355}{6}} = \frac{18}{71} \text{ A}$$

In Fig. 2.21(b), we apply the current-division rule to get

$$i_2 = \frac{50}{50 + 70} \times i_1 = \frac{50}{120} \times \frac{18}{71} = \frac{15}{142} \text{ A}$$

$$i_3 = \frac{70}{50 + 70} \times i_1 = \frac{70}{120} \times \frac{18}{71} = \frac{21}{142} \text{ A}$$

In Fig. 2.21(a), we see that

$$i_4 = \frac{30}{60+30} \times i_3 = \frac{30}{90} \times \frac{21}{142} = \frac{7}{142} \text{ A}$$

$$i_5 = \frac{60}{60+30} \times i_3 = \frac{60}{90} \times \frac{21}{142} = \frac{7}{71} \text{ A}$$

EXERCISE

2.4.4 Determine I and I_1 in the circuit.

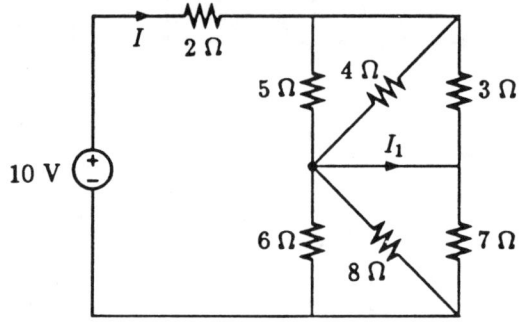

Ans. $\dfrac{17155}{9569}$ A, $-\dfrac{1660}{9569}$ A.

2.5 Ladder and ladder-like analyses

The circuit of Fig. 2.21(a) has a special pattern in that every other element is alternately in series or in parallel as we travel from the right to the left. Such a network is known as a ladder network. Our strategy in Example 10 was to successively combine elements from the right to the left until we have reduced the entire ladder to a single equivalent resistance. Then we successively applied the current-division rule from the left to the right until all currents were found. This process is rather cumbersome. We shall show how such a network can be analyzed much more efficiently by the method in the following example.

EXAMPLE 11. Fig. 2.22 gives a three-rung ladder. We wish to find all currents and voltages in the circuit.

SIMPLE MEMORYLESS CIRCUITS

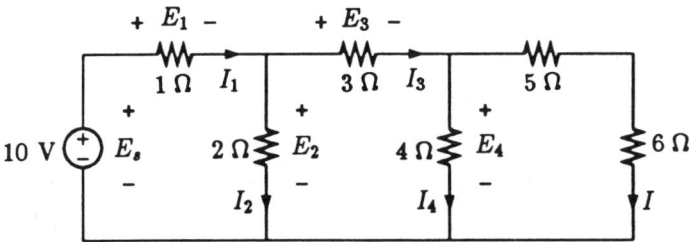

Figure 2.22: Example of the ladder analysis.

SOLUTION Starting from the rightmost branch, we assume the current there to be I as shown. We can successively obtain

$$E_4 = (5+6) \times I = 11I$$

$$I_4 = \frac{E_4}{4} = \frac{11}{4}I$$

$$I_3 = I_4 + I = \left(\frac{11}{4} + 1\right)I = \frac{15}{4}I$$

$$E_3 = 3 \times I_3 = \frac{45}{4}I$$

$$E_2 = E_3 + E_4 = \left(\frac{45}{4} + 11\right)I = \frac{89}{4}I$$

$$I_2 = \frac{E_2}{2} = \frac{89}{8}I$$

$$I_1 = I_2 + I_3 = \left(\frac{89}{8} + \frac{15}{4}\right)I = \frac{119}{8}I$$

$$E_1 = 1 \times I_1 = \frac{119}{8}I$$

$$E_s = E_1 + E_2 = \left(\frac{119}{8} + \frac{89}{4}\right)I = \frac{297}{8}I$$

Now, we have

$$\frac{297}{8}I = 10$$

or

$$I = \frac{80}{297} \text{ A}$$

Once I is found, all other quantities are obtainable from the expressions listed.

When a network includes only one independent source, all quantities in the network are proportional to the strength of that source. In such a case, we may let the unknown assume some convenient value. Then we proceed with the ladder analysis to find the strength of the independent source that would be required to give the unknown its assumed value. The actual value of the unknown corresponding to the actual source strength can be determined simply by proportionality. For example, for the circuit of Fig. 2.22, we can let $I = 1$ A. Then we would find that $E_s = 297/8$ V. By proportionality, we have

$$\frac{I}{10} = \frac{1}{\frac{297}{8}} \tag{2.22}$$

Thence

$$I = \frac{80}{297} \text{ A} \tag{2.23}$$

as before.

The ladder analysis procedure is not limited to ladders. Anytime we can express all quantities of a network in terms of one unknown (or, as we shall see later, a small number of unknowns), we have greatly simplified our task. We can have more than one independent source. We can have controlled sources, as long as the controlled quantities are easily expressible in terms of the unknown.

EXAMPLE 12. Use the procedure similar to the ladder analysis to determine I in the circuit of Fig. 2.23.

SOLUTION We have successively

$$I_1 = I + 3$$

$$E_1 = 3I_1 = 3I + 9$$

SIMPLE MEMORYLESS CIRCUITS

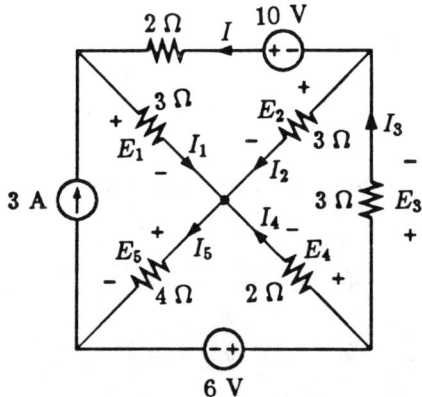

Figure 2.23: Circuit used in Example 12.

$$E_2 = E_1 + 2I - 10 = 5I - 1$$

$$I_2 = \frac{E_2}{3} = \frac{5}{3}I - \frac{1}{3}$$

$$I_3 = I_2 + I = \frac{8}{3}I - \frac{1}{3}$$

$$E_3 = 3I_3 = 8I - 1$$

$$E_4 = E_2 + E_3 = 13I - 2$$

$$I_4 = \frac{E_4}{2} = \frac{13}{2}I - 1$$

$$I_5 = I_1 + I_2 + I_4 = \frac{55}{6}I + \frac{5}{3}$$

$$E_5 = 4I_5 = \frac{110}{3}I + \frac{20}{3}$$

Now we let

$$E_5 + E_4 = 6$$

which results in

$$\frac{110}{3}I + \frac{20}{3} + 13I - 2 = 6$$

Solving yields

$$I = \frac{4}{149} \text{ A}$$

EXAMPLE 13. Determine I in the circuit of Fig. 2.24.

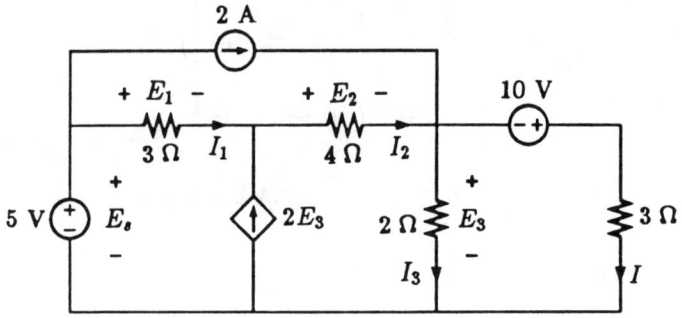

Figure 2.24: Circuit used in Example 13.

SOLUTION We can obtain successively

$$E_3 = 3I - 10$$

$$I_3 = \frac{E_3}{2} = \frac{3}{2}I - 5$$

$$I_2 = I_3 + I - 2 = \frac{5}{2}I - 7$$

$$E_2 = 4I_2 = 10I - 28$$

$$I_1 = I_2 - 2E_3 = -\frac{7}{2}I + 13$$

$$E_1 = 3I_1 = -\frac{21}{2}I + 39$$

$$E_s = E_3 + E_2 + E_1 = \frac{5}{2}I + 1 \ .$$

Now, we make

$$E_s = \frac{5}{2}I + 1 = 5$$

Solving gives

$$I = 1.6 \text{ A}$$

EXAMPLE 14. Use the ladder-like approach to determine I_1 and I_1' in the circuit of Fig. 2.25.

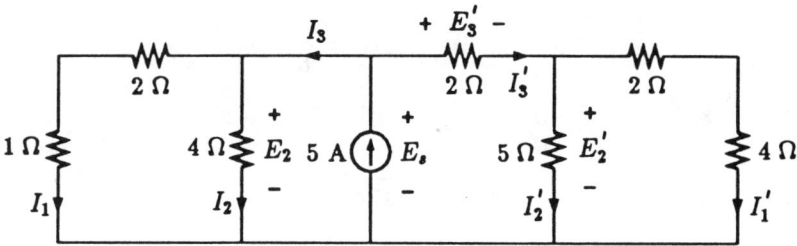

Figure 2.25: Circuit used in Example 14.

SOLUTION This circuit may be viewed as having two ladders joined together—a step ladder. We need two variables, one for each ladder. We proceed with each ladder separately, as in Example 11, until they meet. Thus we have

$$E_2 = 3I_1$$

$$I_2 = \frac{1}{4}E_2 = \frac{3}{4}I_1$$

$$I_3 = I_1 + I_2 = \left(1 + \frac{3}{4}\right)I_1 = \frac{7}{4}I_1$$

and

$$E_2' = 6I_1'$$

$$I_2' = \frac{1}{5}E_2' = \frac{6}{5}I_1'$$

$$I_3' = I_1' + I_2' = \left(1 + \frac{6}{5}\right)I_1' = \frac{11}{5}I_1'$$

$$E_3' = 2I_3' = \frac{22}{5}I_1'$$

$$E_s = E_2' + E_3' = \left(6 + \frac{22}{5}\right)I_1' = \frac{52}{5}I_1'$$

We now concentrate on that part of the network around the current source where the two ladders meet. We have

$$I_3 + I_3' = 5 \quad \text{and} \quad E_s = E_2$$

or

$$\frac{7}{4}I_1 + \frac{11}{5}I_1' = 5 \quad \text{and} \quad 3I_1 = \frac{52}{5}I_1'$$

Solving gives

$$I_1 = \frac{65}{31} \text{ A} \quad \text{and} \quad I_1' = \frac{75}{124} \text{ A}$$

EXERCISE

2.5.1 Use ladder analysis to find E in the circuit.

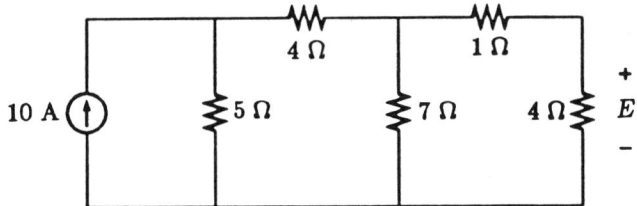

Ans. $\dfrac{1400}{143}$ V.

2.6 Delta-wye and wye-delta transformations

When three resistances are connected in the form of a wye configuration as shown in Fig. 2.26(a), there is an equivalent delta configuration as shown in Fig. 2.26(b); and vice versa. The proof that these two configurations can be made equivalent is best done in the context of three-terminal, two-port networks, which we shall study in Chapter 10. In the meantime, we shall develop the formulas for

SIMPLE MEMORYLESS CIRCUITS

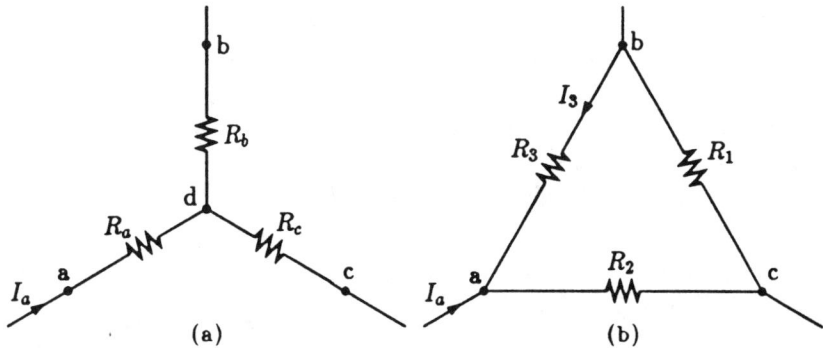

Figure 2.26: Equivalence of delta- and wye-connected resistances.

the relationships among the element values if these networks are to be equivalent.

If these two networks are to be equivalent, their resistances between any two terminals with the third terminal open must be equal. This leads to the following relationships

$$R_a + R_b = \frac{R_3(R_1 + R_2)}{R_1 + R_2 + R_3}$$

$$R_b + R_c = \frac{R_1(R_2 + R_3)}{R_1 + R_2 + R_3} \qquad (2.24)$$

$$R_c + R_a = \frac{R_2(R_3 + R_1)}{R_1 + R_2 + R_3}$$

Solving for R_a, R_b, and R_c from Eq. (2.24), we obtain

$$R_a = \frac{R_2 R_3}{R_1 + R_2 + R_3}$$

$$R_b = \frac{R_1 R_3}{R_1 + R_2 + R_3} \qquad (2.25)$$

$$R_c = \frac{R_1 R_2}{R_1 + R_2 + R_3}$$

Eqs (2.25), therefore, enable us to convert a set of delta-connected resistances into a wye-connected equivalent.

Similarly, we can develop a set of formulas in the opposite direction of the conversion. If the two networks in Fig. 2.26 are to be equivalent, then the resistance between any one of its terminals (say, a) and the terminal made up of the other two terminals short-circuited (b and c) must be equal. There are three such combinations. In this development, it is easier to use the conductances. These equations are

$$G_1 + G_2 = \frac{G_c(G_a + G_b)}{G_a + G_b + G_c}$$

$$G_2 + G_3 = \frac{G_a(G_b + G_c)}{G_a + G_b + G_c} \quad (2.26)$$

$$G_3 + G_1 = \frac{G_b(G_c + G_a)}{G_a + G_b + G_c}$$

Solving for G_1, G_2, and G_3 from Eq. (2.26), we obtain

$$G_1 = \frac{G_b G_c}{G_a + G_b + G_c}$$

$$G_2 = \frac{G_a G_c}{G_a + G_b + G_c} \quad (2.27)$$

$$G_3 = \frac{G_a G_b}{G_a + G_b + G_c}$$

Eqs. (2.27) are more frequently written in terms of the resistances. They read

$$R_1 = \frac{R_a R_b + R_b R_c + R_c R_a}{R_a}$$

$$R_2 = \frac{R_a R_b + R_b R_c + R_c R_a}{R_b} \quad (2.28)$$

$$R_3 = \frac{R_a R_b + R_b R_c + R_c R_a}{R_c}$$

SIMPLE MEMORYLESS CIRCUITS

Eqs. 2.27 and 2.28 enable us to convert a set of wye-connected resistances into its delta-connected equivalent.

At this point it is appropriate to point out that when a delta-connected and a wye-connected networks are made equivalent, they are equivalent only as three-terminal networks connected to terminals a, b, and c. There is no longer any one-to-one correspondence between the innards of these two networks. For example, node d of Fig. 2.26(a) does not exist in Fig. 2.26(b). Current I_3 in Fig. 2.26(b) has no counterpart in Fig. 2.26(a). However, I_a and E_{bc}, and similar quantities, in both networks are identical. The delta-wye and wye-delta transformations are sometimes referred to as the pi-to-tee and tee-to-pi transformations because the resistors in Fig. 2.26 are sometimes drawn as shown in Fig. 2.27. Of course, networks in Figs. 2.26 and 2.27 are electrically identical.

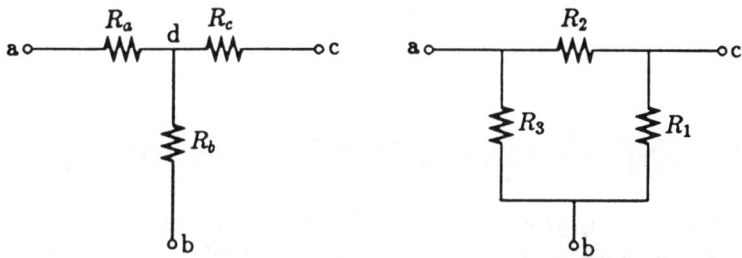

Figure 2.27: Wye- and delta-connected resistors redrawn to appear as tee- and pi-connected networks.

EXAMPLE 15. Evaluate E_{db} in the circuit of Fig. 2.28(a).

SOLUTION We first convert the delta-connected resistances among terminals a, b, and d. The result is shown in Fig. 2.28(b). In this equivalent circuit, we have only series-parallel combinations. We first compute

$$I = \frac{10}{0.5 + (2+2)\|(0.4+5)} = \frac{940}{263}$$

$$I_1 = \frac{5.4}{9.4}I = \frac{540}{263}$$

$$I_2 = \frac{4}{9.4}I = \frac{400}{263}$$

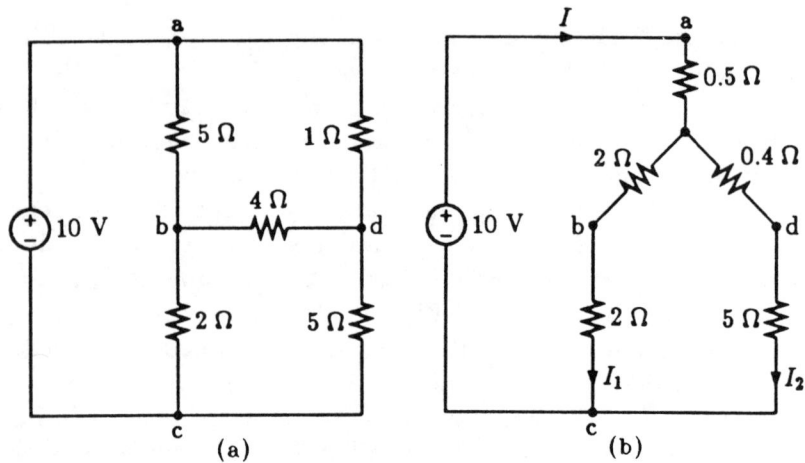

Figure 2.28: Circuits used in Example 15.

$$E_{db} = 5I_2 - 2I_1 = \frac{920}{263} \text{ V}$$

2.7 Circuits with ideal op amps

The ideal op amp was defined in Chapter 1. Here we shall consider several simple circuits containing resistors and ideal op amps.

2.7.1 The noninverting voltage amplifier

Consider the circuit Fig. 2.29. The op amp requires that

$$e_- = e_s \tag{2.29}$$

But the voltage-division rule requires that

$$e_- = \frac{R_2}{R_1 + R_2} e_o \tag{2.30}$$

Hence

$$\frac{e_o}{e_s} = \frac{R_1 + R_2}{R_2} \tag{2.31}$$

Therefore, the circuit of Fig. 2.29 is a voltage-controlled voltage source whose voltage gain is always positive and no less than unity.

SIMPLE MEMORYLESS CIRCUITS

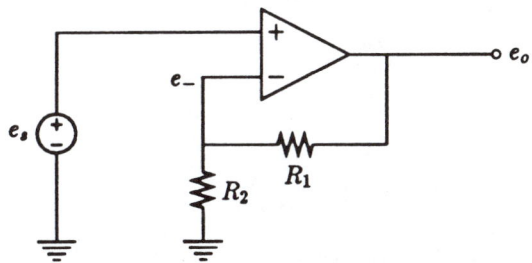

Figure 2.29: The noninverting voltage amplifier.

2.7.2 The voltage follower

If we let $R_2 \to \infty$ in Eq. (2.31), then

$$e_o = e_s \qquad (2.32)$$

Since $R_2 \to \infty$ is an open circuit, the value of R_1 is arbitrary. If we let $R_1 = 0$, we have the circuit in Fig. 2.30.

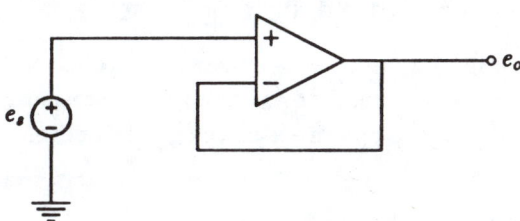

Figure 2.30: The voltage follower.

2.7.3 The inverting voltage amplifier

In the circuit in Fig. 2.31. we have $e_- = 0$. The currents in the two resistors are

$$i_1 = \frac{e_s}{R_1} \quad \text{and} \quad i_2 = \frac{e_s}{R_2} \qquad (2.33)$$

Since the input of the op amp cannot have any current, we must have

$$\frac{e_o}{R_2} + \frac{e_s}{R_1} = 0 \qquad (2.34)$$

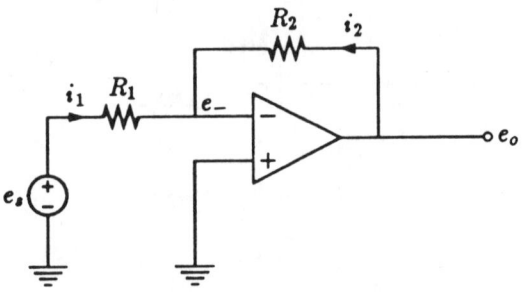

Figure 2.31: The inverting voltage amplifier.

or

$$\frac{e_o}{e_s} = -\frac{R_2}{R_1} \qquad (2.35)$$

The circuit in Fig. 2.31 is a voltage-controlled voltage source whose voltage gain is always negative and is determined by the ratio of the two resistances.

2.7.4 The current-to-voltage converter

The circuit in Fig. 2.32 acts as a current-controlled voltage source. The source current i_s must flow through the resistor. Since the inverting terminal of the op amp is at the ground potential, it follows that

$$e_o = -i_s R \qquad (2.36)$$

Therefore, the output voltage is proportional to the input current.

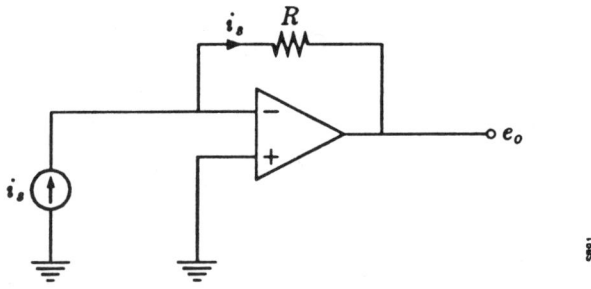

Figure 2.32: A current-to-voltage converter.

2.7.5 A voltage-to-current converter

For the circuit in Fig. 2.33, since the inverting terminal is at ground

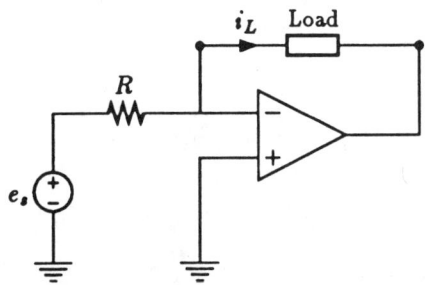

Figure 2.33: A voltage-to-current converter.

potential, the current flowing in R is e_s/R. But this current cannot enter the op amp. It must flow through the load,

$$I_L = \frac{e_s}{R} \tag{2.37}$$

no matter what the load is. This is one way to generate a constant current source if a constant voltage is available. This circuit is also a voltage-controlled current source.

EXERCISES

2.7.1 Calculate I_L in the circuit.

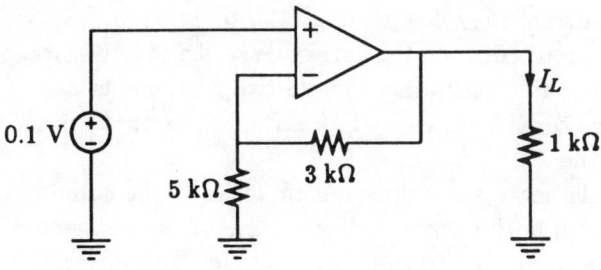

Ans. 0.16 mA.

2.7.2 Obtain the relationship between e_s and e_o, and that between i_1 and i_2.

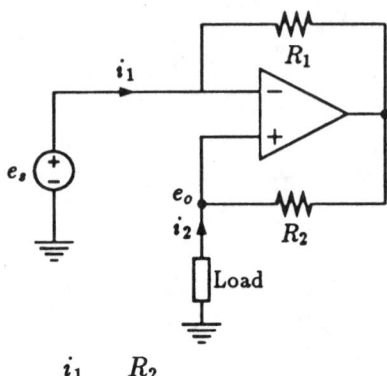

Ans. $e_s = e_o$, $\dfrac{i_1}{i_2} = \dfrac{R_2}{R_1}$.

2.8 Duality

We have now covered enough circuit principles and example problems to observe that there are two parallel sets of proceedings in the material of this chapter. If the student has not noticed this parallelism, now is a good time to review the content covered so far and observe this phenomenon.

To start off, Ohm's law has two forms

$$e = Ri \quad \text{or} \quad i = Ge$$

Hence, if we interchange the roles of the voltage e and the current i, then the role of R in one formula is the same as the role of G in the other. This dualism is call the *principle of duality*.

If the two terminals are short-circuited, then the voltage is zero and any current may flow through the short circuit. If two terminals are open-circuited, then the current is zero and any voltage may exist between the two terminals. These two statements are dual of each other if we consider the *short circuit* and the *open circuit* to be dual of each other.

Two elements are connected in series if the same current must flow through both elements. Two elements are connected in parallel if they are connected to the same voltage. Hence, *series* and *parallel* are dual of each other.

SIMPLE MEMORYLESS CIRCUITS

The total resistance of a series combination of resistances is the sum of all resistances. The total conductance of a parallel combination of conductances is the sum of all conductances. These two statements are dual of each other.

The following is a list of easily observed dual quantities, adjectives, circuit connections, formulas, and devices.

Table 2.1 Dual entities

Voltage	Current
Volt	Ampere
Resistance	Conductance
Ohm	Mho
Current source	Voltage source
KCL	KVL
Series	Parallel
Open	Short
Loop	Closed surface
Voltage-division rule	Current-division rule
Current-controlled current source	Voltage-controlled voltage source
Voltage-controlled current source	Current-controlled voltage source
Delta connection	Wye connection

With these observations, we can see that the two circuits in Fig. 2.34 are dual of each other. In these two circuits, every ele-

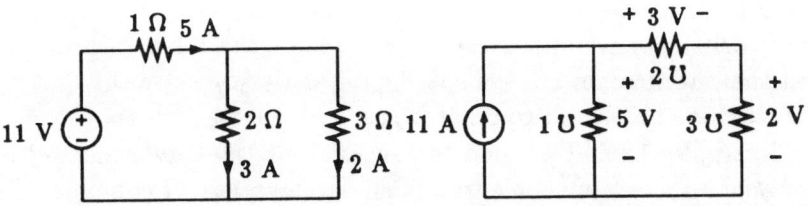

Figure 2.34: Two circuits that are dual of each other.

ment, quantity, and connection has a dual counterpart in the other circuit. The dual elements and dual quantities all have the same numerical value but in dual units. In fact, if one analyzes both circuits, one would go through the same sets of computational steps with exactly the same numbers appearing in steps that are dual of each other.

The principle of duality is indeed very helpful in many situations. For example, the combined conductance of two conductances, G_1 and G_2, in series is $G_1G_2/(G_1 + G_2)$. This is simply the dual of the combination of two resistances in parallel. The voltage-division and current-division rules are dual of each other. We can also observe that the delta-to-wye and the wye-to-delta conversion formulas are dual of each other—if G and R are interchanged. Thus, by observing the duality, we can reduce by half the number of formulas we need to memorize.

Although the principle of duality is sound and helpful, we should not carry this too far, as there are devices and circuits whose dual counterparts do not exist. However, within certain limitations, duality is a very convenient way to aid our understanding of certain circuit behavior and extend our ability to solve numerous problems. We shall continue to point out the duality features as we proceed.

2.9 Summary

At the beginning of this chapter, we defined the characteristics of a linear resistor and stated KVL and KCL. With these basic rules, we then pointed out several special interconnections of elements—the series-parallel connections, the ladder networks, the delta-wye transformations. These analysis techniques only apply to interconnections of elements that have certain specific features. Networks for which these techniques do apply occur quite often in practice, but these techniques are too specific for analyzing networks that do not fit these particular configurations.

Examples 12 and 13 tend to suggest that the ladder-method of analysis may be applicable to a fairly large variety of complex network configurations. However, this is not the case. Except for the true ladder, the ladder-like method of analysis tends to rely more on the ingenuity and the insight of the solver, rather than furnishing us with a straightforward procedure for analyzing an arbitrary network.

SIMPLE MEMORYLESS CIRCUITS

We clearly need more systematic and unrestricted analysis methods that are applicable to any given network, methods that will lead to the solution with a reasonably predictable amount of effort. We shall present several of these more general methods in the next chapter.

Problems

2.1 Obtain the unknown quantity in each of the following diagrams.

2.2 Determine i_1, i_2, i_3, and i_4.

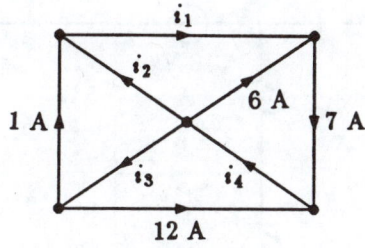

▲ **2.3** Determine i_1, i_2, i_3, i_4, i_5, and i_6.

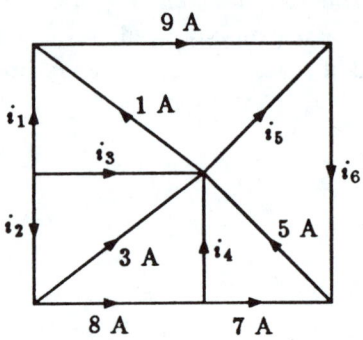

2.4 Determine E and I.

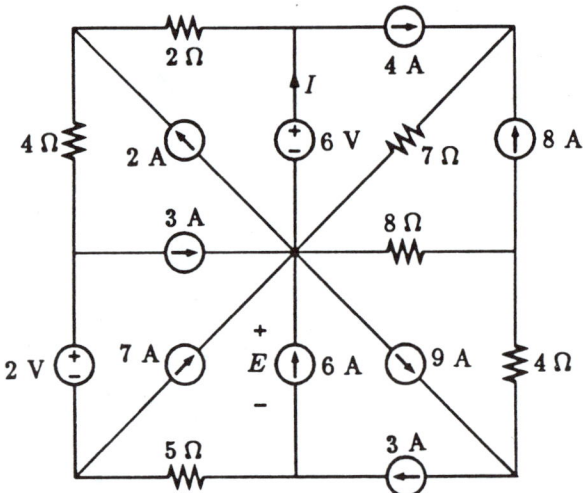

▲▷ **2.5** Determine V.

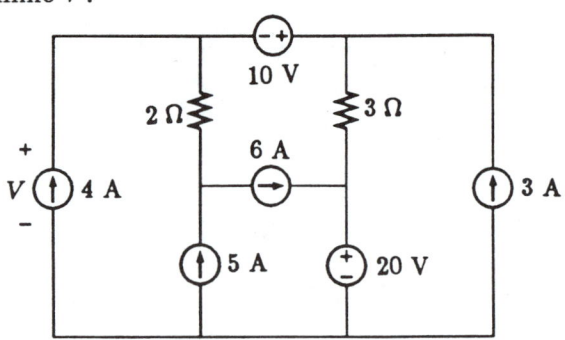

SIMPLE MEMORYLESS CIRCUITS

2.6 Determine E and I.

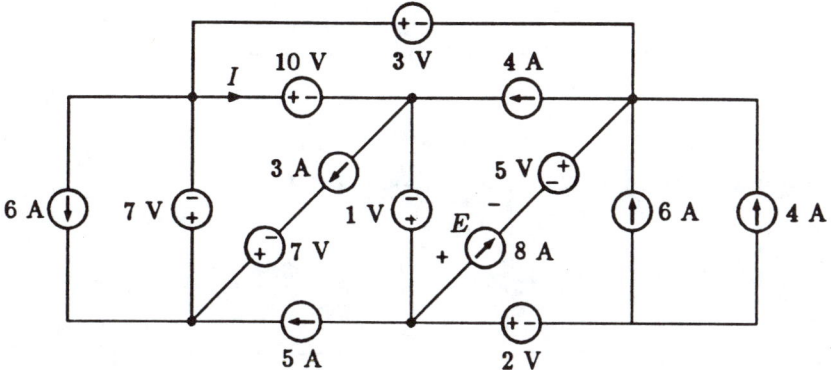

2.7 Determine I.

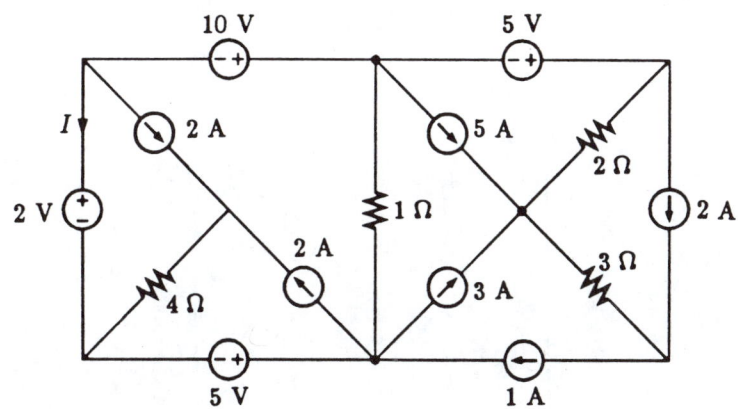

2.8 Determine I.

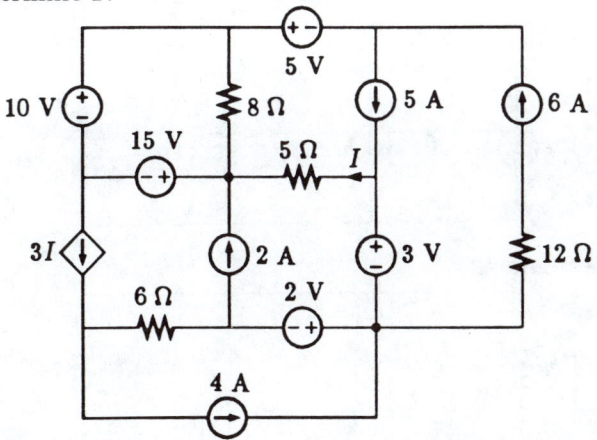

2.9 Determine I_1, I_2, and I_3.

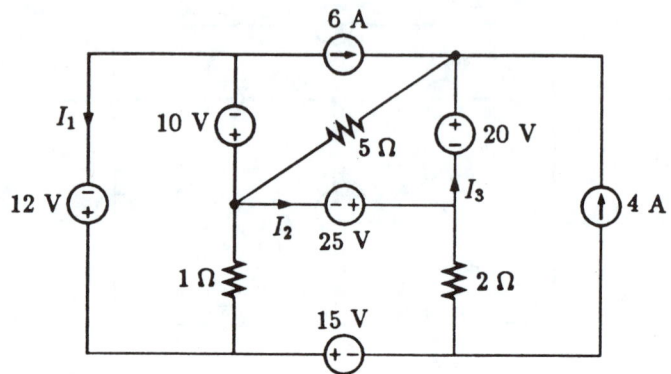

2.10 Determine all E's.

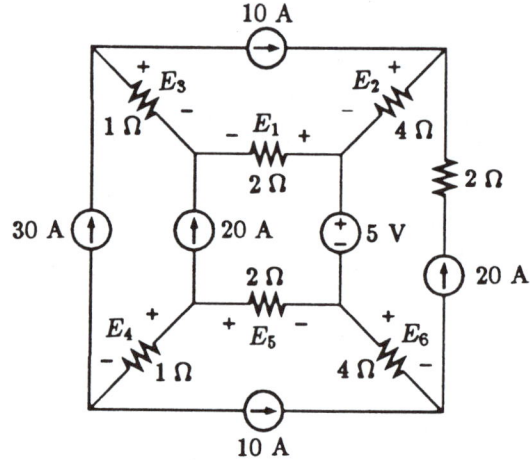

2.11 Determine I.

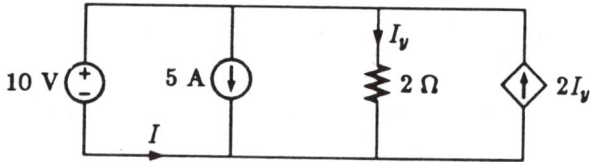

2.12 Determine I.

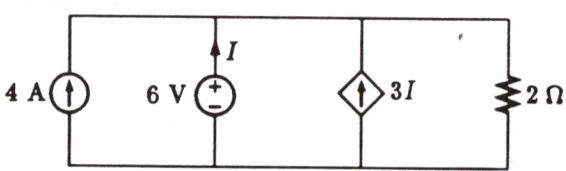

SIMPLE MEMORYLESS CIRCUITS

2.13 Determine E_1.

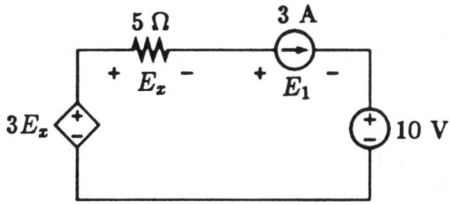

2.14 Determine I

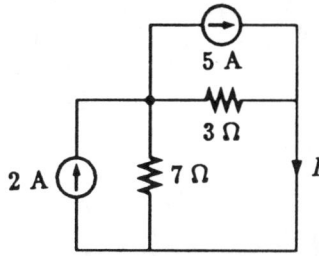

2.15 Determine I.

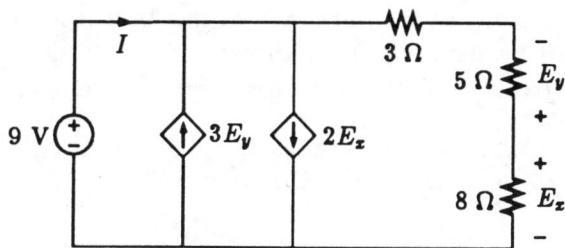

2.16 Determine I.

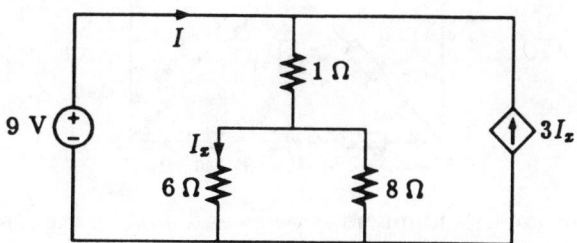

2.17 Determine E.

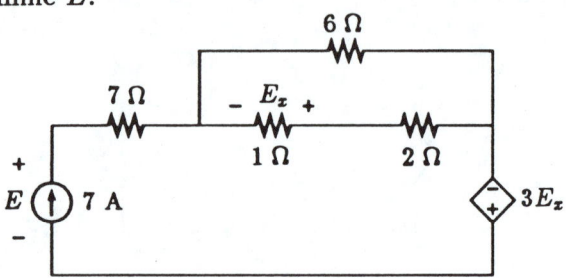

2.18 Determine E and I.

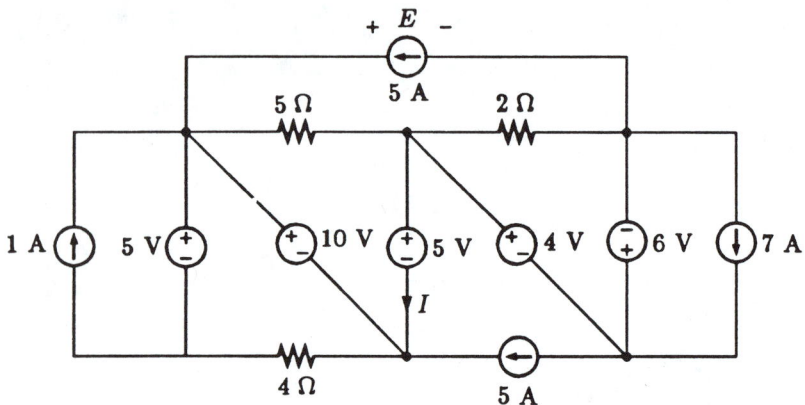

2.19 In the following network, a tree has been chosen and is indicated by heavy lines. Express every link voltage in terms of twig voltages and every twig current in terms link currents *directly*.

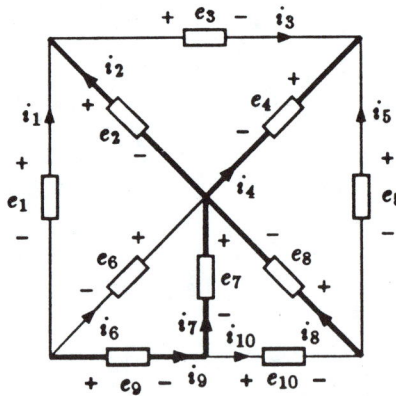

2.20 Determine the numbers of twigs and links in the following net-

SIMPLE MEMORYLESS CIRCUITS

work.

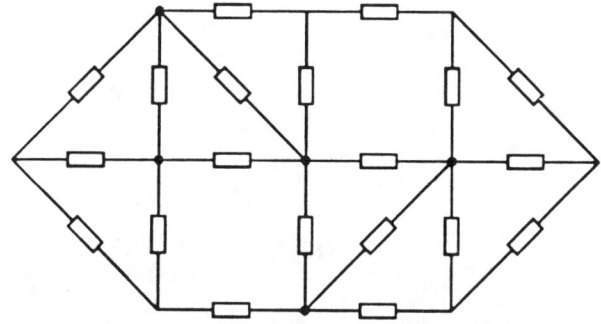

▲ 2.21 Determine the equivalent resistance R_{eq}.

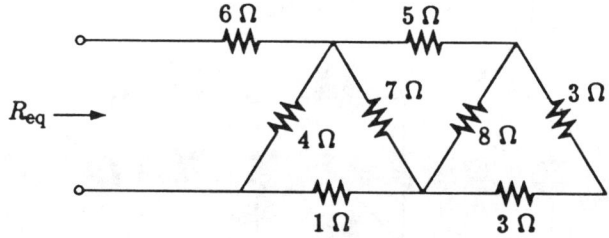

2.22 Determine I.

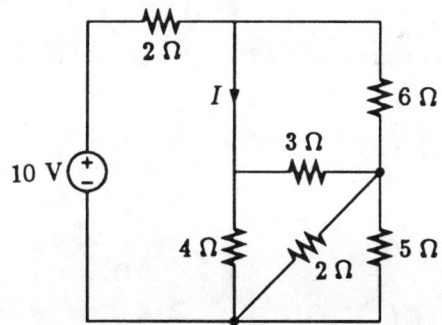

2.23 Determine I.

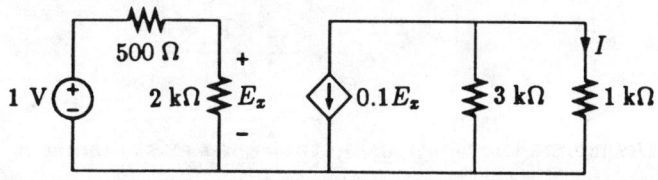

△ **2.24** Determine E.

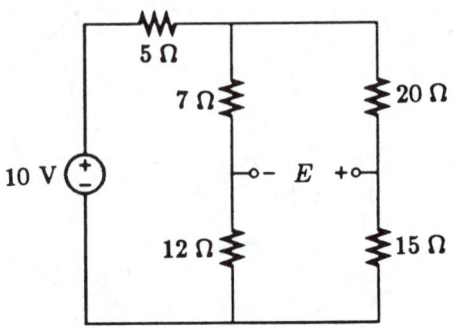

△ **2.25** Determine I.

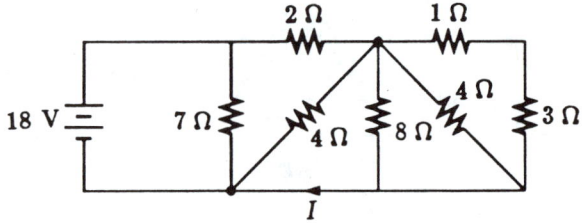

2.26 The power consumed by the 10-Ω resistor is 100 W. Determine the value of E.

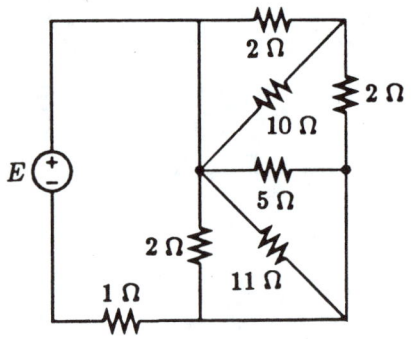

△ **2.27** Determine the relationship that must exist among R_1, R_2, R_3, and R_4 so that the value of I is always zero.

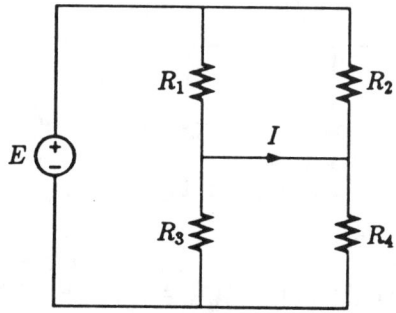

△ **2.28** Determine I.

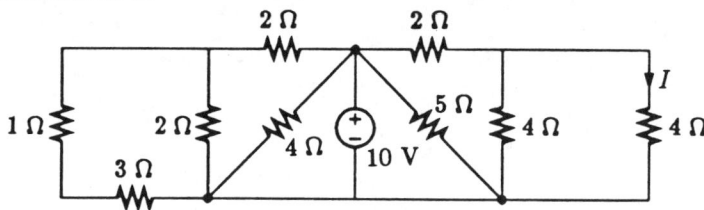

2.29 Determine the value of R.

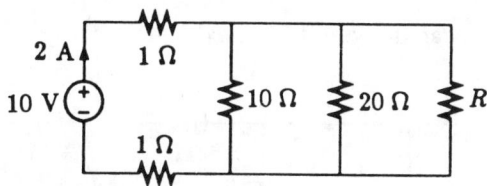

2.30 Determine I.

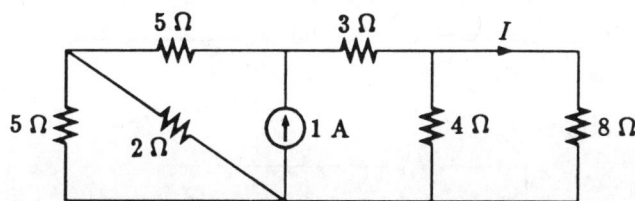

2.31 Determine I_1 and I_2.

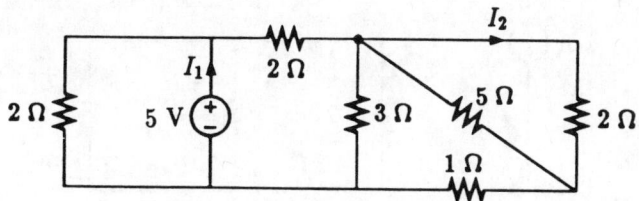

2.32 Determine E_1 and E_2.

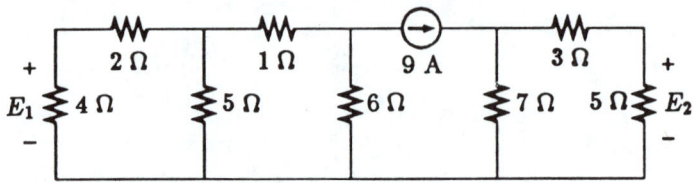

2.33 In the following circuit, $e_s(t) = 10\epsilon^{-t}$ V. Determine all current in the circuit.

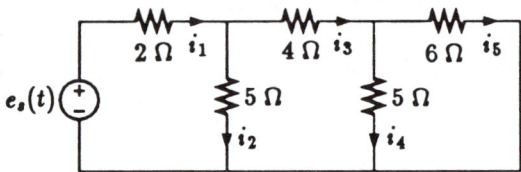

2.34 Find the power delivered to the 100-Ω resistor.

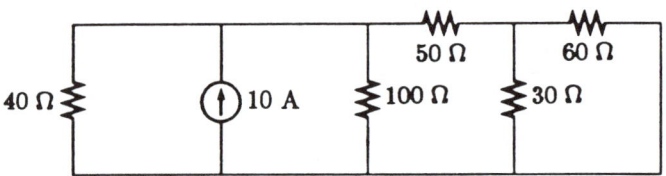

2.35 Determine I.

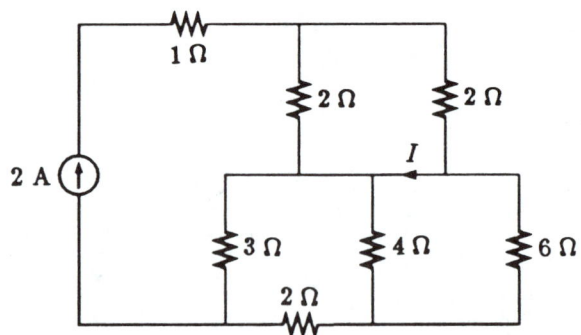

2.36 Determine I.

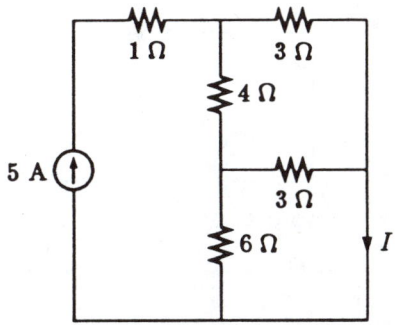

2.37 Determine I.

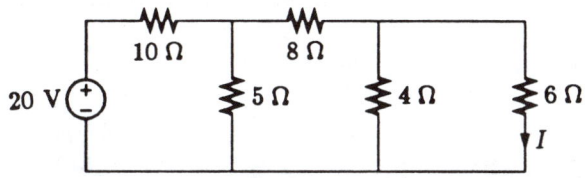

2.38 Determine E.

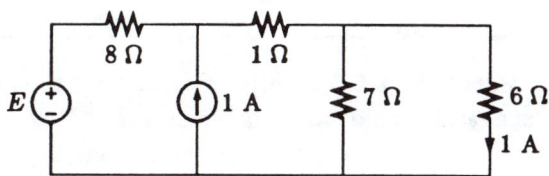

2.39 Determine I.

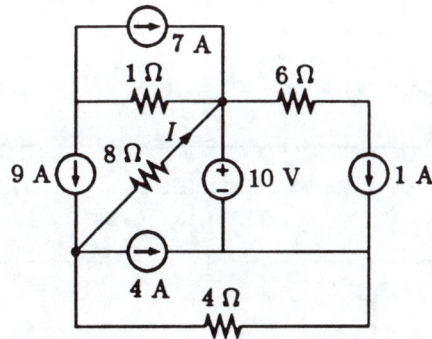

2.40 Use ladder analysis to determine E_2 and the equivalent resistance seen by the voltage source.

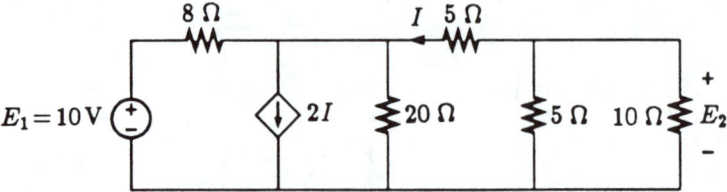

2.41 Use ladder analysis to determine I.

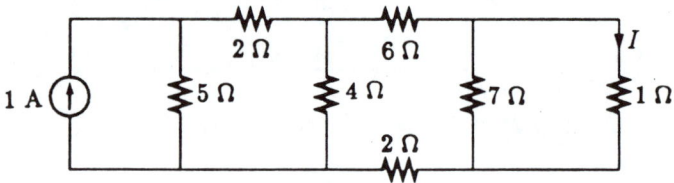

○ **2.42** Use ladder analysis to determine E_x.

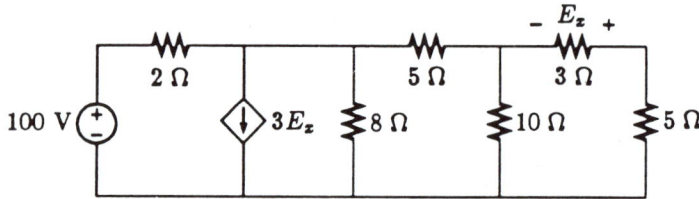

△ **2.43** First, convert the 4-5-6-Ω resistor delta into its equivalent wye. Then use ladder analysis to determine E_y.

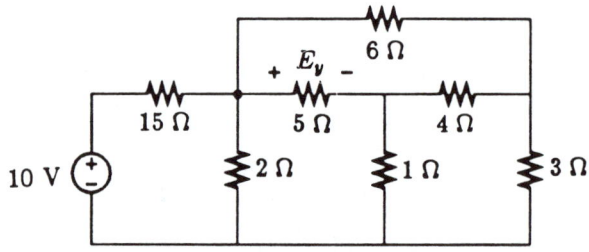

△ **2.44** Use ladder analysis to determine I.

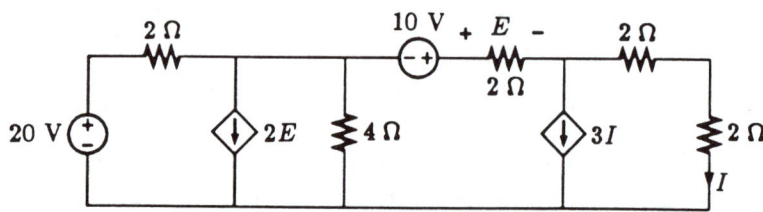

SIMPLE MEMORYLESS CIRCUITS

2.45 Each resistance is 1-Ω. Calculate the value of E.

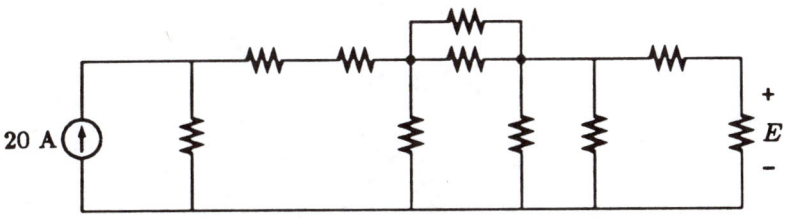

2.46 Use the ladder-like analysis to determine V_1 and V_2.

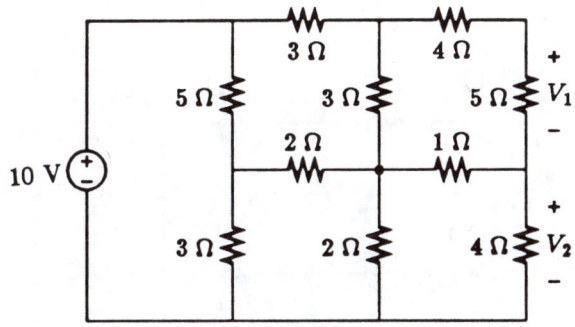

2.47 Determine E_s.

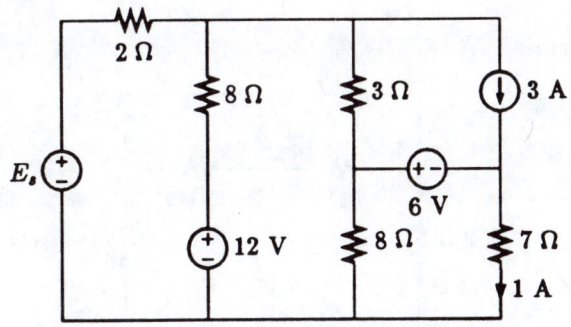

2.48 Determine I.

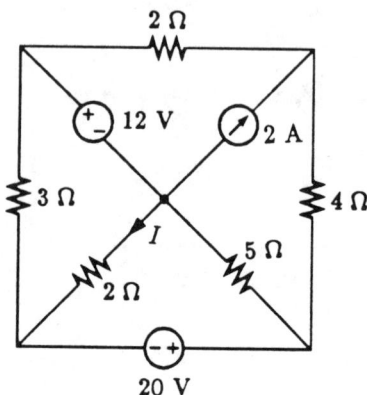

2.49 Determine the value of I_s.

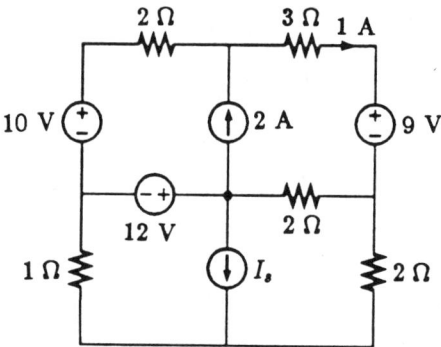

2.50 Determine E.

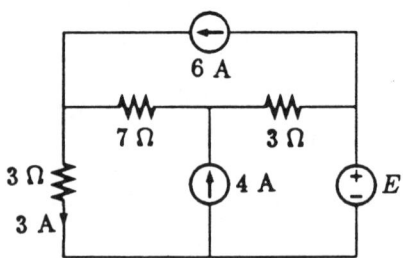

SIMPLE MEMORYLESS CIRCUITS

2.51 Determine I.

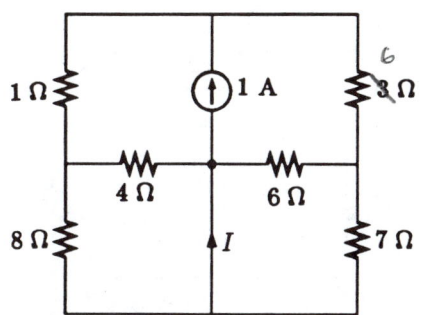

△ **2.52** Determine E.

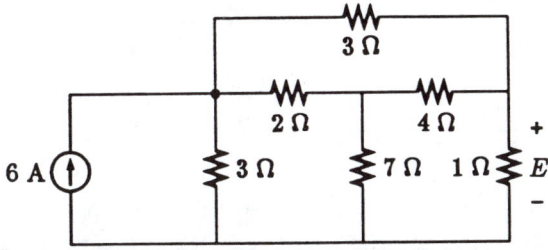

△ **2.53** Determine the value of R_{eq}.

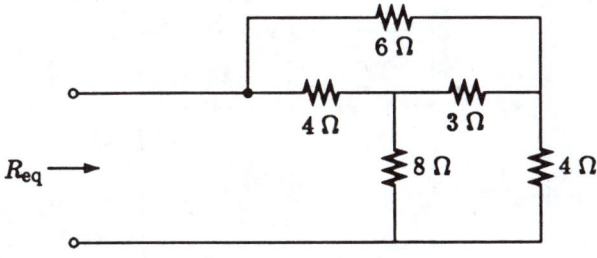

2.54 Determine E.

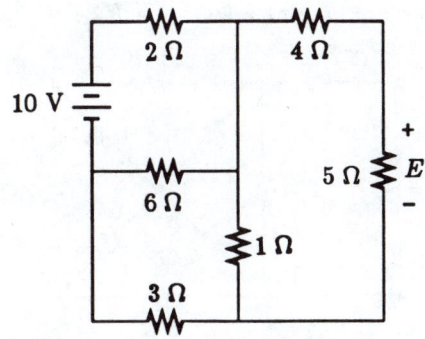

2.55 Determine E.

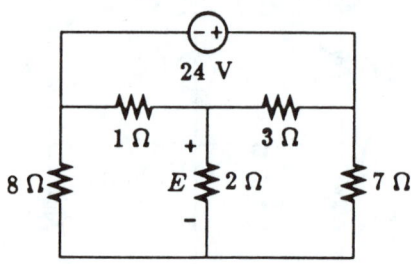

2.56 Determine E.

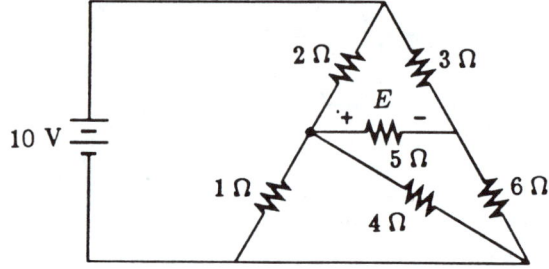

2.57 Determine I.

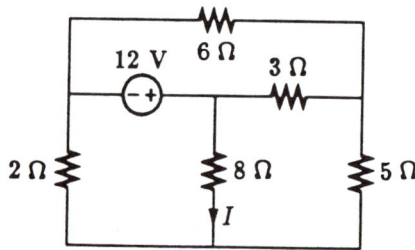

2.58 Determine the equivalent resistance R_{eq}.

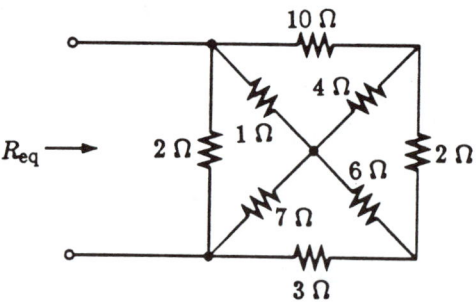

SIMPLE MEMORYLESS CIRCUITS

△ **2.59** Express e_o in terms of e_1 and e_2.

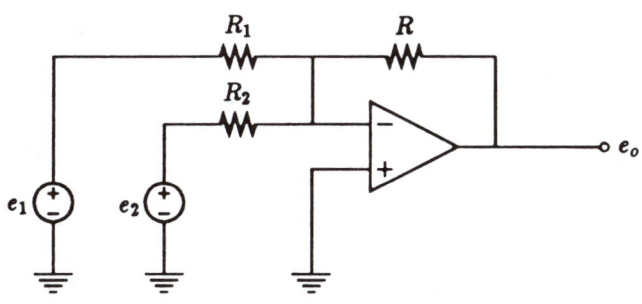

2.60 Determine i.

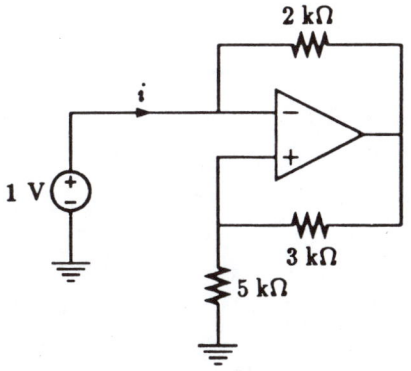

△ **2.61** Express e_o in terms of e_1 and e_2.

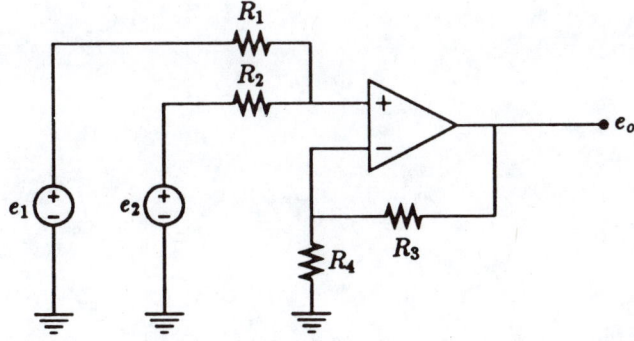

2.62 Determine $i(t)$.

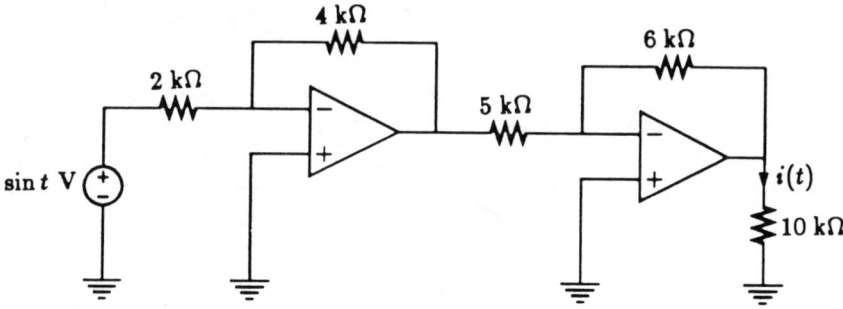

Chapter 3
General Analysis Methods

In Chapter 2, we described several techniques of circuit analysis. Those techniques are applicable only to very simple or special circuit connections, or they require some experience on the solver's part in knowing how to arrive at the solution. The ability to recognize the latter is not easily acquired.

In this chapter, we shall describe several methods by which networks of any complexity can be analyzed in a systematic manner. These methods are well formulated and, once followed, will always lead to the solution of all voltages and currents in a network with a reasonably predictable amount of labor.

3.1 General considerations

The initial consideration in developing totally general methods for analyzing any given network is to choose just enough key variables or unknowns so that these unknowns can be found first. Once these key unknowns are found, the rest of the quantities can be obtained with relative ease. In this chapter, we shall consider the analysis problem to be solved once a set of key unknowns is found.

In Section 2.3, we introduced the concept of network tree and cotree. All voltages in a network are expressible in terms of the twig voltages. All currents in a network are expressible in terms of link currents. Since the voltage sources are either fixed (for independent voltage sources) or expressible in terms of other quantities (for controlled voltage sources), it is only natural that *all voltage sources* (independent or controlled) *should be included in all trees*.

No voltage source should appear in any cotree. Similarly, *all current sources* (independent or controlled) *should be included in all cotrees*. No current source should appear in any tree. These restrictions are automatically understood whenever we choose a tree.

We have indeed two methods that are based on this consideration. In one of these methods, we use the nonsource twig voltages as the key unknowns. We then obtain the KCL equations for these unknowns. Once these voltages are found, the problem is practically solved. This method is known as the *cutset analysis*.

The other method uses the nonsource link currents as unknowns. We then obtain the KVL equations for these unknowns. Again, once these currents are found, the problems is practically solved. This method is known as the *loop analysis*.

3.1.1 Cutset analysis

The procedure for cutset analysis may be outlined as follows:

1. Choose a network tree.
2. Assign an unknown voltage variable to each nonsource twig.
3. Apply KVL to express all link voltages in terms of twig voltages.
4. Express all branch currents (including those of the controlled current sources) in terms of the twig voltage variables.
5. Choose a set of closed surfaces such that each closed surface intersects only one twig. Write a KCL equation for each of these closed surfaces.
6. Solve the set of equations for all the unknown variables.

EXAMPLE 1. Obtain the three twig voltages, e_1, e_2, and e_3, in the circuit in Fig. 3.1.

SOLUTION Steps 1 through 4 are all implemented in Fig. 3.1. The closed surfaces specified in step 5 are also shown (denoted by CS 1, CS 2, and CS 3). Applying KCL to the three closed surfaces, we get

$$2e_1 + 5(e_1 + e_2 - e_3) + 5 = 0$$

$$5(e_1 + e_2 - e_3) + 3e_2 + 4(e_2 - 10) + 5 = 0$$

GENERAL ANALYSIS METHODS

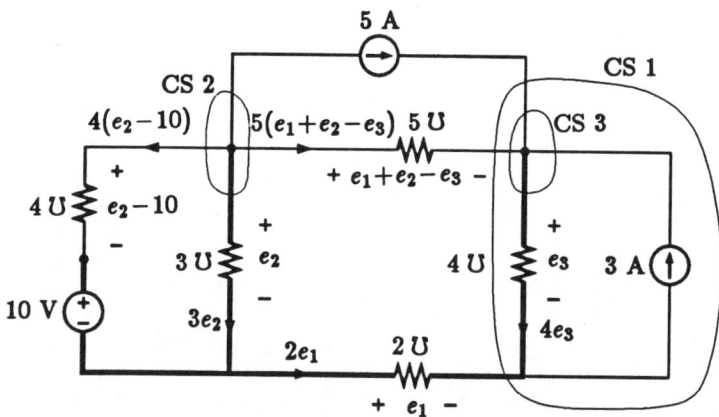

Figure 3.1: A cutset analysis problem.

$$5(e_1 + e_2 - e_3) - 4e_3 + 3 + 5 = 0$$

Solving these three equations simultaneously, we get

$$e_1 = -\frac{835}{306} \text{ V} \qquad e_2 = \frac{755}{153} \text{ V} \qquad e_3 = \frac{647}{306} \text{ V}$$

3.1.2 Loop analysis

In loop analysis, the unknown variables are link currents. From Section 2.3 it is clear that if we start with a tree, the introduction of each link creates a closed loop through some twigs. A convenient notion of this approach is to imagine that each link current "flows" through only those twigs with which the link forms a closed loop. Currents in such a flow pattern are known as *loop currents*. It should be clear that each twig current is simply the sum of all the loop currents that flow through that twig, and that the KCL is automatically satisfied for all closed surfaces since a loop current that enters a closed surface must also leave the closed surface.

The concept of the loop current is illustrated in Fig. 3.2(a) in which there are four links and four loop currents—i_1, i_2, i_3, and i_4. By letting i_1, i_2, i_3, and i_4 flow through only the twigs, we have

$$i_5 = i_2 - i_3 - i_4$$

$$i_6 = i_1 - i_2 + i_3$$

$i_7 = i_1 - i_4$

$i_8 = i_1 - i_2$

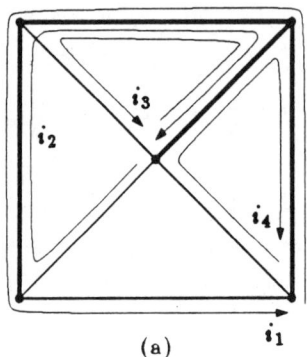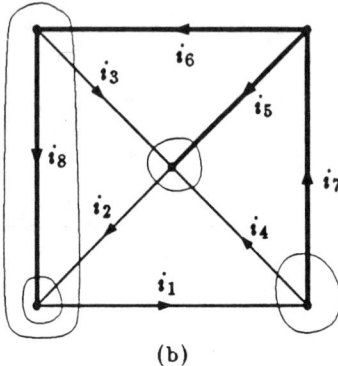

Figure 3.2: The loop currents.

Note that this process is equivalent to expressing all twig currents in terms of the link currents in such a way that KCL is satisfied for the four closed surfaces shown, each of which intersects only one twig.

The procedure for loop analysis may be outlined as follows:

1. Choose a tree and thus its corresponding cotree.
2. Assign a current variable to each nonsource link. (If a link consists of a current source, either independent or controlled, then the link current is either known or dependent.)
3. Express all twig currents in terms of link currents by summing up all loop currents that flow through each branch. This step is equivalent to drawing one closed surface for each twig so that it intersects only the twig (and some links) and then applying KCL to that closed surface.
4. Express all branch voltages (including those of the controlled voltage sources) in terms of link current variables.
5. Write one KVL equation for each loop formed by a nonsource link and part of the tree.
6. Solve these equations for the unknown current variables.

EXAMPLE 2. Determine the two link currents, i_1 and i_2, in the circuit in Fig. 3.3 by loop analysis.

GENERAL ANALYSIS METHODS

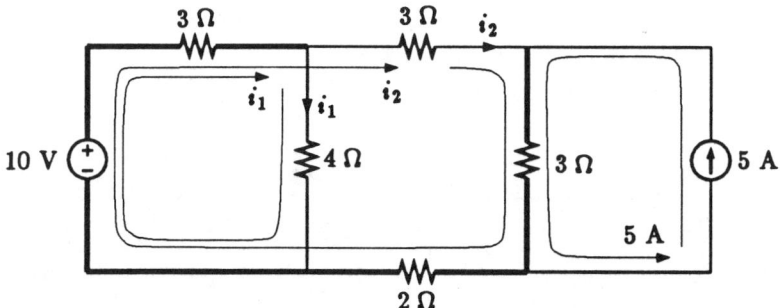

Figure 3.3: Circuits for loop analysis.

SOLUTION Steps 1 through 4 are implemented in the Fig. 3.3. KVL equations read

$$4i_1 - 10 + 3(i_1 + i_2) = 0$$

$$3i_2 + 3(i_2 + 5) + 2i_2 - 10 + 3(i_1 + i_2) = 0$$

Solving gives

$$i_1 = \frac{125}{68} \text{ A} \qquad i_2 = -\frac{65}{68} \text{ A}$$

3.1.3 Alternative methods of choosing voltage variables

The procedure of using twig voltages as our variables of analysis in cutset analysis is a direct and logical consequence when we ask ourselves which set of existing branch voltages constitute a set of independent key unknowns. We are not restricted to using only branch voltages as unknowns. Any voltage between a pair of nodes may also be used to define the voltage between these two nodes whether a branch exists between the pair of nodes or not. Hence, if we choose a set of *node-pair voltages* such that all node voltages are defined, these node-pair voltages may also serve as the key variables of the analysis.

The concept of a tree is still useful in that the set of node-pair voltages should be chosen in such a way that they form an imaginary tree.[1] When this is done, the set of node-pair voltages has all the

---

[1] Sometimes called a *pseudo-tree*.

attributes of a set of tree voltages and, therefore, may be used as the set of key variables in the analysis.

Fig. 3.4 shows a six-node network. We need five key voltage

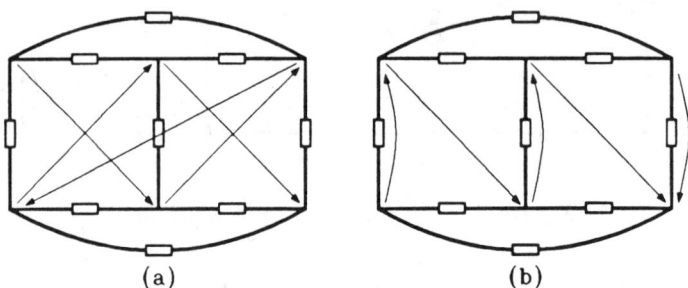

Figure 3.4: Node-pair voltages.

unknowns. Instead of using twig voltages, Fig. 3.4(a) shows five node-pair[2] voltages that have a tree-like structure. However, none of the five node-pair voltages is the voltage of any branch. These five voltages may also be used as the voltage unknowns, and cutset analysis procedure may be followed. In a way, each node-pair voltage arrow may be thought of as a branch that admits no current.

Fig. 3.4(b) shows another choice of five node-pair voltages, some of the which are also branch voltages. This set of five voltages may also be used as a set of independent unknowns.

A particularly useful and advantageous way of choosing a set of node-pair voltages is to choose all voltages to emanate from a common node. Fig. 3.5(a) shows such a choice. When this is done,

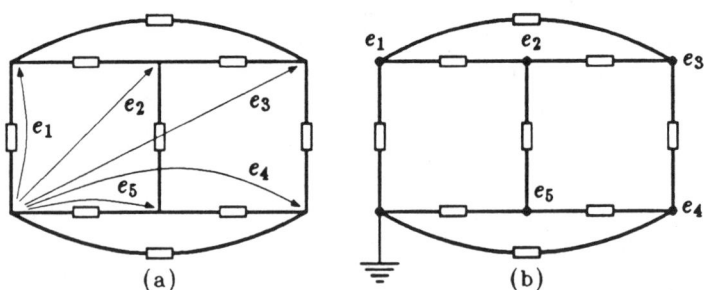

Figure 3.5: A special choice of node-pair voltages.

---

[2] Here, arrows are used to indicate voltage differences between nodes since plus and minus signs would be difficult to follow.

GENERAL ANALYSIS METHODS

all voltage unknowns are referred to the common node. If we ground this common node, then all node-pair voltages are simply the node voltages with reference to ground. This is done in Fig. 3.5(b). Such a choice greatly simplifies the diagram and is the basis on which node analysis in the next section is formulated.

3.1.4 Alternative methods of choosing current variables

The procedure of using link currents as unknowns in loop analysis is also a direct and logical consequence of designating a set of branch currents as the key independent current variables. There is another method of choosing a set of current variables when they are not actual branch currents.

The concept of *circulating currents* is an extension of the idea of loop currents. We may designate a number of fictitious currents flowing along some chosen paths. These currents are fictitious in that they may not be isolated physically or measured by meters. These circulating currents are assumed to flow along complete paths. Therefore, KCL is always satisfied with respect to any closed surface. But these circulating currents can be used as the variables of analysis as we can with loop currents.

Fig. 3.6 is an example of the concept of circulating currents.

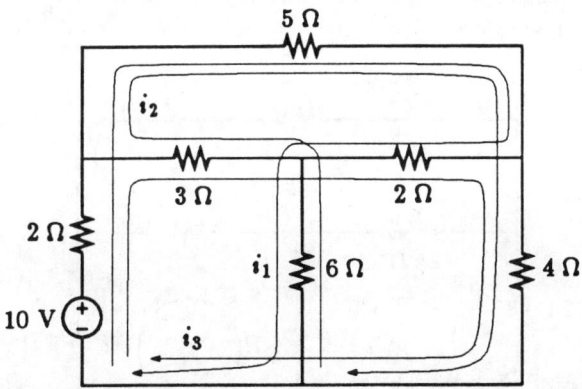

Figure 3.6: Example of circulating currents.

Three circulating currents—i_1, i_2, and i_3—are shown. None of these three circulating currents by themselves can be identified with any branch current. But any branch current can be expressed as some combination of the circulating currents. In Fig. 3.6, writing KVL equations around the three meshes give

$$2(i_1 + i_3) + 3(i_3 - i_2) + 6(i_1 - i_2) = 10$$

$$3(i_2 - i_3) + 5(i_1 + i_2) + 2(i_1 - i_3) = 0$$

$$2(i_3 - i_1) + 4(i_2 + i_3) + 6(i_2 - i_1) = 0$$

Solving gives

$$i_1 = \frac{245}{368} \text{ A} \qquad i_2 = -\frac{5}{368} \text{ A} \qquad i_3 = \frac{335}{368} \text{ A}$$

Once we have calculated all circulating currents, all branch currents can be obtained by properly combining the appropriate circulating currents.

Although circulating currents can be used as the unknowns of analysis and the number of circulating currents required is known in advance, we cannot randomly choose them without advance assurance that they will form an independent set. There is one scheme that will always give us a set of independent circulating currents. This scheme is to assign one circulating current for each mesh. Such circulating currents are known as *mesh currents*. Fig. 3.7 shows the three mesh currents for the circuit in Fig. 3.6. Mesh current are used in mesh analysis in Section 3.3.

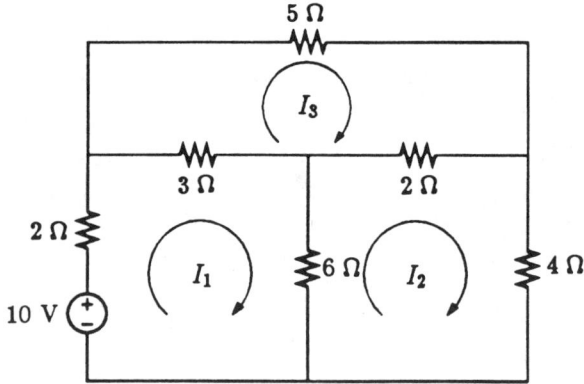

Figure 3.7: The mesh currents.

GENERAL ANALYSIS METHODS 83

There is one complication in using mesh currents as current unknowns. This occurs when a network is not a planar. A network is a planar when it can be drawn on a plane without any of its branches crossing each other. When a network is not a planar, there are more meshes than the number of independent current unknowns. Therefore, it is difficult to tell in advance whether a set of mesh currents are independent or not. Fortunately, nonplanar networks do not occur very often in practice.

3.2 Node analysis

In Section 3.1, we discussed several schemes for choosing a set of independent key voltage variables. The last of those (Subsection 3.1.3) calls for grounding one node. This grounded node is known as *ground*, *reference node*, or *datum node*. The ungrounded node voltages form a set of independent voltage unknowns. Now we shall utilize this method and formulate a general procedure to analyze any given network.

3.2.1 Networks with resistors and independent current sources

Node analysis is particularly simple to follow and understand if the network contains only resistors and independent current sources. The following are the steps for carrying out the node analysis:

1. Choose a reference node and ground that node.
2. Assign a voltage unknown variable to each ungrounded node.
3. Express all branch currents in terms of these node voltages.
4. Write a KCL equation for each node.
5. Solve these equations simultaneously for the node voltages.

EXAMPLE 3. Use node analysis to determine all currents in the circuit of Fig. 3.8.

SOLUTION This is three-node circuit. Recall that all nodes that are short-circuited must have the same potential. The circuit of Fig. 3.8 can be redrawn as shown in Fig. 3.9(a) to show that there are only three nodes electrically. Alternatively, we can take the

84 FUNDAMENTALS OF CIRCUIT ANALYSIS

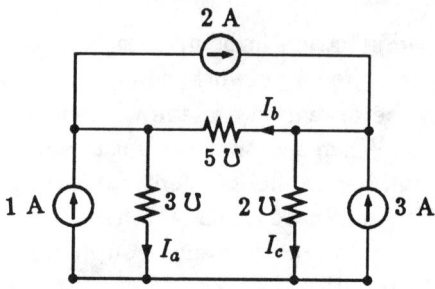

Figure 3.8: Circuit for Example 3.

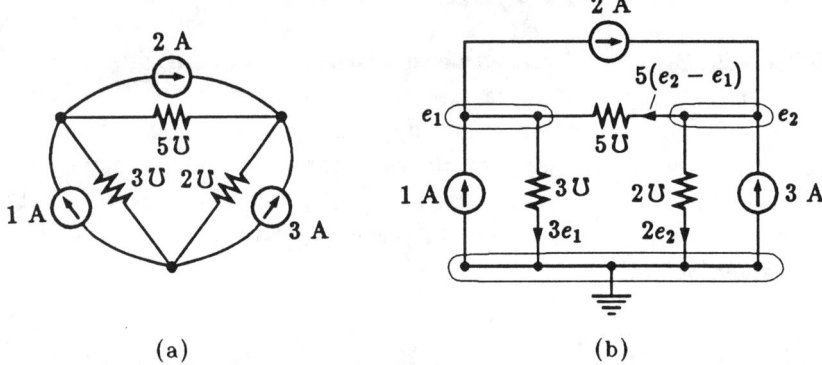

(a)   (b)

Figure 3.9: (a) Circuit of Fig. 3.8 redrawn. (b) Same circuit readied for node analysis.

the original diagram and encircle all points that are at the same potential. The encircled area can be viewed as an expanded node. This is done in Fig. 3.9(b). We will not do this in future circuits, as the student should get used to visualizing these expanded nodes when they occur. In Fig. 3.9(b) we also ground the bottom node and use it as the reference.

The three resistor currents are indicated in Fig. 3.9(b). Next we write one KCL equation for each ungrounded node. For node 1,

$$3e_1 - 5(e_2 - e_1) = 1 - 2 \tag{3.1}$$

For node 2,

$$2e_2 + 5(e_2 - e_1) = 3 + 2 \tag{3.2}$$

Eqs. (3.1) and (3.2) may be rewritten as

GENERAL ANALYSIS METHODS

$$8e_1 - 5e_2 = -1$$
$$-5e_1 + 7e_2 = 5$$

Applying Cramer's rule, we have

$$e_1 = \frac{\begin{vmatrix} -1 & -5 \\ 5 & 7 \end{vmatrix}}{\begin{vmatrix} 8 & -5 \\ -5 & 7 \end{vmatrix}} = \frac{18}{31} \text{ V}$$

$$e_2 = \frac{\begin{vmatrix} 8 & -1 \\ -5 & 5 \end{vmatrix}}{\begin{vmatrix} 8 & -5 \\ -5 & 7 \end{vmatrix}} = \frac{35}{31} \text{ V}$$

Thus

$$I_a = 3e_1 = \frac{54}{31} \text{ A} \qquad I_b = 5(e_2 - e_1) = \frac{51}{31} \text{ A} \qquad I_c = 2e_2 = \frac{70}{31} \text{ A}$$

In writing the node equation for a node, one good scheme to follow is to equate all currents leaving a node through all resistors to all currents entering the node from all current sources. For currents in resistors, let us refer to Fig. 3.10. It is clear that the current leaving node A is $G_{AB}(e_A - e_B)$. The current leaving node B is $G_{AB}(e_B - e_A)$. If we follow this rule Eq. (3.1) will read

$$3e_1 + 5(e_1 - e_2) = 1 - 2 \tag{3.3}$$

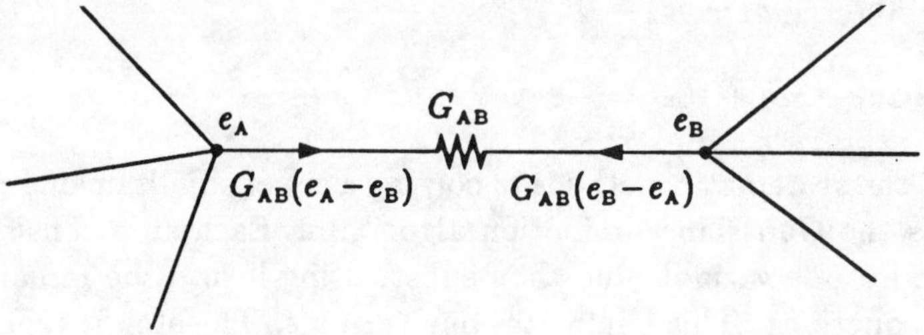

Figure 3.10: Current in a resistor between two nodes.

which, of course, contains the same information as Eq. (3.1). However, in following this convention, we no longer need to indicate the current direction every time.

EXAMPLE 4. Find the node voltages, e_1, e_2, and e_3, in the circuit of Fig. 3.11.

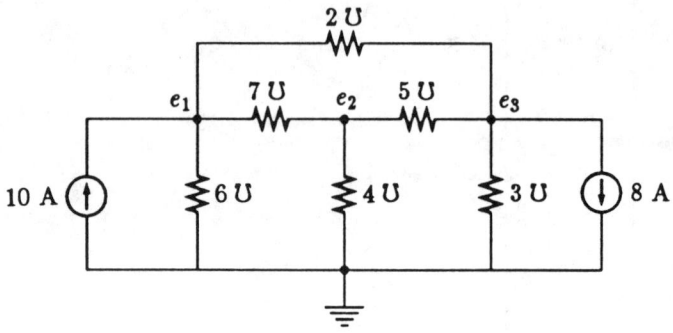

Figure 3.11: Circuit for Example 4.

SOLUTION Following the convention just described, the node equations will read

$$6e_1 + 7(e_1 - e_2) + 2(e_1 - e_3) = 10$$

$$7(e_2 - e_1) + 4e_2 + 5(e_2 - e_3) = 0$$

$$2(e_3 - e_1) + 5(e_3 - e_2) + 3e_3 = -8$$

Collecting like terms, we simplify the node equations to read

$$15e_1 - 7e_2 - 2e_3 = 10 \qquad (3.4)$$

$$-7e_1 + 16e_2 - 5e_3 = 0 \qquad (3.5)$$

$$-2e_1 - 5e_2 + 10e_3 = -8 \qquad (3.6)$$

One of the systematic methods of solving a set of simultaneous equations is the Gaussian elimination algorithm. Each step consists of solving for one variable and then substituting it into the remainder of the equations. This eliminates one variable. The step is repeated

GENERAL ANALYSIS METHODS

until only one variable remains. Once this remaining variable is known, the steps are reversed to find all other variables. We shall use this procedure to solve Eqs. (3.4), (3.5), and (3.6). Solving for e_1 from Eq. (3.4), we get

$$e_1 = \frac{1}{15}(7e_2 + 2e_3 + 10) \tag{3.7}$$

Substituting Eq. (3.7) into Eqs. (3.5) and (3.6), we get

$$191e_2 - 89e_3 = 70 \tag{3.8}$$

$$89e_2 - 146e_3 = 100 \tag{3.9}$$

This eliminates the first variable, e_1. We now solve for e_2 from Eq. (3.8) to get

$$e_2 = \frac{1}{191}(89e_3 + 70) \tag{3.10}$$

Substituting Eq. (3.10) into Eq. (3.9), we eliminate e_2 and obtain

$$e_3 = -\frac{78}{121} \text{ V} \tag{3.11}$$

Now we substitute Eq. (3.11) into Eq. (3.10) to get

$$e_2 = \frac{8}{121} \text{ V} \tag{3.12}$$

Finally, we substitute Eqs. (3.11) and (3.12) into Eq. (3.7) to get

$$e_1 = \frac{74}{121} \text{ V} \tag{3.13}$$

It is helpful to observe certain patterns in node equations written this way.

1. All terms on the left-hand side of each equation carry positive signs.
2. In every voltage term, the first voltage is always the voltage variable of the node for which the node equation is written.
3. In every voltage term, the second voltage is always the voltage variable of the node to which the current is directed. When a resistor

is connected between a node and ground, the second voltage is absent (or, more accurately, equal to zero).

4. Quantities representing the current in a resistor, but appearing in different equations, are always the negative of each other.

A review of a set of node equations written this way will help eliminate errors and give us a chance to double-check our work before proceeding further.

EXAMPLE 5. Analyze the circuit in Fig. 3.12 using node analysis.

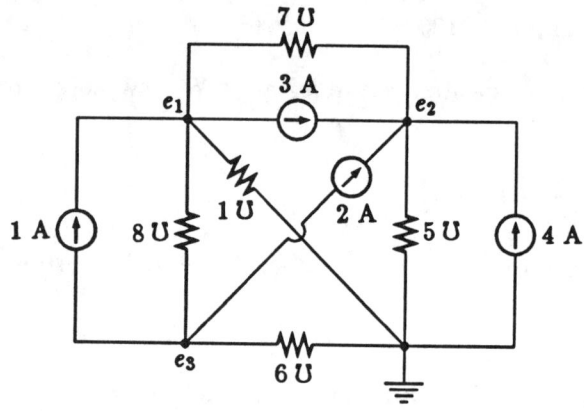

Figure 3.12: Circuit for Example 5.

SOLUTION   The node equations are

$$e_1 + 7(e_1 - e_2) + 8(e_1 - e_3) = 1 - 3$$

$$7(e_2 - e_1) + 5e_2 = 9$$

$$8(e_3 - e_1) + 6e_3 = -1 - 2$$

Simplify these equations to get

$$16e_1 - 7e_2 - 8e_3 = -2$$

$$-7e_1 + 12e_2 = 9$$

$$-8e_1 + 14e_3 = -3$$

GENERAL ANALYSIS METHODS

Write these equations in matrix form

$$\begin{bmatrix} 16 & -7 & -8 \\ -7 & 12 & 0 \\ -8 & 0 & 14 \end{bmatrix} \begin{bmatrix} e_1 \\ e_2 \\ e_3 \end{bmatrix} = \begin{bmatrix} -2 \\ 9 \\ -3 \end{bmatrix}$$

Hence

$$\begin{bmatrix} e_1 \\ e_2 \\ e_3 \end{bmatrix} = \begin{bmatrix} 16 & -7 & -8 \\ -7 & 12 & 0 \\ -8 & 0 & 14 \end{bmatrix}^{-1} \begin{bmatrix} -2 \\ 9 \\ -3 \end{bmatrix} = \frac{1}{1234} \begin{bmatrix} 258 \\ 1076 \\ -117 \end{bmatrix} \text{V}$$

or

$$e_1 = \frac{129}{617} \text{V} \qquad e_2 = \frac{538}{617} \text{V} \qquad e_3 = -\frac{117}{1234} \text{V}$$

EXERCISES

3.2.1 Use node analysis to find e_1 and e_2.

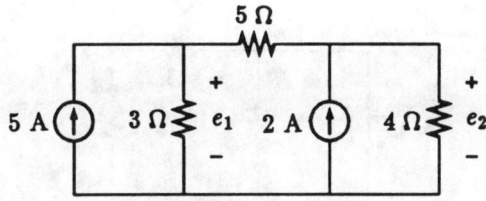

Ans. $e_1 = \dfrac{53}{4}$ V, $e_2 = \dfrac{31}{3}$ V.

3.2.2 Use node analysis to determine the appropriate voltages. Then determine i.

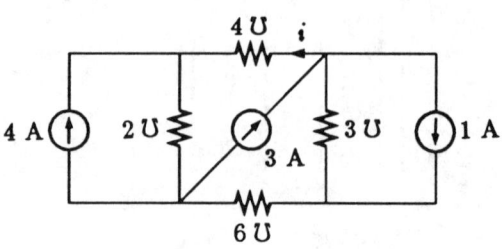

Ans. $-\dfrac{2}{3}$ A.

3.2.2 Networks containing controlled current sources

In node analysis, when controlled current sources are present, they can be treated exactly like independent sources first. Then we need to add the quantitative information regarding these controlled sources. Once we have this information, it can be treated as additional equations. Or else, this information can be incorporated into the node equations.

EXAMPLE 6. Analyze the circuit in Fig. 3.13 using node analysis.

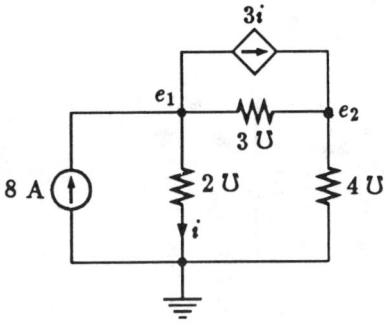

Figure 3.13: Circuit for Example 6.

GENERAL ANALYSIS METHODS

SOLUTION We proceed to write the node equations, treating the controlled source as if it were an independent source. The node equations read

$$2e_1 + 3(e_1 - e_2) = 8 - 3i \tag{3.14}$$

$$4e_2 + 3(e_2 - e_1) = 3i \tag{3.15}$$

For the controlled source, we have

$$i = 2e_1 \tag{3.16}$$

Substitute this into the node equations and rearrange to get

$$11e_1 - 3e_2 = 8$$

$$-9e_1 + 7e_2 = 0$$

Solving these two equations simultaneously, we get

$$e_1 = 1.12 \text{ V} \qquad e_2 = 1.44 \text{ V}$$

An alternative to this procedure is to solve Eqs. (3.14), (3.15), and (3.16) simultaneously. The values of e_1 and e_2 should check with the answers above, and the value for i will be obtained at the same time.

EXAMPLE 7. Determine the values of i and e in the circuit of Fig. 3.14.

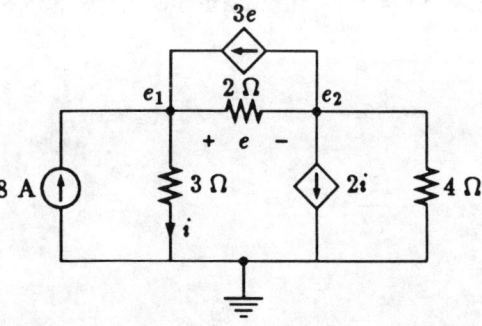

Figure 3.14: Circuit for Example 7.

SOLUTION The node equations are

$$\frac{1}{3}e_1 + \frac{1}{2}(e_1 - e_2) = 8 + 3e$$

$$\frac{1}{4}e_2 + \frac{1}{2}(e_2 - e_1) = -3e - 2i$$

The controlling quantities are

$$e = e_1 - e_2$$

$$i = \frac{1}{3}e_1$$

Substitute these controlling quantities into the node equations to get

$$\frac{1}{3}e_1 + \frac{1}{2}(e_1 - e_2) = 8 + 3(e_1 - e_2)$$

$$\frac{1}{4}e_2 + \frac{1}{2}(e_2 - e_1) = -8(e_1 - e_2) - \frac{2}{3}e_1$$

Rearrange to get

$$-\frac{13}{6}e_1 + \frac{5}{2}e_2 = 8$$

$$\frac{19}{6}e_1 - \frac{9}{4}e_2 = 0$$

Solving these two equations simultaneously, we get

$$e_1 = \frac{432}{73} \quad \text{and} \quad e_2 = \frac{608}{73}$$

Hence

$$e = e_1 - e_2 = -\frac{176}{73} \text{ V}$$

$$i = \frac{1}{3}e_1 = \frac{144}{73} \text{ A}$$

GENERAL ANALYSIS METHODS

EXERCISES

3.2.3 Determine i in the circuit using node analysis.

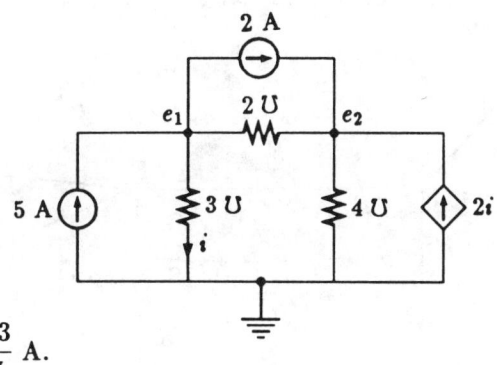

Ans. $\dfrac{33}{7}$ A.

3.2.4 Determine e and i in the circuit.

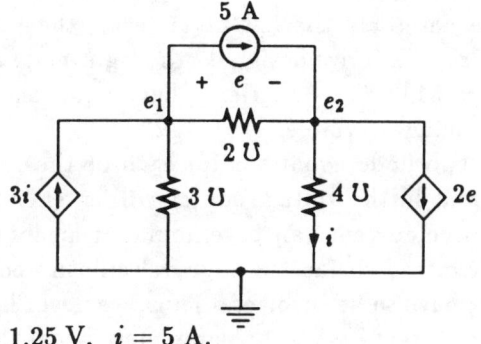

Ans. $e = 1.25$ V, $i = 5$ A.

3.2.3 Networks containing voltage sources

At a first glance, we might suspect that the presence of voltage sources complicates node analysis. But upon closer examination, we will find that voltage sources actually simplify node analysis. Two factors are in our favor in this consideration. First, a voltage source constrains the voltage difference between the two terminals to which it is connected. Therefore, once we know the voltage at one terminal, we also know the voltage at the other terminal of the voltage source.

The other factor is that a voltage source can sustain any current. Therefore, we are not concerned about the amount of currents flowing through the voltage sources.

We can take advantage of these two factors by considering nodes connected by voltage sources as a group. This is illustrated in Fig. 3.15, in which two nodes are constrained by the voltage source

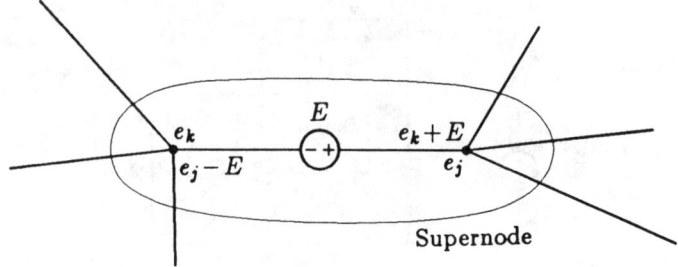

Figure 3.15: A voltage source connected between two nodes.

E. We can choose either of the two node voltages as our voltage variable. If we designate the left node as e_k, then the right node voltage will be $e_k + E$. If we designate the right node as e_j, then the left node voltage will be $e_j - E$. Hence between these two nodes, we only need one voltage variable.

If we write two node equations for each or the two constrained nodes, the current in the voltage source will enter both node equations. This source current can be eliminated algebraically. But if we consider the closed surface that encircles both nodes, the source current will not have to be involved. Thus, we never have to address what the source current is. We can completely bypass the source current in our analysis. The closed surfaces that encircle constrained nodes have been given different names, such as *generalized nodes, constrained nodes, extended nodes*, and more recently, *supernodes*. We shall adopt the term "supernode."

GENERAL ANALYSIS METHODS

EXAMPLE 8. Apply node analysis to the circuit of Fig. 3.16.

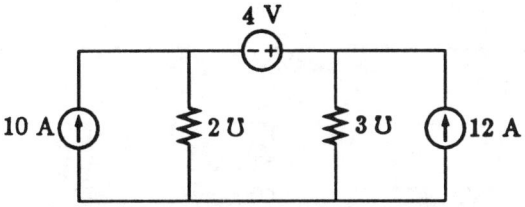

Figure 3.16: Circuit for Example 8.

SOLUTION Choose the bottom node as the reference and the voltage e_1 as the unknown as shown in the following diagram.

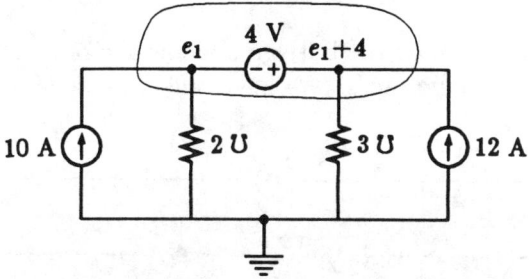

There is only one unknown node voltage. Therefore we have only one KCL equation. Writing the node equation for the supernode shown, we get

$$2e_1 + 3(e_1 + 4) = 12 + 10$$

Solving, we get

$$e_1 = 2 \text{ V}$$

Once e_1 is known, all other quantities including the current through the voltage source can be found with relative ease.

If a node is connected to ground through a voltage source, its voltage is known. That node actually becomes part of the supernode that includes the ground.

EXAMPLE 9. Find i_1, i_2, i_3, and i_4 in the circuit in Fig. 3.17.

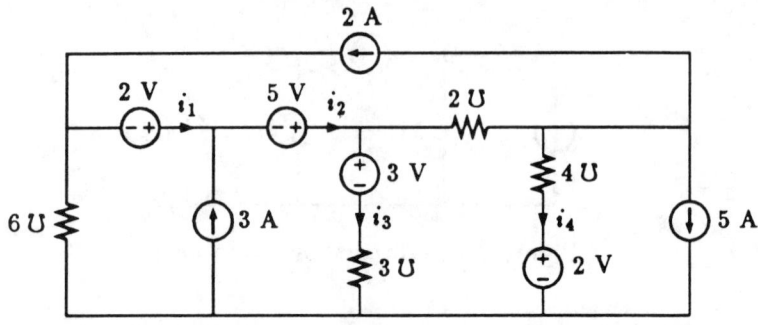

Figure 3.17: Circuit for Example 9.

SOLUTION Choose the bottom node as the reference. We have two unknown node voltages. We have one supernode and one ordinary node.

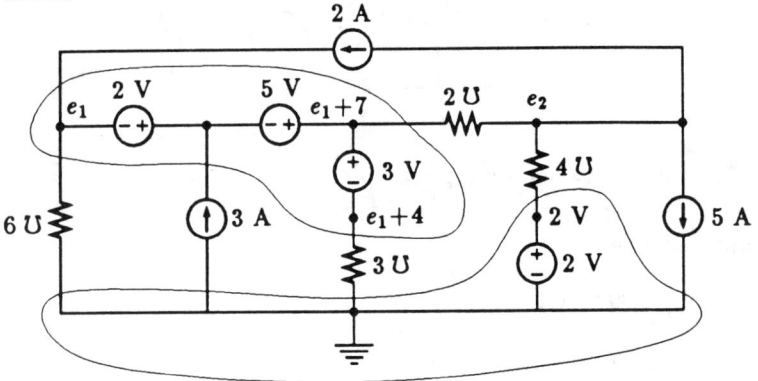

The KCL equations are

$$6e_1 + 3(e_1 + 4) + 2(e_1 + 7 - e_2) = 2 + 3$$

$$2(e_2 - e_1 - 7) + 4(e_2 - 2) = -2 - 5$$

Solving these two equations simultaneously, we get

$$e_1 = -\frac{48}{31} \text{ V} \qquad e_2 = \frac{123}{62} \text{ V}$$

Hence

GENERAL ANALYSIS METHODS 97

$$i_1 = 2 - 6e_1 = \frac{350}{31} \text{ A}$$

$$i_2 = i_1 + 3 = \frac{443}{31} \text{ A}$$

$$i_3 = 3(e_1 + 4) = \frac{228}{31} \text{ A}$$

$$i_4 = 4(e_2 - 2) = -\frac{2}{31} \text{ A}$$

EXERCISE

3.2.5 Use node analysis to find i in the circuit.

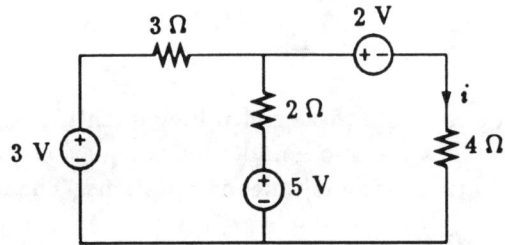

Ans. $\frac{11}{26}$ A.

If a network contains controlled voltage sources, we can first treat them as if they were independent voltages sources. Then we need to express the strengths of these voltage sources in terms of the unknown node voltages, much like we did with the controlled current sources.

EXAMPLE 10. Determine e_1 and e_2 in the circuit in Fig. 3.18.

SOLUTION We have two voltage unknowns. The KCL equations read

$$2e + 3(e_1 + 6) + 2(e_1 + 6 + 3i - 5e) + 3(e_1 + 6 + 3i - e_2) = 2$$

$$3(e_2 - e_1 - 6 - 3i) + 4e_2 = -2$$

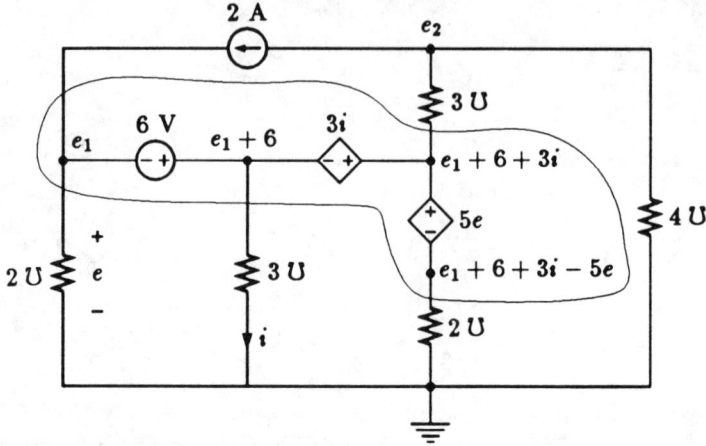

Figure 3.18: Circuit for Example 10.

For the controlled sources, we have

$$e = e_1$$

$$i = 3(e_1 + 6)$$

We could either substitute these controlling quantities into the KCL equations and then solve two simultaneous equations, or we could treat all four equations as simultaneous equations. The final answers are

$$e_1 = -\frac{1678}{225} \text{ V} \qquad e_2 = -\frac{98}{15} \text{ V}$$

EXERCISES

3.2.6 Determine e in the circuit.

Ans. $e = \dfrac{9}{25}$ V.

GENERAL ANALYSIS METHODS 99

3.2.7 Determine e in the circuit.

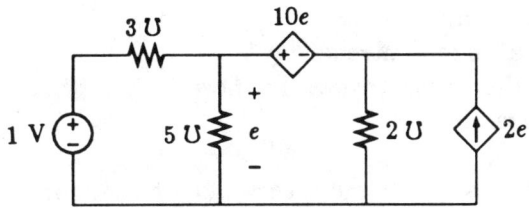

Ans. -0.25 V.

3.2.8 Determine i and e in the circuit.

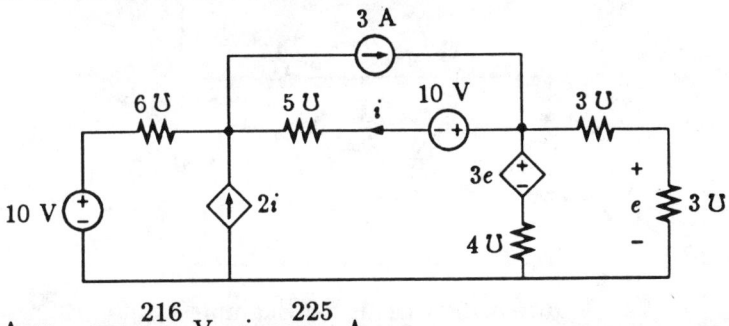

Ans. $e = \dfrac{216}{13}$ V, $i = \dfrac{225}{13}$ A.

3.3 Mesh Analysis

In Section 3.1, we discussed several schemes of choosing a set of key independent current variables. The last of those (Subsection 3.1.4) calls for assigning one mesh current for each interior mesh (excluding the outermost mesh). This is the basis of the mesh analysis we are going to develop in this section. We shall assume that every network we deal with is a planar network.

3.3.1 Networks with resistors and independent voltage sources

When a network contains only resistors and independent voltage sources, the procedure for mesh analysis is as follows:

1. Assign a mesh-current variable to each mesh.

2. Write one KVL equation for each mesh, considering the mesh currents as circulating currents.
3. Solve for all mesh currents.
4. Combine the mesh currents that flow by each branch to obtain each branch current.

EXAMPLE 11. With the mesh currents assigned for the network in Fig. 3.19, determine I_a, I_b, and I_c.

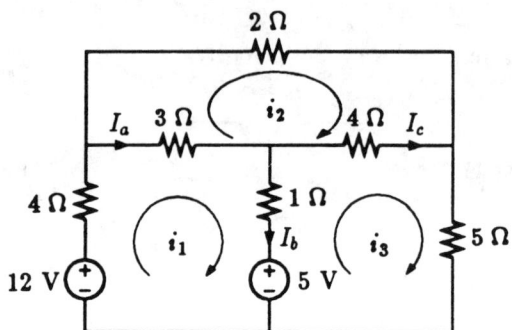

Figure 3.19: Circuit for Example 11.

SOLUTION KVL equations for the three meshes are

$$4i_1 + 3(i_1 - i_2) + 1(i_1 - i_3) = 12 - 5$$

$$2i_2 + 4(i_2 - i_3) + 3(i_2 - i_1) = 0$$

$$1(i_3 - i_1) + 4(i_3 - i_2) + 5i_3 = 5$$

Solving the three simultaneous equations, we get

$$i_1 = \frac{89}{67} \text{ A} \qquad i_2 = \frac{59}{67} \text{ A} \qquad i_3 = \frac{66}{67} \text{ A}$$

Hence

$$I_a = i_1 - i_2 = \frac{30}{67} \text{ A}$$

$$I_b = i_1 - i_3 = \frac{23}{67} \text{ A}$$

$$I_c = i_3 - i_2 = \frac{7}{67} \text{ A}$$

GENERAL ANALYSIS METHODS

In Example 11, we have assigned the directions of all mesh currents to be clockwise. There are some advantages in doing this. As Example 11 illustrates, the following patterns exist and are worth observing.

1. The exterior branch current is equal to a mesh current.
2. The interior branch current is always the difference between the two adjacent mesh currents.
3. We have placed all voltage drops across all resistors around a mesh on the left-hand side of each mesh equation. We also placed all voltage rises around a mesh on the right-hand side of each mesh equation.
4. All left-hand side terms have positive signs and all current quantities are the differences of two mesh currents. They all begin with the mesh current of the mesh whose KVL equation is being written.
5. The right-hand side of each equation is the sum of the voltage rises of the mesh in the direction of the mesh current.

This pattern of assigning the mesh currents and writing KVL equations is not the only good way of implementing mesh analysis. There is nothing sacred about this convention. One really doesn't have to follow any set convention as long as the basic rules are followed—Ohm's law and Kirchhoff's laws. Students are encouraged to form their own habits of doing mesh analysis.

EXERCISES

3.3.1 Find the two mesh currents in the circuit.

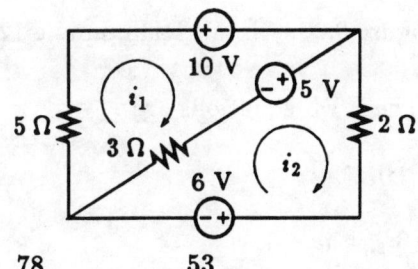

Ans. $i_1 = -\dfrac{78}{31}$ A, $i_2 = -\dfrac{53}{31}$ A.

102 FUNDAMENTALS OF CIRCUIT ANALYSIS

3.3.2 Determine the three mesh currents in the circuit.

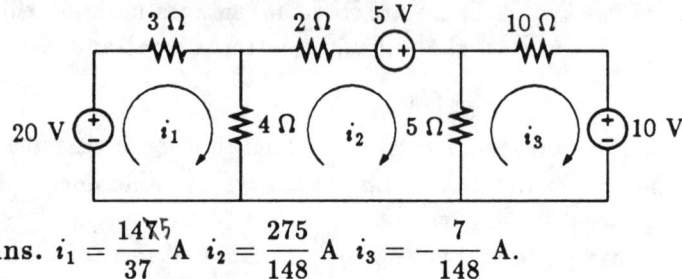

Ans. $i_1 = \dfrac{1475}{37}$ A $i_2 = \dfrac{275}{148}$ A $i_3 = -\dfrac{7}{148}$ A.

3.3.2 Networks containing controlled voltage sources

In mesh analysis, if controlled voltage sources are present, they can be treated as if they were independent voltage sources first. Then, in addition to KVL equations, we need to furnish the expressions for the controlling quantities in terms of the mesh currents. Once we have this information, it can be treated as additional equations. Or, this information can be incorporated into the KVL equations.

EXAMPLE 12. Obtain the mesh currents in the circuit of Fig. 3.20.

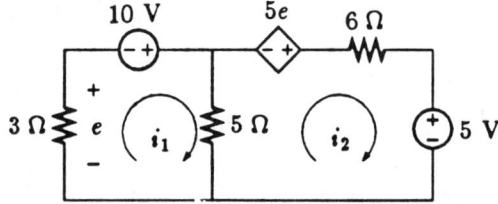

Figure 3.20: Circuit for Example 12.

SOLUTION The KVL equations are

$$3i_1 + 5(i_1 - i_2) = 10$$

$$5(i_2 - i_1) + 6i_2 = 5e - 5$$

For the controlled source,

GENERAL ANALYSIS METHODS

$$e = -3i_1$$

Substitute this into the second KVL equation and we get

$$5(i_2 - i_1) + 6i_2 = -15i_1 - 5$$

Solving this equation with the first KVL equation simultaneously gives

$$i_1 = \frac{85}{138} \text{ A} \qquad i_2 = -\frac{70}{69} \text{ A}$$

EXAMPLE 13. Obtain the mesh currents in the circuit of Fig. 3.21.

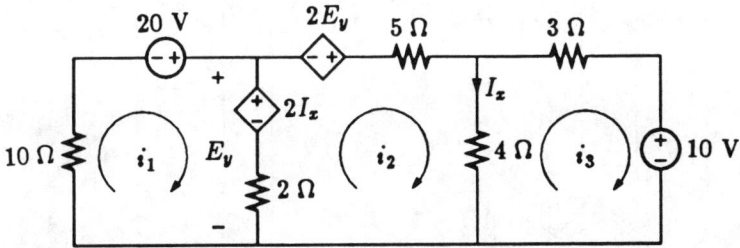

Figure 3.21: Circuit for Example 13.

SOLUTION KVL equations are

$$10i_1 + 2(i_1 - i_2) = 20 - 2I_x$$

$$2(i_2 - i_1) + 5i_2 + 4(i_2 - i_3) = 2I_x + 3E_y$$

$$4(i_3 - i_2) + 3i_3 = -10$$

For the controlling quantities,

$$E_y = 2(i_1 - i_2) + 2I_x$$

$$I_x = i_2 - i_3$$

We can treat these relationships as five simultaneous equations. Writing them in matrix form, we have

$$\begin{bmatrix} 12 & -2 & 0 & 0 & 2 \\ -2 & 11 & -4 & -2 & -2 \\ 0 & -4 & 7 & 0 & 0 \\ 2 & -2 & 0 & -1 & 2 \\ 0 & 1 & -1 & 0 & -1 \end{bmatrix} \begin{bmatrix} i_1 \\ i_2 \\ i_3 \\ E_y \\ I_z \end{bmatrix} = \begin{bmatrix} 20 \\ 0 \\ -10 \\ 0 \\ 0 \end{bmatrix}$$

Hence

$$\begin{bmatrix} i_1 \\ i_2 \\ i_3 \\ E_y \\ I_z \end{bmatrix} = \begin{bmatrix} 12 & -2 & 0 & 0 & 2 \\ -2 & 11 & -4 & -2 & -2 \\ 0 & -4 & 7 & 0 & 0 \\ 2 & -2 & 0 & -1 & 2 \\ 0 & 1 & -1 & 0 & -1 \end{bmatrix}^{-1} \begin{bmatrix} 20 \\ 0 \\ -10 \\ 0 \\ 0 \end{bmatrix}$$

$$= \frac{1}{201} \begin{bmatrix} 310 \\ 240 \\ -150 \\ 920 \\ 390 \end{bmatrix}$$

We have

$$i_1 = \frac{310}{201} \text{ A} \qquad i_2 = \frac{80}{67} \text{ A} \qquad i_3 = -\frac{50}{67} \text{ A}$$

GENERAL ANALYSIS METHODS

EXERCISES

3.3.3 Obtain the value of E in the circuit using mesh analysis.

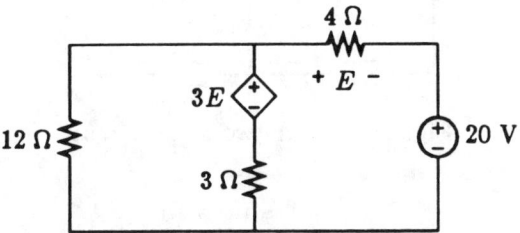

Ans. 25 V.

3.3.4 Determine I_x and E_y in the circuit using mesh analysis.

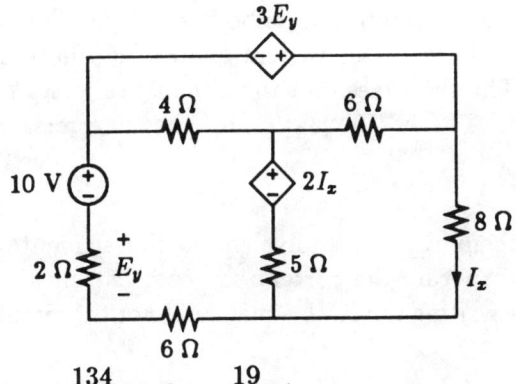

Ans. $E_y = -\dfrac{134}{109}$ V $I_x = \dfrac{19}{109}$ A.

3.3.3 Networks containing current sources

When current sources are present in a network, they have two effects on mesh analysis. One is that a current source constrains some mesh currents. The other is that the voltages across these current sources are not relevant to the KVL equations since current sources can sustain any voltages.

Take the two-mesh circuit of Fig. 3.22. The current source I_s forces

FUNDAMENTALS OF CIRCUIT ANALYSIS

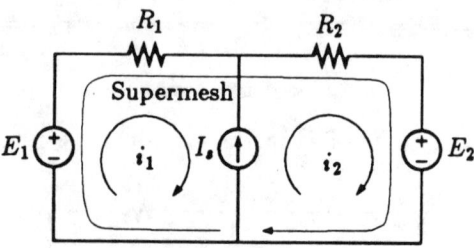

Figure 3.22: A two-mesh circuit.

$$i_2 - i_1 = I_s \qquad (3.17)$$

It is not necessary to write two mesh equations. If we did, each equation would have to include a term representing the voltage across the current source. Upon eliminating this voltage, we would obtain the KVL equation for the loop that embodies both meshes. Such a loop that ignores the presence of current sources can logically be called a *supermesh*. The supermesh for the circuit in Fig. 3.22 is indicated on the diagram. The KVL equation for the supermesh is

$$i_1 R_1 + i_2 R_2 = E_1 - E_2 \qquad (3.18)$$

Another scheme that is equivalent to the elementary approach above is to incorporate the current source constraint into the mesh currents before writing the KVL equations. For the circuit of Fig. 3.22, we have

$$i_2 = I_s + i_1 \qquad (3.19)$$

If we use $I_s + i_1$ in place of i_2 as is done in Fig. 3.23, the constraint of the current source is already satisfied. We can then proceed with the KVL equation for the supermesh, which would read

$$i_1 R_1 + (I_s + i_1) R_2 = E_1 - E_2 \qquad (3.20)$$

for this particular circuit.

There is still another variation on the two methods described above. We can direct all mesh currents around supermeshes only. At the same time, we also let all source currents circulate around the supermeshes. This way, only the minimum number of mesh currents

GENERAL ANALYSIS METHODS   107

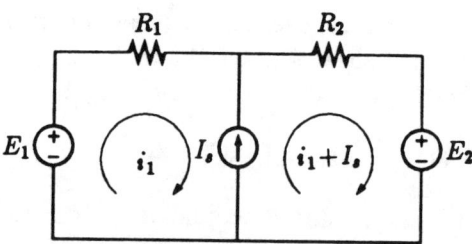

Figure 3.23: Mesh currents that satisfies the current source constraint.

will be introduced and all source currents retain their original values. For the circuit in Fig. 3.22, this scheme will result in the current pattern of Fig. 3.24. The KVL equation for this circuit is obviously the same as Eq. (3.20).

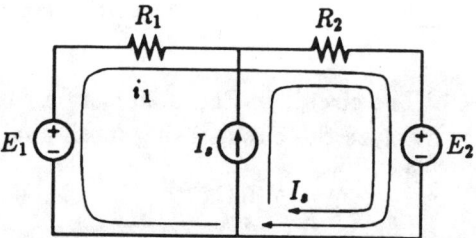

Figure 3.24: Another scheme of directing currents.

EXAMPLE 14. Determine I in the circuit in Fig. 3.25.

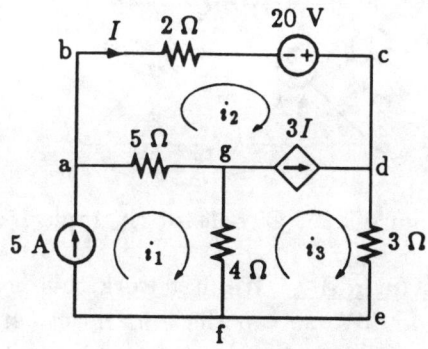

Figure 3.25: Circuit for Example 14.

108   FUNDAMENTALS OF CIRCUIT ANALYSIS

SOLUTION There are three mesh currents. But there are two current sources. Therefore, there is only one independent mesh current. The two current sources place the following constraints on the three mesh currents.

$$i_1 = 5 \quad \text{and} \quad i_3 - i_2 = 3I = 3i_2 \implies i_2 = \frac{1}{4}i_3$$

We next write a KVL equation around the supermesh abcdefga. It reads

$$2i_2 + 3i_3 + 4(i_3 - i_1) + 5(i_2 - i_1) = 20$$

We can now either substitute the constraining equations into the mesh equation or solve the three equations simultaneously. Either approach gives

$$I = i_2 = \frac{13}{7} \text{ A}$$

EXAMPLE 15. The circuit in Fig. 3.26 has four meshes and two current sources. Analyze the circuit using mesh analysis.

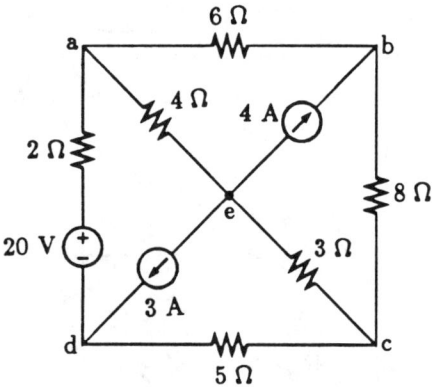

Figure 3.26: Circuit for Example 15.

SOLUTION **Method 1**. We first work this problem using the elementary method. We set up the four mesh currents as shown. The current sources constrain the mesh currents as follows

$$i_1 - i_4 = 3 \qquad i_3 - i_2 = 4$$

GENERAL ANALYSIS METHODS 109

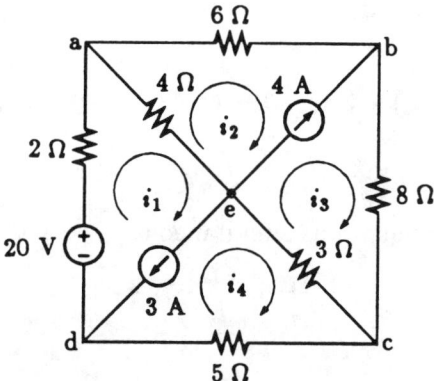

We next obtain the two supermesh KVL equations. Around daecd, we have

$$2i_1 + 4(i_1 - i_2) + 3(i_4 - i_3) + 5i_4 = 20$$

Around abcea, we have

$$4(i_2 - i_1) + 6i_2 + 8i_3 + 3(i_3 - i_4) = 0$$

Solving yields

$$i_1 = \frac{23}{7} \text{ A} \qquad i_2 = -\frac{10}{7} \text{ A}$$

$$i_3 = \frac{18}{7} \text{ A} \qquad i_4 = \frac{2}{7} \text{ A}$$

Method 2. We now impose the current source constraints in advance. The setup is shown below.

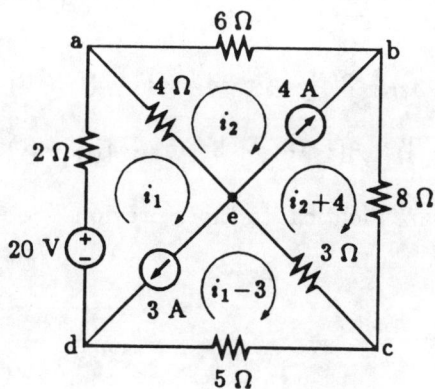

FUNDAMENTALS OF CIRCUIT ANALYSIS

We next obtain the two KVL equations around the two supermeshes. We get

$$2i_2 + 4(i_1 - i_2) + 3(i_1 - 3 - 4 - i_2) + 5(i_1 - 3) = 20$$

$$6i_2 + 8(i_2 + 4) + 3(i_2 + 4 - i_1 + 3) + 4(i_2 - i_1) = 0$$

Solving these two equations simultaneously, we get

$$i_1 = \frac{23}{7} \text{ A} \qquad i_2 = -\frac{10}{7} \text{ A}$$

Method 3. We now direct the mesh currents around the supermesh only and let the source currents circulate around the supermesh. The current pattern is shown in the following diagram.

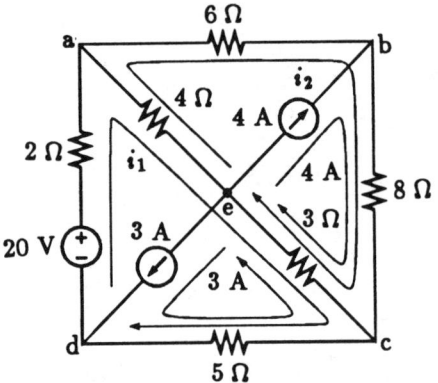

Now we write the KVL equations around the two supermeshes. They are

$$2i_1 + 4(i_1 - i_2) + 3(i_1 - i_2 - 3 - 4) + 5(i_1 - 3) = 20$$

$$6i_2 + 8(i_2 + 4) + 3(i_2 - i_1 + 3 + 4) + 4(i_2 - i_1) = 0$$

These equations are identical to those previously obtained by using Method 2.

GENERAL ANALYSIS METHODS 111

3.4 Duality

The student might have noticed through the chapter that there are two parallel tracks of material that are similar to each other. In fact, if we put them side-by-side, we will see that they are exactly alike except for the interchange of certain key words. This parallelism is a direct extension of the principle of duality we introduced in Chapter 2. As an example, we can see that the following two statements are dual of each other:

In *node* analysis, we use the *ungrounded node voltages* as unknowns. The number of *KCL* equations is equal to the number of nonsource *twigs*.

In *mesh* analysis, we use the *interior mesh currents* as unknowns. The number of *KVL* equations is equal to the number of nonsource *links*.

Therefore, we can add the following to our collection of dual entities:

Branch voltage	Branch current
Tree	Cotree
Twig	Link
Twig voltage	Link current
Cutset or closed surface	Loop
Cutset analysis	Loop analysis
Node	Mesh
Ungrounded node	Interior mesh
Node-pair voltage	Circulating current
Node voltage	Mesh current
Grounded node	Exterior mesh
Supernode	Supermesh
Node analysis	Mesh analysis

The student should be cautioned that there are some complications in carrying this duality to its limit with general network configurations. Although duality can be extended to all networks, a number of fine points should be considered and some exceptions accounted for. This is studied in network topology. However, with our limited experience with circuits and without claiming too much generality, the observation of duality remains both useful and helpful in appreciating the general approach to most practical circuit problems.

To illustrate the duality of node analysis and mesh analysis, we superimpose two dual circuits in Fig. 3.27.

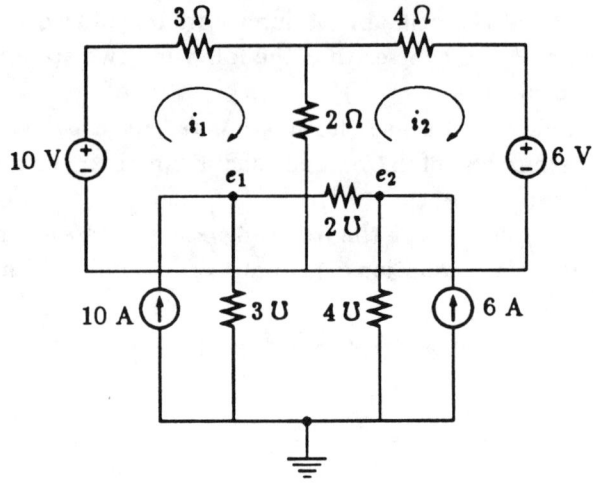

Figure 3.27: Two dual circuits.

For the circuit with voltage sources, we use mesh analysis. The mesh equations are

$$3i_1 + 2(i_1 - i_2) = 10$$

$$4i_2 + 2(i_2 - i_1) = 6$$

For the circuit with current sources, we use node analysis. The node equations are

$$3e_1 + 2(e_1 - e_2) = 10$$

$$4e_2 + 2(e_2 - e_1) = 6$$

These two sets of equations are clearly dual of each other.

3.5 Some concluding remarks

In this chapter, we introduced four general systematic methods of analyzing networks of any complexity. The cutset analysis uses twig voltages as unknown variables, and the equations are KCL equations for a set of appropriately chosen closed surfaces (or cutsets). The loop analysis uses link currents as unknown variables, and the equations are KVL equations around a set of appropriately chosen loops. These two methods are logical and are direct results of trying to develop general methods for analyzing networks using actual branch voltages or branch currents as unknowns. Although they are not particularly difficult to apply, they are not as efficient and streamlined as node and mesh analyses.

We introduced cutset and loop analyses here primarily for intellectual reasons. We placed no emphasis on the details of these methods and their application to the analysis of actual networks. These two methods do serve as stepping stones for the development of node and mesh analyses as there is a great deal of similarity between the two pairs of analysis methods. Either pair of methods may be used to arrive at the basic conclusions. In other words, we regard node and mesh analyses as extensions and refinements of cutset and loop analyses. From a practical point of view, cutset and loop analyses are almost never utilized because they are much more cumbersome to apply, compared with node and mesh analyses.

Between node and mesh analyses, node analysis is usually the method of choice for the majority of circuit analysis problems. There are several reasons for this preference. First is the relative simplicity of implementing node analysis.

The other reason that node analysis is the preferred method is that networks can be described verbally in terms of connections among nodes with no ambiguity and no confusion. Imagine that you want to describe a network to another person over the telephone. You tell that person how many nodes you have and what is connected between each pair of nodes. No matter how that person lays out the nodes, if that person has followed your description correctly, the network will be completely equivalent to yours electrically regardless of how the nodes are laid out. The two networks may be different in appearance, but they are electrically identical.

This is not the case if the same thing is attempted with meshes.

For this reason, almost all computer-aided network analysis and design programs use node analysis.

There are some situations in which mesh analysis might be preferred to node analysis. One such situation is when a network has far fewer mesh equations then node equations. The other situation is when a network contains mutual inductances—the element we will treat in Chapter 11.

We can determine the number of node equations and the number of mesh equations for a given network in advance. The number of node (or cutset) equations is equal to $N-1$, where N is the number of nonsource tree branches or the number of ungrounded nodes or supernodes. The number of mesh (or loop) equations is equal to the number of nonsource links or the number of interior meshes or supermeshes.

Problems

3.1 Use cutset analysis to determine all voltages in the circuit. Then compute the power delivered to each resistor.

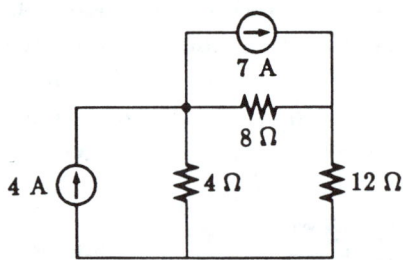

3.2 Use cutset analysis to determine E_1 and E_2.

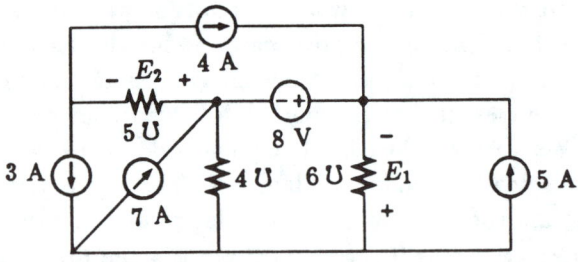

GENERAL ANALYSIS METHODS

3.3 Use cutset analysis to determine E_1 and E_2.

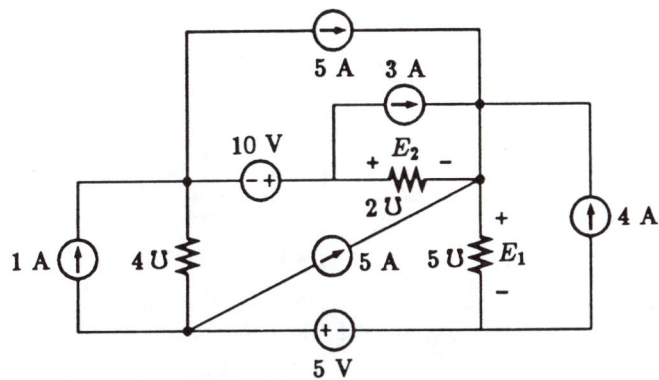

3.4 Use cutset analysis to determine E_1, E_2, and E_3.

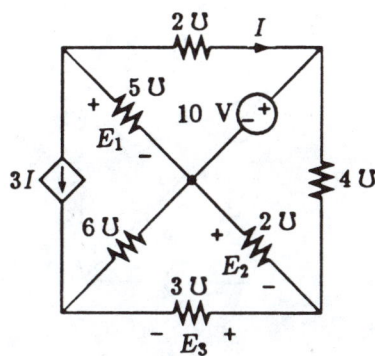

3.5 Use loop analysis to determine I_1, I_2, and I_3.

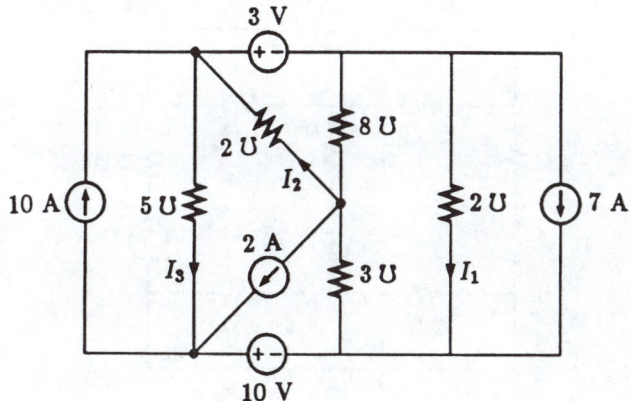

116 FUNDAMENTALS OF CIRCUIT ANALYSIS

▲▲ **3.6** Use loop analysis to determine I_1, I_2, and I_3.

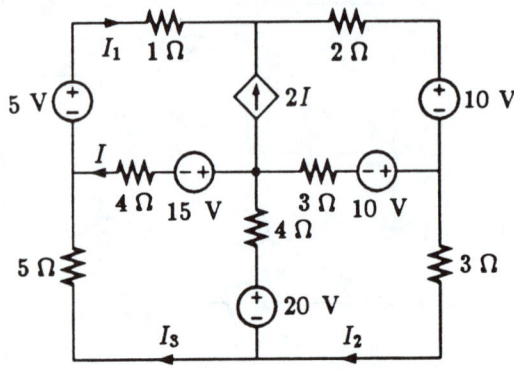

△ **3.7** Use loop analysis to determine I_1 and I_2.

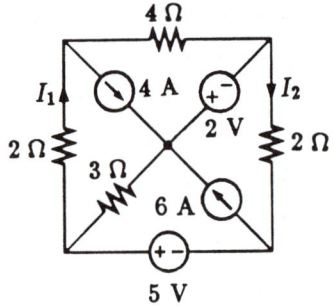

3.8 Use loop analysis to determine I_1 and I_2.

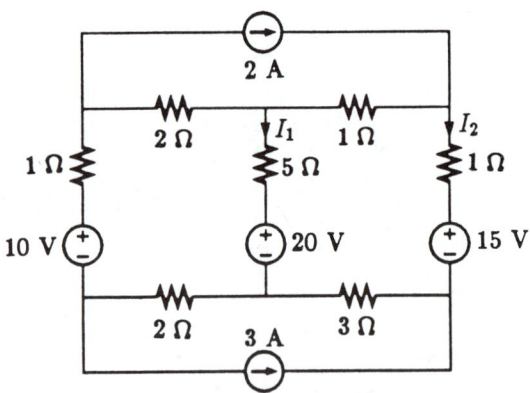

GENERAL ANALYSIS METHODS

△ **3.9** Use node analysis to determine E_1, E_2, and E_3.

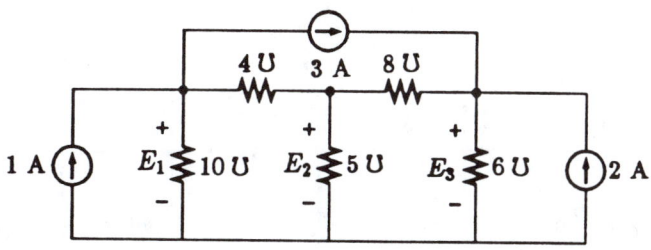

3.10 First perform node analysis on the circuit. Then determine all e's.

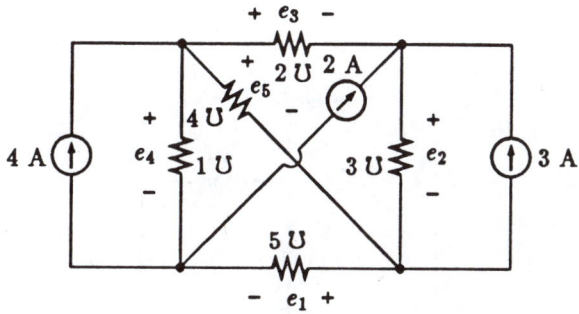

△△ **3.11** Use node analysis to determine e_1 and e_2.

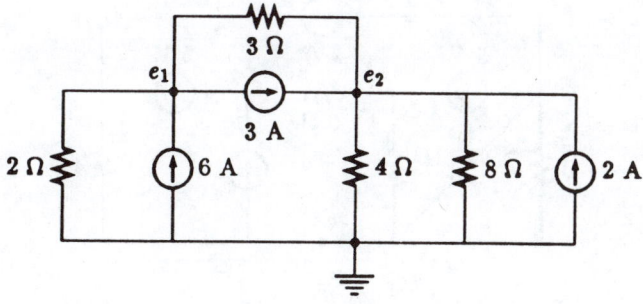

3.12 Use node analysis to determine e_1, e_2, and e_3.

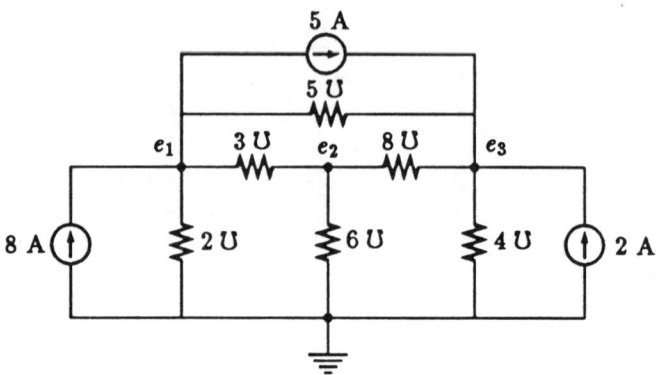

3.13 First perform node analysis on the circuit. Then determine I.

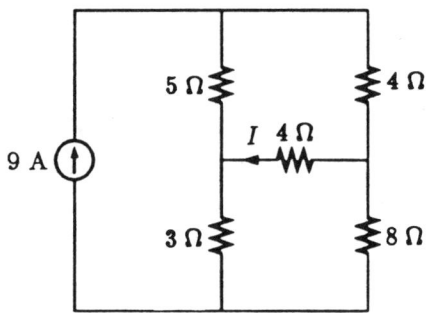

3.14 Use node analysis to determine E_1 and E_2.

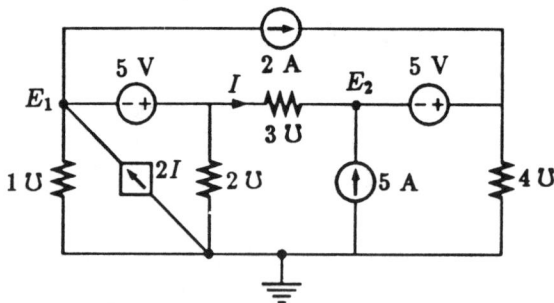

GENERAL ANALYSIS METHODS

3.15 Use node analysis to determine E.

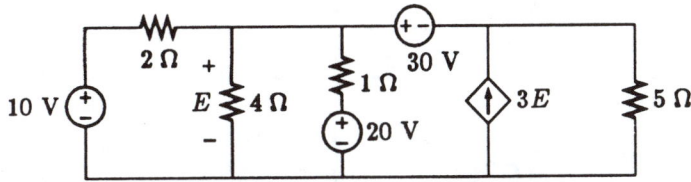

3.16 First perform node analysis on the circuit. Then determine E.

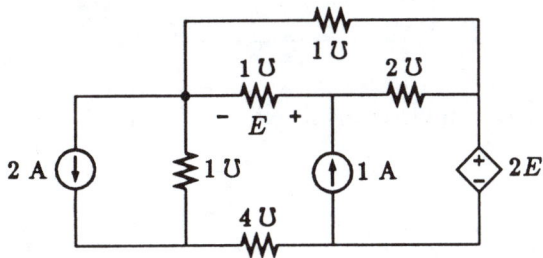

3.17 Use node analysis to determine e_1 and e_2.

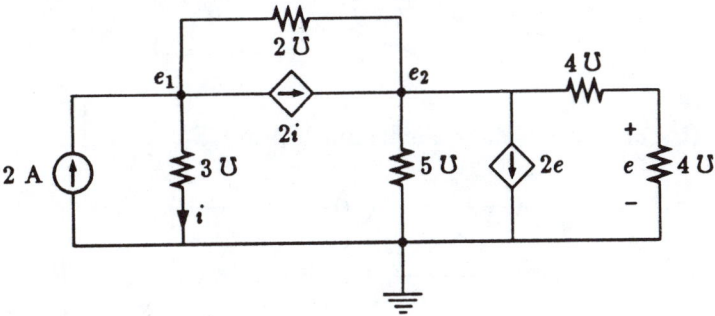

3.18 Use node analysis to determine v_1 and v_2.

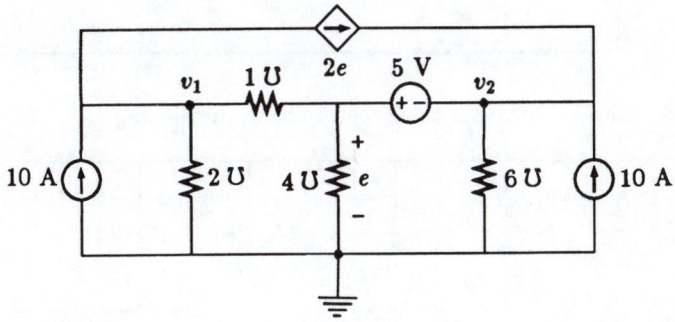

3.19 Use node analysis to determine E.

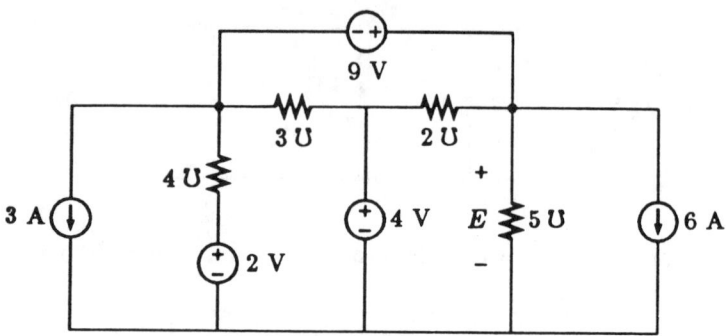

3.20 First perform node analysis on the circuit. Then determine I.

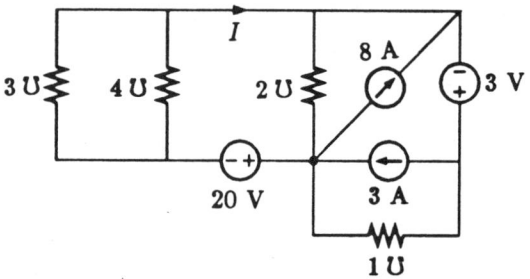

3.21 Use node analysis to determine E_1 and E_2.

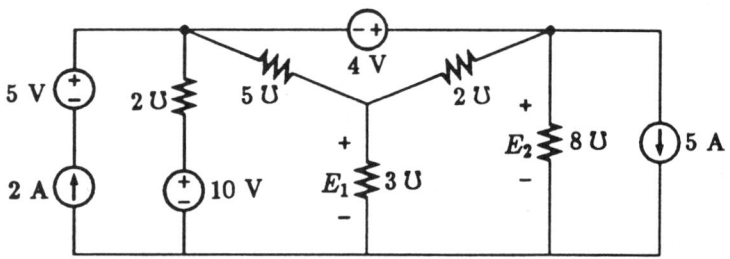

3.22 Use node analysis to determine e and v.

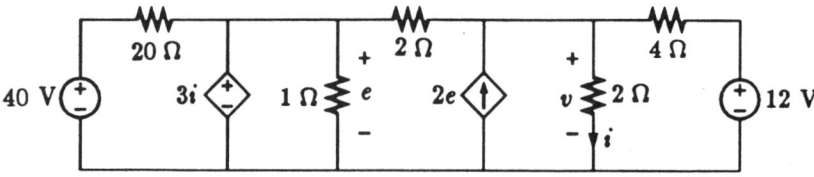

GENERAL ANALYSIS METHODS

3.23 Use node analysis to determine E_1 and E_2.

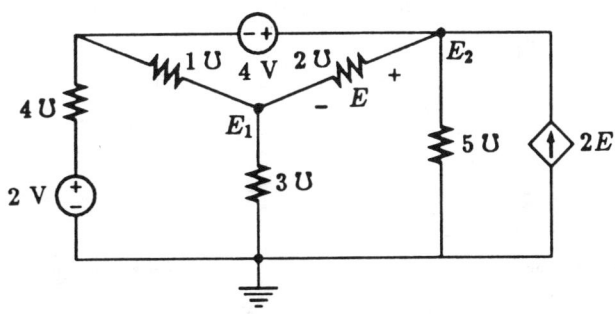

3.24 Use node analysis to determine E_1 and E_2.

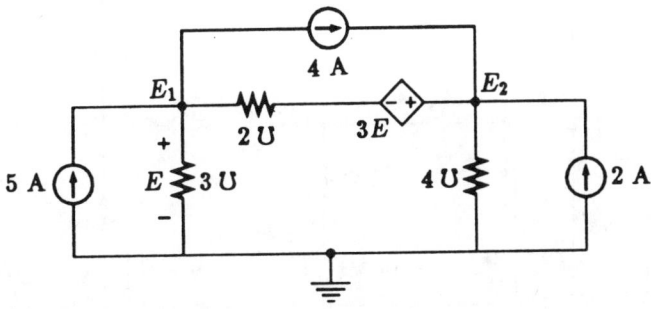

3.25 Use node analysis to determine E_1 and E_2.

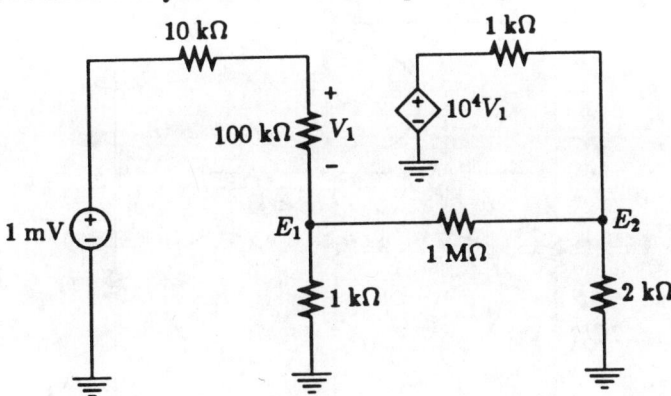

3.26 Use node analysis to determine E_1 and E_2.

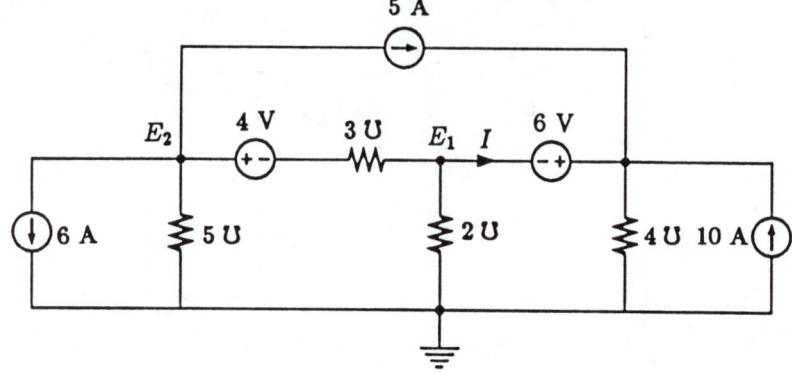

3.27 Use mesh analysis to solve for the three circulating currents.

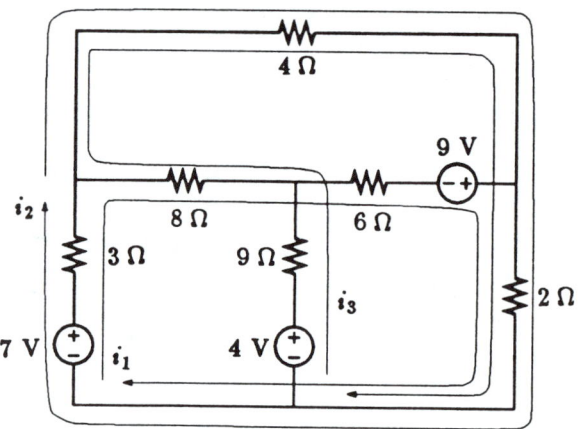

▷ **3.28** Use mesh analysis to determine I_1, I_2, and I_3.

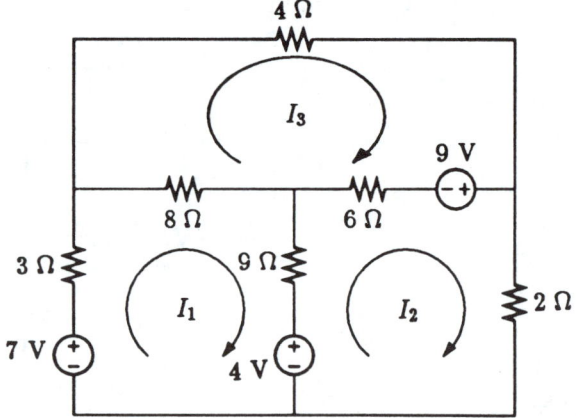

GENERAL ANALYSIS METHODS

3.29 Use mesh analysis to determine I.

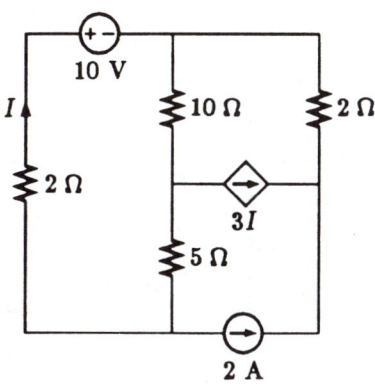

3.30 Use mesh analysis to determine I.

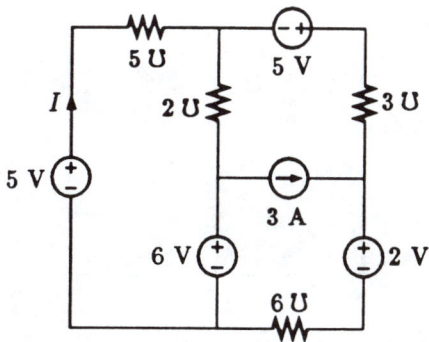

3.31 Use mesh analysis to determine I.

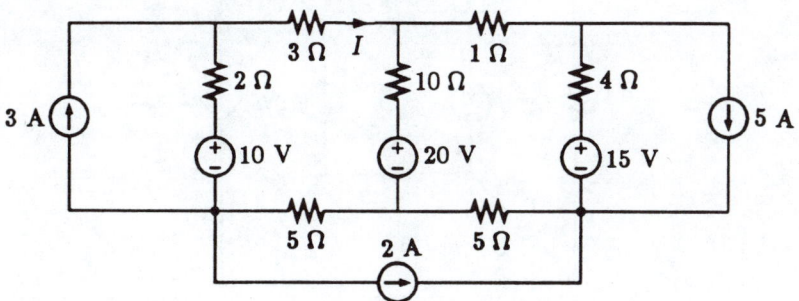

3.32 Use mesh analysis to determine I.

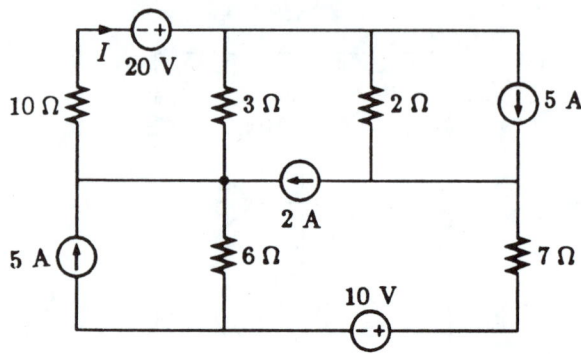

3.33 Use mesh analysis to determine I.

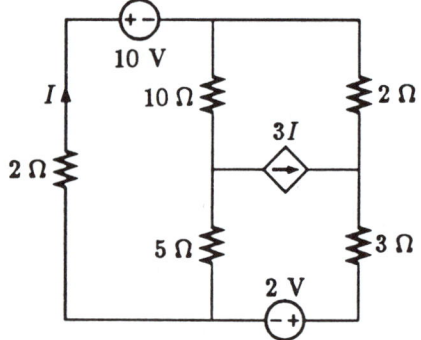

3.34 Use mesh analysis to determine I.

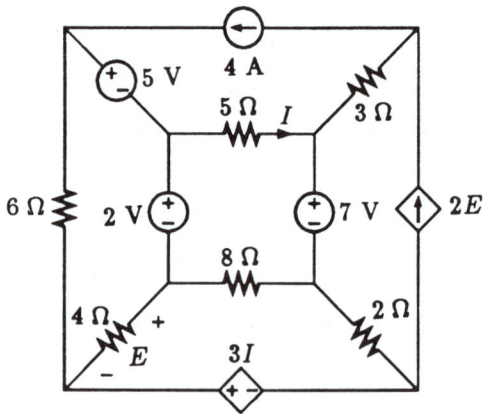

GENERAL ANALYSIS METHODS   125

3.35 Use mesh analysis to determine i.

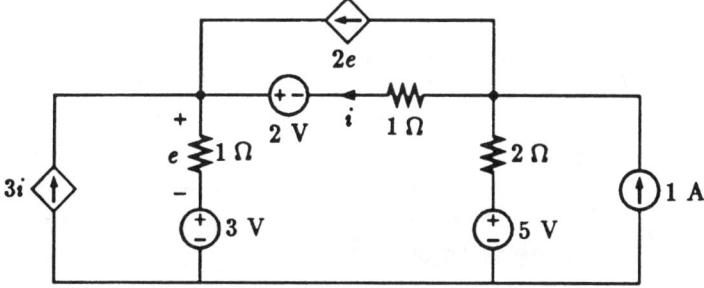

3.36 Use mesh analysis to determine I.

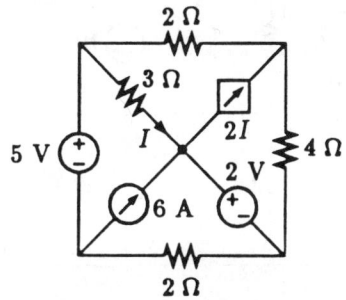

3.37 Use mesh analysis to determine i_1 and i_2.

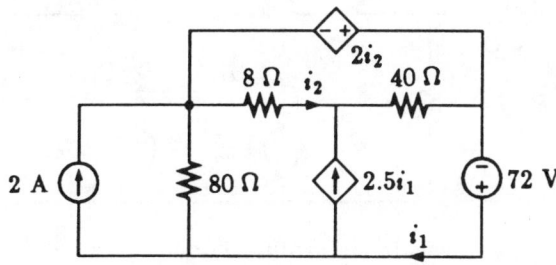

3.38 Use mesh analysis to determine I_x.

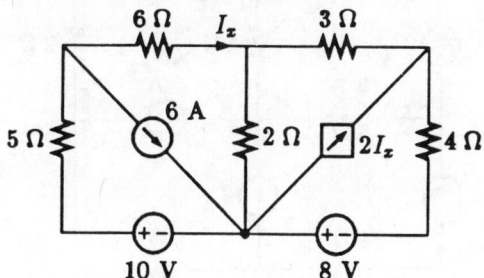

3.39 Use mesh analysis to determine I.

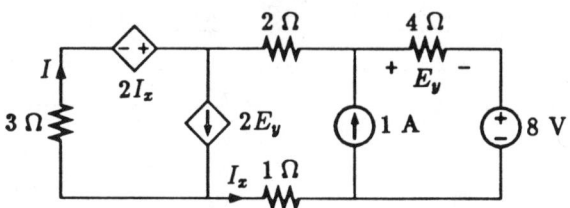

3.40 Use mesh analysis to determine I.

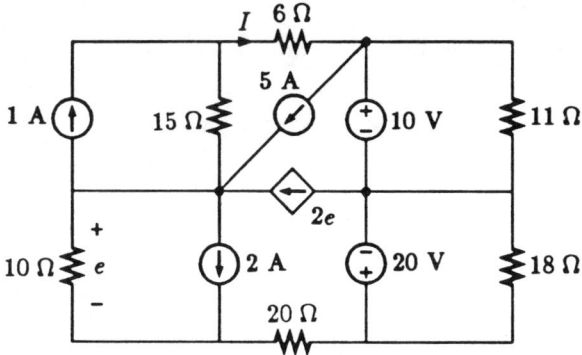

3.41 First perform mesh analysis. Then determine E.

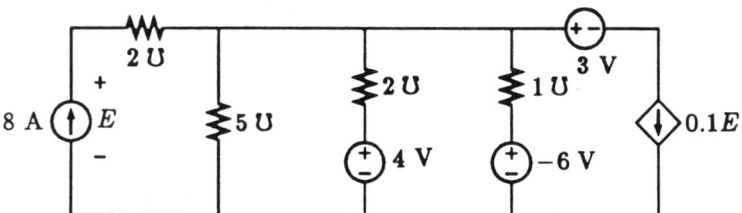

3.42 Use mesh analysis to determine I.

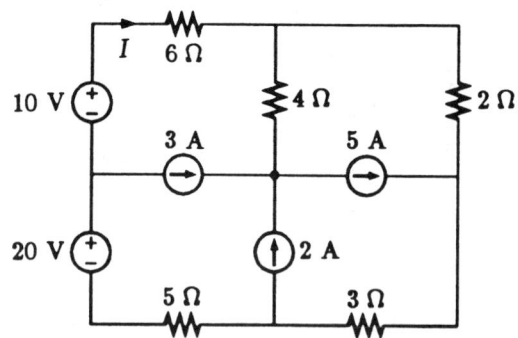

3.43 Use mesh analysis to determine i.

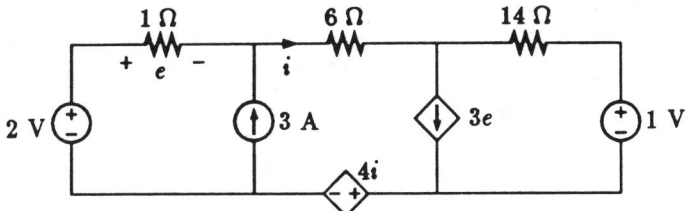

3.44 Use mesh analysis to determine I_1 and I_2.

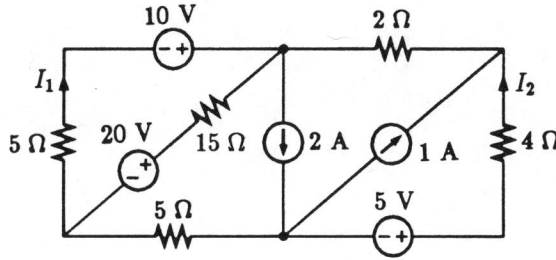

Chapter 4

Network Properties and Theorems

In this chapter, we shall describe several properties of certain classes of networks. A few of the properties are stated in the form of theorems, partly for convenience and partly for historical reasons. We find some properties helpful in simplifying the solution of particular network problems. Others are used to derive additional theorems or to explain certain phenomena in networks.

4.1 Linear elements and networks

In Chapter 2, we described the relationship between the voltage and current in a resistor. They must obey Ohm's law. That is

$$e = Ri \tag{4.1}$$

Such a resistor is a *linear* element in that, if we plot an *e*-versus-*i* curve in the *e-i* plane, it is a straight line passing through the origin.

In Chapter 2, where we described the four types of controlled sources, we also assumed that the controlled quantity in each is proportional to the controlling quantity. Or they all have the form

$$y = a x \tag{4.2}$$

where y represents the controlled quantity and x represents the controlling quantity. Such sources are also *linear* elements if a is a constant. A controlled source would not be linear if, for instance, $y = ax^2$ or $y = a\sqrt{x}$.

FUNDAMENTALS OF CIRCUIT ANALYSIS

A network that consists of only linear elements and is excited by some independent sources has several simple and important properties. We shall describe these properties without proof here. In stating these properties, we shall call the independent sources *excitations*. We shall focus our attention on certain voltages and/or currents in the network. We shall call these voltages and currents *responses*.

4.1.1 Proportionality property

The responses in a network containing only linear resistors and linear controlled sources is proportional to the excitation. In other words, if excitation x produces a response y, then excitation kx will produce a response ky, for any constant k. This property is also known as the scaling property or homogeneity.

EXAMPLE 1. Show that I_1 is proportional to E in the circuit in Fig. 4.1.

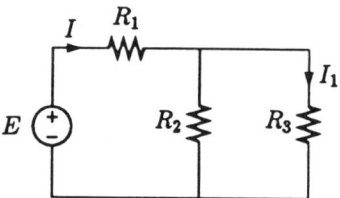

Figure 4.1: Circuit for Example 1.

SOLUTION

$$I = \frac{E}{R_1 + \dfrac{R_2 R_3}{R_2 + R_3}}$$

$$I_1 = \frac{R_2}{R_2 + R_3} \times I = \frac{R_2}{R_1 R_2 + R_2 R_3 + R_3 R_1} \times E \propto E$$

EXAMPLE 2. (a) In the ladder network in Fig. 4.2, assume $E_o = 1$ V. Calculate E_s. (b) Calculate E_o if $E_s = 5$ V.

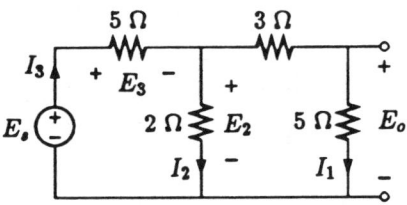

Figure 4.2: A ladder network.

SOLUTION (a) We obtain successively

$$I_1 = \frac{1}{5}$$

$$E_2 = 8 \times I_1 = \frac{8}{5}$$

$$I_2 = \frac{E_2}{2} = \frac{4}{5}$$

$$I_3 = I_1 + I_2 = 1$$

$$E_3 = 5 \times I_3 = 5$$

$$E_s = E_2 + E_3 = \frac{33}{5} = 6.6 \text{ V}$$

(b) By proportionality

$$\frac{1}{6.6} = \frac{E_o}{5}$$

$$E_o = \frac{5}{6.6} = 0.7576 \text{ V}$$

4.1.2 Additivity property

If the response of a network is y_1 when its excitation is x_1, and if its response is y_2 when its excitation is x_2, then the network is *additive* if its response is $y_1 + y_2$ when its excitation is $x_1 + x_2$.

For example, if in the circuit in Fig. 4.3

FUNDAMENTALS OF CIRCUIT ANALYSIS

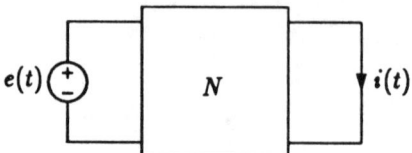

Figure 4.3: A network and its response.

$i(t) = 3$ A for $e(t) = 10$ V

$i(t) = 30 \cos t$ A for $e(t) = 100 \sin t$ V

and it is additive, then

$i(t) = 3 + 30 \cos t$ A for $e(t) = 10 + 100 \sin t$ V

Note that this network is not proportional.[1]

If a network contains only linear resistors and linear controlled sources, then the network is additive.

4.1.3 Linearity property

The linearity[2] property implies, and is implied by, both proportionality and additivity. If the response of a network is y_1 when its excitation is x_1, and if its response is y_2 when its excitation is x_2, then the network is *linear* if its response is $ay_1 + by_2$ when the excitation is $ax_1 + bx_2$, where a and b are any constants. A network that contains only linear resistors and linear controlled sources is linear.

EXAMPLE 3. If the response y of a network is related to its excitation x by the relationship

$$y = A\frac{dx}{dt}$$

is this network linear?

SOLUTION From the given relationship, the proportionality property is satisfied since

---

[1] Proportionality is not a prerequisite for additivity.

[2] The student should be careful to distinguish between an element being linear and a network being linear.

NETWORK PROPERTIES AND THEOREMS

$$ky = kA\frac{dx}{dt} = A\frac{d(kx)}{dt}$$

The additivity is also satisfied since

$$y_1 + y_2 = A\frac{dx_1}{dt} + A\frac{dx_2}{dt} = A\frac{d}{dt}(x_1 + x_2)$$

Hence, the network is linear.

4.2 Superposition theorem

If a network is linear, then the current in any branch (or the voltage between any two points) of the network is the sum of all the currents in that branch (or of all the voltages between those two points), each of which is calculated with only one independent source active at a time. While one independent source is active, all other independent sources are idle: Voltage sources are reduced to short circuits and current sources are reduced to open circuits.

Fig. 4.4 illustrates the application of this theorem. In Fig. 4.4(a), there are three independent sources connected to N. In each of parts (b) through (d) in Fig. 4.4, the network is excited by only one of the three sources. Two points and one branch are singled out to demonstrate the application of the superposition theorem at those localities.

The following are based on the superposition theorem

$$E_a = E_{a1} + E_{a2} + E_{a3} \qquad (4.3)$$

$$I_b = I_{b1} + I_{b2} + I_{b3} \qquad (4.4)$$

The distinction between the superposition and the additivity properties of a linear network lies in the fact that additivity pertains to only the sum of the values of a single source, while superposition pertains to the total effect of several sources, typically scattered over different parts of a network. In a way, the additivity property might be regarded as a special case of the superposition property.

As an example of the superposition theorem, the circuit in Fig. 4.5(a) could be replaced by the two separate circuits shown in Fig. 4.5(b) and (c). The currents and voltages in (a) would be the sums of the corresponding currents and voltages in parts (b) and (c). If we obtain the mesh equations for (b) and (c), they would read

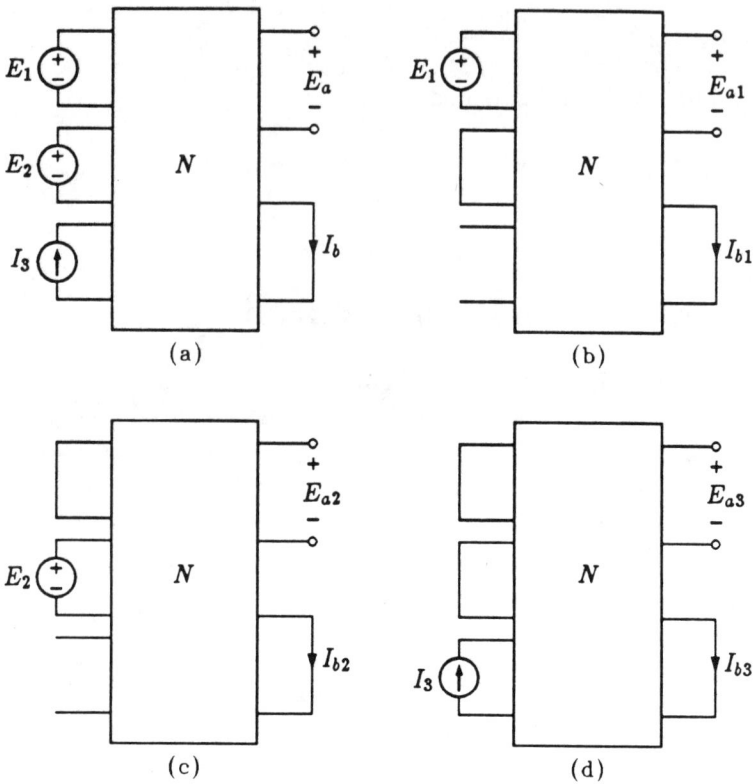

Figure 4.4: The application of the superposition theorem.

$$5I_{1a} - 3I_{2a} = 10 \qquad -3I_{1a} + 8I_{2a} = 0 \qquad (4.5)$$

and

$$5I_{1b} - 3I_{2b} = 0 \qquad -3I_{1b} + 8I_{2b} = -5 \qquad (4.6)$$

If we wrote the mesh equations for the circuit of Fig. 4.5(a) directly, they would read

$$5I_1 - 3I_2 = 10 \qquad -3I_1 + 8I_2 = -5 \qquad (4.7)$$

A simple manipulation shows that, if I_{1a}, I_{1b}, I_{2a}, and I_{2b} satisfy Eqs. (4.5) and (4.6), then the addition of Eqs. (4.5) and (4.6) will result in Eq. (4.7) if

$$I_1 = I_{1a} + I_{1b} \qquad I_2 = I_{2a} + I_{2b} \qquad (4.8)$$

NETWORK PROPERTIES AND THEOREMS

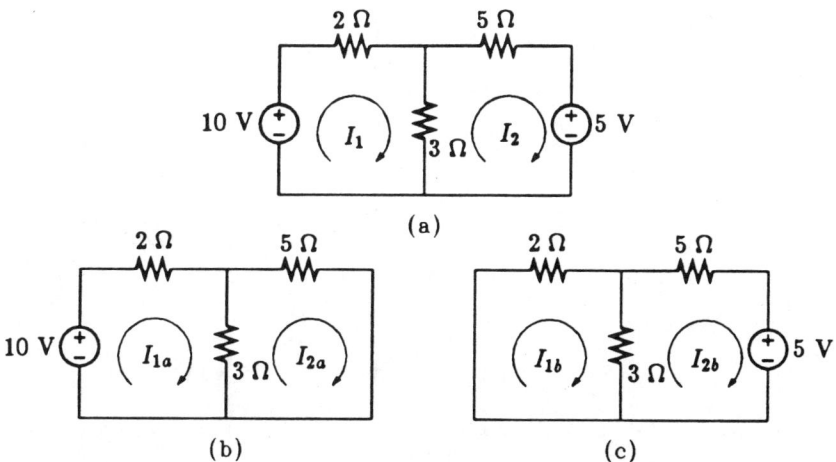

Figure 4.5: An example of superposition.

A generalization of this procedure would constitute a rigorous proof of the superposition theorem. We will not do this in this volume.

EXAMPLE 4. Analyze the circuit in Fig. 4.6(a) using superposition.

SOLUTION The circuit has three independent sources. We can analyze the three separate circuits shown in Fig. 4.6(b), (c), and (d). Each of these circuits can be analyzed by using series or parallel combination of resistances and voltage- or current-division rules. All currents are indicated on the circuit diagrams. The current in the original network can be obtained by summing up the currents in the corresponding branches in each constituent network. Or

$$I_1 = 1 - 3 - 1 = -3 \text{ A}$$

$$I_2 = 1 + 2 - 1 = 2 \text{ A}$$

$$I_3 = \frac{1}{2} - \frac{1}{3} + 2 = 1 \text{ A}$$

$$I_4 = \frac{1}{2} + \frac{7}{2} - 3 = 1 \text{ A}$$

$$I_5 = -\frac{1}{2} + \frac{3}{2} + 3 = 4 \text{ A}$$

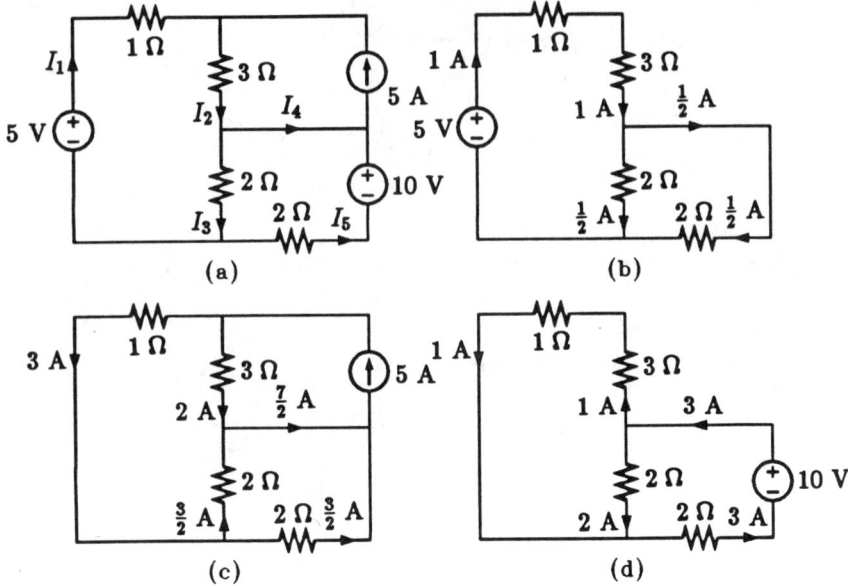

(a) (b) (c) (d)

Figure 4.6: Circuit for Example 4

EXAMPLE 5. Determine the output voltage E_o of the circuit in Fig. 4.7, which contains a transistor equivalent circuit, using superposition. $V_s = 1$ mV. $I_s = 0.1$ mA.

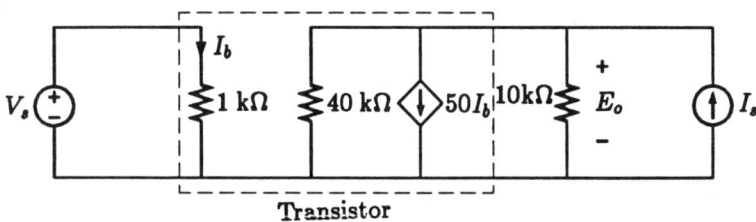

Figure 4.7: A transistor equivalent circuit.

SOLUTION Since the circuit is excited by two independent sources, it can be analyzed as two separate circuits each with one source

NETWORK PROPERTIES AND THEOREMS

active. With only the voltage source active, we have[3]

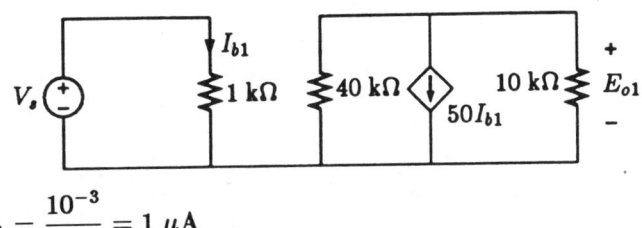

$$I_{b1} = \frac{10^{-3}}{1\,k} = 1\,\mu A$$

$$E_{o_l} = -50 \times I_{b1} \times \frac{40\,k \times 10\,k}{50\,k} = -0.4\,V$$

With only the current source active, we have

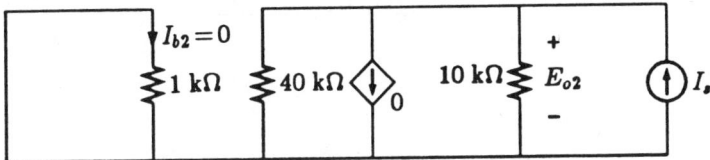

We find that $I_{b2} = 0$ and the current in the controlled current source happens to have a zero value. Thus

$$E_{o2} = 0.1 \times 10^{-3} \times 8\,k = 0.8\,V$$

Hence the output voltage is

$$E_o = E_{o1} + E_{o2} = -0.4 + 0.8 = 0.4\,V$$

EXERCISES

4.2.1 Find the value of I by superposition.

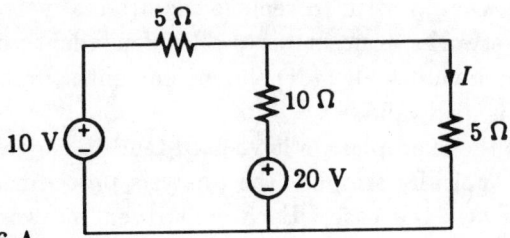

Ans. 1.6 A.

---
[3] Here k stands for 10^3.

4.2.2 Determine the value of e by superposition.

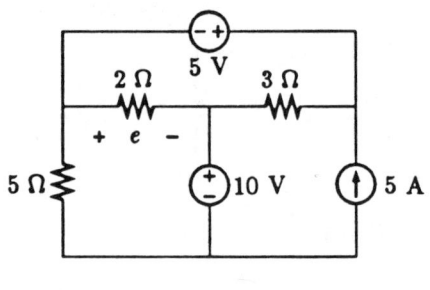

Ans. $\dfrac{40}{31}$ V.

4.2.3 Determine the value of i by superposition.

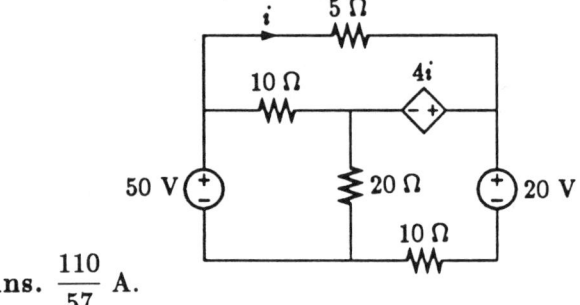

Ans. $\dfrac{110}{57}$ A.

Although the superposition theorem states that a network with several independent sources may be analyzed by considering one independent source active in each constituent network, there is no reason why some of the constituent networks cannot be combined. In other words, we may group some of the independent sources in some of the constituent networks if we wish, as long as each independent source is included in some constituent network once, and once only. For example, we may wish to replace the original network with two constituent networks—one includes all independent voltage sources and the other includes all independent current sources. Any other combination is also valid.

Several of the examples we have used tend to suggest that superposition will typically simplify the analysis procedure. In general, this is not always the case. Each constituent network may entail just as much labor to analyze as the original network. The thrust of superposition theorem is that we *can* replace the original network

NETWORK PROPERTIES AND THEOREMS

with some constituent networks such that the independent sources are somehow distributed throughout the constituent networks. The purpose of using superposition is not always to make the task of analysis simpler. When superposition does simplify the analysis, it is usually just a coincidence. The key is that the effects of the independent sources can be superimposed in a linear network. One situation in which superposition is necessary is when we wish to know the effect of each independent source on a certain electrical quantity in a network.

4.3 Equivalence of a linear two-terminal network

If a network contains only resistors, controlled sources, independent sources and is connected externally through two terminals, the network puts a very simple constraint between its terminal voltage and current. We shall develop this relationship here.

Let this two-terminal network be N and its two terminals be A and B, as in Fig. 4.8. We wish to examine what must happen

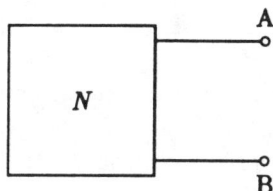

Figure 4.8: A two-terminal network.

when this network is connected to "something." We shall call this "something" a load, as shown in Fig. 4.9.

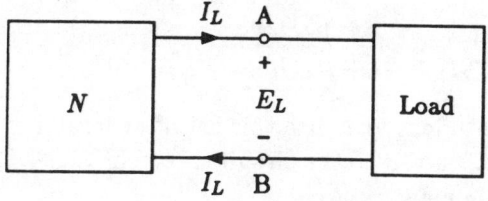

Figure 4.9: A two-terminal network connected to a load.

140   FUNDAMENTALS OF CIRCUIT ANALYSIS

The identity of the load is of no concern to us here. It can be a single resistor, linear or nonlinear, or it can be another complicated network containing an arbitrary number of branches of whatever type. Our purpose here is to assert how I_L and E_L must be related to each other as dictated by N. What happens in the load is not an issue at all.

We shall assume that all independent sources are voltage sources. We shall apply mesh analysis to N. In this particular formulation of mesh analysis, we shall set up the analysis such that I_L and E_L are part of mesh 1 as represented in Fig. 4.10.

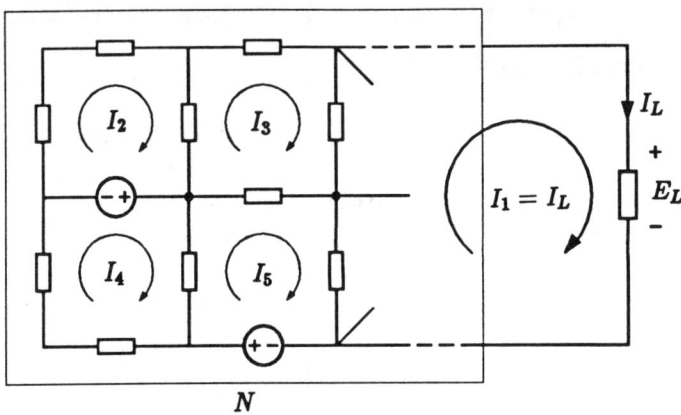

Figure 4.10: Mesh analysis of a general network with a load.

Direct application of mesh analysis in the routine manner will result in the following mesh equations.

$$R_{11}I_L + R_{12}I_2 + \cdots + R_{1n}I_n = -E_L + E_{s1}$$

$$R_{21}I_L + R_{22}I_2 + \cdots + R_{2n}I_n = E_{s2} \qquad (4.9)$$

$$\cdots\cdots\cdots\cdots\cdots$$

$$R_{n1}I_L + R_{n2}I_2 + \cdots + R_{nn}I_n = E_{sn}$$

To help the student visualize this set of general mesh equations, let us apply this analysis to the circuit shown in Fig. 4.11. The mesh equations for this network are

NETWORK PROPERTIES AND THEOREMS

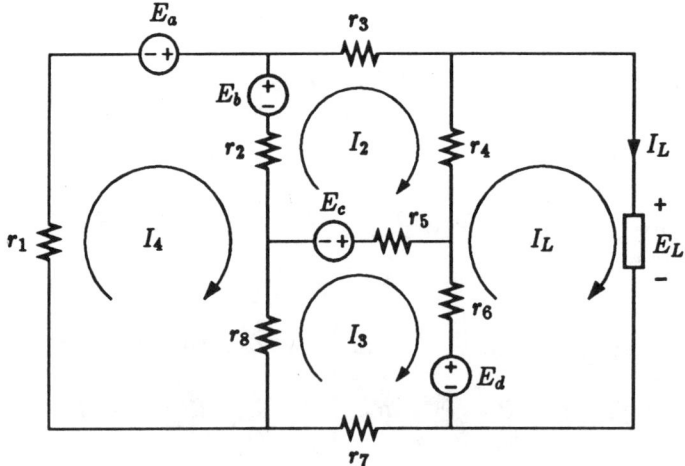

Figure 4.11: An example of mesh analysis.

$$(r_4 + r_6)I_L - r_4 I_2 - r_6 I_3 = -E_L + E_d$$

$$-r_4 I_L + (r_2 + r_3 + r_4 + r_5)I_2 - r_5 I_3 - r_2 I_4 = E_b - E_c$$
(4.10)
$$-r_6 I_L - r_5 I_2 + (r_5 + r_6 + r_7 + r_8)I_3 - r_8 I_4 = E_c - E_d$$

$$-r_2 I_2 - r_8 I_3 + (r_1 + r_2 + r_8)I_4 = E_a - E_b$$

So, the R's in Eq. (4.9) represent the coefficients of the various current variables, the E_s's represent the quantities stemming from the various independent voltage sources, I_L occupies the places one would normally put the quantity I_1, and E_L is one of the voltages in the equation associated with mesh No. 1.

In this example, as well as in Fig. 4.10, no current source or controlled sources are present. As we showed in Section 3.3, these sources will only affect the coefficients and constants of the mesh equations of a network. Thus Eq. (4.9) is entirely general in form when we apply mesh analysis to the general network as configured in Fig. 4.10.

Solving for I_L in Eq. (4.9), we have

$$I_L = \frac{\begin{vmatrix} -E_L + E_{s1} & R_{12} & \cdots & R_{1n} \\ E_{s2} & R_{22} & \cdots & R_{2n} \\ \vdots & \vdots & & \vdots \\ E_{sn} & R_{n2} & \cdots & R_{nn} \end{vmatrix}}{\Delta}$$

$$= \frac{\begin{vmatrix} -E_L & R_{12} & \cdots & R_{1n} \\ 0 & R_{22} & \cdots & R_{2n} \\ \vdots & \vdots & & \vdots \\ 0 & R_{n2} & \cdots & R_{nn} \end{vmatrix} + \begin{vmatrix} E_{s1} & R_{12} & \cdots & R_{1n} \\ E_{s2} & R_{22} & \cdots & R_{2n} \\ \vdots & \vdots & & \vdots \\ E_{sn} & R_{n2} & \cdots & R_{nn} \end{vmatrix}}{\Delta}$$

$$= -E_L \frac{\Delta_{11}}{\Delta} + I_{\text{sc}} \qquad (4.11)$$

where

$$\Delta = \begin{vmatrix} R_{11} & R_{12} & \cdots & R_{1n} \\ R_{21} & R_{22} & \cdots & R_{2n} \\ \vdots & \vdots & & \vdots \\ R_{n1} & R_{n2} & \cdots & R_{nn} \end{vmatrix} \qquad (4.12)$$

$$\Delta_{11} = \begin{vmatrix} R_{22} & R_{23} & \cdots & R_{2n} \\ R_{32} & R_{33} & \cdots & R_{3n} \\ \vdots & \vdots & & \vdots \\ R_{n2} & R_{n3} & \cdots & R_{nn} \end{vmatrix} \qquad (4.13)$$

NETWORK PROPERTIES AND THEOREMS

$$I_{sc} = \frac{\begin{vmatrix} E_{s1} & R_{12} & \cdots & R_{1n} \\ E_{s2} & R_{22} & \cdots & R_{2n} \\ \vdots & \vdots & & \vdots \\ E_{sn} & R_{n2} & \cdots & R_{nn} \end{vmatrix}}{\Delta} \tag{4.14}$$

For a given network N, I_{sc} is a constant and has the dimension of the current. Since Δ_{11} is $(n-1) \times (n-1)$ while Δ is $n \times n$, the quantity Δ/Δ_{11} will be another constant and has the dimension of the resistance. We shall define

$$R_{eq} = \frac{\Delta}{\Delta_{11}} \tag{4.15}$$

and Eq. (4.11) reads

$$I_L = I_{sc} - \frac{E_L}{R_{eq}} \tag{4.16}$$

Eq. (4.16) asserts that the current I_L and the terminal voltage E_L of a two-terminal network must satisfy a linear equation. This is the major conclusion we are seeking. It may be rearranged to read

$$E_L = I_{sc} R_{eq} - I_L R_{eq} \tag{4.17}$$

The quantity $I_{sc} R_{eq}$ is another constant. We shall designate it by

$$E_{oc} = I_{sc} R_{eq} \tag{4.18}$$

and Eq. (4.17) becomes

$$E_L = E_{oc} - I_L R_{eq} \tag{4.19}$$

Eq. (4.16) and Eq. (4.19) both state that, graphically, I_L and E_L must vary along a straight line in the I_L-E_L coordinate plane as shown in Fig. 4.12, in which a positive R_{eq} is assumed. The slope of the straight line and the two intercepts with the axes have some fairly simple interpretations which will be dealt with later.

144 FUNDAMENTALS OF CIRCUIT ANALYSIS

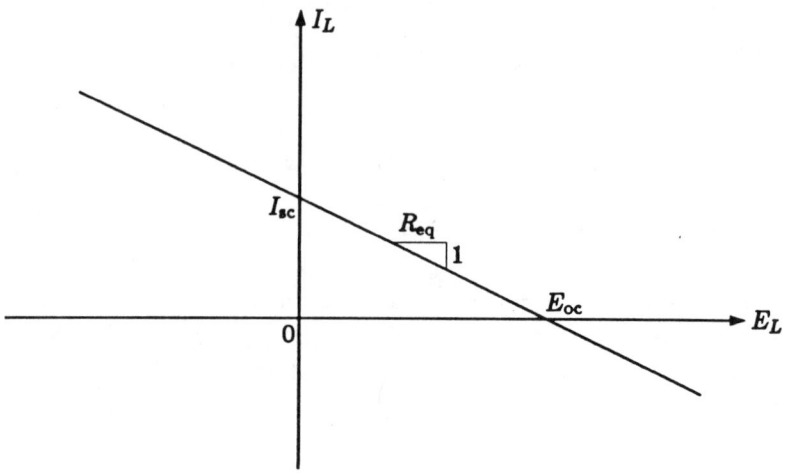

Figure 4.12: I_L-E_L relationship of a linear two-terminal network.

EXAMPLE 6. We wish to determine the relationship between E_L and I_L that must be satisfied at the two terminals A and B of network N in Fig. 4.13. Using node analysis, we set up the circuit

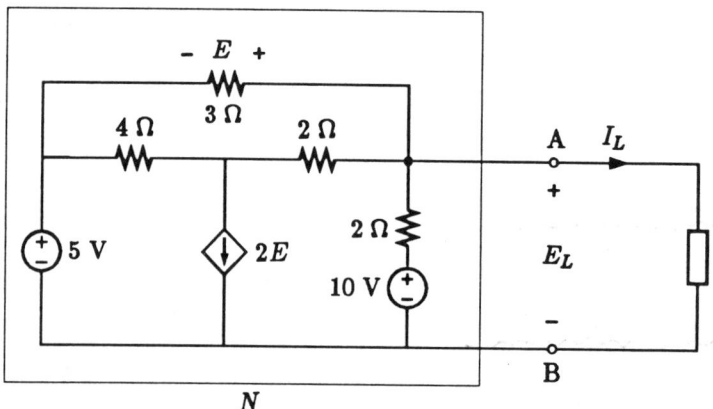

Figure 4.13: Circuit for Example 6.

as shown in Fig. 4.14. The node equations are

$$\frac{E_1 - 5}{4} + 2E + \frac{E_1 - E_L}{2} = 0$$

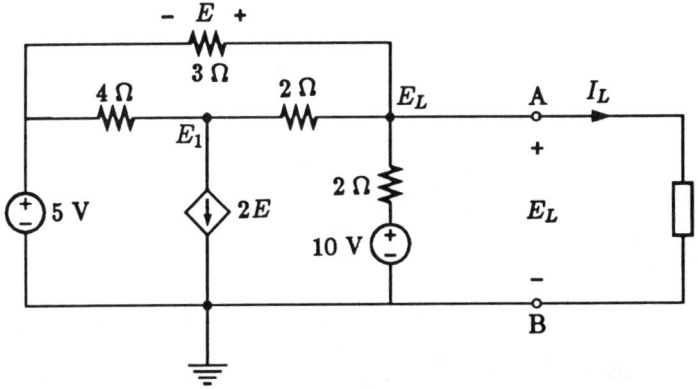

Figure 4.14: Circuit of Fig. 4.13 readied for node analysis.

$$\frac{E_L - E_1}{2} + \frac{E_L - 10}{2} + I_L + \frac{E_L - 5}{3} = 0$$

The controlling quantity is

$$E = E_L - 5$$

Eliminating E and E_1 from these three equations, we obtain the equation

$$14 E_L + 6 I_L = 85 \tag{4.20}$$

The usefulness of this type of relationship is that no matter how complex network N is, once we have found its I_L-E_L relationship, this relationship represents network N completely, as far as its effects on the outside world is concerned. If we are not interested in what goes on inside N, we never need to know any more about N than this relationship.

Returning to the circuit in Fig. 4.13, if the load is a resistor of, say, 5 Ω, then

$$E_L = 5 I_L$$

Combining this relationship with Eq. (4.20), we get

$$14 \times 5 I_L + 6 I_L = 85$$

146 FUNDAMENTALS OF CIRCUIT ANALYSIS

$$76 I_L = 85$$

$$I_L = \frac{85}{76} \text{ A}$$

$$E_L = \frac{425}{76} \text{ V}$$

If, instead, the load is a voltage source with

$$E_L = 10 \text{ V}$$

then

$$14 \times 10 + 6 I_L = 85$$

$$I_L = -\frac{55}{6} \text{ A}$$

If the load connected to N is a more complex arrangement than a single element, the effect of N is still described by Eq. (4.20). For instance, if N is a part of the circuit shown in Fig. 4.15, we can write

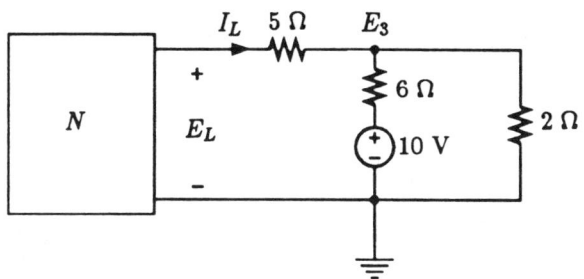

Figure 4.15: Network N as a part of another network.

a node equation for the node whose voltage is E_3 to get

$$\frac{E_3 - 10}{6} + \frac{E_3 - E_L}{5} + \frac{E_3}{2} = 0$$

We also have

$$I_L = \frac{E_L - E_3}{5}$$

NETWORK PROPERTIES AND THEOREMS

These two equations coupled with Eq. (4.20) enable us to solve for E_3, E_L, and I_L. The results are

$$E_3 = \frac{635}{194} \text{ V} \qquad E_L = \frac{1135}{194} \text{ V} \qquad I_L = \frac{50}{97} \text{ A}$$

EXERCISES

4.3.1 Determine the relationship between E_L and I_L in the circuit.

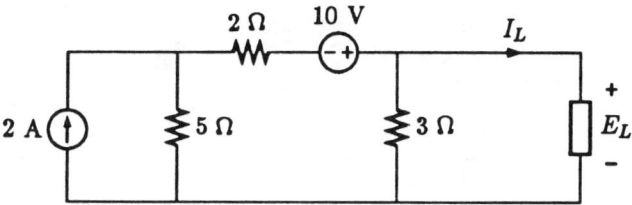

Ans. $E_L = 6 - 2.1 I_L$.

4.3.2 Determine the relationship between E_L and I_L in the circuit.

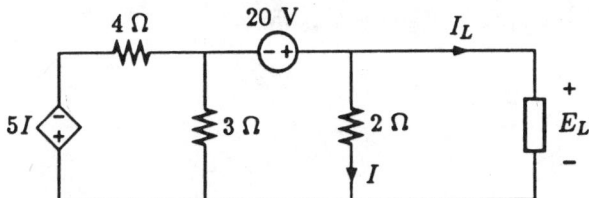

Ans. $41 E_L = 280 - 24 I_L$.

4.3.1 Determination of parameters of the equivalent circuits of a two-terminal network

Given a two-terminal linear network, it is sometimes more convenient to determine the coefficients that appear in Eqs. (4.16) and (4.19) individually. Let us examine the significance of these coefficients.

From Eq. (4.16), we can say that

$$I_{sc} = I_L \bigg|_{E_L = 0} \tag{4.21}$$

This means that the quantity I_{sc} is the value of I_L when the load is a short circuit—hence the subscript "sc". Let's use the circuit of Example 6 (Fig. 4.13) again. To determine its I_{sc}, we shall use node analysis to analyze the circuit in Fig. 4.16. Writing a node equation

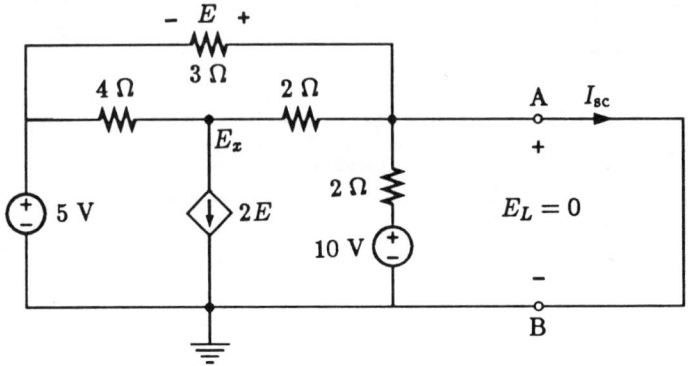

Figure 4.16: Finding the I_{sc} of network N of Fig. 4.13.

for the node whose voltage is E_x, we get

$$\frac{E_x - 5}{4} + 2E + \frac{E_x}{2} = 0$$

$$E = -5$$

Solving, we get

$$E_x = 15$$

and

$$I_{sc} = \frac{E_x}{2} + \frac{5}{3} + \frac{10}{2} = \frac{85}{6} \text{ A}$$

From Eq. (4.19), we can say that

$$E_{oc} = E_L \bigg|_{I_L = 0} \tag{4.22}$$

NETWORK PROPERTIES AND THEOREMS

Thus, the quantity E_{oc} is the voltage rise from B to A when the two-terminal network is left open-circuited—hence the script "oc".

Again, we shall use the circuit of Example 6 (Fig. 4.13) as an illustration. To determine E_{oc} for network N, we shall analyze the circuit in Fig. 4.17. This time, for variety, we shall use mesh analysis.

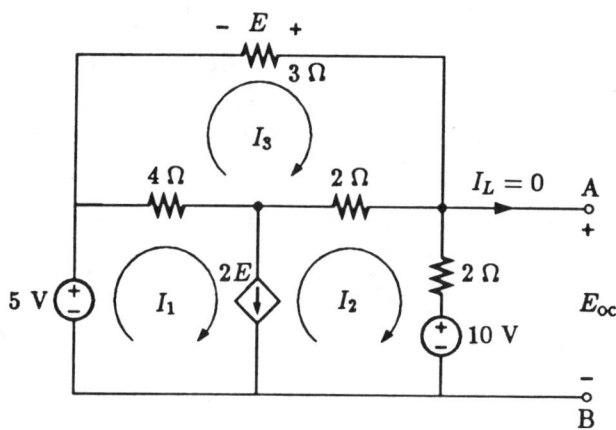

Figure 4.17: Finding the E_{oc} of network N of Fig. 4.13.

The mesh equations are

$$4(I_1 - I_3) + 2(I_2 - I_3) + 2I_2 = 5 - 10$$

$$3I_3 + 2(I_3 - I_2) + 4(I_3 - I_1) = 0$$

$$I_1 - I_2 = 2E = -6I_3$$

Solving gives

$$I_1 = \frac{5}{28} \qquad I_2 = -\frac{55}{28} \qquad I_3 = -\frac{5}{14}$$

Whence,

$$E_{oc} = 10 + 2I_2 = 10 - \frac{55}{14} = \frac{85}{14} \text{ V}$$

Once we have I_{sc} and E_{oc} for a two-terminal network, its R_{eq} can be obtained by Eq. (4.18). For the circuit in Fig. 4.13, we have

$$R_{eq} = \frac{E_{oc}}{I_{sc}} = \frac{85}{14} \times \frac{6}{86} = \frac{3}{7} \, \Omega$$

The quantity R_{eq} can also be obtained for a two-terminal network directly. Let us examine the significance of this quantity as given in Eq. (4.15). We go back to the general set of mesh equations as expressed in Eq. (4.9). If we set all E_s's to zero, we then have

$$R_{11}I_L + R_{12}I_2 + \cdots + R_{1n}I_n = -E_L$$

$$R_{21}I_L + R_{22}I_2 + \cdots + R_{2n}I_n = 0 \qquad (4.23)$$

$$\cdots\cdots\cdots\cdots\cdots\cdots$$

$$R_{n1}I_L + R_{n2}I_2 + \cdots + R_{nn}I_n = 0$$

$$I_L = \frac{\begin{vmatrix} -E_L & R_{12} & \cdots & R_{1n} \\ 0 & R_{22} & \cdots & R_{2n} \\ \vdots & \vdots & & \vdots \\ 0 & R_{n2} & \cdots & R_{nn} \end{vmatrix}}{\Delta} = -E_L \frac{\Delta_{11}}{\Delta} = -R_{eq} E_L \qquad (4.24)$$

where Δ and Δ_{11} are defined in Eqs. (4.12) and (4.13) respectively.

Eqs. (4.23) are the same mesh analysis equations when the original network N of Eq. (4.10) is modified. The modification is that all voltage sources are idled—set to zero. We shall denote this modified network by N_0, and it is sometimes referred to as the "dead" network of N. Fig. 4.18(a) represents such a situation. If we compare

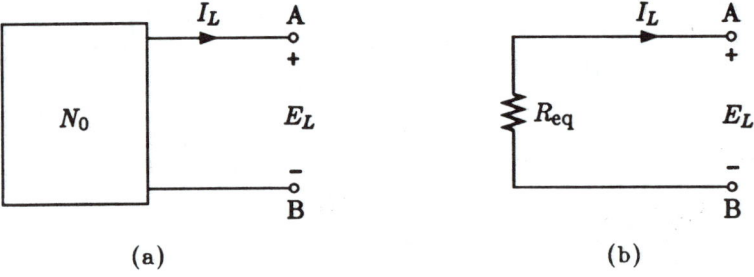

Figure 4.18: Equivalent resistance of the "dead" network.

NETWORK PROPERTIES AND THEOREMS

this arrangement with that shown in Fig. 4.18(b), in which

$$E_L = -R_{eq} I_L \qquad (4.25)$$

it is seen that R_{eq} is the equivalent resistance of the "dead" network N_0 between terminals A and B.

Again, let us return to Example 6 and Fig. 4.13. R_{eq} is the resistance between terminals A and B in Fig. 4.19.

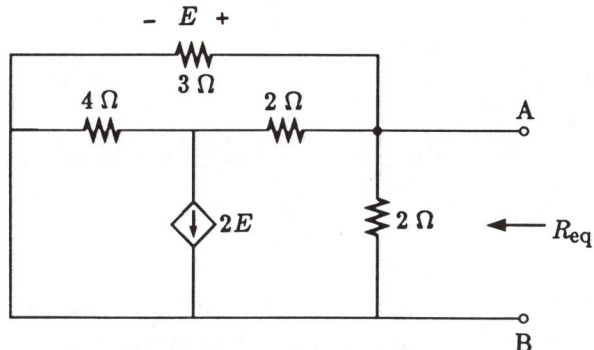

Figure 4.19: Equivalent resistance of N_0.

To obtain R_{eq} in Fig. 4.19, we apply a 1-volt source between terminals A and B as in Fig. 4.20. Then we calculate the current I_t. We write a node equation for the node whose voltage is E_x.

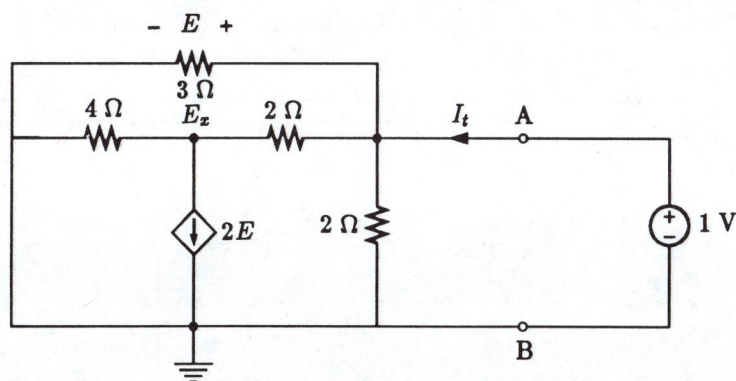

Figure 4.20: Applying a voltage to determine R_{eq}.

$$\frac{E_x}{4} + 2 \times 1 + \frac{E_x - 1}{2} = 0$$

which yields

$$E_x = -2 \text{ V}$$

Whence,

$$I_t = \frac{1}{3} + \frac{1-(-2)}{2} + \frac{1}{2} = \frac{7}{3} \text{ A}$$

Thus,

$$R_{eq} = \frac{1}{I_t} = \frac{3}{7} \, \Omega$$

4.3.2 Norton's and Thévenin's equivalent circuits

Although Eqs. (4.16) and (4.19) are technically complete and mathematically sound in representing a two-terminal network, they are sometimes awkward and inconvenient to use in conjunction with the network to which N is connected. Although the information conveyed by either of these two equations is accurate and sufficient, the equations are not always the most appropriate way to represent N. It would be desirable to develop circuit models to take the place of these equations.

There are two simple equivalent circuits that are commonly used to replace a two-terminal network. These circuits are shown in Fig. 4.21. Each of these circuits contains exactly the same information contained in Eqs. (4.16) and (4.19).

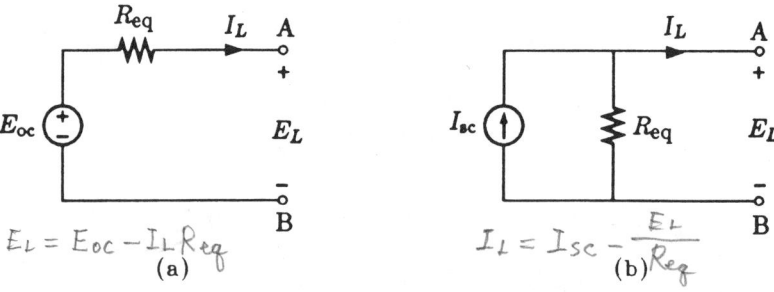

Figure 4.21: (a) Thévenin's and (b) Norton's equivalent circuits.

NETWORK PROPERTIES AND THEOREMS

The equivalent circuit in Fig. 4.21(a) is known as the *Thévenin's* equivalent circuit. The circuit in Fig. 4.21(b) is known as the *Norton's* equivalent circuit. The reader can easily verify that the circuit in Fig. 4.21(a) is simply the circuit model of Eq. (4.19), while that in Fig. 4.21(b) is that of Eq. (4.16). *Do this as an exercise.*

Since the two circuits of Fig. 4.21 are both equivalent to N, they are obviously equivalent to each other also. This observation is sometimes useful in itself without regard to their parent circuit N. As an example, the two branches in Fig. 4.22 are clearly equivalent to each other when these branches are both taken as a whole. This equivalence is useful when we come to the subject of source transformation. These branches are sometimes referred to as the *Thévenin's branch* and the *Norton's branch*.

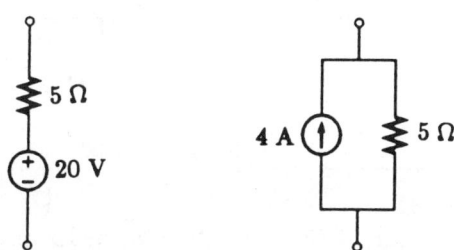

Figure 4.22: Equivalence of Thévenin's and Norton's branches.

EXAMPLE 7. Obtain the Thévenin's and Norton's equivalent circuits for the network in Fig. 4.23(a).

SOLUTION E_{oc} is simply the voltage rise from B to A under the open-circuit condition. From Fig. 4.23(a), it is clear that

$$E_{oc} = E_{BA} = 10 \times \frac{10}{5+10} = \frac{20}{3} \text{ V}$$

To obtain I_{sc}, we analyze the circuit in Fig. 4.23(b). We have

$$I_1 = \frac{10}{5 + \dfrac{2 \times 10}{2 + 10}} = \frac{3}{2}$$

$$I_{sc} = \frac{10}{12} I_1 = \frac{5}{4} \text{ A}$$

154 FUNDAMENTALS OF CIRCUIT ANALYSIS

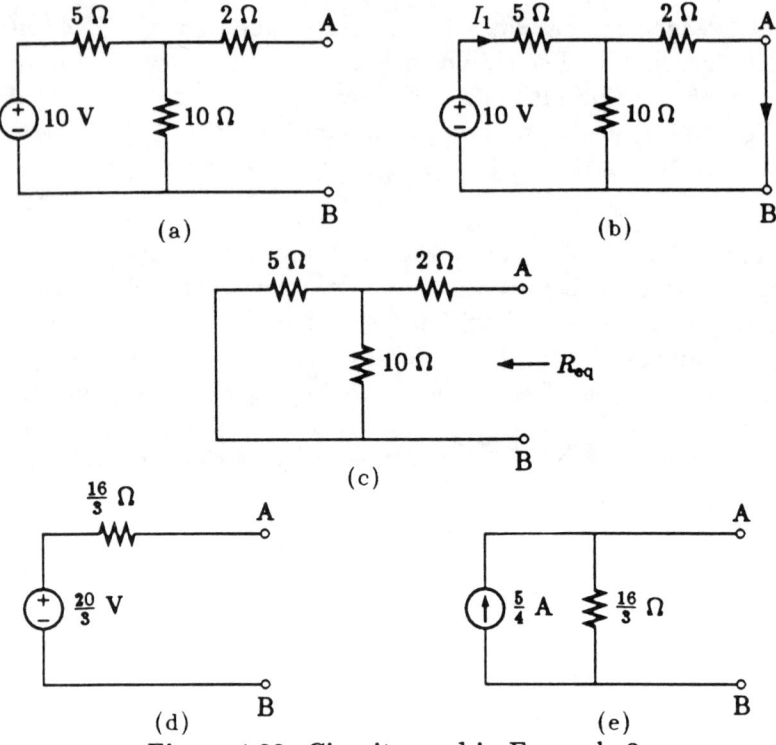

Figure 4.23: Circuits used in Example 2

To find R_{eq}, we look at Fig. 4.23(c). Since it is simply a series-parallel combination, we have

$$R_{eq} = 2 + \frac{5 \times 10}{5 + 10} = \frac{16}{3} \, \Omega$$

Hence, the Thévenin's and Norton's equivalent circuits are shown in Fig. 4.23(d) and (e), respectively. To check

$$I_{sc} R_{eq} = \frac{5}{4} \times \frac{16}{3} = \frac{20}{3} = E_{oc}$$

EXAMPLE 8. Obtain the Thévenin's and Norton's equivalent circuits for the network in Fig. 4.24(a).

SOLUTION To obtain E_{oc}, we analyze the circuit of Fig. 4.24(b). Several of the currents can be expressed in terms of E as shown.

NETWORK PROPERTIES AND THEOREMS

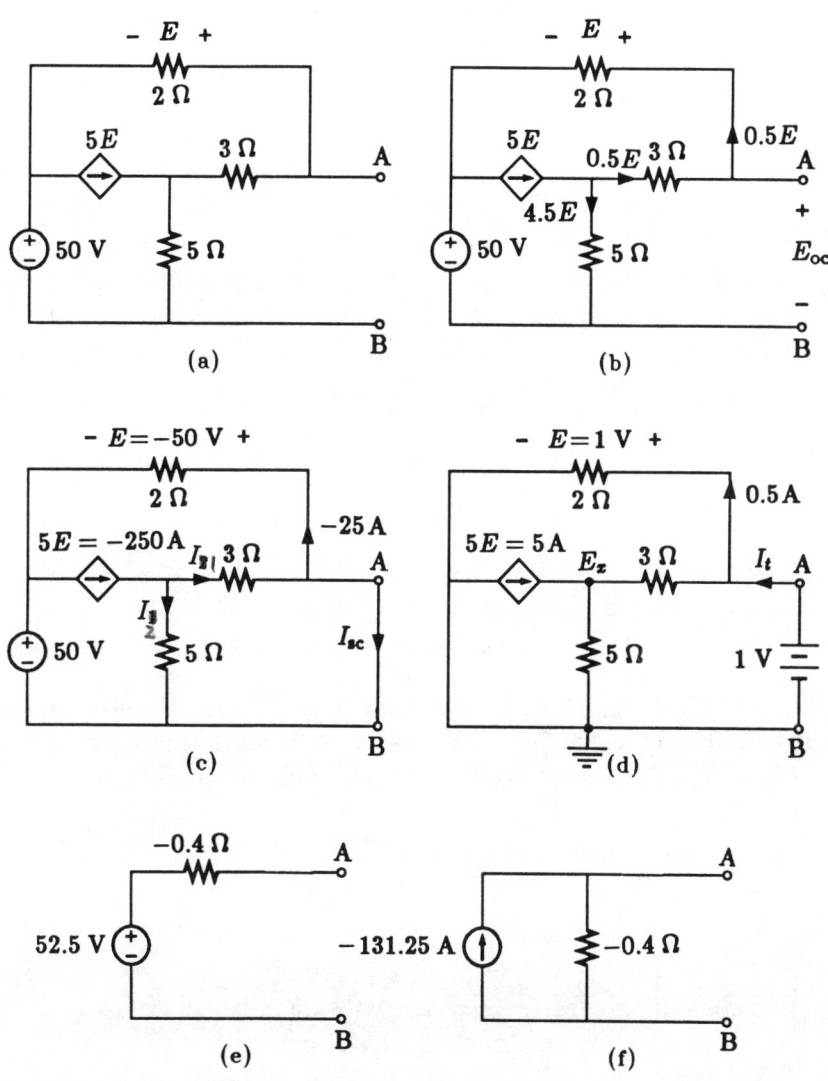

Figure 4.24: Circuits used in Example 8.

Then we write a mesh equation around the only supermesh, bypassing the controlled current source. We get

$$50 + E + 1.5 E - 5 \times 4.5 E = 0$$

Solving we get

$$E = 2.5 \text{ V}$$

and

$$E_{oc} = E_{AB} = 50 + E = 52.5 \text{ V}$$

To obtain I_{sc}, we analyze the circuit under the short-circuit condition as shown in Fig. 4.24(c). Because of the short circuit, the analysis of this circuit is particularly simple. We have

$$E = -50 \text{ V} \qquad 5E = -250 \text{ A}$$

The 3-Ω and the 5-Ω resistances are connected in parallel. Hence

$$I_1 = -250 \times \frac{5}{8} = -156.25 \text{ A} \qquad I_2 = -250 \times \frac{3}{8} = -93.75 \text{ A}$$

and

$$I_{sc} = I_1 - \frac{E}{2} = -156.25 + 25 = -131.25 \text{ A}$$

To obtain R_{eq} directly, we apply a 1-V source between A and B as shown in Fig. 4.24(d). We use node analysis and observe that $E = 1$. Write the node equation for the node whose voltage is E_x to get

$$\frac{E_x - 1}{3} + \frac{E_x}{5} = 5 E = 5$$

Solving to get

$$E_x = 10 \qquad I_t = \frac{1 - E_x}{3} + \frac{E}{2} = -2.5 \text{ A}$$

Hence

$$R_{eq} = \frac{1}{I_t} = -0.4 \text{ }\Omega$$

The Thévenin's and Norton's equivalent circuits are shown in Fig. 4.24(e) and (f) respectively.

NETWORK PROPERTIES AND THEOREMS

4.3.3 Thévenin's and Norton's Theorems

The thrust of this section is frequently stated in the form of two theorems. These two theorems are essentially a verbal summary of what we have already established.

Thévenin's Theorem

> A network N that contains only linear resistors, linear controlled sources, and independent sources, and is connected to the outside at two terminals, can be replaced by a voltage source in series with another network N_0. The strength of the voltage source is equal to and in the same direction as the voltage rise from one of its two terminals to the other when they are left open-circuited. The series network N_0 is network N with all its independent sources idled—voltage sources replaced by short circuits and current sources replaced by open circuits.

In the theorem, the "dead" network N_0 is left as another network. This is formally correct. As a practical matter, it is usually desirable to replace it by its equivalent resistance. Sometimes, it is possible to reduce N_0 step-by-step until it is simplified to a single resistance. At other times, it may require considerably more elaborate analysis to calculate this equivalent resistance.

Norton's Theorem.

> A network N, that contains only linear resistors, linear controlled sources, and independent sources, and is connected to the outside at two terminals, can be replaced by a current source in parallel with another network N_0. The strength of the current source is the same as the current that would flow in the short circuit when the latter is placed across the two terminals. The direction of the source current is such that it produces a current in the same direction in the short circuit for both N and its equivalent. The parallel network N_0 is network N with all its independent sources idled.

Since N_0 in both theorems is exactly the same, the same remarks that follow Thévenin's threorem also apply here.[4]

EXERCISES

4.3.3 Obtain the Thévenin's equivalent circuit of the network with respect to terminals A and B.

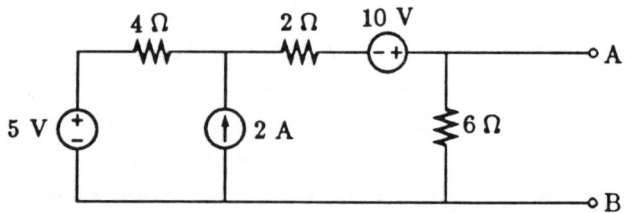

Ans. $E_{oc} = 11.5$ V, $R_{eq} = 3\ \Omega$.

4.3.4 Obtain the Norton's equivalent circuit of the network with respect to terminals A and B.

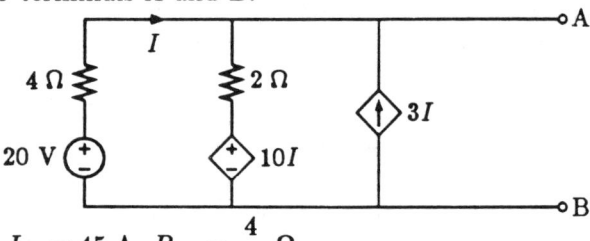

Ans. $I_{sc} = 45$ A, $R_{eq} = \dfrac{4}{11}\ \Omega$.

4.3.5 Obtain the Thévenin's equivalent circuit of the network with respect to terminals A and B.

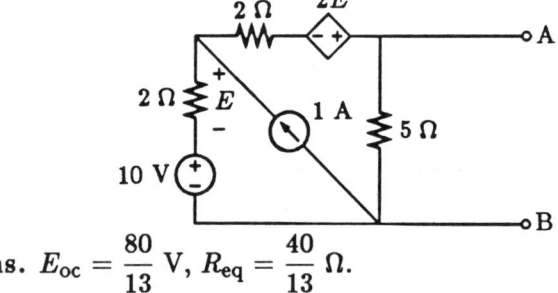

Ans. $E_{oc} = \dfrac{80}{13}$ V, $R_{eq} = \dfrac{40}{13}\ \Omega$.

---

[4]These theorems are adequate when the networks contains no energy-storing elements such as capacitors and inductors. When energy-storing elements are present, it is necessary to further state that in N_0, all initial energy be removed.

NETWORK PROPERTIES AND THEOREMS 159

4.3.4 Determination of Norton's and Thévenin's equivalent circuits by external means

The three key quantities—I_{sc}, E_{oc}, and R_{eq}— that serve to quantify the external characteristics of a two-terminal network, as interpreted in the previous subsection, are not always easy to determine directly. If the circuit of N is known, it is a matter of analyzing two or three circuits to determine these quantities. This is, however, not always the case.

The concept of these equivalent circuits is not limited to dc situations. Hence, the basic principle of the equivalence is extremely important for reasons other than simplifying N into a two-element equivalent circuit. Network N could represent a huge piece of machinery, a powerful transmitter feeding a radio station antenna, a long power transmission line, or a telephone or television cable reaching a household. Any time we have access to a circuit at two terminals, we can use the idea that everything behind these two terminals, can be represented by a two-element equivalent circuit. Everything that is connected to the two terminals in a household electric receptacle—which may include all the wiring, meters, circuits breakers, transformers, substation equipment, power plants miles away, as well as all other consumers' circuits—can be totally summarized by a simple two-terminal equivalent circuit.

Therefore, it is not always practical to determine the equivalent circuits directly. Open-circuiting a piece of huge machinery will require the shut-down of a great deal of services. Short-circuiting a transmitter may damage the electronic parts of the equipment. It is not always feasible to shut down all power sources, some of them may be miles away to determine the equivalent resistance of the "dead" network.

As explained earlier, the basis of these two equivalent circuits is really a model of the I_L-E_L relationship of a two-terminal linear network. Since this relationship must be a straight line in the I_L-E_L coordinate plane, we only need to have the knowledge of any two pairs of corresponding values of I_L and E_L. Once we have these data, the equation for the straight line can be obtained and the equivalent circuits constructed.

EXAMPLE 9. We wish to determine the Thévenin's equivalent circuit of a two-terminal network N without knowing its interior except that it contains only resistors, controlled sources, and independent sources. In this example, we connect N to two different load resistances (R_L) and measure the current I_L in each case, as shown in Fig. 4.25(a). When $R_L = 10\ \Omega$, $I_L = 7$ A. When $R_L = 20\ \Omega$, $I_L = 5$ A.

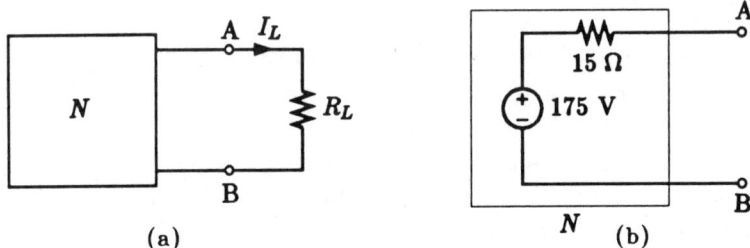

Figure 4.25: Circuits used in Example 9.

SOLUTION From the given data and using Eq. (4.19), we have

$$7 \times 10 = E_{oc} - 7 R_{eq}$$

$$5 \times 20 = E_{oc} - 5 R_{eq}$$

Solving, we get

$$E_{oc} = 175\ \text{V} \quad \text{and} \quad R_{eq} = 15\ \Omega$$

The equivalent circuit is shown in Fig. 4.25(b).

EXAMPLE 10. In this example, we apply a voltage source and a current source to obtain the Norton's equivalent circuit for N. The two arrangements and their results are shown in Fig. 4.26(a) and (b). Using Eq. (4.16), we have

$$-2 = I_{sc} - \frac{10}{R_{eq}}$$

$$-4 = I_{sc} + \frac{40}{R_{eq}}$$

NETWORK PROPERTIES AND THEOREMS

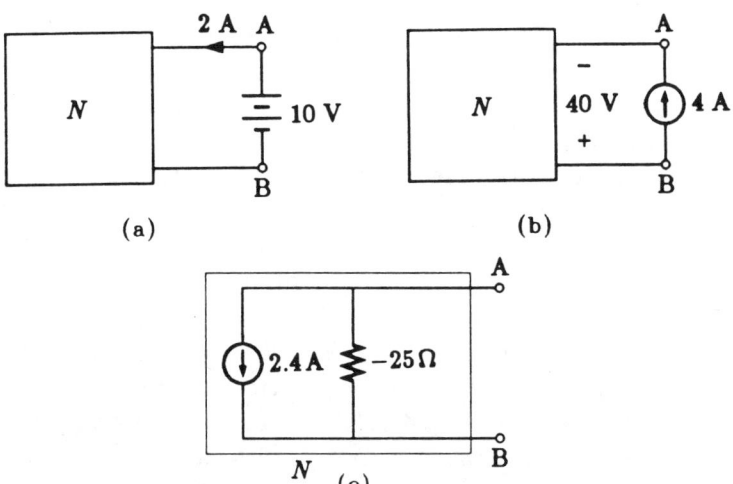

Figure 4.26: Circuits used in Example 10.

Solving, we obtain

$$R_{eq} = -25 \text{ }\Omega \quad \text{and} \quad I_{sc} = -2.4 \text{ A}$$

The Norton's equivalent circuit for N is shown in Fig. 4.26(c).

EXERCISES

4.3.6 Obtain the Thévenin's equivalent circuit of the two-terminal network N.

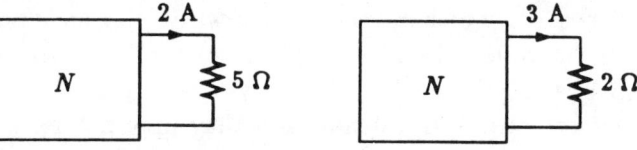

Ans. $E_{oc} = 18$ V, $R_{eq} = 4 \text{ }\Omega$.

4.3.7 Obtain the Thévenin's equivalent circuit of the two-terminal network N.

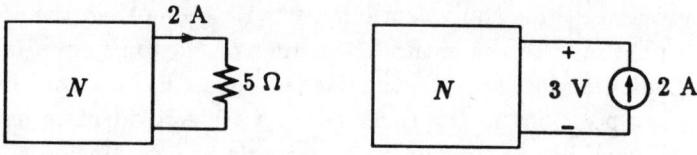

Ans. $E_{oc} = 6.5$ V, $R_{eq} = -1.75 \text{ }\Omega$.

4.4 Source transformation

The equivalence of the Thévenin's circuit and the Norton's circuit of Fig. 4.21 can sometimes be used to good advantage in circuit analysis. Inasmuch as these two circuits are equivalent to each other, whenever we have a voltage source (independent or controlled) in series with a resistor, this combination can be replaced by a current source in parallel with the same resistor, and vice versa.

We shall illustrate this application with an example. We shall start with the circuit in Fig. 4.27(a). The two branches with the voltage sources can be converted into their Norton's equivalent. The result is the network in Fig. 4.27(b), in which the two new Norton's equivalent circuits are enclosed in dashed boxes.

In an analogous manner, the two current sources, in conjunction with their respective parallel resistances in Fig. 4.27(a), can be replaced by their Thévenin's equivalent circuits. The result is the network in Fig. 4.27(c), in which the two new Thévenin's equivalent circuits are enclosed in dashed boxes.

The student may wonder why we want to perform these transformations. At this point, the important thing is that these transformations are valid and can always be effected as we wish. Whether or not these transformations are advantageous will depend largely on the particular situation at hand; the same applies to all network theorems we have learned thus far. In the example given in Fig. 4.27, we may observe that the circuit in Fig. 4.27(b) contains only current sources, which is easy to use if we employ node analysis. On the other hand, the circuit in Fig. 4.27(c) contains only voltage sources, which is easy to use if mesh analysis is employed. But these advantages are strictly incidental and they may not be attractive in all situations. Some examples will illustrate how this transformation can be advantageous.

At this point, it is extremely important to point out that when we replace these equivalent circuits with each other, the equivalence applies only to the entire Thévenin's or Norton's circuits as respective entities. In Fig. 4.27, the dashed boxes that occupy the corresponding positions in the three circuits are equivalent to each other only if each box is considered as a single branch. Inside these boxes, we must be careful not to over-interpret their equivalence. For example, the current through the 2-Ω resistors in (a) and that in (b) are not

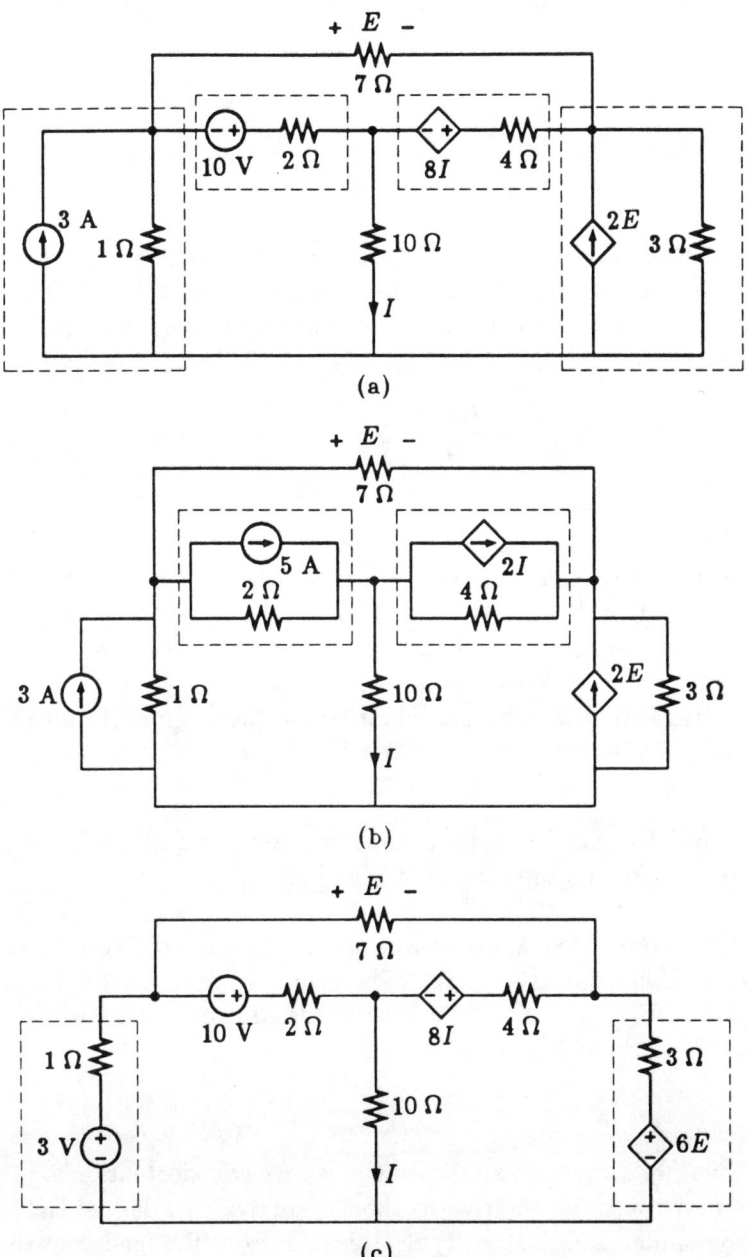

Figure 4.27: Example illustrating source transformations.

164   FUNDAMENTALS OF CIRCUIT ANALYSIS

the same. Likewise, the voltages across the 3-Ω resistor in (a) and that in (c) are not the same. Voltage E and current I are unchanged in all three circuits.

EXAMPLE 11. Use source transformation to find E_{ab} for the circuit in Fig. 4.28(a).

SOLUTION We first transform the two voltage sources into their Norton's equivalent. We have the circuit in Fig. 4.28(b). Then we combine the two current sources and the three resistances in parallel. We then have the circuit in Fig. 4.28(c). There, we readily find

$$E_{ab} = 6.5 \times \frac{20}{19} = \frac{130}{19} \text{ V}$$

Note in Example 11 that the only structural items in Fig. 4.28(c) that is readily identifiable with the circuit in Fig. 4.28(a) are nodes a and b. Everything else has been either altered or combined with something else. However, once we have found E_{ab}, which is the same in all three circuits, all other quantities in the original circuit can be calculated with relative ease.

EXAMPLE 12. Use source transformation to find the Thévenin's equivalent for the top circuit in Fig. 4.29.

SOLUTION We apply source transformation and series-parallel combination from left to right. The result is the sequence of circuits shown in Fig. 4.29. The final Thévenin's equivalent circuit is shown at the bottom of Fig. 4.29.

Another important application of source transformation is to provide us with an alternative method of analyzing a ladder network. In the ladder analysis, we typically work from the load toward the source in a step-by-step manner. Source transformation enables us to work from the source toward the load. Fig. 4.30 illustrates how

NETWORK PROPERTIES AND THEOREMS

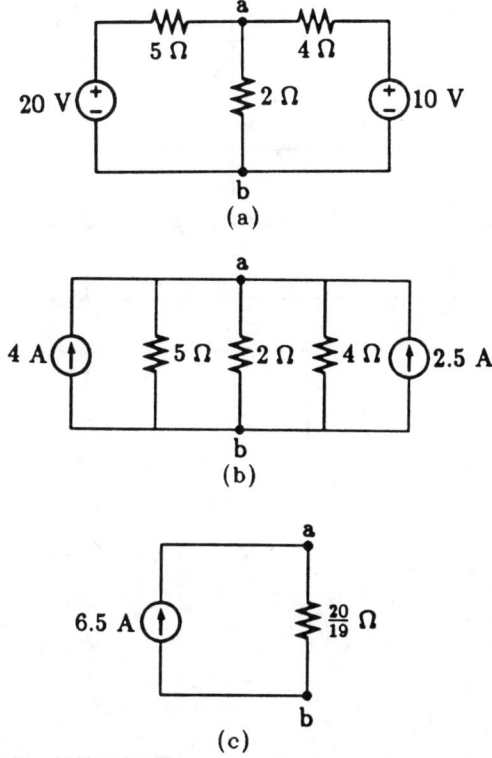

Figure 4.28: Circuit used in Example 11.

FUNDAMENTALS OF CIRCUIT ANALYSIS

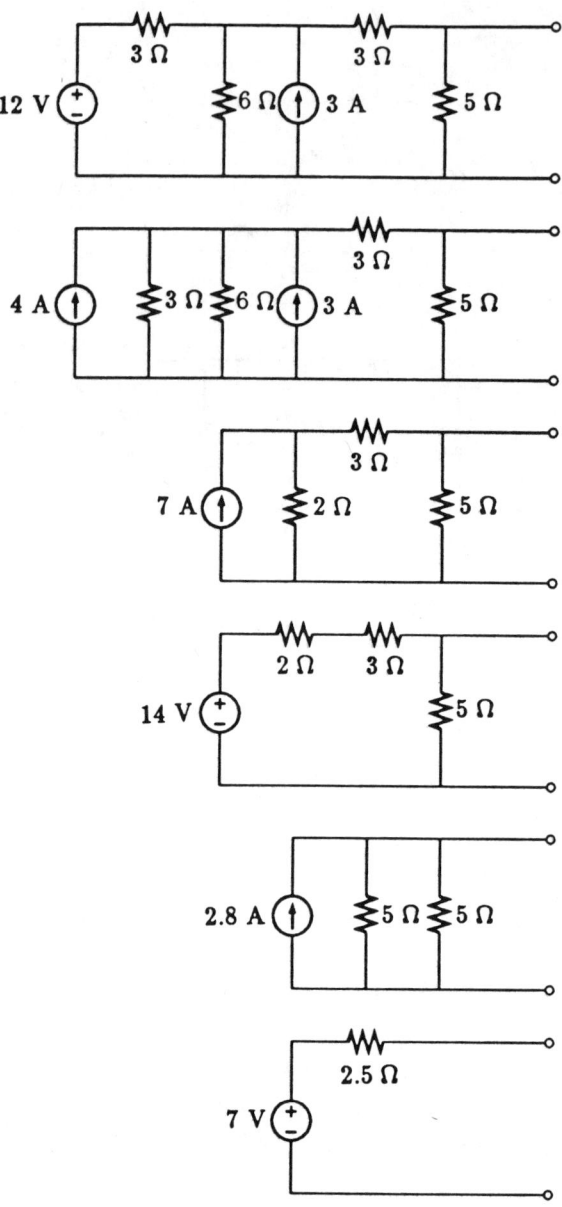

Figure 4.29: Circuits used in Example 12.

NETWORK PROPERTIES AND THEOREMS

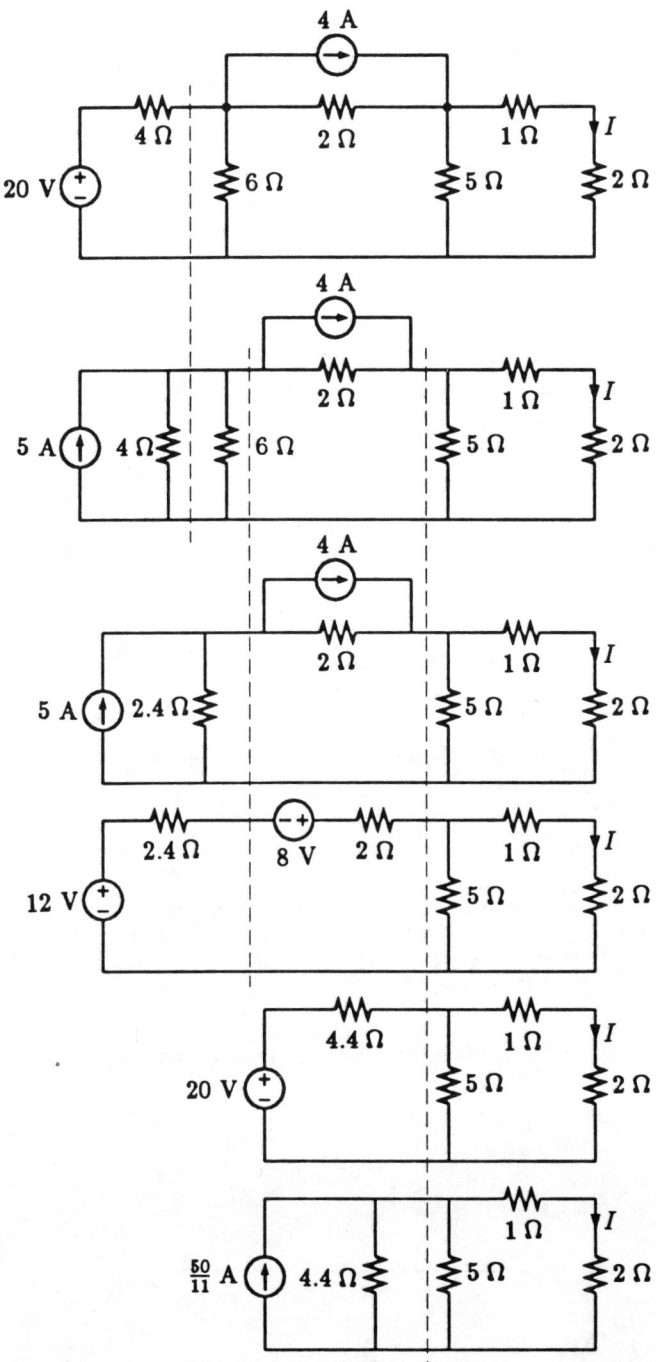

Figure 4.30: Analysis of a network using source transformation.

168 FUNDAMENTALS OF CIRCUIT ANALYSIS

a network is transformed step by step from left to right. We have been given the top network and current I is to be determined. These steps are shown in the sequence of equivalent networks from top to bottom. The vertical dashed lines indicate that those parts of the networks to the left are equivalent in terms of how they appear electrically at the dashed borders. After the bottom network has been reached, we can apply the current division rule to obtain

$$I = \frac{50}{11} \times \frac{\frac{1}{3}}{\frac{1}{4.4} + \frac{1}{5} + \frac{1}{3}} = \frac{500}{251} \text{ A}$$

EXERCISES

4.4.1 Obtain the Thévenin's equivalent circuit for the circuit using source transformation.

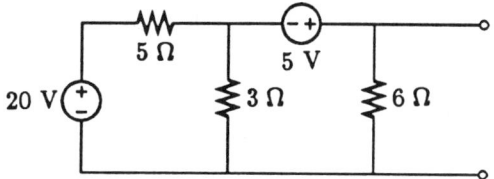

Ans. $E_{oc} = \frac{200}{21}$ V, $R_{eq} = \frac{10}{7}$ Ω.

4.4.2 Use source transformation to determine i in the circuit.

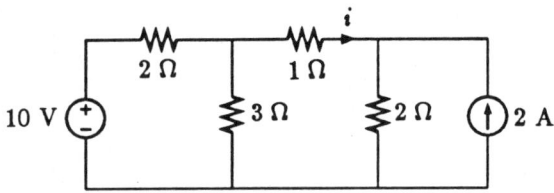

Ans. $\frac{10}{21}$ A.

4.5 Source shifting

The methods described in the previous section are limited to situations where a voltage source is connected in series with, or a current source is connected in parallel with, a resistance. Actually, these techniques can also be employed in other situations if a slight modification is effected to the network beforehand. This modification is known as *source shifting*.

There are two types of source shifting—the E-shift and the I-shift.

The E-shift enables us to "push a voltage through a node." This source shifting is indicated in Fig. 4.31. When we have a situation

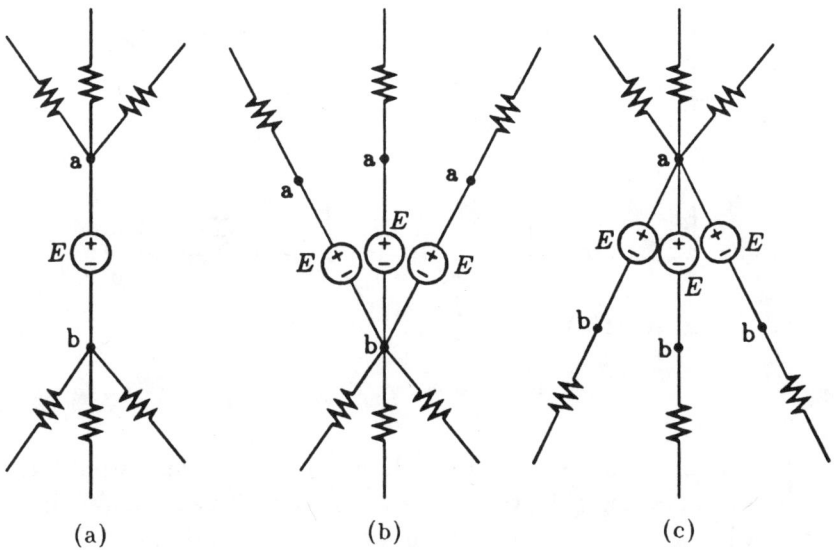

Figure 4.31: The E-shift.

as shown in part (a), we can push the source either through node a or node b. The results are shown in parts (b) and (c), respectively. It is easy to see that the voltage difference among all points with the same designation are unchanged in all three diagrams. Therefore, no electrical change has taken place. The only differences among these three arrangements are the currents in the various sources. Since voltage sources can sustain any currents, what currents flow in the various sources make little difference to the sources or to the rest of the network.

The E-shift, in effect, splits a node into several nodes with the same potential. This procedure is also known as *node splitting*. Another way to look at this procedure is to reverse the shifting process. Specifically, if we have the situation in (b) to start with, since all three nodes marked "a" are all at the same potential, they may be connected together. When we do this, the arrangement in part (a) results.

The I-shift enables us to "wrap a current source along a continuous path." This procedure is illustrated in Fig. 4.32. We start with

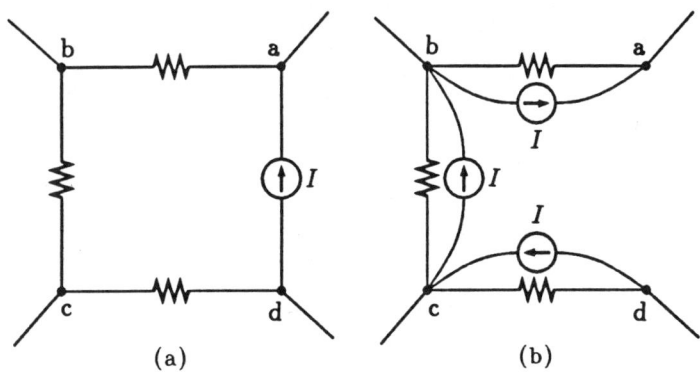

Figure 4.32: The I-shift.

the situation as shown in part (a). In (b), we have wrapped the current source originally connected between terminals a and d along the path abcd. It is clear that the net currents entering each node by all current sources remain unchanged before and after the I-shift. Since current sources can sustain any voltages, the replacement of one current source by a number of sources makes little difference to the sources or to the rest of the network.

An application of the source-shifting technique is illustrated in Fig. 4.33. We have the circuit in (a) given. We wish to find an equivalent circuit that contains either only voltage sources or only current sources.

To eliminate the voltage sources, we could first replace ahg and

NETWORK PROPERTIES AND THEOREMS

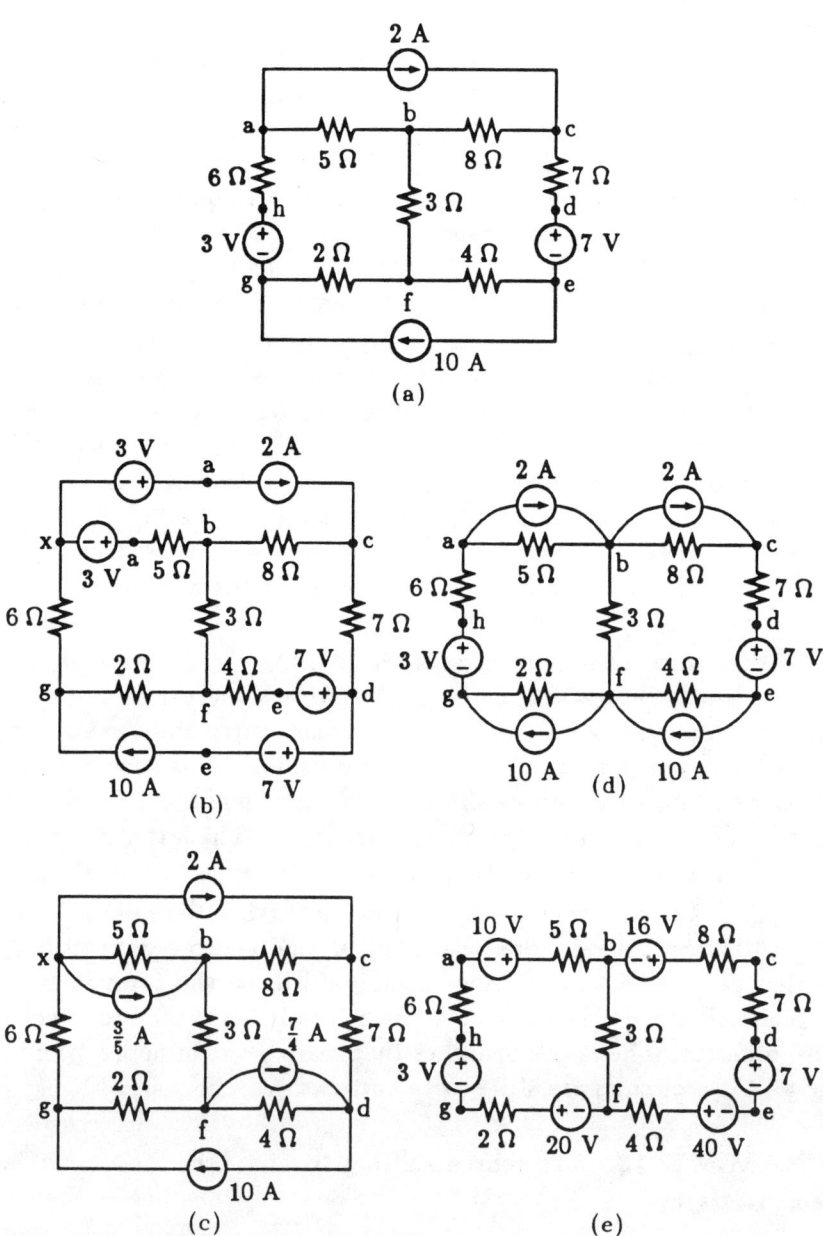

Figure 4.33: A example of source shifting.

cde by their Norton's equivalent. But we choose to do it in a different way. We push the 3-V source through node a (after first interchanging its position with the 6-Ω resistor) and the 7-V source through node e. The result is the circuit in (b). Note that node h no longer appears in (b) because of the positional exchange of the 6-Ω resistor and the 3-V source. Instead, a new node x appears. However, this does not alter the entire branch ahg or its equivalent—branch axg. After these E-shifts, branches xab and fed are transformed into their Norton's equivalent as shown in (c). The two voltage sources in series with the current sources are immaterial and are dropped.

Although (c) is equivalent to (a), one needs to exercise caution in identifying the results obtained in (c) with quantities in (a). Branches bf, gf, and bc are, of course, unaltered. Currents in the 6-Ω and the 7-Ω resistors are also unchanged. All other quantities may or may not be obviously identifiable. For example, the current in the two 5-Ω resistors in the two circuits are not the same; likewise with the currents in the 4-Ω resistors. But E_{fd}, as well as many other voltages with the same subscripts, will remain the same in all three diagrams.

To eliminate the current sources in (a), we can wrap the 2-A source along path abc and the 10-A source along efg as shown in (d). Each parallel combination of a current source and a resistance in (d) can be replaced by its Thévenin's equivalent as shown in (e). In performing these source shiftings, we have modified a four-mesh network in (a) into a two-mesh network in (e). The latter is considerably simpler to analyze using, say, mesh analysis.

Again, we must exercise caution in identifying the quantities in (e) with those in (a). Branches ahg, bf, and cde are undisturbed. Other voltages with the same subscripts will be the same in both circuits. However, the currents in the 2-Ω resistors in the two circuits are different. The entire branches that exist between nodes g and f in all three circuits are still equivalent.

EXAMPLE 13. Use source shifting to find I in the circuit in Fig. 4.34(a).

SOLUTION First we split node a as shown in (b). Then we perform source transformations resulting in Norton's equivalent and then transform back to Thévenin's equivalent. The final network is

NETWORK PROPERTIES AND THEOREMS

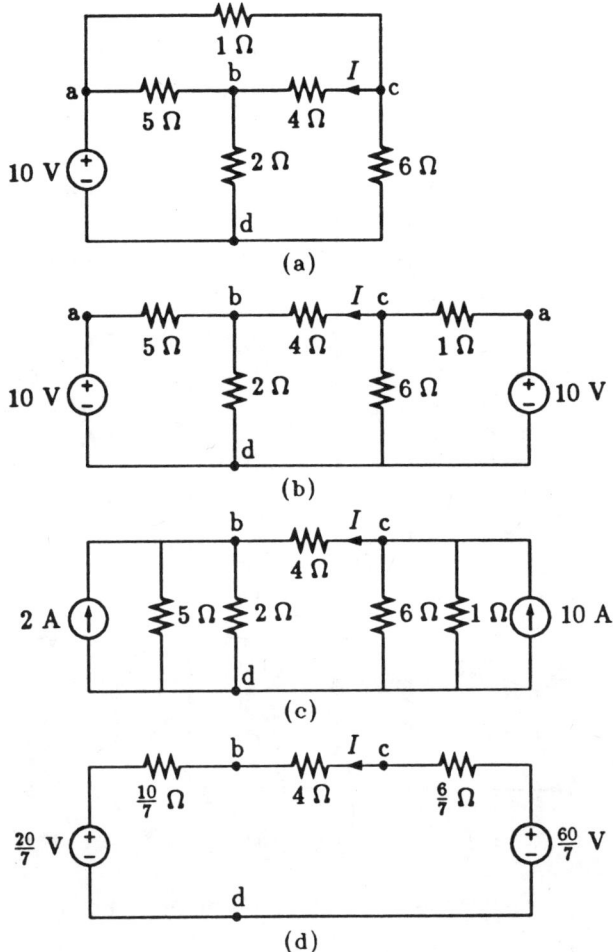

Figure 4.34: Circuits used in Example 13.

shown in (d). Throughout these source-shiftings and transformations, the 4-Ω resistor has not been disturbed. Hence, I is the same in all four circuits. From (d)

$$I = \frac{\frac{60}{7} - \frac{20}{7}}{\frac{10}{7} + 4 + \frac{6}{7}} = \frac{10}{11} \text{ A}$$

EXAMPLE 14. Use source shifting to find E_{bd} in the circuit of Fig. 4.35(a).

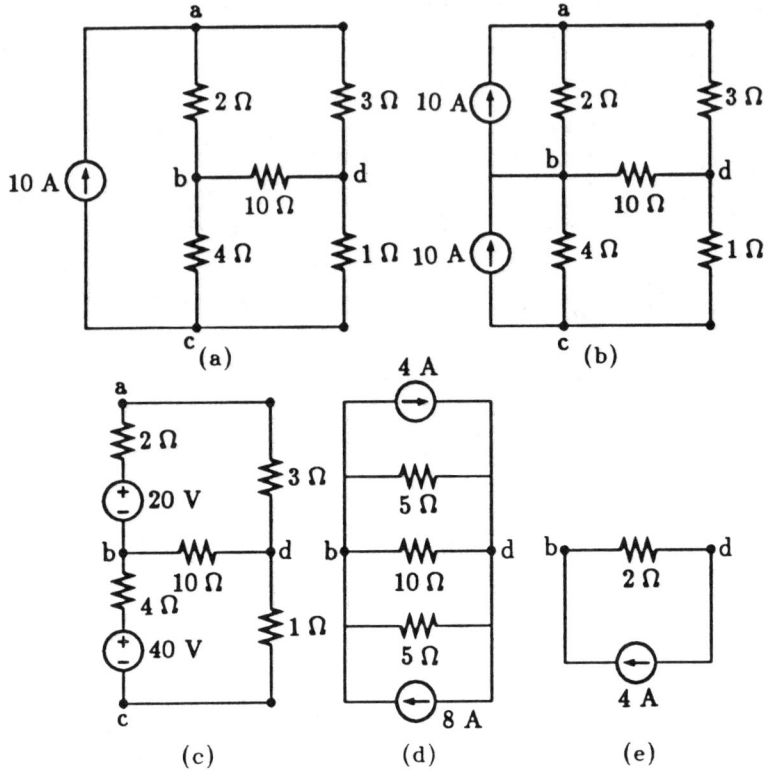

Figure 4.35: Circuit used in Example 14.

SOLUTION First we wrap the current source along path abc as shown in (b). Then we convert the parallel combination of each current source and its parallel resistance into its Thévenin's equivalent as shown in (c). Then we combine the series resistances and transform each Thévenin's branch into a Norton's branch. Finally, combine the parallel current sources and resistances to obtain the circuit in (e). From (e)

$$E_{bd} = 4 \times 2 = 8 \text{ V}$$

EXERCISES

4.5.1 Use source shifting and source transformation to obtain the Norton's equivalent of the circuit with respect to terminals A and B.

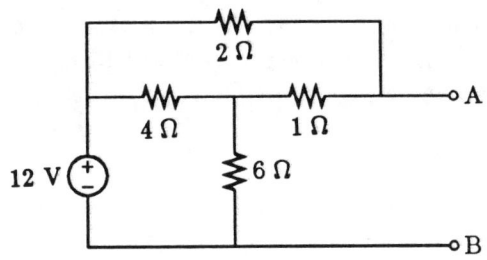

Ans. $I_{sc} = \dfrac{138}{17}$ A, $R_{eq} = \dfrac{34}{27}$ Ω.

4.5.2 Use source shifting and source transformation to obtain the Thévenin's equivalent of the circuit with respect to terminals A and B.

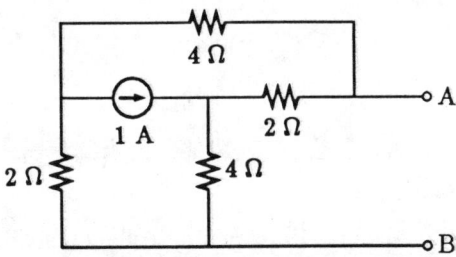

Ans. $E_{oc} = 1$ V, $R_{eq} = 3$ Ω.

FUNDAMENTALS OF CIRCUIT ANALYSIS

4.6 Tellegen's theorem

Suppose we have two networks N and N' with the same number of branches (n) and identical interconnections. We denote the voltages in N by e_j $(j = 1, 2, \ldots n)$ and the current in the corresponding branches in N' by i'_j $(j = 1, 2, \ldots n)$. The element or elements contained in a branch of N need not have any similarity to those in the corresponding branch of N'. We further require that the direction of each voltage in N relative to its corresponding current in N' be made consistent throughout both networks. That is to say, i'_j must be either all pointed from the minus to plus or all pointed from the plus to minus signs of the corresponding e_j. As long as all e's satisfy KVL and all i''s satisfy KCL, *Tellegen's theorem* states that

$$\sum_{j=1}^{n} e_j i'_j = 0 \qquad (4.26)$$

The two general circuits in Fig. 4.36 illustrate how Tellegen's

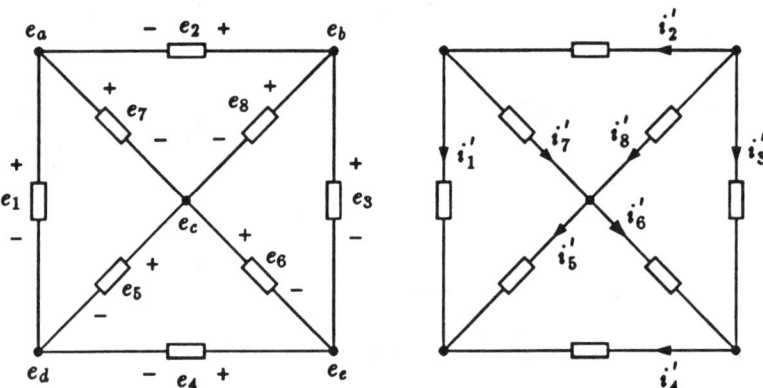

Figure 4.36: Circuits illustrating Tellegen's theorem.

theorem applies to two networks of similar topology, as well as the reason that the theorem is true. For the circuit in Fig. 4.36, we have

$$e_1 i'_1 + e_2 i'_2 + e_3 i'_3 + e_4 i'_4 + e_5 i'_5 + e_6 i'_6 + e_7 i'_7 + e_8 i'_8$$

$$= (e_a - e_d)i'_1 + (e_b - e_a)i'_2 + (e_b - e_e)i'_3 + (e_e - e_d)i'_4$$

$$+ (e_c - e_d)i'_5 + (e_c - e_e)i'_6 + (e_a - e_c)i'_7 + (e_b - e_c)i'_8$$

$$= e_a(i'_1 - i'_2 + i'_7) + e_b(i'_2 + i'_3 + i'_8) + e_c(i'_5 + i'_6 - i'_7)$$

$$+ e_d(-i'_1 - i'_4 - i'_5) + e_e(-i'_3 + i'_4 - i'_6) = 0 \quad (4.27)$$

It can be observed that each of the primed current quantities that multiply every node voltage is the sum of the current leaving the respective nodes. By KCL, each of these quantities is equal to zero. Hence, the sum in Eq. (4.27) is zero.

Although this example does not constitute a rigorous proof of the theorem, the student should be able to project that a generalization of this formalism will constitute a proof of the theorem.

As an example, the two circuits in Fig. 4.37 are totally unrelated.

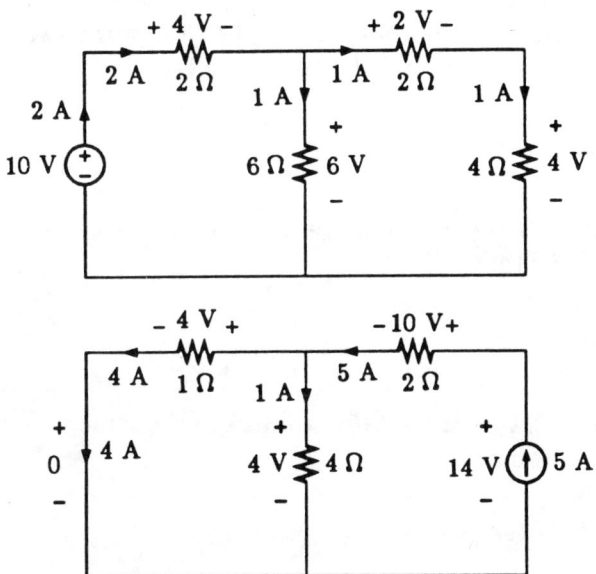

Figure 4.37: A example of Tellegen's theorem.

They have the same topology. Thus, each voltage in one circuit may be paired with the current in the other circuit branch by branch according to their locations. Tellegen's theorem asserts that

$$10 \times 4 + 4 \times (-4) + 6 \times 1 + 2 \times (-5) + 4 \times (-5) = 0 \quad (4.28)$$

$$0 \times 2 + 4 \times (-2) + 10 \times (-1) + 4 \times 1 + 14 \times 1 = 0 \quad (4.29)$$

Note that in writing out Eqs. (4.28) and (4.29), we have taken each voltage rise in one circuit and multiplied it by the corresponding current in the other circuit in the standard Ohm's law direction (from plus to minus). Of course, we could have taken all currents in the opposite directions.

Tellegen's theorem is a very elementary theorem. The only requirement is that KVL and KCL are satisfied. Hence, it is an extension of Kirchhoff's laws. There is no restriction as to what each branch is. It can be a source, independent or controlled; a resistor, linear or nonlinear; or a composite of a number of elements. If the voltages and currents are time functions and the elements are time-varying, Tellegen's theorem must be satisfied at every instant. In fact, we can point out the following ramifications of the theorem:

1. Tellegen's theorem may be applied to two networks at two different instants. That is

$$\sum_{j=1}^{n} e_j(t_1) i_j'(t_2) = 0 \qquad (4.30)$$

2. Tellegen's theorem may be applied to the same network at two different instants. That is

$$\sum_{j=1}^{n} e_j(t_1) i_j(t_2) = 0 \qquad (4.31)$$

3. Tellegen's theorem may be applied to the same network at the same instant. That is

$$\sum_{j=1}^{n} e_j(t_1) i_j(t_1) = 0 \qquad (4.32)$$

4.6.1 Conservation of power

If we apply Eq. (4.32) to a network and, in doing so, adopt the standard Ohm's law directions for all voltages and currents, Eq. (4.32) states that the algebraic sum of all powers delivered to all branches at any instant must add up to zero. This is the *law of conservation of power*.

Sometimes, we prefer to state the law of conservation of power by saying that the total power received by one part of a network must be equal to the total power supplied by the remainder of the network.

EXAMPLE 15. Analyze the circuit in Fig. 4.38 and verify that the power is conserved.

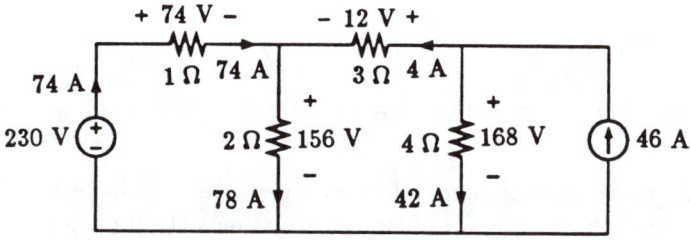

Figure 4.38: Circuit used in Example 15

SOLUTION Routine analysis will lead to the currents and voltages shown in Fig. 4.38. The powers delivered to the five branches are listed as follows.

Power delivered to the 1-Ω resistor $= 74 \times 74 = 5,476$ W

Power delivered to the 2-Ω resistor $= 156 \times 78 = 12,168$ W

Power delivered to the 3-Ω resistor $= 12 \times 4 = 48$ W

Power delivered to the 4-Ω resistor $= 168 \times 42 = 7,056$ W

Power delivered to the voltage source $= -230 \times 74 = -17,020$ W

Power delivered to the current source $= -168 \times 46 = -7,728$ W

The sum of these six powers is indeed equal to zero.

4.6.2 Tellegen's corollary

A special application of Tellegen's theorem pertains to the situation in which two networks contain an identical linear resistive subnetwork. Typically, this application becomes useful when two networks are largely identical except for a small number of branches.

We have two networks shown in Fig. 4.39. Subnetwork N con-

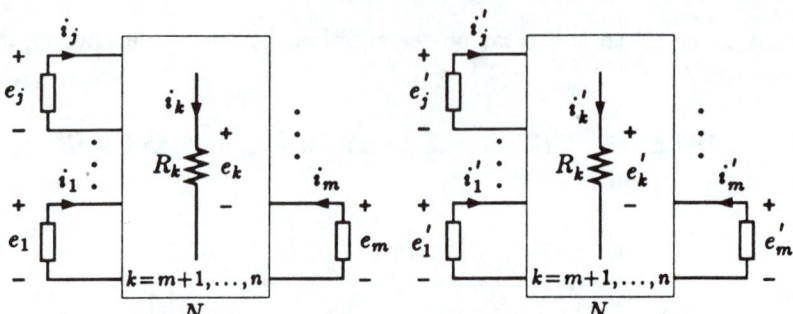

Figure 4.39: Two networks with a common resistive subnetwork.

tains only linear resistors and they are identical. We have a typical branch—the kth branch—of N shown, together with its voltages and currents.

Outside of the two N's there are m branches, a typical branch is numbered the jth branch. We represent these branches by m boxes. These branches need not have any similarity between the two networks. Tellegen's theorem requires that

$$\sum_{j=1}^{m} e_j i_j' + \sum_{k=m+1}^{n} e_k i_k' = 0 \qquad (4.33)$$

$$\sum_{j=1}^{m} e_j' i_j + \sum_{k=m+1}^{n} e_k' i_k = 0 \qquad (4.34)$$

But

$$e_k = R_k i_k \quad \text{and} \quad e_k' = R_k i_k', \quad k = m+1, \ldots, n \qquad (4.35)$$

Therefore,

$$e_k i_k' = i_k R_k i_k' = e_k' i_k, \quad k = m+1, \ldots, n \qquad (4.36)$$

Hence, the second summation terms in Eqs. (4.33) and (4.34) are identical and we may conclude that

$$\sum_{j=1}^{m} e_j i_j' = \sum_{j=1}^{m} e_j' i_j \qquad (4.37)$$

We shall call the relationship in Eq. (4.37) *Tellegen's corollary*. It should be emphasized that the two summations in Eq. (4.37) are equal to each other. Each summation is, in general, not equal to zero itself.

EXAMPLE 16. When the four-terminal resistive network N of Fig. 4.40 is subjected to various external connections, the results are

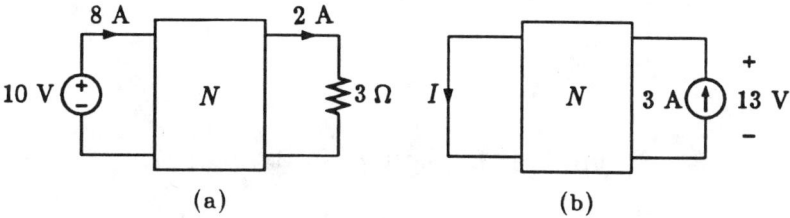

(a)   (b)

Figure 4.40: Circuit used in Example 16.

indicated in the figure. Determine current I.

SOLUTION Using Tellegen's corollary, we have

$$10 \times I + 6 \times (-3) = 0 \times (-8) + 13 \times 2$$

Solving, we obtain

$$I = 4.4 \text{ A}$$

EXAMPLE 17. When the eight-terminal resistive network N in Fig. 4.41 is subjected to various external connections, the results are indicated in the figure. Determine current I.

SOLUTION Applying Tellegen's corollary, we have

$$5 \times (-2) + 6 \times I + 5 \times 2 + 0 \times 0$$

$$= 3 \times (-3) + 0 \times 2 + 0 \times (-2) + 2 \times 3$$

Solving yields

$$I = -\frac{1}{2} \text{ A}$$

FUNDAMENTALS OF CIRCUIT ANALYSIS

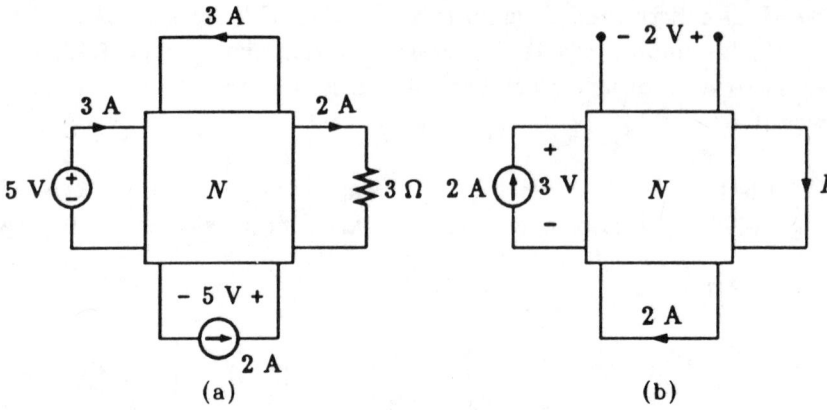

Figure 4.41: Circuit used in Example 17.

4.6.3 Reciprocity theorem

In a network that contains linear resistors and one independent voltage (or current) source, the ratio of a current (or voltage) response in one part of the network to the strength of the voltage (or current) source is the same if the locations of the source and the response are interchanged.

Reciprocity theorem can best be viewed as the consequence of Tellegen's corollary. The two versions of the reciprocity theorem are represented in Figs. 4.42 and 4.43, in which N is a network containing only linear resistors. In Fig. 4.42, the sources are voltage

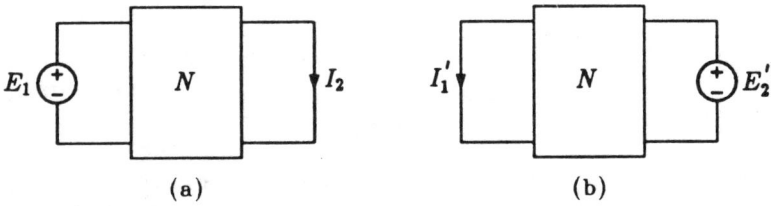

Figure 4.42: Relationship in a reciprocal network.

sources and the responses are currents. Application of Eq. (4.37) leaves only two non-zero terms, or

$$E_1 I_1' = E_2' I_2 \qquad (4.38)$$

since $E_2 = 0$ and $E_1' = 0$. Eq. (4.38) is more commonly presented as

NETWORK PROPERTIES AND THEOREMS

$$\frac{I_2}{E_1} = \frac{I_1'}{E_2'} \tag{4.39}$$

In Fig. 4.43, the sources are current sources and the responses

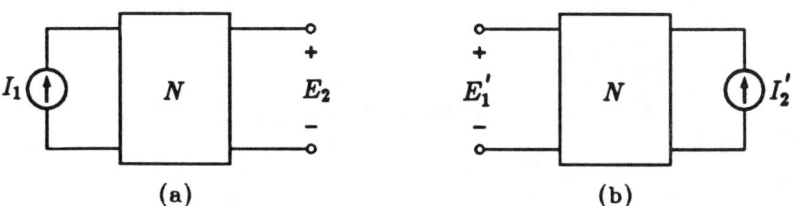

(a)   (b)

Figure 4.43: Another relationship in a reciprocal network.

are voltages. Application of Eq. (4.37) gives

$$E_1' I_1 = E_2 I_2' \tag{4.40}$$

since $I_2 = 0$ and $I_1' = 0$. Eq. (4.40) is more commonly presented as

$$\frac{E_2}{I_1} = \frac{E_1'}{I_2'} \tag{4.41}$$

Eqs. (4.39) and (4.41) show that, in reciprocity, the response and excitation must be of the opposite type (one a current and the other a voltage). The student should be cautioned that reciprocity theorem does not apply to the ratio of a voltage response to a voltage source, or a current response to a current source. Also, the relative directions of these quantities should follow the general scheme of Tellegen's theorem.

Finally, reciprocity is an important property of a special class of networks. Usually, reciprocity is not useful as a tool of analysis. Rather, it is a special property that is possessed by a restricted class of networks. It is important for us to recognize when a network is reciprocal and when it is not. For example, a network that contains controlled sources is, in general, not reciprocal.

EXAMPLE 18. By direct analysis of the two circuits of Fig. 4.44, verify that the network is reciprocal. $E_1 = 100$ V and $E_2' = 50$ V.

SOLUTION By direct analysis, we find

$$I_1' = 2 \text{ A} \quad \text{and} \quad I_2 = 4 \text{ A}$$

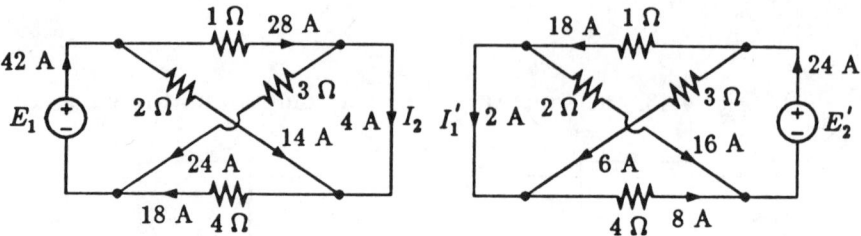

Figure 4.44: Circuits used in Example 18.

Hence,

$$\frac{I_2}{E_1} = \frac{4}{100} = \frac{I_1'}{E_2'} = \frac{2}{50}$$

The reciprocity property is satisfied.

4.7 Maximum power transfer

Situations often arise in practical engineering in which we wish to determine how much power we can draw from a given piece of apparatus or equipment. Obviously, if the question is about an ideal voltage or current source, the amount of power available is unlimited. However, no practical device is ever ideal, so the amount of power available from a device and how to attain this power are two elementary questions. An example of such a situation is when an audio amplifier is connected to a loudspeaker. The amplifier could be represented by the linear two-terminal network N in Fig. 4.45(a) and the resistance of the loudspeaker by R_L.

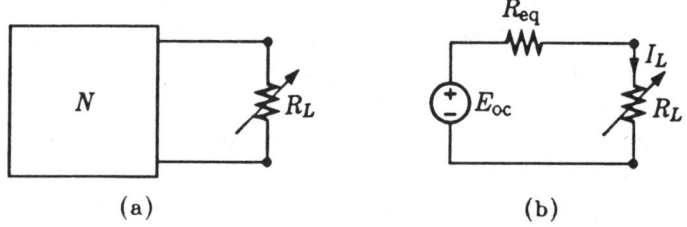

Figure 4.45: A two-terminal network connected to a load.

Thévenin's and Norton's theorems provide us with a convenient way to formulate the answers to these questions. Given a two-termi-

NETWORK PROPERTIES AND THEOREMS

nal network, which represents the device in question, we can replace it with its Thévenin's equivalent as shown in Fig. 4.45(b). At the two extremes when $R_L = 0$ and $R_L = \infty$, the power received by R_L is zero. As the value of R_L is varied from 0 to ∞, there is a maximum value of the power P_L received by R_L. From Fig. 4.45(b), it is easily seen that

$$P_L = I_L^2 R_L = \left[\frac{E_{oc}}{R_{eq} + R_L}\right]^2 \times R_L = \frac{R_L}{(R_{eq} + R_L)^2} \times E_{oc}^2 \quad (4.42)$$

We now differentiate P_L with respect to R_L and set the derivative equal to zero. Solving for R_L leads to the result that P_L is maximum when

$$R_L = R_{eq} \quad (4.43)$$

Thus, the power delivered to R_L is maximum when the load resistance R_L is equal to the Thévenin's equivalent resistance R_{eq}. When this condition is met, we say that the load resistance is matched with the source resistance. The maximum power delivered to R_L is

$$P_{\max} = \frac{E_{oc}^2}{4R_{eq}} \quad (4.44)$$

This maximum power is often referred to as the *available power*.

EXAMPLE 19. Determine the available power to R_L in the arrangement shown in Fig. 4.46.

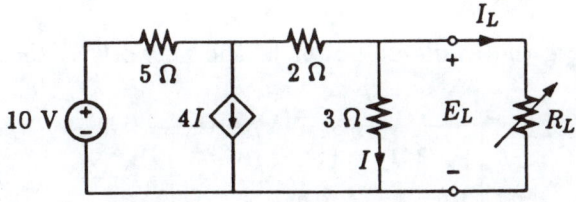

Figure 4.46: Circuit used in Example 19.

SOLUTION We shall use node analysis to first find the relationship between E_L and I_L. The arrangement is shown below.

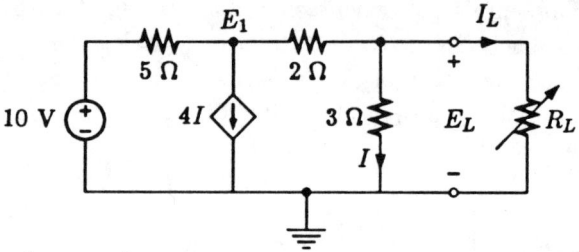

The node equations are

$$\frac{E_1 - 10}{5} + \frac{E_1 - E_L}{2} + 4I = 0$$

$$\frac{E_L - E_1}{2} + \frac{E_L}{3} + I_L = 0$$

We also have

$$I = \frac{E_L}{3}$$

Elimination of E_1 and I gives

$$E_L = 1 - 0.7 I_L$$

Comparison with Eq. (4.19) gives

$$E_{oc} = 1 \text{ V} \quad \text{and} \quad R_{eq} = 0.7 \text{ }\Omega$$

From Eq. (4.44), we have

$$\text{Available power} = \frac{1}{4 \times 0.7} = 0.3571 \text{ W}$$

EXERCISE

4.7.1 What is the available power of the network at terminals A and B?

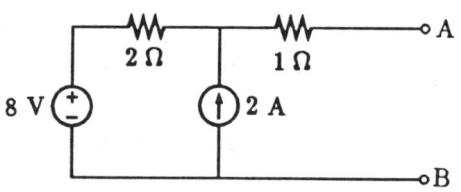

Ans. 12 W.

NETWORK PROPERTIES AND THEOREMS

4.8 Conclusions

In this chapter, we have presented several important network properties and theorems. The additivity property, the proportionality property, the linearity property, the superposition theorem, Thévenin's theorem, Norton's theorem, and the maximum-power theorem apply to networks containing independent sources, linear resistors, and linear controlled sources.

Tellegen's theorem is not limited to linear networks. It is one of the most elementary and general relationships in network theory. It applies to any network or networks with the same topology, regardless of the types of elements included. It finds several special applications in modern network theory. One of the special interpretations of this theorem is Tellegen's corollary.

One of the applications of Tellegen's corollary leads to the conclusion that a network containing only linear resistors must be reciprocal. Networks containing controlled sources are, in general, not reciprocal.

We have presented these properties and theorems using only memoryless network elements, chiefly because this makes the statements and derivations simple and the ideas easier to grasp. In fact, many of these properties and theorems are also applicable to many broader classes of networks. For example, many of the theorems also apply to ac circuits, which we will take up in Chapter 9. Thévenin's and Norton's theorems occur also in numerous situations in energy-storing elements in Chapter 6. Reciprocity is important in the discussion of reciprocal two-ports (Chapter 10) and multiterminal networks (Chapter 12). Hence, it is highly desirable that the properties and theorems presented in their present form be fully understood and appreciated. Their extension to other situations will be noted as occasions arise.

Problems

4.1 Determine the contribution of each source to I. Then calculate the value of I.

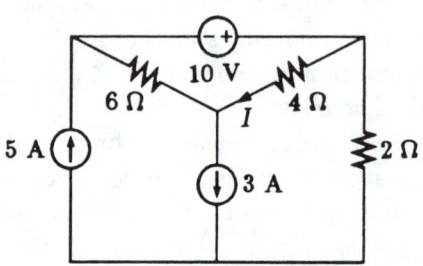

△ **4.2** Determine the contribution of each source to E. Then calculate E.

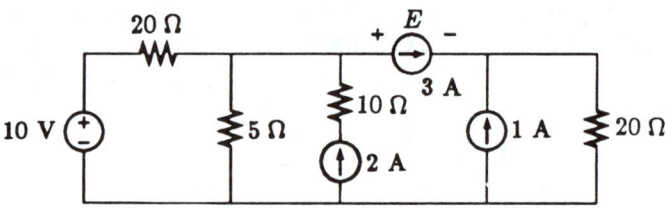

4.3 In the circuit, it is known that if $E_1 = 10$ V and $I_2 = -1$ A, then $I_x = 2$ A; and if $E_1 = -7$ V and $I_2 = 4$ A, then $I_x = -1.5$ A. Determine I_x if $E_1 = -1$ V and $I_2 = 1$ A.

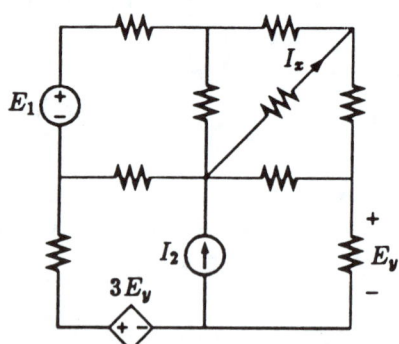

4.4 Determine I.

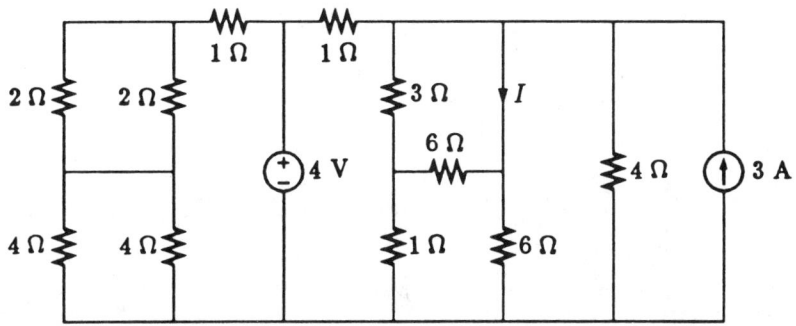

4.5 Use superposition theorem to calculate $i(t)$.

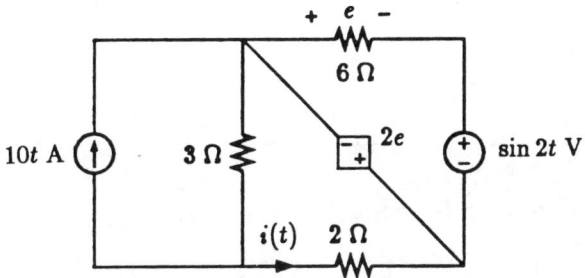

4.6 Use the result of Problem 4.5 to obtain $i(t)$.

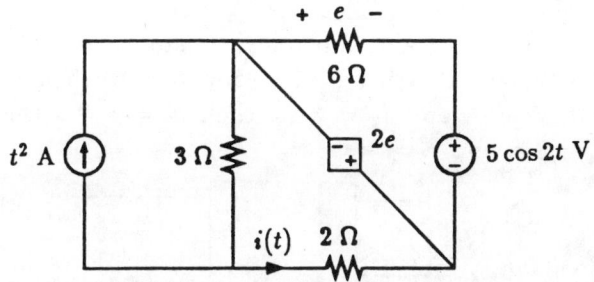

4.7 In the circuit, it is known that if $E_1 = 20$ V and $I_1 = 5$ A, then $I = 3$ A; and if $E_1 = -5$ V and $I_1 = 15$ A, then $I = 1$ A.

Calculate I if $E_1 = 12$ V and $I_1 = 10$ A.

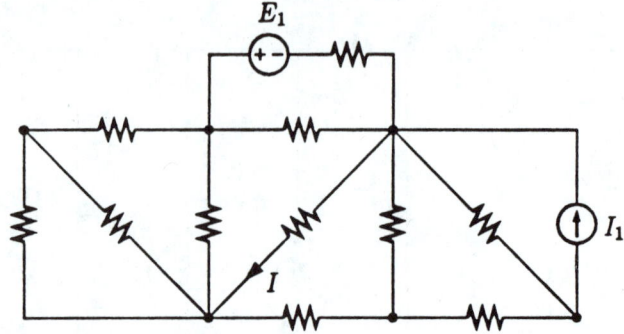

○○ **4.8** Determine the contribution of each source to E. Then calculate E.

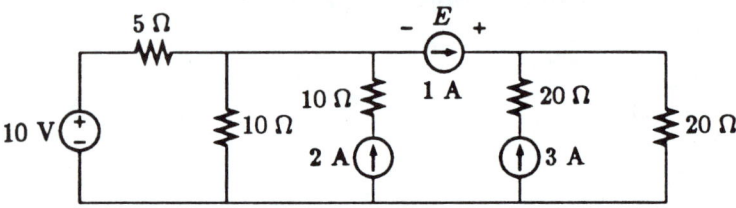

≫ **4.9** Express E in terms of E_1, I_1, and I_2.

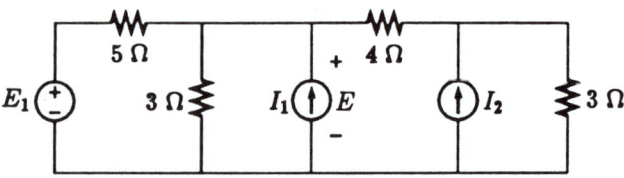

4.10 Network N contains only linear resistors and controlled sources. It is known that if $I_1 = 5$ A and $I_2 = 0$, then $E = 3$ V; and if $I_1 = 1$ A and $I_2 = 3$ A, then $E = 5$ V. Determine I_2 if $I_1 = 3$ A and $E = 5$ V.

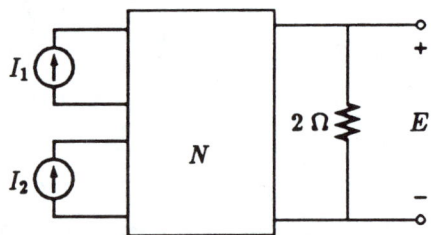

NETWORK PROPERTIES AND THEOREMS 191

4.11 Determine the relationship between E_L and I_L.

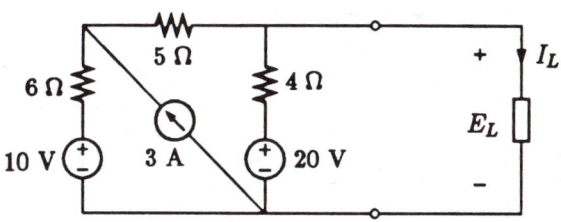

4.12 Determine the relationship between E_L and I_L.

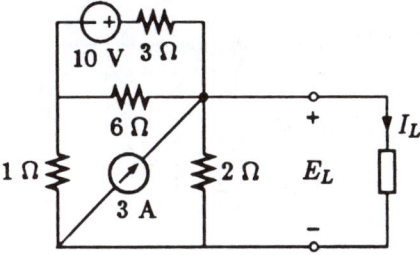

4.13 Determine the relationship between E_L and I_L.

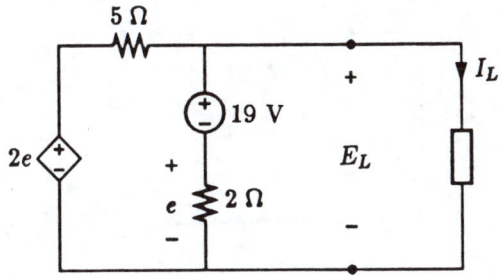

4.14 Determine the relationship between E_L and I_L.

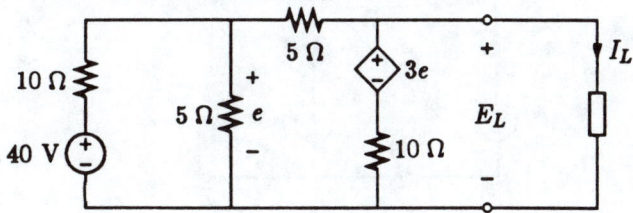

4.15 Determine the relationship between E_L and I_L.

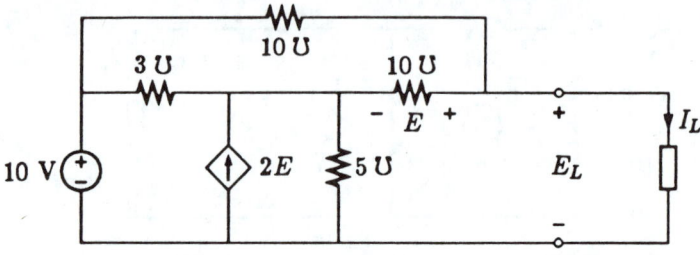

4.16 Determine the Thévenin's and Norton's equivalent circuits of the following circuit with respect to terminals A and B.

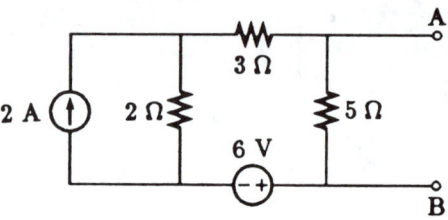

4.17 Determine the Thévenin's and Norton's equivalent circuits of the following circuit with respect to terminals A and B.

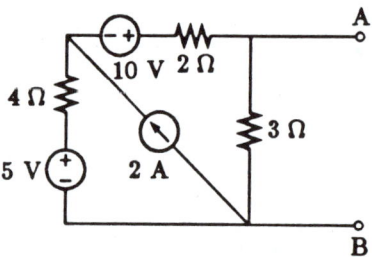

4.18 Determine the Thévenin's and Norton's equivalent circuits of the following circuit with respect to terminals A and B.

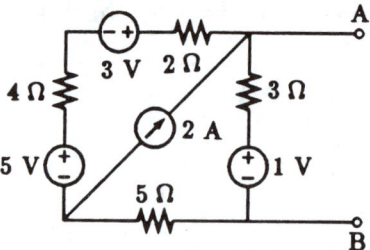

4.19 Determine the Thévenin's and Norton's equivalent circuits of the following circuit with respect to terminals A and B.

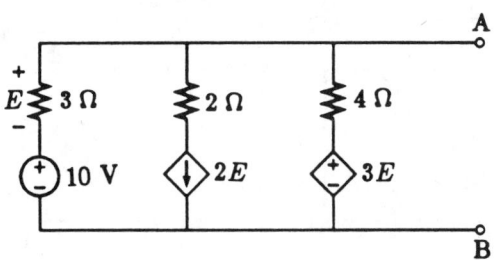

4.20 Determine the Thévenin's and Norton's equivalent circuits of the following circuit with respect to terminals A and B.

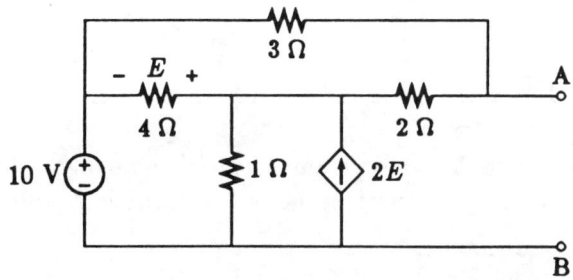

4.21 Determine the Thévenin's and Norton's equivalent circuits of the following circuit with respect to terminals A and B.

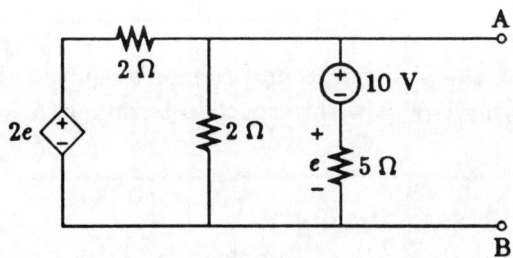

4.22 Determine the Thévenin's and Norton's equivalent circuits of the following circuit with respect to terminals A and B.

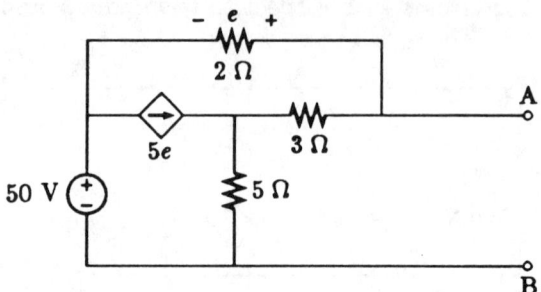

4.23 Determine the Thévenin's and Norton's equivalent circuits of the following circuit with respect to terminals A and B.

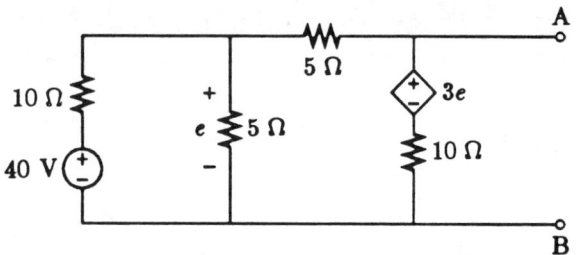

4.24 Determine the Thévenin's and Norton's equivalent circuits of the following circuit with respect to terminals A and B.

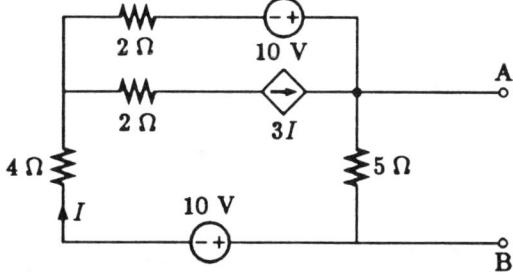

4.25 Determine the Thévenin's and Norton's equivalent circuits of the following circuit with respect to terminals A and B.

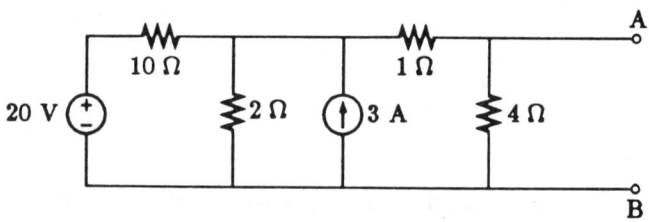

4.26 Determine the Thévenin's and Norton's equivalent circuits of the following circuit with respect to terminals A and B.

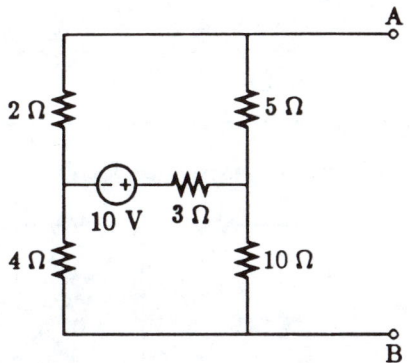

4.27 Determine the Thévenin's and Norton's equivalent circuits of the following circuit with respect to terminals A and B.

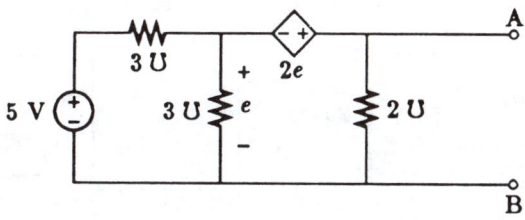

4.28 Determine the equivalent resistance R_{eq}.

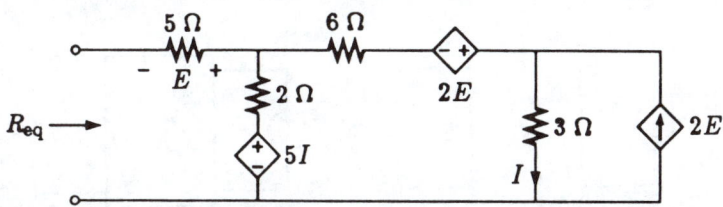

4.29 Determine the equivalent resistance R_{eq}.

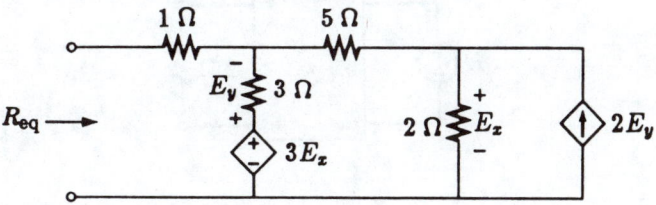

4.30 Determine the equivalent resistance R_{eq}.

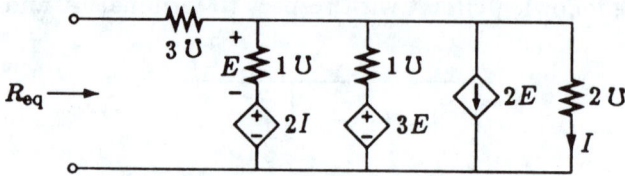

4.31 Determine the equivalent resistance R_{eq}.

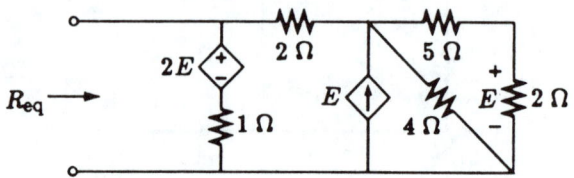

4.32 Network N is linear and contains independent sources. Determine its Thévenin's equivalent circuit with respect to terminals A and B.

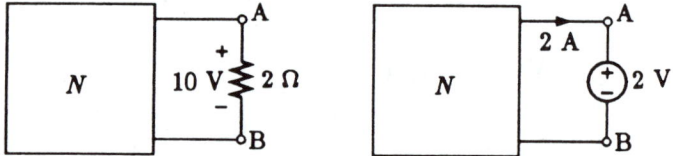

4.33 Network N contains only linear resistors and controlled sources. Determine E.

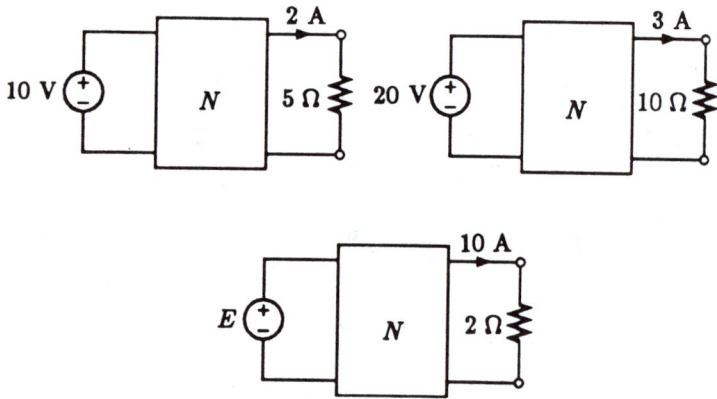

4.34 It is known that if $R = 5\,\Omega$, then $I = 3$ A; and if $R = 10\,\Omega$, then $I = 2$ A. What is the value of I if $R = 2\,\Omega$?

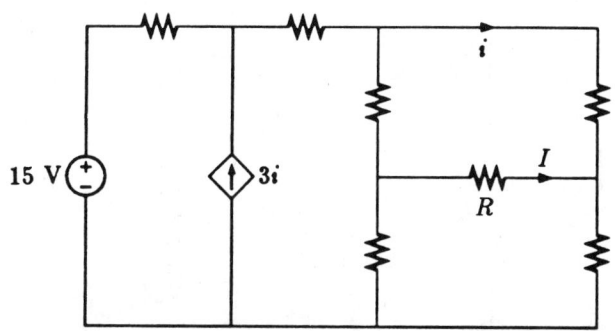

4.35 Network N contains only linear resistors and controlled sources. Determine I.

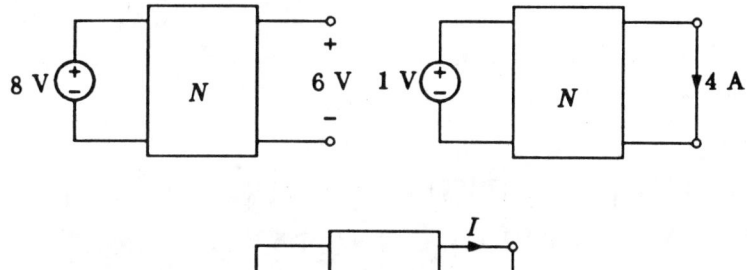

4.36 It is known that if $E = 10$ V and $R = 5\,\Omega$, then $I = 3$ A; and if $E = 20$ V and $R = 10\,\Omega$, then $I = 5$ A. Calculate I for $E = 1$ V and $R = 2\,\Omega$.

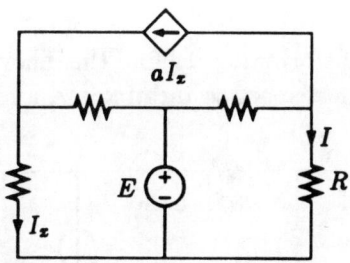

4.37 Network N contains only linear resistors and controlled sources. Determine I.

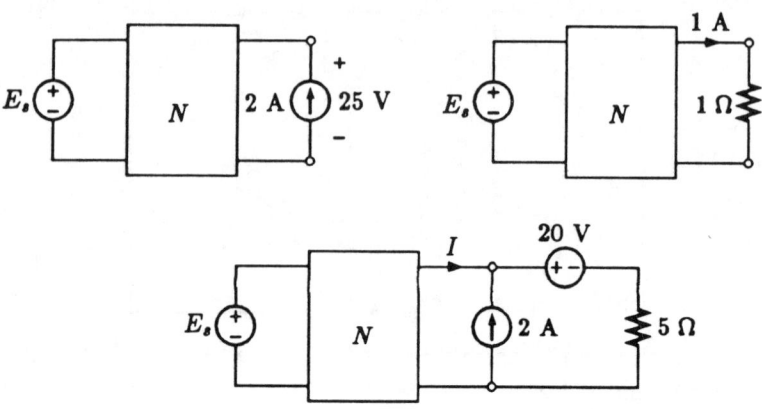

4.38 Network N contains only linear resistors and controlled sources. Determine I.

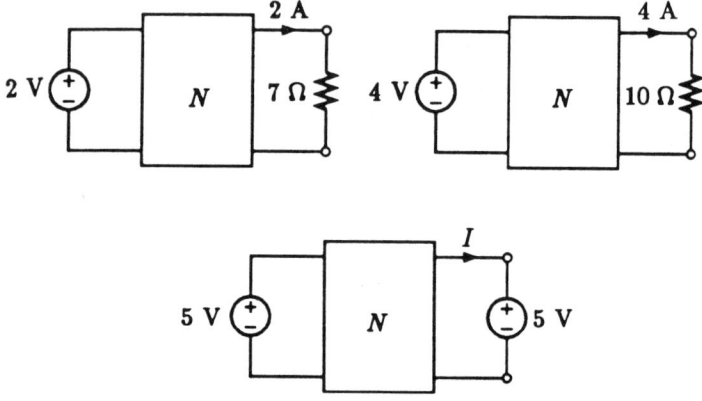

4.39 Use source transformation to find the Thévenin's equivalent of the circuit with respect to terminals A and B.

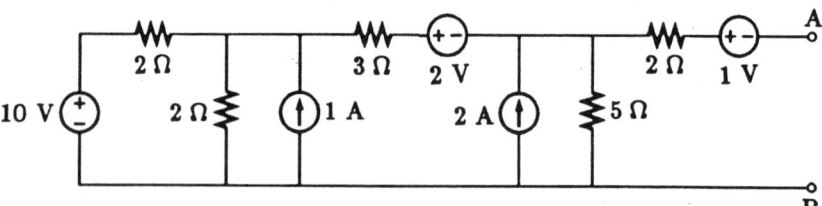

NETWORK PROPERTIES AND THEOREMS

4.40 Use source transformation to find the value of I.

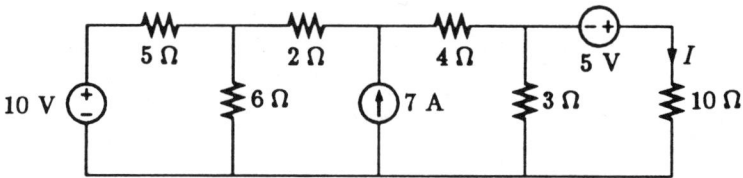

4.41 Use source transformation to determine E.

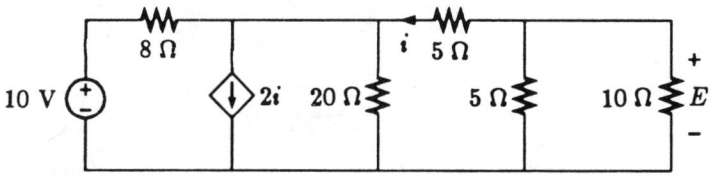

4.42 Apply source transformation until everything is in parallel with the 5-A source. Then determine e.

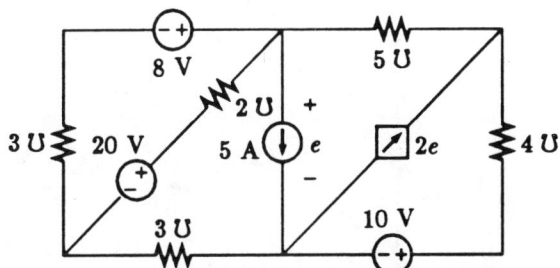

4.43 Apply I-shift and source transformation to reduce the circuit to a single-loop circuit. Do this without disturbing the branch with current i. Then calculate i.

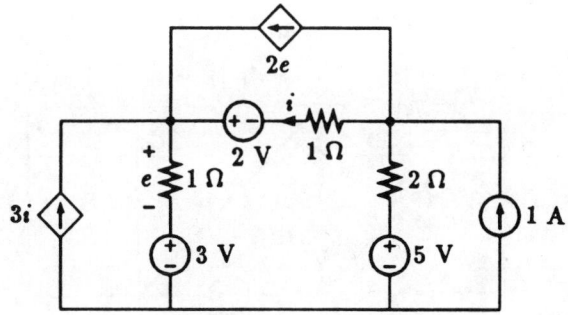

4.44 Apply E-shift to push the two voltage sources through nodes A and B. Then transform every Thévenin's branch into a Norton's branch. Finally, perform node analysis to determine I.

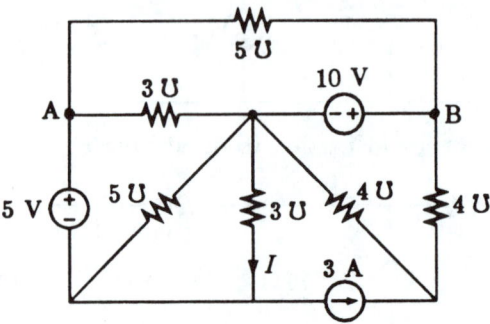

4.45 Use source shifting and source transformation to reduce the circuit to a single-loop circuit leaving the 2-Ω branch undisturbed. Then determine I.

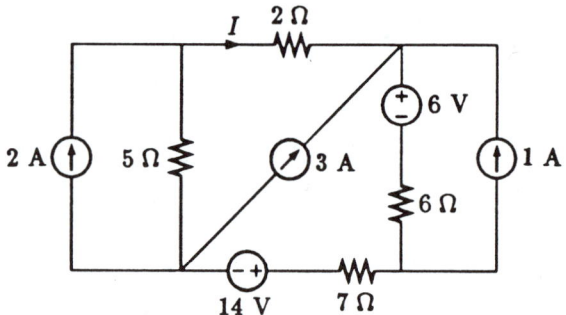

4.46 Apply source shifting and source transformation to eliminate all current sources. Do this without disturbing the two branches with I_1 and I_2. Then use mesh analysis to determine I_1 and I_2.

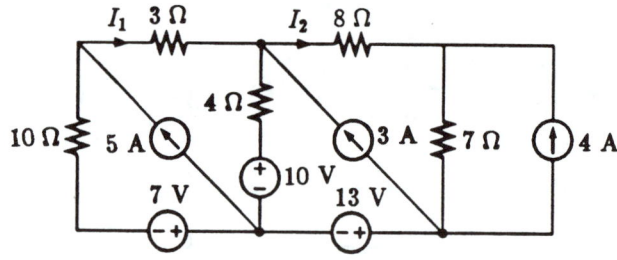

4.47 Apply source shifting and source transformation to eliminate all current sources. Do this without disturbing the two branches with I_1 and I_2. Then use mesh analysis to determine I_1 and I_2.

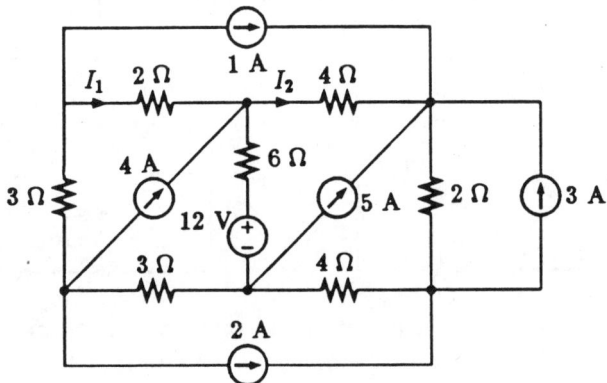

4.48 First, write every current and every voltage in terms of i. Then apply Tellegen's theorem to determine i.

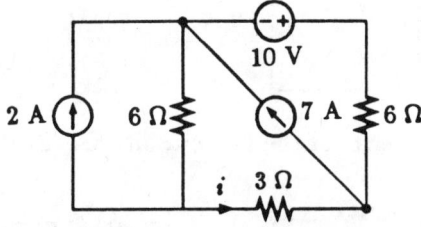

4.49 Network N contains only linear resistors. Determine I.

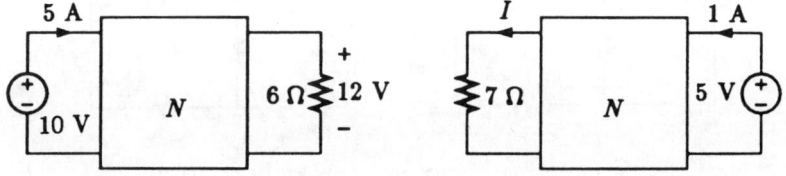

4.50 The resistive parts of the two circuits are identical. Determine I.

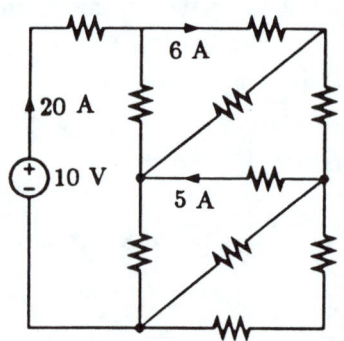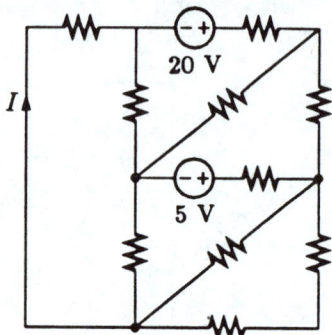

4.51 First, obtain all currents and voltages in both circuits. Then verify Tellegen's theorem by using voltages from one circuit and currents from the other, and vice versa.

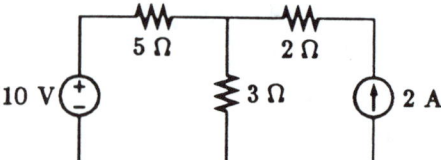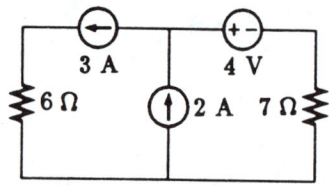

4.52 The resistive parts of the two circuits are identical. Determine I.

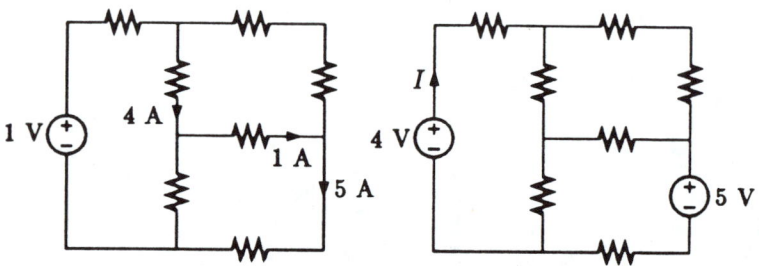

4.53 Network N contains linear resistors only. Determine I.

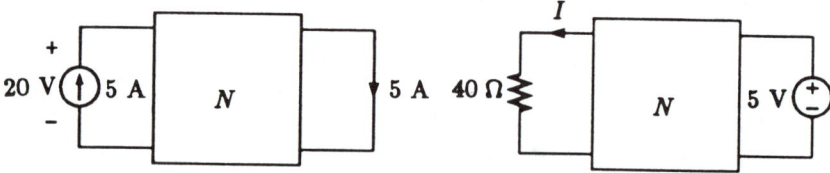

4.54 The resistive parts of the two circuits are identical. Determine E.

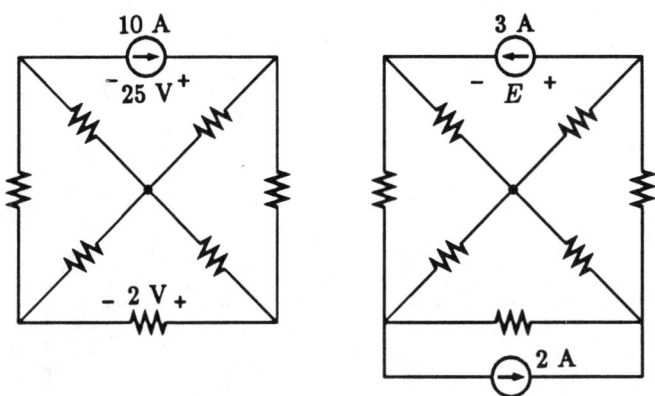

4.55 Network N contains only linear resistors. Determine the value of R.

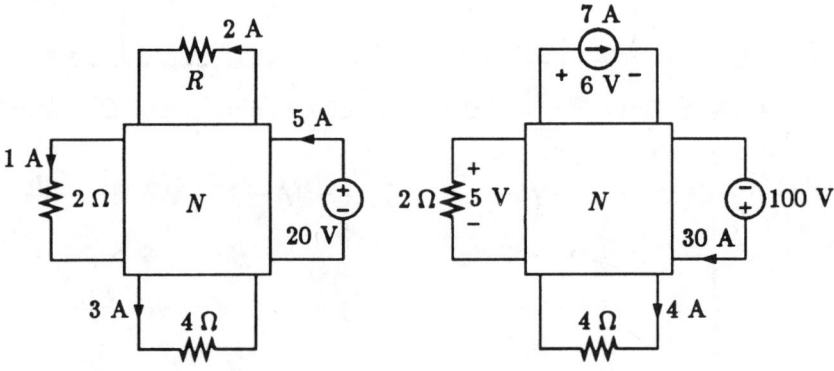

4.56 Network N contains only linear resistors. Determine E.

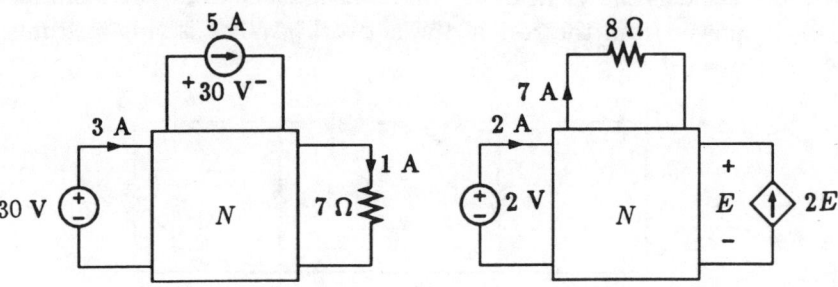

4.57 Network N contains only linear resistors. Determine I.

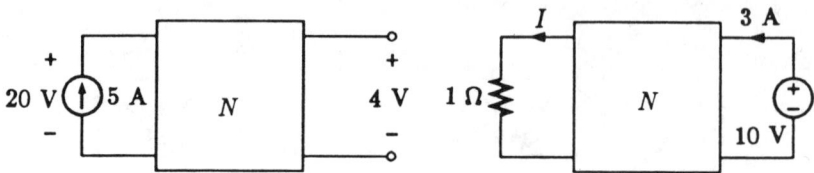

4.58 The resistive parts of the two circuits are identical. Determine E.

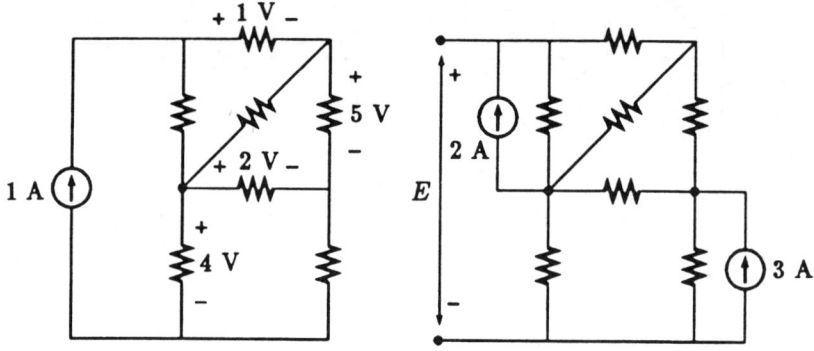

4.59 The resistive parts of the two circuits are identical. Determine E.

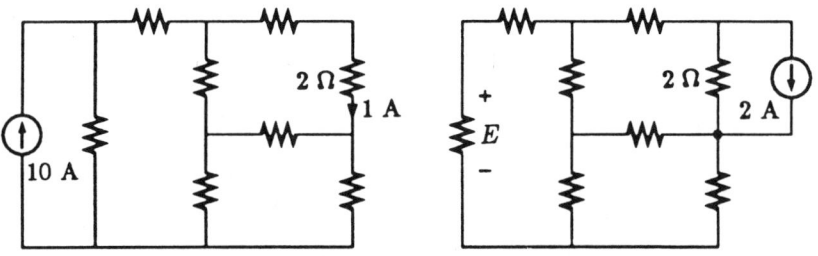

4.60 What is the value of R_L that will make it draw the maximum power from the rest of the network? What is this maximum power?

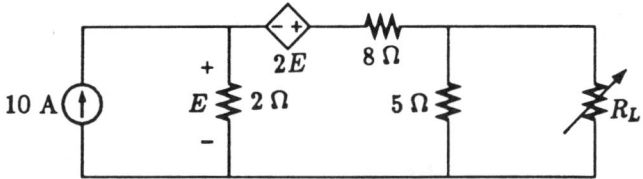

4.61 What is the power available from the network at terminals A and B?

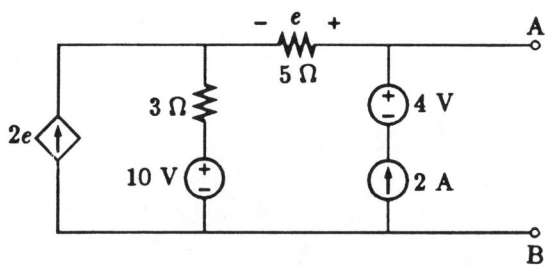

4.62 Calculate the available power of the circuit at terminals A and B.

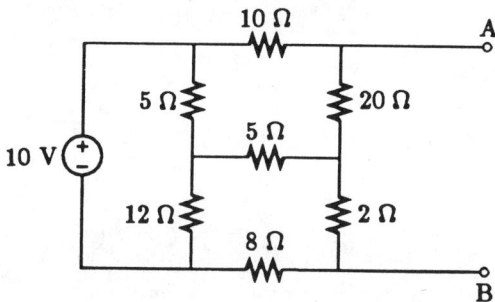

Chapter 5

Memory Elements

The past four chapters have been devoted to networks containing linear resistors, linear controlled sources, ideal op amps, and independent sources. These networks are memoryless. In these networks, the response at any time depends only on the values of the independent sources at that time. As soon as the independent sources change their strengths, the responses change immediately in accordance with the new conditions.

In this chapter we introduce two new ideal elements—the capacitor and the inductor—which are capable of storing energy in their electric and magnetic fields, respectively. In circuits that contain these devices, the response at any given time depends not only on the strengths of the independent sources, but also on the energy stored in the capacitors and inductors; this stored energy depends on the past history of the circuit.

Because networks containing capacitors and inductors are memory networks, their analysis generally requires the solution of some differential and integral relationships instead of just algebraic equations. At the same time, these networks offer a great deal more in their performance and application.

5.1 The capacitor

A *capacitor* is an electrostatic energy-storage device that stores the energy in an electric field and can be directly related to the electric charge stored in the device. The symbol for a capacitor is shown in Fig. 5.1(a). The external electrical behavior of a capacitor can be

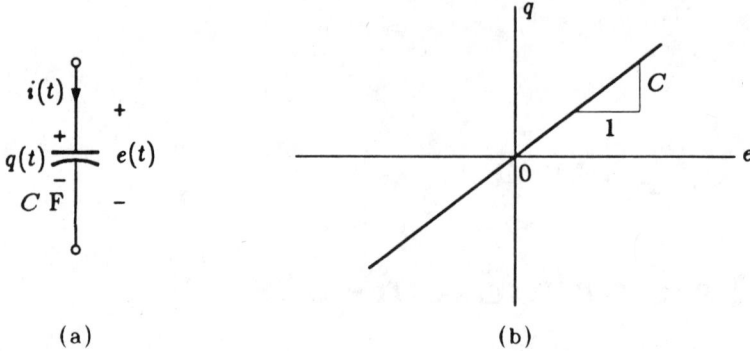

(a)   (b)

Figure 5.1: A linear capacitor and its q-e characteristic.

completely characterized by the charge-versus-voltage (q-e) curve. A linear capacitor has a linear q-e curve, as shown in Fig. 5.1(b). Analytically, such a device is characterized by either of the following two equations:

$$q(t) = C\, e(t) \tag{5.1}$$

$$e(t) = S\, q(t) \tag{5.2}$$

in which C is a constant known and the *capacitance* and S is another constant known as the *elastance*. When q is in coulombs and e is in volts, C has a unit of *farads* (represented by the letter F). Clearly

$$CS = 1 \tag{5.3}$$

The simplest example of a linear capacitor is the parallel-plate capacitor illustrated in Fig. 5.2. Its capacitance is

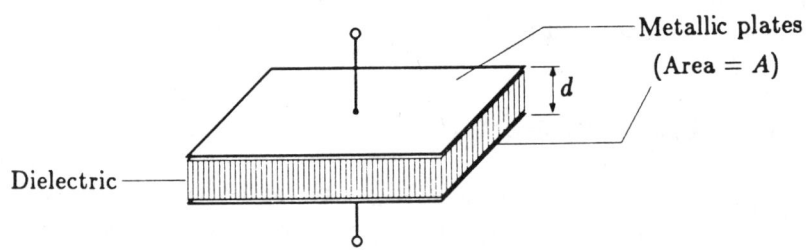

Figure 5.2: A parallel-plate capacitor.

MEMORY ELEMENTS

$$C = \epsilon \frac{A}{d} \tag{5.4}$$

where ϵ is the permittivity of the dielectric material that occupies the space between the plates, A is the area of each plate, and d is the distance between the plates.

Capacitors can be made of metal foil and paper, special plastics, or ceramics. In a modern, completely integrated circuit, a capacitor may simply be the characteristic of a certain part of a complicated structure and never appear in isolated form.

In most of our problems, we wish to find the voltage-current relationship of an element. Eqs. (5.1) and (5.2) are not suitable for these purposes. Since the current is also the time rate of change of the stored charge, we have

$$i(t) = \frac{d\,q(t)}{dt} \tag{5.5}$$

Conversely,

$$q(t) = \int_{-\infty}^{t} i(\lambda)\, d\lambda \tag{5.6}$$

or

$$q(t) = q(t_0) + \int_{t_0}^{t} i(\lambda)\, d\lambda, \qquad t > t_0 \tag{5.7}$$

Hence the e-i relationship of a capacitor is

$$e(t) = \frac{1}{C} \int_{-\infty}^{t} i(\lambda)\, d\lambda \tag{5.8}$$

or

$$e(t) = \frac{q(t_0)}{C} + \frac{1}{C} \int_{t_0}^{t} i(\lambda)\, d\lambda = e(t_0) + \frac{1}{C} \int_{t_0}^{t} i(\lambda)\, d\lambda \tag{5.9}$$

for $t > t_0$, and

$$i(t) = C \frac{d\,e(t)}{dt} \tag{5.10}$$

EXAMPLE 1. Refer to Fig. 5.1(a) with $C = 0.1$ F. Obtain current $i(t)$ for the voltage $e(t)$ shown in Fig. 5.3.

210   FUNDAMENTALS OF CIRCUIT ANALYSIS

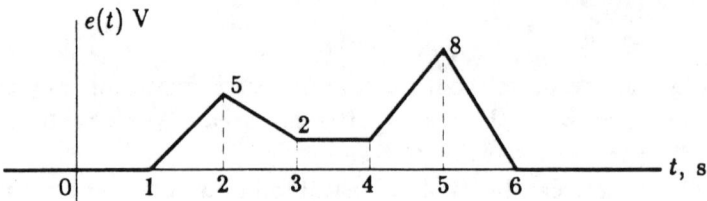

Figure 5.3: Voltage $e(t)$ of Example 1.

SOLUTION Since $e(t)$ consists of broken straight lines, its derivative is shown below.

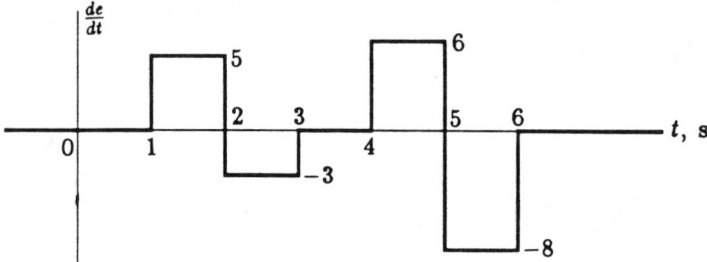

Hence, the current $i(t)$ is shown in the following diagram.

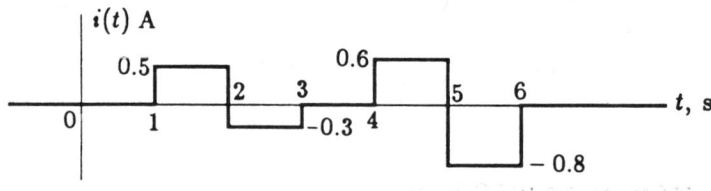

EXAMPLE 2. Refer to Fig. 5.1(a) with $C = 1$ F. Obtain voltage $e(t)$ for the current $i(t)$ shown in Fig. 5.4.

SOLUTION We have

$$i(t) = 0, \quad t < 0$$

$$i(t) = t, \quad 0 < t < 2$$

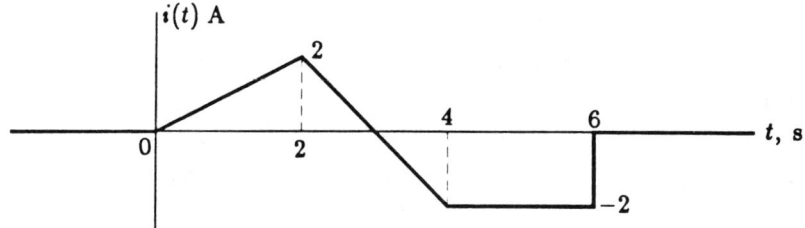

Figure 5.4: Current $i(t)$ of Example 2.

$i(t) = -2(t-3), \quad 2 < t < 4$

$i(t) = -2, \quad 4 < t < 6$

$i(t) = 0, \quad 6 < t$

Thus,

$e(t) = 0, \quad t < 0$

$e(t) = \int_0^t \lambda \, d\lambda = \dfrac{t^2}{2}, \quad 0 < t < 2$

$e(2) = 2$

$e(t) = 2 + \int_2^t -2(\lambda - 3) \, d\lambda = -t^2 + 6t - 6, \quad 2 < t < 4$

$e(4) = 2$

$e(t) = 2 + \int_4^t (-2) \, d\lambda = 10 - 2t, \quad 4 < t < 6$

$e(t) = e(6) = -2, \quad 6 < t$

The variation of $e(t)$ is shown below.

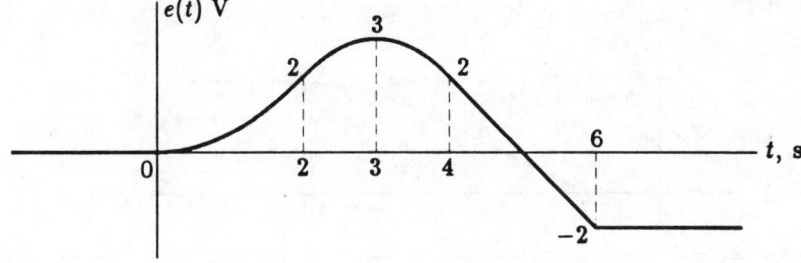

EXERCISES

5.1.1 In Fig. 5.1(a), $C = 0.5$ F and $e(t) = 5\cos 3t$ V. Find $i(t)$.

Ans. $-7.5 \sin 3t$ A.

5.1.2 In Fig 5.1(a), $C = 1$ F and $e(t)$ is given below. Find $i(t)$.

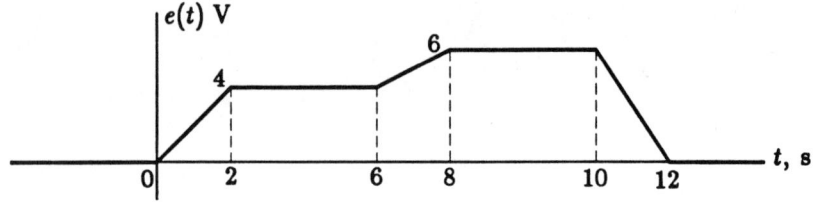

Ans.

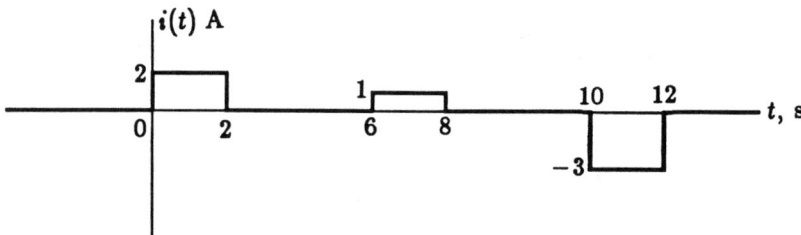

5.1.3 In Fig. 5.1(a), $C = 10$ F and $i(t)$ is given below. Find $e(t)$.

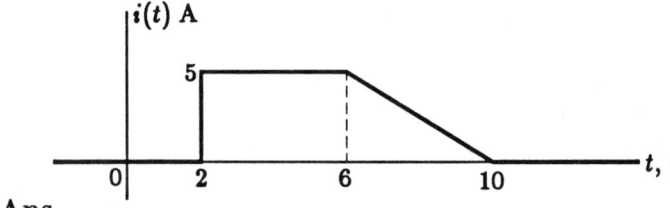

Ans.

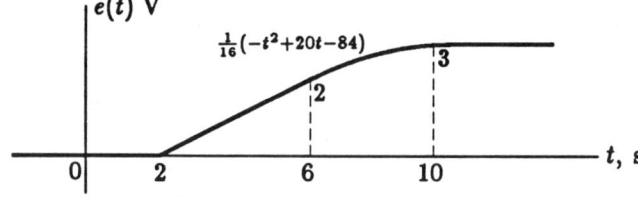

5.2 Energy stored in a capacitor

When a capacitor has a charge stored in it, it represents a certain amount of stored energy. The energy delivered to a device at any instant is the integral of the power, $p(t)$, delivered to that device up to that time. For a linear capacitor, this energy is

$$w(t) = \int_{-\infty}^{t} p(\lambda)\, d\lambda = \int_{-\infty}^{t} e(\lambda) i(\lambda)\, d\lambda = \int_{-\infty}^{t} \frac{q(\lambda)}{C} \cdot \frac{dq(\lambda)}{d\lambda}\, d\lambda$$

$$= \frac{1}{C} \int_{\lambda=-\infty}^{\lambda=t} q(\lambda)\, dq = \frac{1}{2C} q^2(t) = \frac{1}{2} C\, e^2(t) \qquad (5.11)$$

if we accept that $q(-\infty) = 0$ and $e(-\infty) = 0$. We could also write the energy in a capacitor as

$$w(t) = w(t_0) + \int_{t_0}^{t} p(\lambda)\, d\lambda, \quad t > t_0 \qquad (5.12)$$

Thus, if we know the current $i(t)$, we can find the voltage $e(t)$. Then we can calculate the power delivered to the capacitor. If we integrate this power, the integral represents the energy stored in the capacitor. On the other hand, in a linear capacitor, the energy is also given by Eq. (5.11). Naturally, these two methods of calculating the energy should give the same result.

EXAMPLE 3. In Fig. 5.1(a), $C = 0.2$ F. and $i(t)$ is given in Fig. 5.5. Obtain the voltage, $e(t)$; the power delivered to the capacitor, $p(t)$; and the energy stored in the capacitor, $w(t)$.

SOLUTION For $t < 0$, everything is zero. We have

$$i(t) = \frac{5}{2} t, \quad 0 < t < 2$$

$$i(t) = \frac{5}{2}(t-4), \quad 2 < t < 4$$

$$i(t) = 0, \quad 4 < t$$

Thus

$$e(t) = \frac{1}{0.2} \int_0^t \frac{5}{2} \lambda\, d\lambda = \frac{25}{4} t^2, \quad 0 < t < 2$$

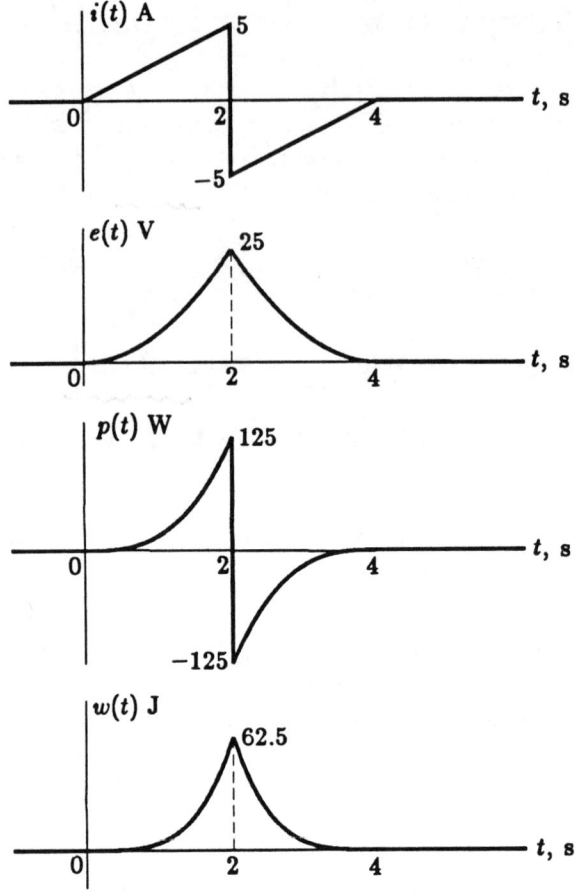

Figure 5.5: Quantities for Example 3.

$e(2) = 25$

$$e(t) = 25 + 5 \int_2^t \frac{5}{2}(\lambda - 4) \, d\lambda = \frac{25}{4}(t-4)^2, \quad 2 < t < 4$$

$e(t) = e(4) = 0, \quad 4 < t$

and

$$p(t) = \frac{5}{2}t \times \frac{25}{4}t^2 = \frac{125}{8}t^3, \quad 0 < t < 2$$

$$p(t) = \frac{5}{2}(t-4) \times \frac{25}{4}(t-4)^2 = \frac{125}{8}(t-4)^3, \quad 2 < t < 4$$

$$p(t) = 0, \quad 4 < t$$

Also

$$w(t) = \frac{125}{32}t^4, \quad 0 < t < 2$$

$$w(t) = \frac{125}{32}(t-4)^4, \quad 2 < t < 4$$

$$w(t) = 0, \quad 4 < t$$

The variations of $e(t)$, $p(t)$, and $w(t)$ are shown in Fig. 5.5.

EXERCISE

5.2.1 In Fig. 5.1(a), $C = 1$ F and $i(t)$ is given below. Find the energy stored in the capacitor.

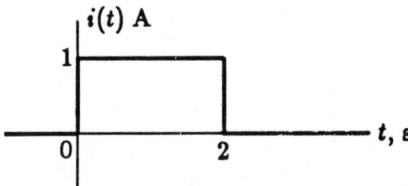

Ans.

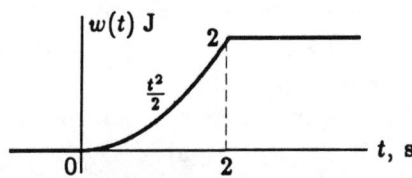

5.3 Series and parallel combinations of capacitors, voltage- and current-division rules

When a number of capacitors are connected in series as shown in Fig. 5.6, we have

$$e = e_1 + e_2 + \cdots + e_n$$
$$= \frac{1}{C_1} \int_{-\infty}^{t} i(\lambda)\,d\lambda + \frac{1}{C_2} \int_{-\infty}^{t} i(\lambda)\,d\lambda + \cdots + \frac{1}{C_n} \int_{-\infty}^{t} i(\lambda)\,d\lambda$$
$$= \left(\frac{1}{C_1} + \frac{1}{C_2} + \cdots + \frac{1}{C_n}\right) \int_{-\infty}^{t} i(\lambda)\,d\lambda \tag{5.13}$$

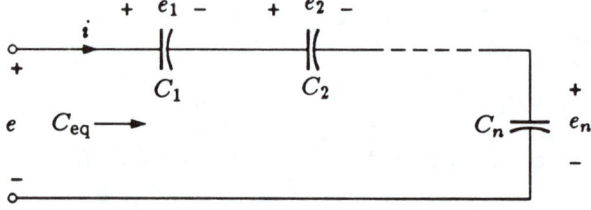

Figure 5.6: Capacitors in series.

If we replace the series combination by an equivalent capacitance, C_{eq}, then

$$e(t) = \frac{1}{C_{eq}} \int_{-\infty}^{t} i(\lambda)\,d\lambda \tag{5.14}$$

A comparison of Eq. (5.14) and Eq. (5.13) gives

$$\frac{1}{C_{eq}} = \frac{1}{C_1} + \frac{1}{C_2} + \cdots + \frac{1}{C_n} \tag{5.15}$$

Thus, when a number of capacitances are connected in series, the sum of their reciprocals (the elastances) is the reciprocal of the equivalent capacitance of the combination.

When two capacitances are connected in series, the combination is equivalent to a capacitance whose value is

$$C_{eq} = \frac{C_1 C_2}{C_1 + C_2} \tag{5.16}$$

From Eq. (5.13), it is easy to see how a voltage is divided among a number of capacitances in series. We have

$$e_i = \frac{\dfrac{1}{C_i}}{\dfrac{1}{C_1} + \dfrac{1}{C_2} + \cdots + \dfrac{1}{C_n}} \times e, \quad i = 1, 2, \ldots, n \tag{5.17}$$

MEMORY ELEMENTS

Eq. (5.17) is the voltage-division rule among a number of capacitors connected in series.

EXAMPLE 4. In the arrangement of Fig. 5.7, determine the equivalent capacitance seen by the source and, e_1, e_2, and e_3.

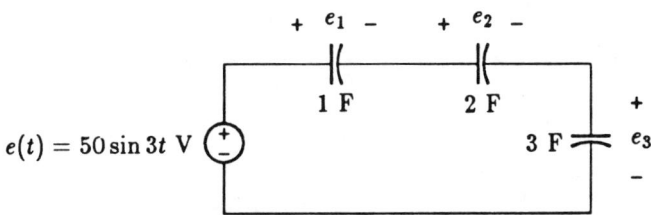

Figure 5.7: Circuit for Example 4.

SOLUTION

$$C_{eq} = \frac{1}{\frac{1}{1} + \frac{1}{2} + \frac{1}{3}} = \frac{6}{11} \text{ F}$$

$$e_1(t) = \frac{\frac{1}{1}}{\frac{1}{1} + \frac{1}{2} + \frac{1}{3}} \times e(t) = \frac{300}{11} \sin 3t \text{ V}$$

$$e_2(t) = \frac{\frac{1}{2}}{\frac{1}{1} + \frac{1}{2} + \frac{1}{3}} \times e(t) = \frac{150}{11} \sin 3t \text{ V}$$

$$e_3(t) = \frac{\frac{1}{3}}{\frac{1}{1} + \frac{1}{2} + \frac{1}{3}} \times e(t) = \frac{100}{11} \sin 3t \text{ V}$$

When a number of capacitances are connected in parallel as shown in Fig. 5.8, it is easy to show that

$$C_{eq} = C_1 + C_2 + \cdots + C_n \qquad (5.18)$$

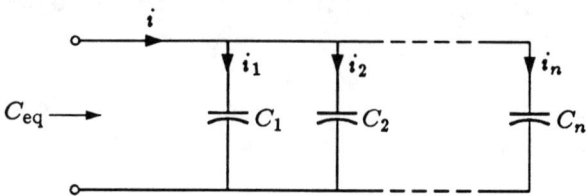

Figure 5.8: Capacitors connected in parallel.

and

$$i_i = \frac{C_i}{C_1 + C_2 + \cdots + C_n} \times i, \quad i = 1, 2, \ldots, n \quad (5.19)$$

Eq. (5.19) is the current-division rule among a number of capacitances connected in parallel.

EXAMPLE 5. In the circuit in Fig. 5.9, determine C_{eq} and the three currents in the three capacitances.

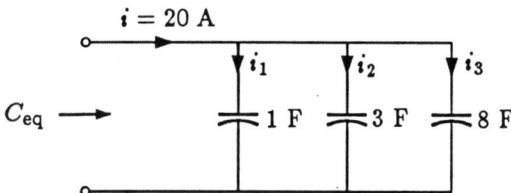

Figure 5.9: Circuit for Example 5.

SOLUTION

$$C_{eq} = 1 + 3 + 8 = 12 \text{ F}$$

$$i_1 = \frac{1}{1+3+8} \times 20 = \frac{5}{3} \text{ A}$$

$$i_2 = \frac{3}{1+3+8} \times 20 = 5 \text{ A}$$

$$i_3 = \frac{8}{1+3+8} \times 20 = \frac{40}{3} \text{ A}$$

MEMORY ELEMENTS

The student should note the similarity between the series-parallel formulas for capacitors and those for resistors. If we employ the elastance for every capacitor, the formulas for series-parallel combinations for capacitors are analogous to those for resistors. This is true even though capacitors are energy-storing elements. The reason is that although we are dealing with memory elements, all elements are of the same type. Therefore, the way they combine or mete out voltages and currents to various elements turn out to be exactly as if they were memoryless. This is true even if the electric quantities are time functions. In fact, one could indeed treat a capacitive network as if it were a resistive network if every capacitor is given in terms of its elastance.

However, one should not take this similarity too far. Our observation in the previous paragraph is sound only if (1) the time interval under consideration extends from $-\infty$ to $+\infty$ and (2) the network or part of a network is totally made up of capacitors.

EXERCISES

5.3.1 Determine C_{eq}.

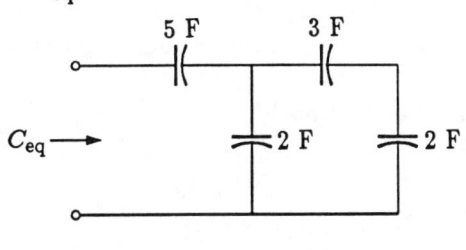

Ans. $\dfrac{80}{41}$ F.

5.3.2 Determine i_1.

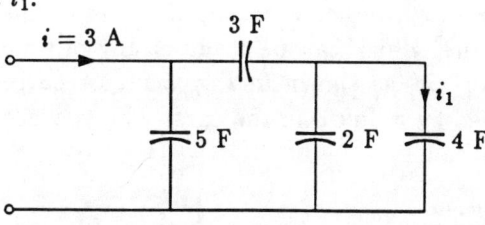

Ans. $\dfrac{4}{7}$ A.

5.4 The inductor

An *inductor* is an electromagnetic energy-storage device whose stored energy is in the form of a magnetic field that can be directly related to the magnetic flux ϕ (usually in webers). The electrical properties of an inductor can be characterized by the flux-versus-current (ϕ-i) curve. The simplest type of inductor is the linear inductor, which has a linear ϕ-i curve, shown in Fig. 5.10. Analytically such

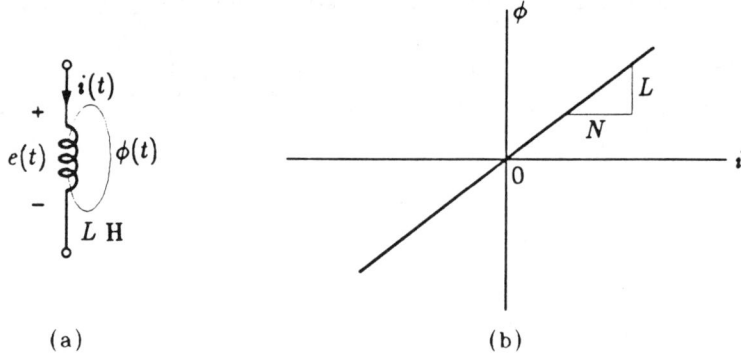

(a) (b)

Figure 5.10: A linear inductor and it ϕ-i characteristic.

an inductor has the relationship

$$N\phi(t) = Li(t) \tag{5.20}$$

where N is the number of turns of the inductor and L is the *inductance* of the inductor. When ϕ is in webers and i is in amperes, the inductance is in *henrys* (represented by the letter H).

The simplest inductor arrangement is a coil wound around a magnetic core, which can be made of any of a number of magnetic materials or air, as shown in Fig. 5.11. A current i in the coil is accompanied by a flux ϕ in the core. The two quantities are related by the equation

$$\phi = \frac{\mu N A}{\ell} i \tag{5.21}$$

where A is the cross-sectional area of the core, ℓ is the average length of the magnetic path, N is the number of turns in the coil, and μ is

MEMORY ELEMENTS

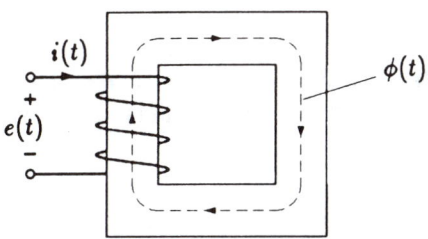

Figure 5.11: A typical inductor arrangement.

the permeability of the core material. The flux ϕ linking an N-turn coil produces a *flux linkage* equal to

$$\Phi = N\phi \tag{5.22}$$

We have

$$\Phi = \frac{\mu N^2 A i}{\ell} \tag{5.23}$$

For such a coil, the inductance is

$$L = \frac{\mu N^2 A}{\ell} \tag{5.24}$$

In an equation similar to Eq. (5.5), we find that the voltage across an inductor and the flux linkage are related by

$$e(t) = \frac{d\,\Phi(t)}{dt} \tag{5.25}$$

Hence the voltage-current relationship in an inductor is

$$e(t) = L\frac{di}{dt} \tag{5.26}$$

Conversely,

$$i(t) = \frac{1}{L}\int_{-\infty}^{t} e(\lambda)\,d\lambda \tag{5.27}$$

or

$$i(t) = i(t_0) + \frac{1}{L}\int_{t_0}^{t} e(\lambda)\,d\lambda, \quad t > t_0 \tag{5.28}$$

222 FUNDAMENTALS OF CIRCUIT ANALYSIS

EXAMPLE 6. Refer to Fig. 5.10(a) with $L = 2$ H. Current $i(t)$ is shown in Fig. 5.12, give a dimensioned sketch of $e(t)$.

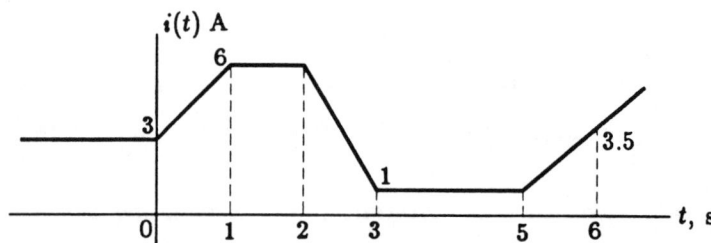

Figure 5.12: Current $i(t)$ for Example 6.

SOLUTION Using Eq. (5.26), we get the following graph for $e(t)$.

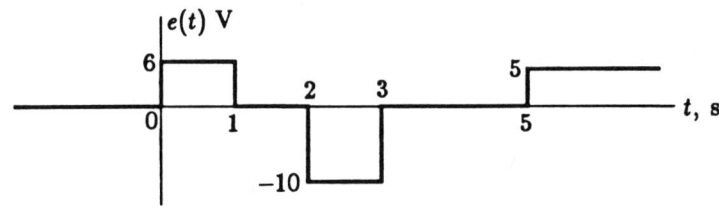

EXAMPLE 7. An initially unenergized 10-H inductor is connected to a 12-V battery for 2 seconds and short-circuited afterwards. What is the current $i(t)$?

SOLUTION

$$i(t) = 0, \quad t < 0$$

$$i(t) = \frac{1}{10} \int_0^t 12 \, d\lambda = 1.2t, \quad 0 < t < 2$$

$$i(t) = i(2) = 2.4 \text{ A}, \quad 2 < t$$

MEMORY ELEMENTS 223

5.5 Energy stored in an inductance

When a magnetic flux is linking an inductor, it represents a certain amount of energy stored in it. The accumulated energy delivered to an inductance at any time is the integral of the power, $p(t)$, delivered to the inductor up to that time. Or

$$w(t) = \int_{-\infty}^{t} p(\lambda)\,d\lambda = \int_{-\infty}^{t} \frac{\Phi(\lambda)}{L} \cdot \frac{d\,\Phi(\lambda)}{d\lambda}\,d\lambda$$

$$= \frac{1}{L} \int_{\lambda=-\infty}^{\lambda=t} \Phi(\lambda)\,d\Phi = \frac{1}{2L}\Phi^2(t) = \frac{1}{2}Li^2(t) \qquad (5.29)$$

if we accept that $\Phi(-\infty) = 0$ and $i(-\infty) = 0$. We could also write

$$w(t) = \int_{-\infty}^{t} p(\lambda)\,d\lambda \qquad (5.30)$$

or

$$w(t) = w(t_0) + \int_{t_0}^{t} p(\lambda)\,d\lambda, \quad t > t_0 \qquad (5.31)$$

EXAMPLE 8. Refer to Fig. 5.10(a), with $L = 4$ H. Let $i(t)$ be that given in Fig. 5.13. Give a dimensioned sketch for $e(t)$, the power delivered to the inductance, $p(t)$, and the energy stored in the inductance, $w(t)$.

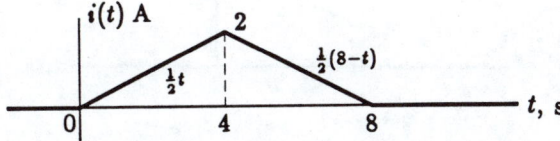

Figure 5.13: The current for Example 8.

SOLUTION The quantities $e(t)$, $p(t)$, and $w(t)$ are shown in the following diagrams together with the equation for each segment of the curves.

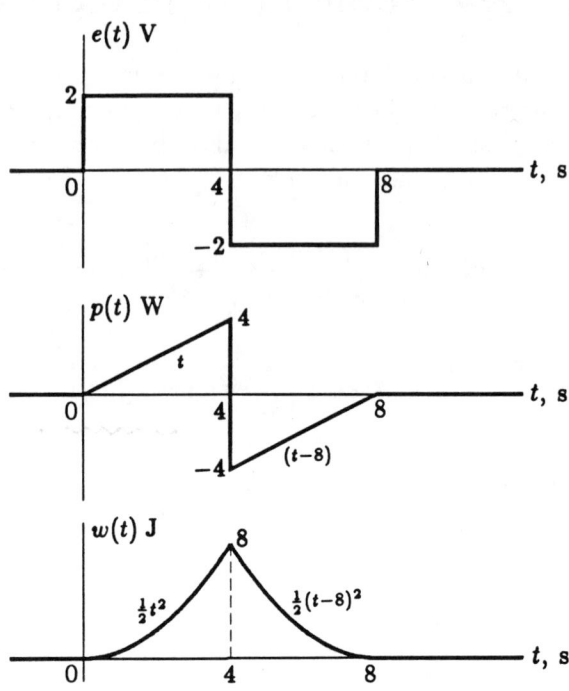

EXERCISE

5.5.1 Refer to Fig. 5.10(a). If $e(t)$ is that given below and $L = 1$ H, give a dimensioned sketch of the energy $w(t)$ stored in the inductance.

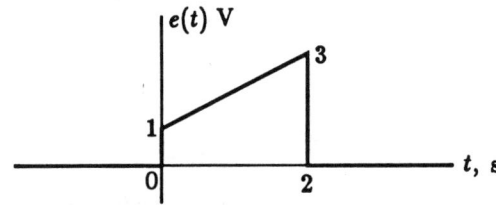

MEMORY ELEMENTS

Ans.

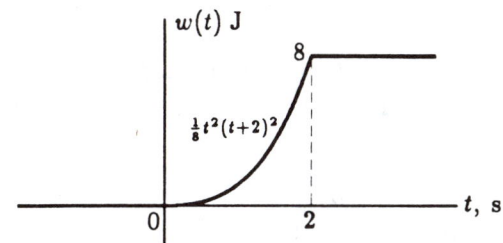

5.6 Series and parallel combinations of inductors, voltage- and current-division rules

When a number of linear inductors are connected in series as shown in Fig. 5.14, we have

$$e = e_1 + e_2 + \cdots + e_n = L_1\frac{di}{dt} + L_2\frac{di}{dt} + \cdots + L_n\frac{di}{dt}$$

$$= (L_1 + L_2 + \cdots + L_n)\frac{di}{dt} \qquad (5.32)$$

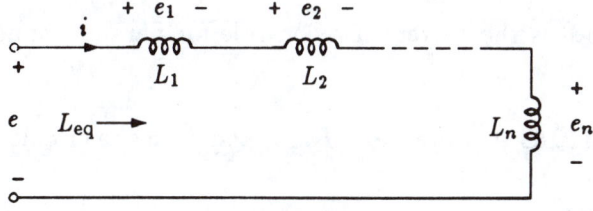

Figure 5.14: Inductors in series.

If we replace the series combination by an equivalent inductance, L_{eq}, then

$$L_{eq} = L_1 + L_2 + \cdots + L_n \qquad (5.33)$$

From Eq. (5.32), it is seen that

$$e_i = L_i\frac{di}{dt} = \frac{L_i}{L_1 + L_2 + \cdots + L_n} \times e, \quad i = 1, 2, \ldots, n \qquad (5.34)$$

Eq. (5.34) is the voltage-division rule for series-connected inductances.

If we have a number of inductors connected in parallel as shown in Fig. 5.15, it is easy to show that

$$\frac{1}{L_{eq}} = \frac{1}{L_1} + \frac{1}{L_2} + \cdots + \frac{1}{L_n} \tag{5.35}$$

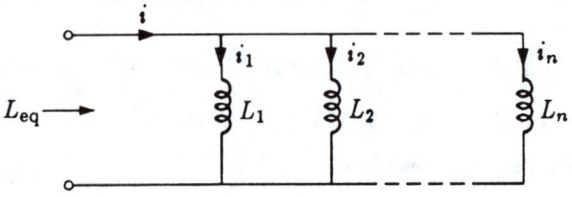

Figure 5.15: Inductors connected in parallel.

and

$$i_i = \frac{\frac{1}{L_i}}{\frac{1}{L_1} + \frac{1}{L_2} + \cdots + \frac{1}{L_n}} \times i \tag{5.36}$$

Eq. (5.36) is the current-division rule for parallel connected inductances.

EXAMPLE 9. Determine I_1, I_2, and I_3 in the circuit of Fig. 5.16.

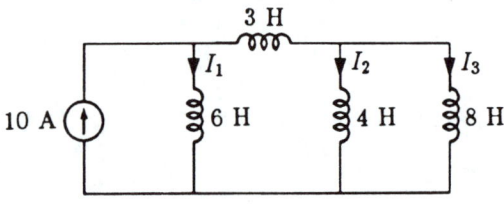

Figure 5.16: Circuit used in Example 9

MEMORY ELEMENTS

SOLUTION

$$4 \| 8 = \frac{4 \times 8}{4 + 8} = \frac{8}{3}$$

$$\frac{8}{3} + 3 = \frac{17}{3}$$

$$I_1 = \frac{\frac{17}{3}}{6 + \frac{17}{3}} \times 10 = \frac{34}{7} \text{ A}$$

$$I_2 + I_3 = 10 - I_1 = \frac{36}{7}$$

$$I_2 = \frac{8}{4 + 8} \times \frac{36}{7} = \frac{24}{7} \text{ A}$$

$$I_3 = \frac{4}{4 + 8} \times \frac{36}{7} = \frac{12}{7} \text{ A}$$

EXERCISE

5.6.1 Determine e in the circuit.

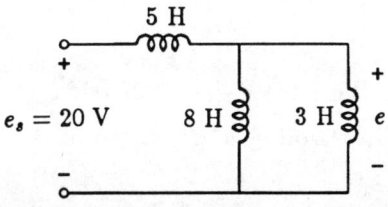

Ans. $\frac{480}{79}$ V.

It is readily seen that series and parallel combinations of inductors and current- and voltage-division rules among inductors are completely analogous to those when the elements are resistors.

5.7 Simple circuits containing inductors and capacitors

When a circuit contains more than one type of element or several independent sources, the analysis of these circuits is usually quite laborious. However, they are still governed by KCL and KVL as well as the voltage-current relationships just described for energy storing elements. We shall illustrate the solution of some of these simple circuits with a few examples.

EXAMPLE 10. Determine $e(t)$ for the circuit in Fig. 5.17.

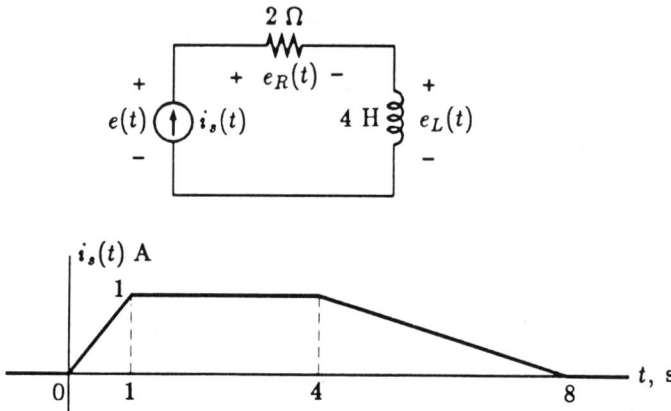

Figure 5.17: Circuit and source current for Example 10.

SOLUTION We have $e_R(t) = 2i_s(t)$, $e_L(t) = 4\dfrac{di_s}{dt}$, and $e(t) = e_R(t) + e_L(t)$. These voltages are shown in the following diagrams.

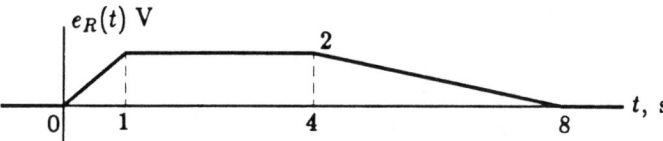

MEMORY ELEMENTS

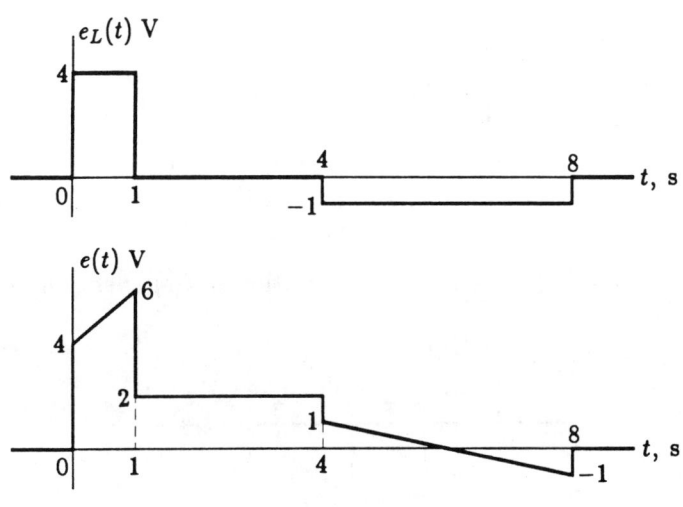

EXAMPLE 11. The source voltage, $e_s(t)$, and the source current, $i_s(t)$, are given for the circuit of Fig. 5.18. Give a dimensioned sketch of $i(t)$.

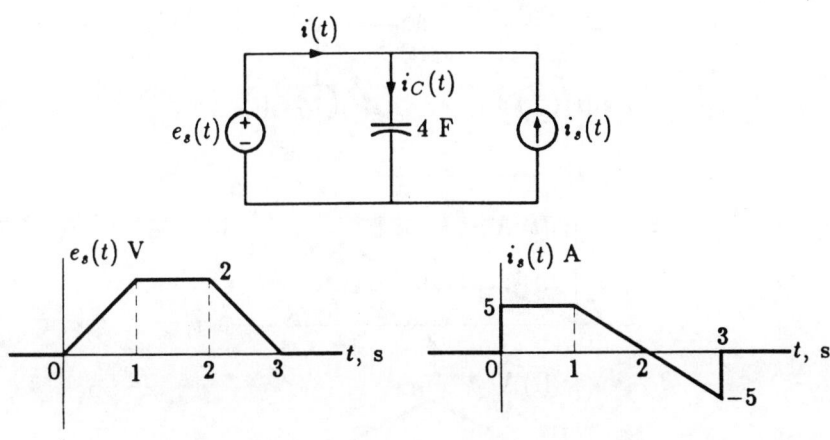

Figure 5.18: Arrangement for Example 11.

SOLUTION The voltage across the capacitor is $e_s(t)$ and its current, $i_C(t)$, is shown below.

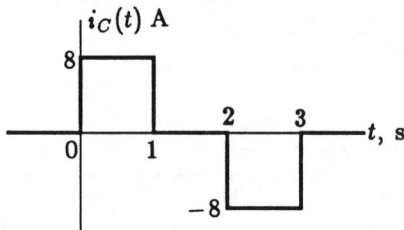

Since $i(t) = i_C(t) - i_s(t)$, the variation of $i(t)$ is shown below.

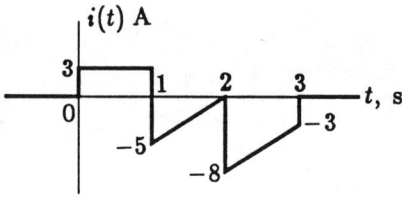

EXAMPLE 12. The source voltage, $e_s(t)$, and the source current, $i_s(t)$, are given for the circuit of Fig. 5.19. Give a dimensioned sketch of $e(t)$.

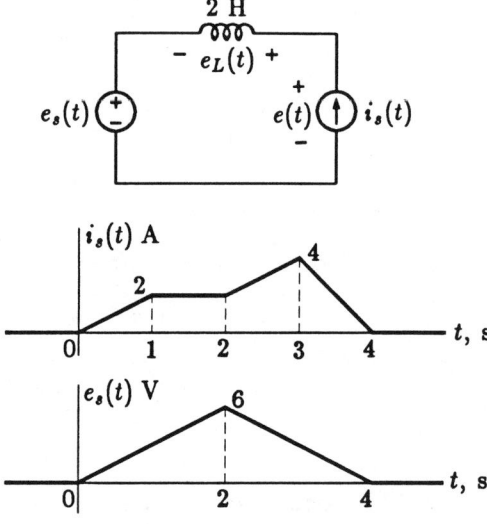

Figure 5.19: Arrangement for Example 12.

MEMORY ELEMENTS

SOLUTION The voltage across the inductor, $e_L(t) = 2\dfrac{di_s}{dt}$, is readily found and is shown below.

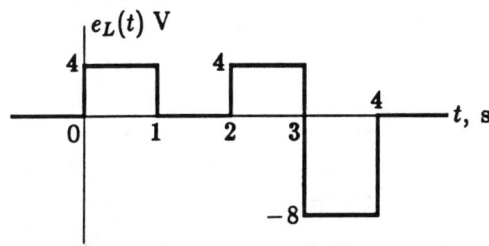

Since $e(t) = e_L(t) + e_s(t)$, the variation of $e(t)$ is shown below.

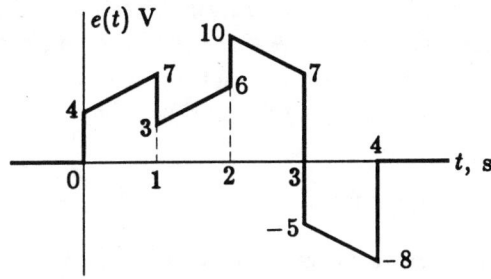

EXERCISES

5.7.1 Obtain $i(t)$.

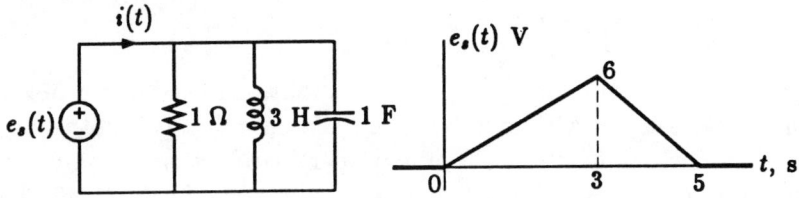

Ans.

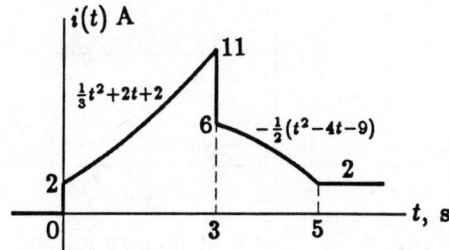

FUNDAMENTALS OF CIRCUIT ANALYSIS

5.8 Singularity functions

We now turn our attention to a class of functions, known as *singularity functions*, that are very useful in representing certain occurrences in circuits and in modeling signals. These functions are singular because they are not ordinary, well-behaved functions. Treating these functions on a sound mathematical basis will require the study of an entirely separate discipline known as *distribution theory*. Our approach here is heuristic in nature and, to the limited extent that we employ these functions and with the aid of certain physical interpretations, we shall proceed to use these functions. One should always keep in mind that these functions are not ordinary functions and should exercise extreme caution when dealing with this class of functions. In this context, these functions are very convenient to use in many areas of electrical engineering well beyond circuit theory.

5.8.1 The unit step function

A *unit step function*, $u(t)$, is defined to be zero for negative t and unity for positive t. Or analytically,

$$u(t) = 0, \quad t < 0$$
$$= 1, \quad t > 0 \tag{5.37}$$

Hence $u(t)$ is a discontinuous function and, when used in a problem, should be handled with great care to make sure that it is not treated as a continuous function without proper justification. A unit step function may be represented graphically as shown in Fig. 5.20. The

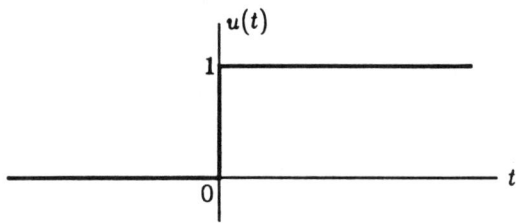

Figure 5.20: The unit step function.

value of $u(t)$ at $t = 0$ is left undefined.[1]

One of the applications of the step function is to represent the switching action in a circuit. In Fig. 5.21, we have a two-terminal

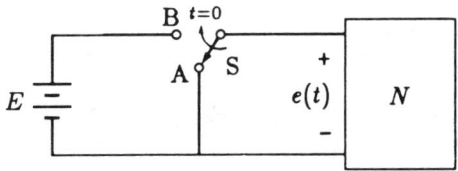

Figure 5.21: Switching action that results in a step voltage to N.

network N which is short-circuited for $t < 0$. At $t = 0$, switch S is moved from position A to position B and connect the battery voltage E to the network. Hence, for network N,

$$e(t) = Eu(t) \tag{5.38}$$

Another application of the step function is to synthesize a piecewise-constant function. From the definition of the unit step function of Eq. (5.37), we see that $u(t - t_0)$ is a step whose step occurs at $t = t_0$, at which time the argument $t - t_0$ is zero. This is illustrated in Fig. 5.22, in which the unit step function is multiplied by A and delayed by t_0.

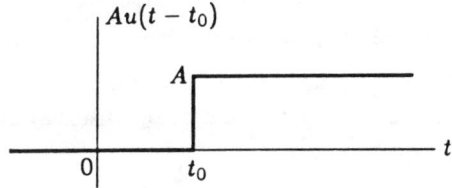

Figure 5.22: A delayed step function.

EXAMPLE 13. Write $f(t)$ of Fig. 5.23 in terms of the sum of step functions.

---

[1] Some authors define $u(0) = 1$. Others define $u(0) = 0$. Still others define $u(0) = \frac{1}{2}$. They do this for good reasons in the context of their applications. We prefer to leave it undefined because it is immaterial to our usage of the function.

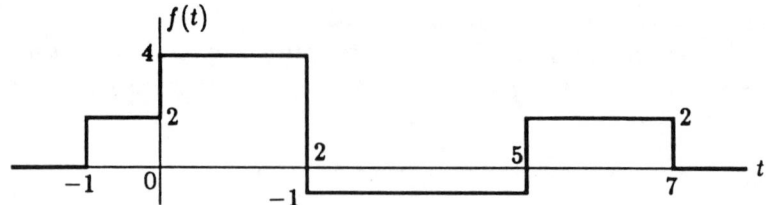

Figure 5.23: A piece-wise constant function.

SOLUTION

$$f(t) = 2\,u(t+1) + 2\,u(t) - 5\,u(t-2) + 3\,u(t-5) - 2\,u(t-7)$$

EXERCISE

5.8.1 Write $g(t)$ as the sum of step functions.

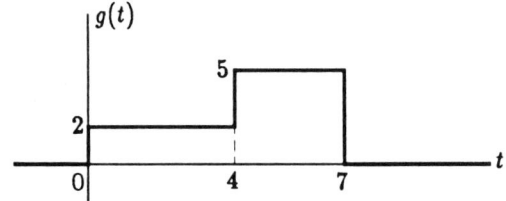

Ans. $g(t) = 2\,u(t) + 3\,u(t-4) - 5\,u(t-7)$.

Another usage of the unit step function is to limit a continuous function to remain as it is inside a certain interval and force it to be zero outside. For example, if

$$f_1(t) = \cos t\, u(t) \tag{5.39}$$

Eq. (5.39) is equivalent to a longer statement that reads

$$\begin{aligned} f_1(t) &= 0, \quad t < 0 \\ &= \cos t, \quad t > 0 \end{aligned} \tag{5.40}$$

A plot of $f_1(t)$ is shown in Fig. 5.24(a).

A function multiplied by a delayed unit step is cut off at a later time. The function is not delayed. Thus the graph of the function

MEMORY ELEMENTS

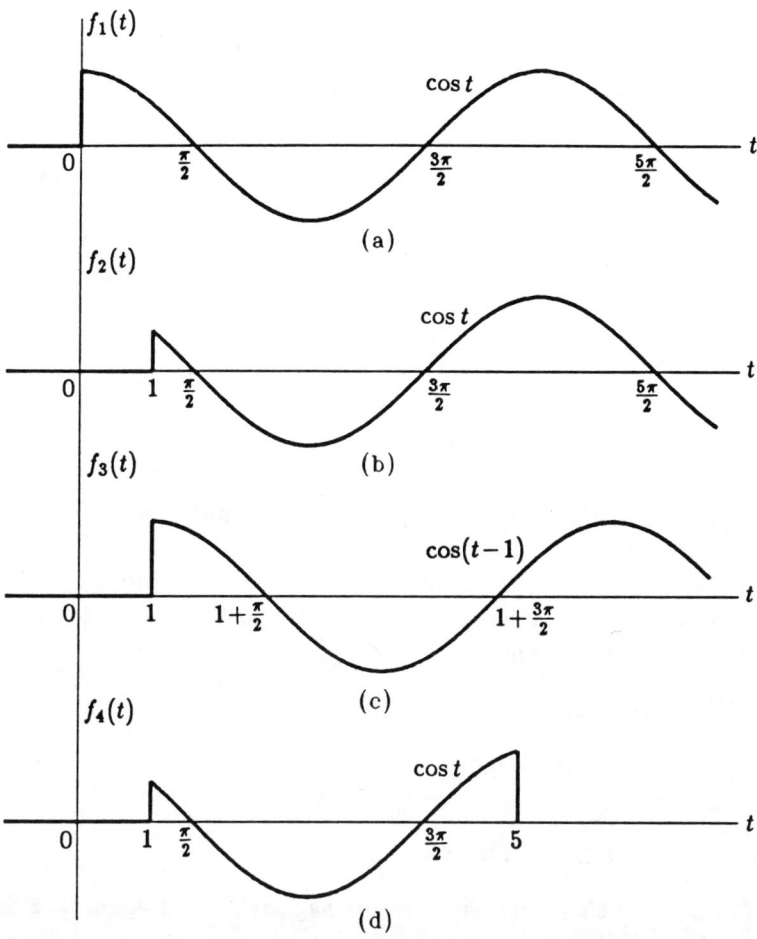

Figure 5.24: Effects of the step function on a continuous function.

$$f_2(t) = \cos t\, u(t-1) \tag{5.41}$$

is shown in Fig. 5.24(b). However, the function

$$f_3(t) = \cos(t-1)\, u(t-1) \tag{5.42}$$

is $f_1(t)$ delayed by one unit of time as shown in Fig. 5.24(c).

A rectangular pulse of a finite duration may be written as the difference of two unit step functions. The function

$$G(t) = u(t-t_1) - u(t-t_2) \tag{5.43}$$

is shown in Fig. 5.25. When a continuous function is multiplied by

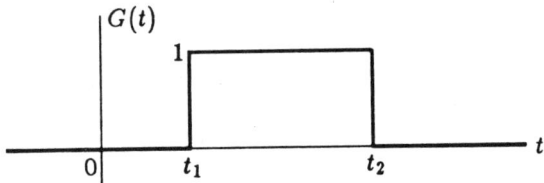

Figure 5.25: A gate or window function.

$G(t)$, only that part of the continuous function in the range $t_1 < t < t_2$ is retained. The function $G(t)$ is often called the *gate function* or the *window function*. Thus

$$f_4(t) = \cos t[u(t-1) - u(t-5)] \tag{5.44}$$

varies as shown in Fig. 5.24(d).

5.8.2 The unit ramp function

A *unit ramp function*, $r(t)$, is zero for negative t and equal to t for positive t. Analytically

$$\begin{aligned} r(t) &= 0, \quad t < 0 \\ &= t, \quad t > 0 \end{aligned} \tag{5.45}$$

Fig. 5.26 shows the variation of a unit ramp function. A unit ramp function is not discontinuous, but its derivative is discontinuous. In fact, its derivative is a unit step function, or

MEMORY ELEMENTS

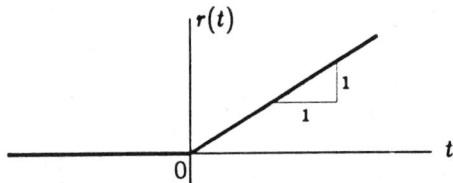

Figure 5.26: A unit ramp function.

$$u(t) = \frac{d\,r(t)}{dt} \tag{5.46}$$

Conversely, a unit ramp function is the integral of a unit step function, or

$$r(t) = \int_{-\infty}^{t} u(\lambda)\,d\lambda \tag{5.47}$$

One of the applications of the ramp function is to synthesize piecewise linear functions. An example is given in Fig. 5.27, in which

$$f(t) = r(t) - 2r(t-1) + r(t-2) \tag{5.48}$$

The three individual terms of Eq. (5.48) are shown in Fig. 5.27(a). The composite of the three terms is shown in Fig. 5.27(b).

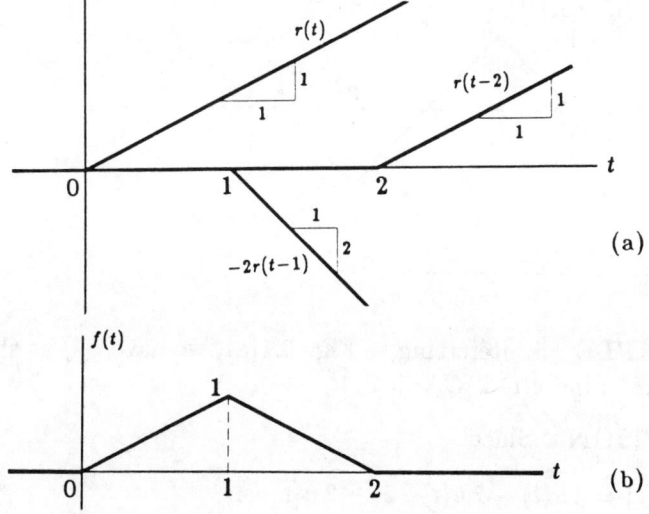

Figure 5.27: Combination of ramp functions.

EXAMPLE 14. Give a dimensioned sketch for each of the following functions.

$$f_a(t) = 4\,r(t-1) - 4\,r(t-2)$$

$$f_b(t) = 2\,r(t) - 2\,r(t-2) + 2\,r(t-4) - 3.5\,r(t-6) + 1.5\,r(t-8)$$

$$f_c(t) = 5\,r(t) - 10\,r(t-1) + 10\,r(t-3) - 5\,r(t-4)$$

SOLUTION

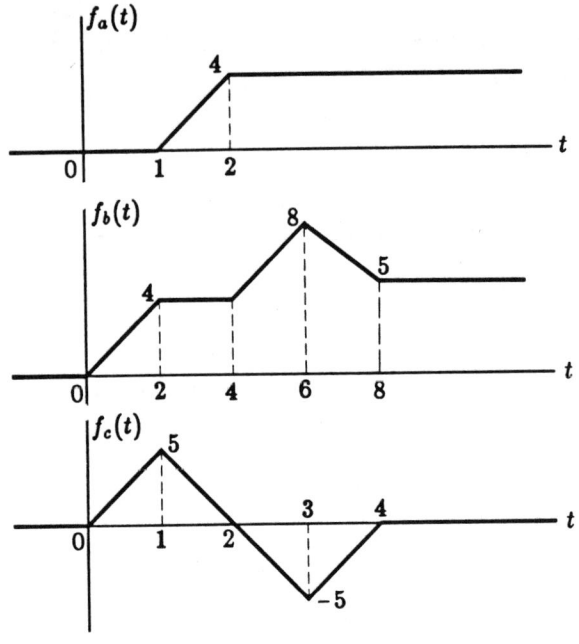

EXAMPLE 15. Referring to Fig. 5.1(a), we have $i(t)$ as shown in Fig. 5.28. Find $e(t)$ if $C = 5$ F.

SOLUTION Since

$$i(t) = 4\,u(t) - 2\,u(t-2) - 2\,u(t-4)$$

$$e(t) = \frac{1}{5}\int_{-\infty}^{t} i(\lambda)\,d\lambda$$

MEMORY ELEMENTS

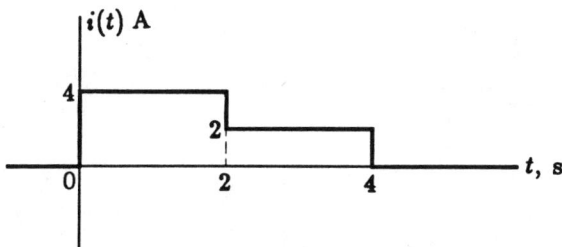

Figure 5.28: Variation of $i(t)$ of Example 15.

$$= 0.8\,r(t) - 0.4\,r(t-2) - 0.4\,r(t-4) \text{ V}$$

Graphically, $e(t)$ is shown below.

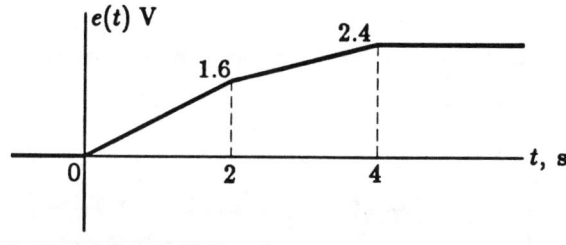

EXAMPLE 16. Referring to Fig. 5.10(a), we have the current $i(t)$ shown in Fig. 5.29. Find $e(t)$ if $L = 2$ H.

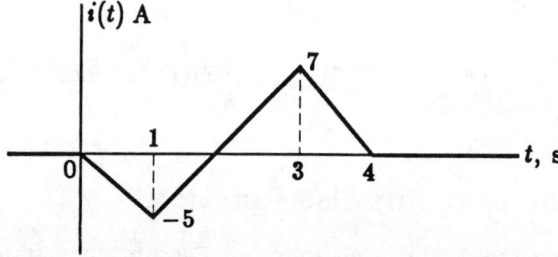

Figure 5.29: Variation of $i(t)$ of Example 16.

SOLUTION We have

$$i(t) = -5\,r(t) + 11\,r(t-1) - 13\,r(t-3) + 7\,r(t-4)$$

Hence

$$e(t) = 2\frac{di}{dt} = -10\,u(t) + 22\,u(t-1) - 28\,u(t-3) + 14\,u(t-4)$$

Graphically, $e(t)$ is shown below.

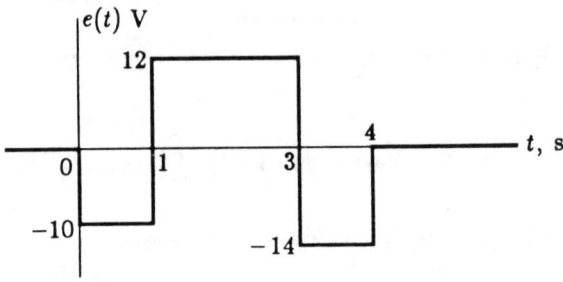

EXERCISE

5.8.2 Write the following functions as the sum of step and ramp functions.

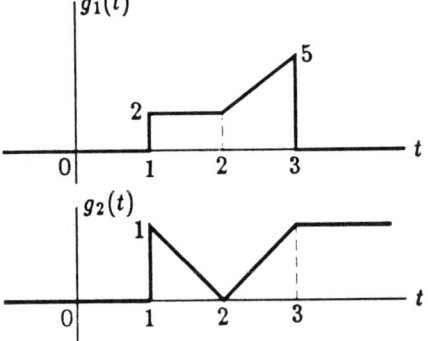

Ans. $g_1(t) = 2\,u(t-1) + 3\,r(t-2) - 3\,r(t-3) - 5\,u(t-3)$, $g_2(t) = u(t-1) - r(t-1) + 2\,r(t-2) - r(t-3)$.

5.8.3 The unit impulse function

If we differentiate a unit step function, we obtain a function that is zero everywhere except at $t = 0$. At $t = 0$, the derivative is infinity. We call this derivative of the unit step function a *unit impulse function*.[2] It is denoted by

$$\delta(t) = \frac{d\,u(t)}{dt} \qquad (5.49)$$

---

[2] It is also known as the delta function.

MEMORY ELEMENTS

Conversely,

$$u(t) = \int_{-\infty}^{t} \delta(\lambda)\, d\lambda \qquad (5.50)$$

Eq. (5.50) states that the area under the impulse is equal to the height of the step function—unity. A unit impulse function is symbolically represented by the vertical arrow as shown in Fig. 5.30, in which the label "1" represents the area contained in the impulse. When we differentiate an A-unit step function, $Au(t)$, we will get an

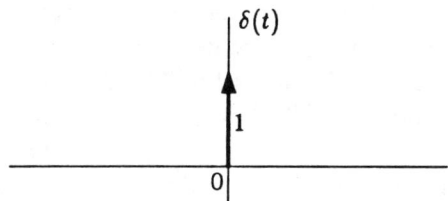

Figure 5.30: The unit impulse function.

A-unit impulse function—$A\delta(t)$. We shall represent such an impulse function by indicating its area to be A units by a label A.

In order to gain some insight into the significance of an impulse function, let's look at the arrangement in Fig. 5.31(a). An ideal

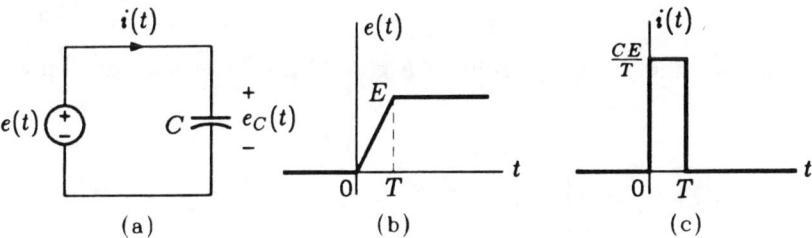

Figure 5.31: Charging of a capacitor by a voltage source.

voltage source is connected directly across a capacitance of C farads. It is obvious that $e_C(t) = e(t)$. Suppose the source voltage $e(t)$ varies with time as shown in Fig. 5.31(b). Since

$$i(t) = C \frac{de_C}{dt} \qquad (5.51)$$

the variation of $i(t)$ is as shown in Fig. 5.31(c). We see that the current is a constant pulse. The area under the pulse is CE coulombs, which remains unchanged even if T is changed.

A moment's reflection will reveal that the voltage source is changing the voltage across the capacitor at a constant rate until it reaches E volts. Since the voltage is proportional to the charge in a capacitor, the source is also charging the capacitor at a constant rate until the final charge reaches CE coulombs. A constant charging rate means a constant current. After the charge has reached CE coulombs, it stays constant and the current is zero.

So T is the duration of charging. If T is decreased, the rate of charging (CE/T) will increase proportionally. However, the total charge transferred is still CE coulombs, no matter what T is.

Now suppose we let T approach zero. This means that $e(t)$ is suddenly changed from 0 to E volts at $t = 0$. The voltage of Fig. 5.31(b) approaches an E-volt step. The current $i(t)$ now behaves rather dramatically in that it approaches a pulse that is infinite in height but lasts for an infinitesimally short instant. The area of this pulse remains CE coulombs. In the limit, we have a current that is a CE-coulomb impulse occurring at $t = 0$. Physically, this simply means that a lump charge of CE coulombs is being transferred from the source to the capacitor instantaneously. We describe this charge transfer by saying that the current is a CE-unit impulse function, that is

$$i(t) = CE\,\delta(t) \text{ A} \qquad (5.52)$$

The switching arrangement in Fig. 5.32 produces the same cur-

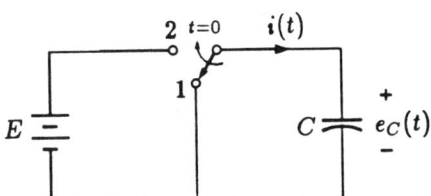

Figure 5.32: The sudden charging or a capacitor.

rent. The capacitor is short-circuited before $t = 0$. There is no voltage across it. Hence, there is no charge in the capacitor. At $t = 0$, the switch is moved to position 2. The voltage $e_C(t)$ is changed suddenly from 0 to E volts. Its charge jumped from 0 to CE coulombs instantaneously, resulting in a current given by Eq. (5.52). As far as the capacitor is concerned, this switching arrangement is identical

MEMORY ELEMENTS

to the arrangement in Fig. 5.31(a)—which involves no switching—if we assume that $e(t)$ is an E-unit step function. In this example, singularity functions enable us to describe notationally the events in a circuit in a relatively simple manner.

EXAMPLE 17. Give a plot of

$$f(t) = 2u(t-1) + r(t-1) - r(t-3) - 2u(t-3) - 2u(t-5)$$

and its derivative $f'(t)$.

SOLUTION

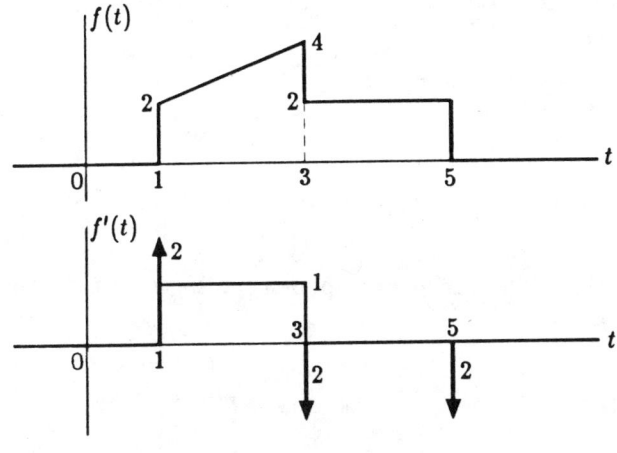

EXAMPLE 18. In Fig. 5.31(a),

$$i(t) = u(t) + \delta(t-1) - u(t-2) \text{ A}$$

Obtain $e(t)$ if $C = 5$ F.

SOLUTION The capacitor is being charged at a constant rate of one coulomb per second between $t = 0$ and $t = 2$. Also, at $t = 1$, a one-coulomb charge is delivered to the capacitor instantaneously. Analytically, we can write

$$e(t) = 0.2[r(t) + u(t-1) - r(t-2)]$$

Graphically, we have the following.

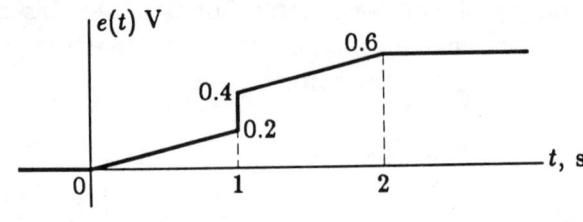

EXAMPLE 19. In the arrangement in Fig. 5.33, obtain $e(t)$.

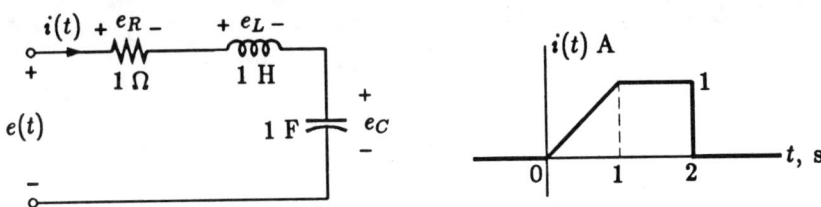

Figure 5.33: Arrangement used in Example 19.

SOLUTION The various component voltages and the total voltage are shown below.

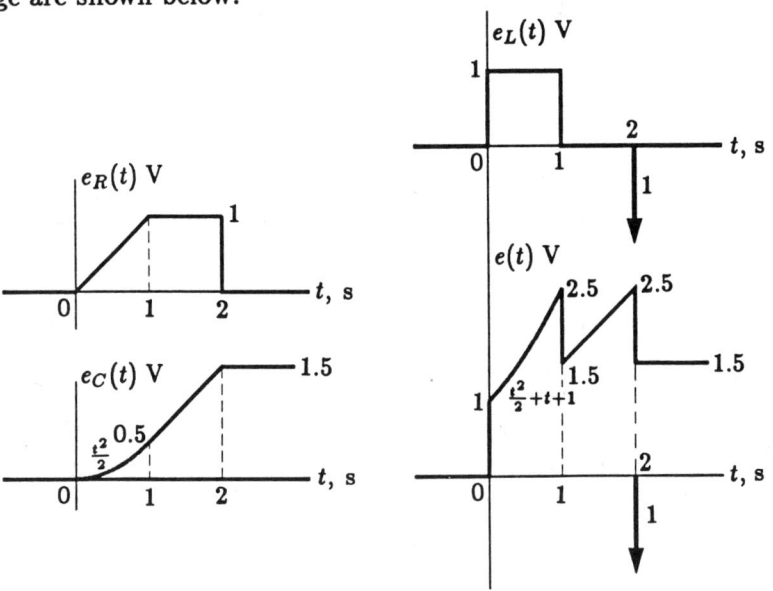

EXERCISES

5.8.3 Obtain current $i(t)$.

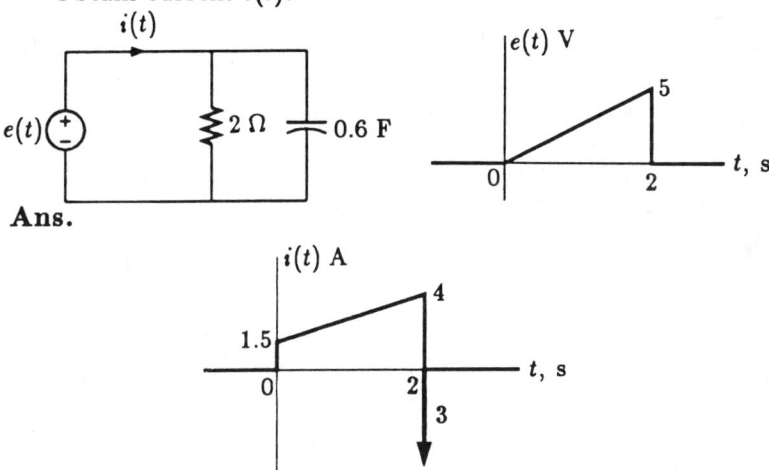

Ans.

5.8.4 Refer to Fig. 5.10(a). Obtain $i(t)$ if $L = 5$ H and $e(t) = 10\delta(t-1)$ V.

Ans. $2u(t-1)$ A.

5.8.4 Other singularity functions

Thus far, we have defined three singularity functions—the step, the ramp, and the impulse functions. These function are useful in representing certain switching actions, describing certain circuit phenomena, and modeling certain signal variations. In working with these functions, care should be exercised because they are singularity functions. We shall present some examples of how, with proper precaution, these functions can be handled with orthodox mathematics.

If we want to integrate a unit step

$$\int_{-\infty}^{t} u(\lambda)\, d\lambda$$

we need to do it in steps. For $t < 0$,

$$\int_{-\infty}^{t} u(\lambda)\,d\lambda = \int_{-\infty}^{t} 0 \cdot d\lambda = 0 \tag{5.53}$$

For $t > 0$,

$$\int_{-\infty}^{t} u(\lambda)\,d\lambda = 0 + \int_{0}^{t} 1 \cdot d\lambda = t \tag{5.54}$$

Hence, the integral is the unit ramp function.

EXAMPLE 20. Evaluate

$$f(t) = \int_{-\infty}^{t} \epsilon^{-\lambda} u(\lambda - 1)\,d\lambda$$

SOLUTION For $t < 1$, $f(t) = 0$ because the integrand is equal to zero for $-\infty < t < 1$. For $t > 1$,

$$f(t) = \int_{1}^{t} \epsilon^{-\lambda}\,d\lambda = \epsilon^{-1} - \epsilon^{-t}$$

Combining the results, we can write a single expression for $f(t)$ for all t. Namely,

$$f(t) = \left(\epsilon^{-1} - \epsilon^{-t}\right) u(t - 1)$$

Similarly, let us evaluate the integral of a unit impulse function

$$\int_{-\infty}^{t} \delta(\lambda)\,d\lambda$$

We have, for $t < 0$,

$$\int_{-\infty}^{t} \delta(\lambda)\,d\lambda = 0 \tag{5.55}$$

because the integrand is equal to zero in the interval of integration. For $t > 0$,

$$\int_{-\infty}^{t} \delta(\lambda)\,d\lambda = 1 \tag{5.56}$$

MEMORY ELEMENTS

because the unit impulse lies within the interval of integration and the area contained in the unit impulse function is unity. Thus, the integral of a unit impulse function is a unit step function.

An important operation of the unit impulse function is

$$f(t)\,\delta(t - t_0) = f(t_0)\,\delta(t - t_0) \tag{5.57}$$

Eq. (5.57) states that the product of a unit impulse function with another function is another impulse function, the area under which is equal to the value of the function at the point where the impulse occurs. In the product of Eq. (5.57), all values of $f(t)$ other than that at $t = t_0$ are multiplied by zero. Note that, in Eq. (5.57), $f(t)$ is a function while $f(t_0)$ is a number for a given $f(t)$.

EXAMPLE 21. The following are a few more examples of how some operations involving singularity functions are handled:

(a) $\displaystyle\int_{-10}^{20} u(t)\,dt = \int_{0}^{20} 1\,dt = 20$

(b) $\displaystyle\int_{5}^{10} \delta(t - 2)\,dt = 0$

(c) $\displaystyle\int_{-2}^{5} \sin t\,\delta(t)\,dt = \int_{-2}^{5} 0 \cdot \delta(t)\,dt = 0$

(d) $\displaystyle\int_{-1}^{1} \cos t\,\delta(t)\,dt = \int_{-1}^{1} \delta(t)\,dt = 1$

(e) $\displaystyle\int_{2}^{t} u(\lambda)\,d\lambda = \int_{-\infty}^{t} u(\lambda)\,d\lambda - \int_{-\infty}^{2} u(\lambda)\,d\lambda = r(t) - 2$

(f) $\displaystyle\frac{d}{dt}[\cos t\,u(t)] = \cos t\,\delta(t) - \sin t\,u(t) = \delta(t) - \sin t\,u(t)$

The three singularity functions we have defined thus far are related to one another by the integral and differential relationships. That is,

$$\delta(t) = \frac{d}{dt}u(t) \quad \text{or} \quad u(t) = \int_{-\infty}^{t} \delta(\lambda)\,d\lambda \tag{5.58}$$

$$u(t) = \frac{d}{dt}r(t) \quad \text{or} \quad r(t) = \int_{-\infty}^{t} u(\lambda)\, d\lambda \tag{5.59}$$

Hence, there is some advantage in employing the subscript notation for these functions. When this set of notations is preferred, we define

$$\delta(t) = u_0(t) \tag{5.60}$$

$$u(t) = u_{-1}(t) \tag{5.61}$$

$$r(t) = u_{-2}(t) \tag{5.62}$$

In this notational convention, each time a function is differentiated, its subscript is incremented by one; each time it is integrated, the subscript is decremented by one. If we follow this pattern, then we will have

$$u_{-3}(t) = \int_{-\infty}^{t} u_{-2}(\lambda)\, d\lambda \tag{5.63}$$

and explicitly

$$\begin{aligned} u_{-3}(t) &= 0, \quad t < 0 \\ &= \frac{t^2}{2}, \quad t > 0 \end{aligned} \tag{5.64}$$

The function $u_{-3}(t)$ is known as the *unit parabola function*. In general,

$$\begin{aligned} u_{-k}(t) &= 0, \quad t < 0 \\ &= \frac{t^{k-1}}{(k-1)!}, \quad t > 0 \end{aligned} \tag{5.65}$$

for $k \geq 1$.

This notational convention also results in the function

$$u_1(t) = \frac{d}{dt}u_0(t) = \frac{d}{dt}\delta(t) \tag{5.66}$$

The function $u_1(t)$ is known as a *unit doublet*. Also,

$$u_2(t) = \frac{d}{dt}u_1(t) \tag{5.67}$$

MEMORY ELEMENTS

The function $u_2(t)$ is known as a *unit triplet*. These functions are not identifiable with any physical quantities or occurrences. They are philosophical in nature. They are used primarily as temporary functions in certain theoretical or analytical developments. Eventually, these functions are either integrated or combined with other functions in some way so as to result in physically meaningful functions.

5.9 Concluding remarks

In this chapter, we have introduced two new linear elements—the inductor and the capacitor. These elements are memory elements because they are energy-storing elements. The analysis of circuits containing these two types of elements is typically quite difficult as we shall see in the next chapter.

However, certain properties of linear networks remain valid when networks contain linear capacitors and inductors. The additivity, the proportionality, the linearity, and the superposition properties still apply.[3]

In addition, two more properties become meaningful when capacitors and inductors are present—differentiability and integrability.

The *differentiability* property of a linear network implies that, if the excitation $x(t)$ is replaced by its derivative $x'(t)$, then the response $y(t)$ is replaced by $y'(t)$.

The *integrability* property of a linear network implies that, if the input $x(t)$ is replaced by its integral,

$$x^{(-1)}(t) = \int_{-\infty}^{t} x(\lambda)\, d\lambda$$

then the response $y(t)$ is replaced by $y^{(-1)}(t)$.

Another observation that should be pointed out is the duality between the inductor and the capacitor. If we retrace some of the development of these two elements, it can be seen that what is said for one element, can also be said for the other if their dual entities

---

[3] This is true only if the capacitors and inductors contain no initial energy. Energy-storing elements with initial energy are equivalent to combinations of independent sources and linear elements and, therefore, are no longer linear.

250   FUNDAMENTALS OF CIRCUIT ANALYSIS

are interchanged. We list a few dual entities as examples.

Inductor	Capacitor
Flux linkage, Φ	Charge, q
$e_L = \dfrac{d\Phi}{dt}$	$i_C = \dfrac{dq}{dt}$
Inductance (in H)	Capacitance (in F)
$e_L = L\dfrac{di}{dt}$	$i_C = C\dfrac{de}{dt}$
$w(t) = \tfrac{1}{2}Li^2$	$w(t) = \tfrac{1}{2}Ce^2$

Finally, we introduced several singularity functions and studied some of their properties. These functions will be used in several situations, one of which will be the treatment of dynamic behavior of circuits containing energy-storing elements dealt with in the next chapter.

Problems

5.1 Refer to Fig. 5.1(a) with $C = 8$ F. For $e(t)$ given, give a dimensioned sketch for $i(t)$.

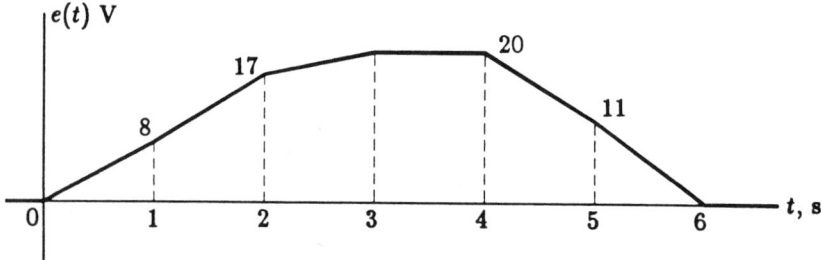

5.2 Refer to Fig. 5.1(a) with $C = 4$ F. For $e(t)$ given, give a dimensioned sketch for $i(t)$.

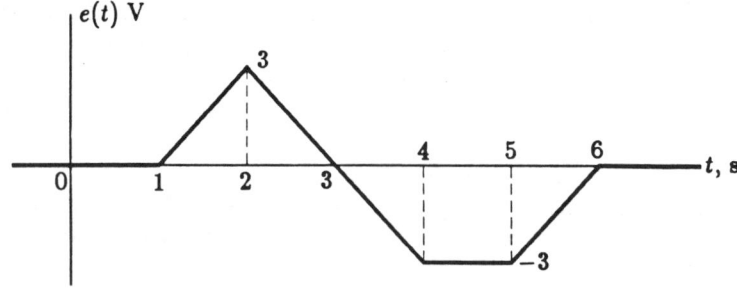

MEMORY ELEMENTS

5.3 Refer to Fig. 5.1(a) with $C = 4$ F. For $i(t)$ given, give a dimensioned sketch for $e(t)$.

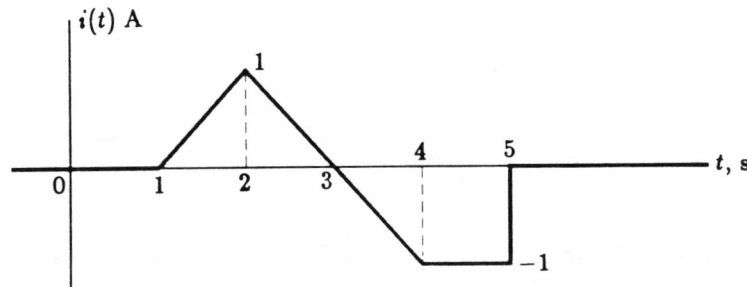

5.4 Refer to Fig. 5.1(a) with $C = 1$ F. For $i(t)$ given, give a dimensioned sketch for $e(t)$.

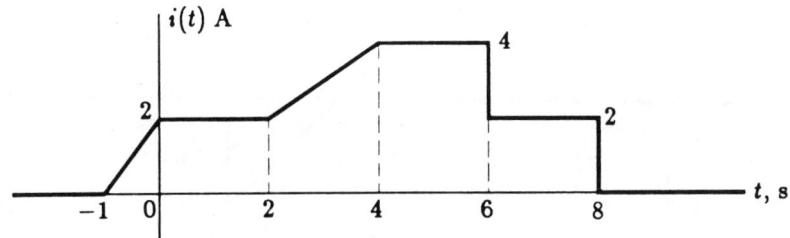

5.5 Refer to Fig. 5.1(a) with $C = 4$ F. For $i(t)$ given, give a dimensioned sketch for $e(t)$.

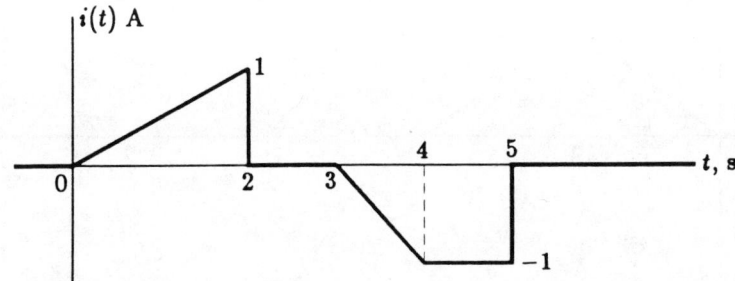

5.6 Refer to Fig. 5.1(a) with $C = 4$ F. For $e(t)$ given, give dimensioned sketches for $i(t)$, the power delivered to the capacitor, and the energy stored in the capacitor.

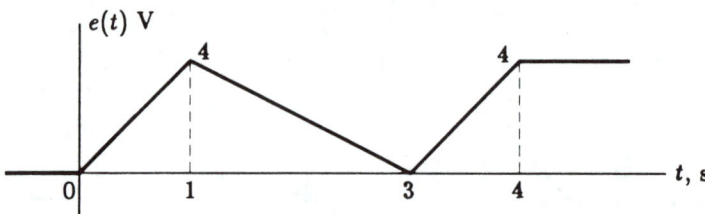

5.7 Refer to Fig. 5.1(a) with $C = 3$ F. For $i(t)$ given, give dimensioned sketches for $e(t)$, the power delivered to the capacitor, and the energy stored in the capacitor.

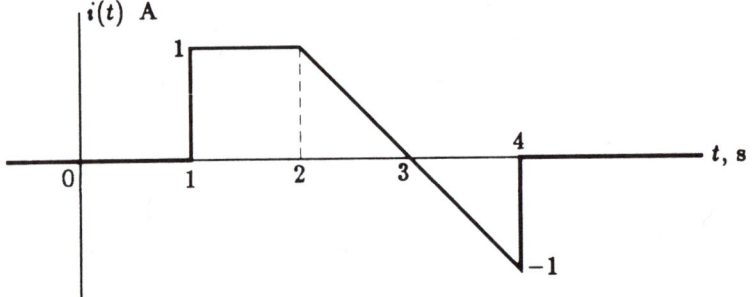

5.8 Refer to Fig. 5.10(a) with $L = 2$ H. For $i(t)$ given, give a dimensioned sketch for $e(t)$.

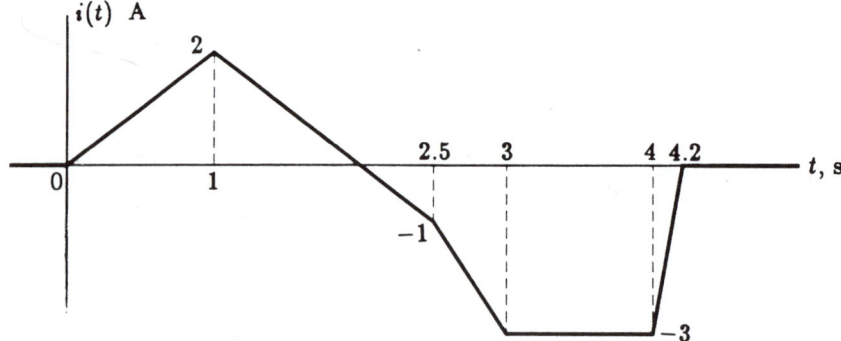

5.9 Refer to Fig. 5.10(a) with $L = 5$ H. For $e(t)$ given, give a

MEMORY ELEMENTS

dimensioned sketch for $i(t)$.

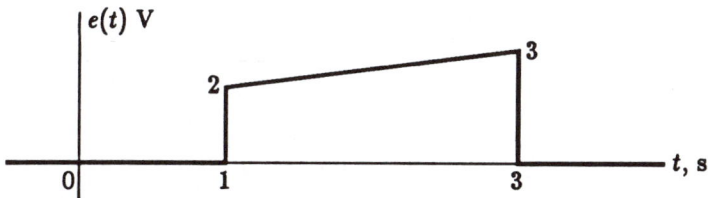

5.10 Refer to Fig. 5.10(a) with $L = 10$ H. For $e(t)$ given, give a dimensioned sketch for $i(t)$.

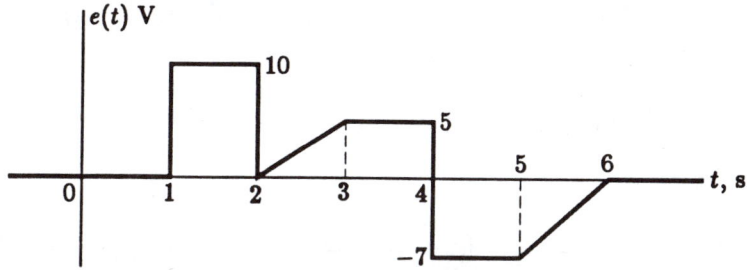

5.11 Refer to Fig. 5.10(a) with $L = 2$ H. For $e(t)$ given, give dimensioned sketches for $i(t)$, the power delivered to the inductor, and the energy stored in the inductor.

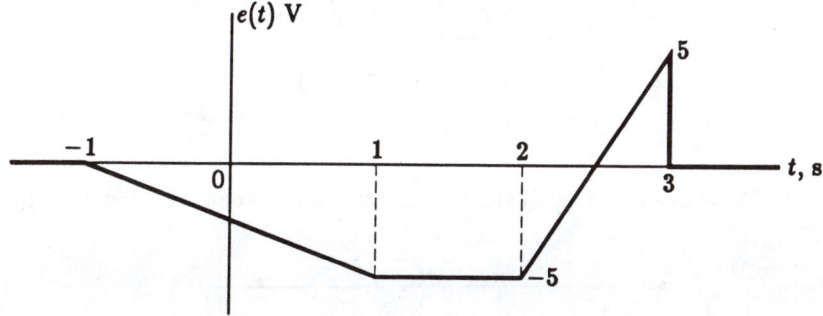

5.12 Find the equivalent inductance between terminals A and B.

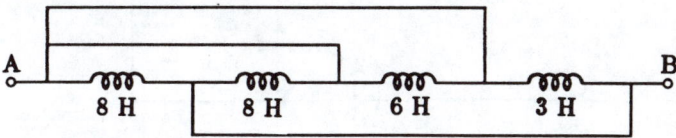

5.13 Determine e.

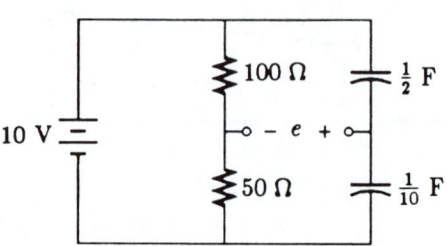

5.14 Determine the time function of $e(t)$.

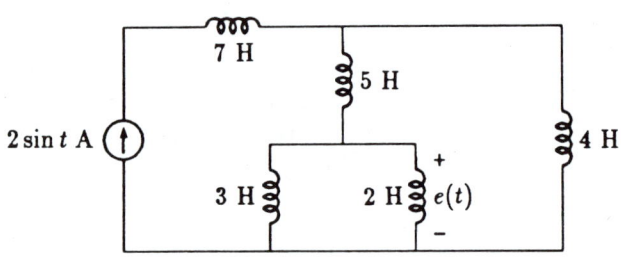

5.15 Determine the equivalent capacitance between terminals A and B.

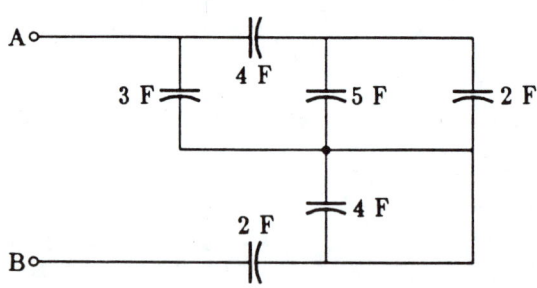

MEMORY ELEMENTS

5.16 Give a dimensioned sketch of $i(t)$.

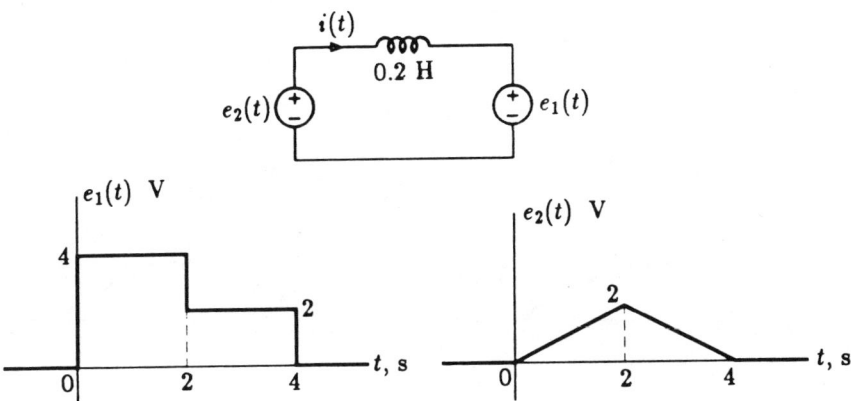

5.17 Give a dimensioned sketch of $i_s(t)$.

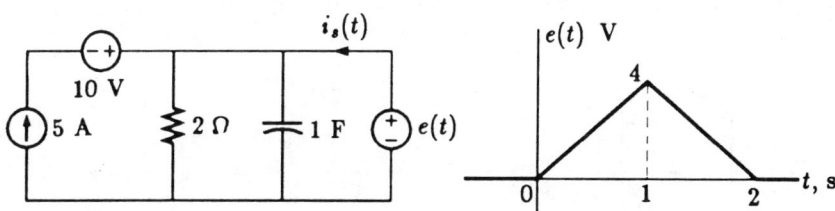

5.18 Give a dimensioned sketch of $e(t)$.

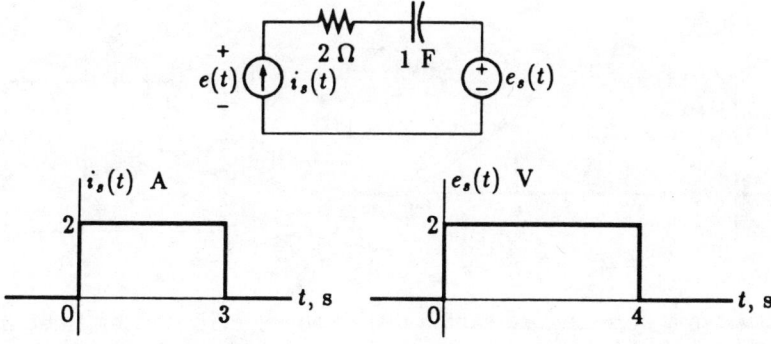

5.19 Give a dimensioned sketch of $e_i(t)$.

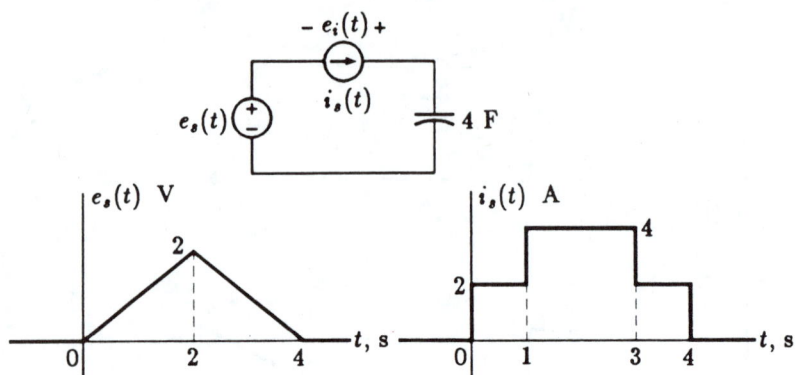

5.20 Give a dimensioned sketch of $e(t)$.

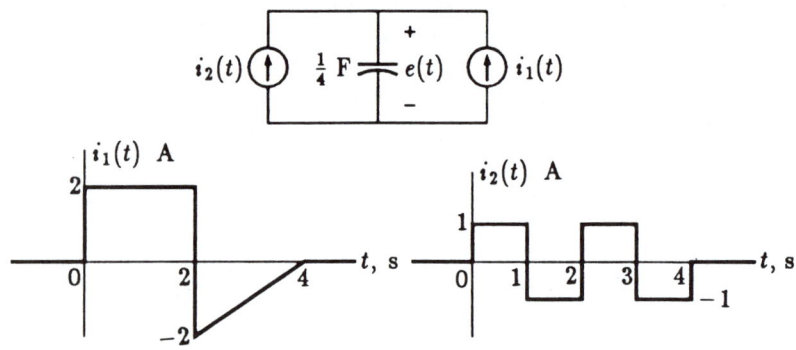

5.21 Give a dimensioned sketch of $i_s(t)$.

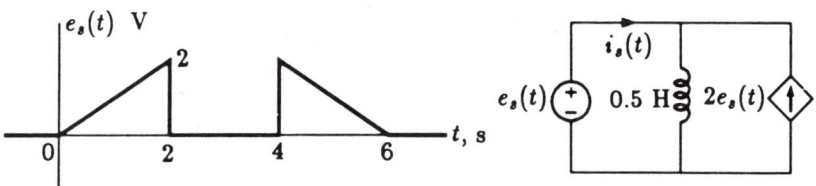

5.22 Give a dimensioned sketch of the power delivered by the current source.

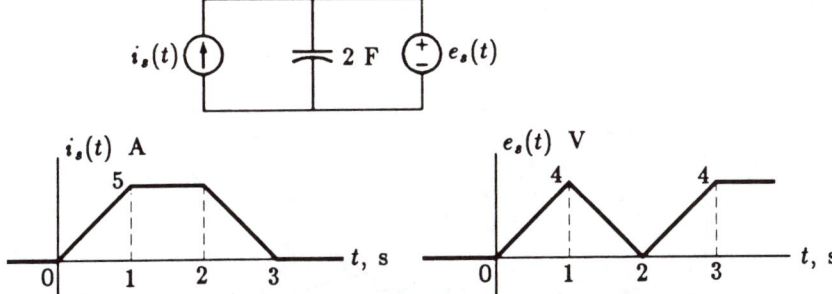

5.23 Give a dimensioned sketch of the power delivered by the current source.

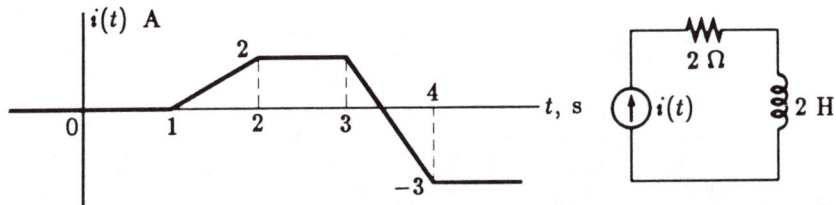

5.24 Give a dimensioned sketch of each of the following functions.

(a) $f_1(t) = u(t) + 2u(t-2) + 4u(t-3) - 5u(t-5)$

(b) $f_2(t) = u(t+1) \times u(1-t)$

(c) $f_3(t) = \epsilon^{-0.2t}[u(t-1) - u(t-4)]$

(d) $f_4(t) = u(t+2) + r(t-2) + u(t-3) - 2r(t-4)$

(e) $f_5(t) = [5u(t+1) - 2u(t-2)] \times [u(t+2) - u(t-5)]$

5.25 Express the following functions as sums of singularity functions.

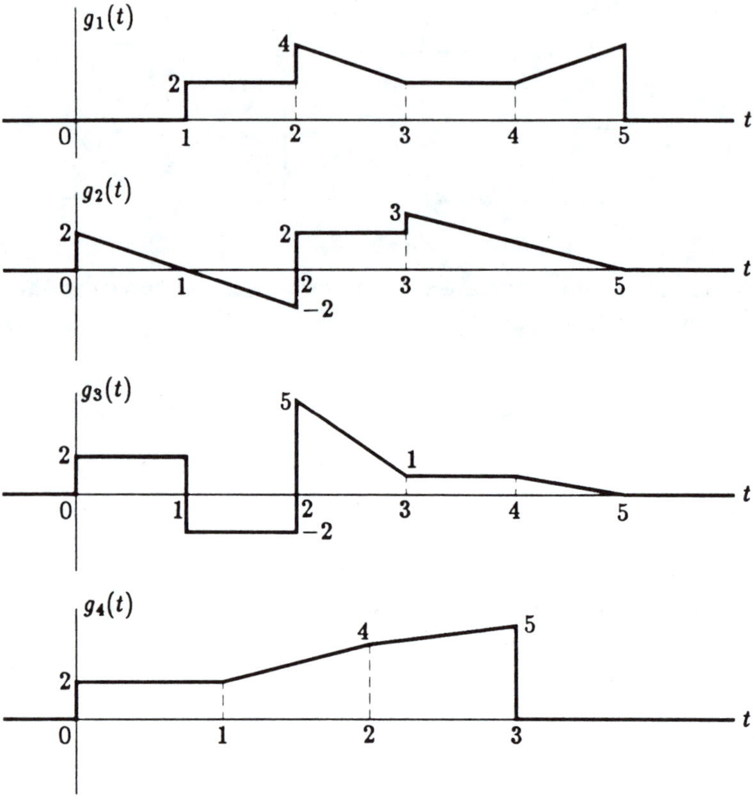

5.26 Give a dimensioned sketch of $i(t)$.

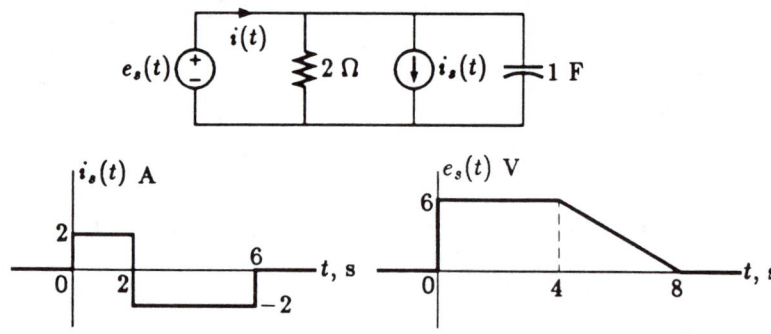

5.27 Give a dimensioned sketch of $i(t)$.

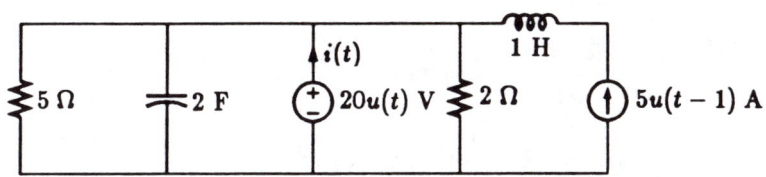

5.28 Give a singularity-function expression and a dimensioned sketch of $i(t)$.

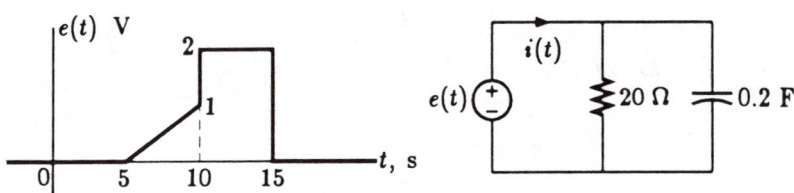

5.29 Give a dimensioned sketch of $i(t)$.

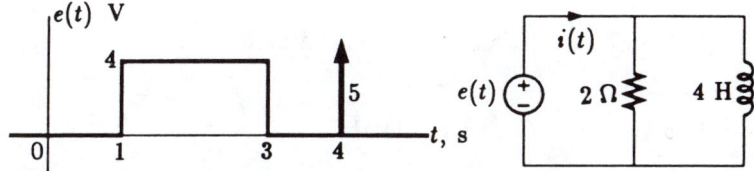

5.30 Evaluate the following integrals. Note that $-\infty < t < \infty$.

(a) $\int_0^5 u(t-2)\, dt$

(b) $\int_{-1}^2 r(t)\, dt$

(c) $\int_0^5 r(t-1)\, dt$

(d) $\int_0^4 [u(t-2) + r(t-3)]\, dt$

(e) $\int_2^t u(x)\, dx$

(f) $\int_2^{10} u_{-3}(\tau - 4)\, d\tau$

(g) $\int_2^t \delta(\lambda - 5)\, d\lambda$

o (h) $\int_1^{10} (x-9)^2\, \delta(x-2)\, dx$

(i) $\int_{-2}^3 [u(t-1)\, u(2-t)]\, dt$

(j) $\int_0^6 r(\tau - 5)\, d\tau$

(k) $\int_{-\infty}^t x^2\, u(x-2)\, dx$

(ℓ) $\int_0^2 \cos 2t\, \delta(t-3)\, dt$

(m) $\int_{-\infty}^t 10(\cos 2t + \sin 3t)\, \delta(t)\, dt$

o (n) $\int_{-\infty}^{10} 5\epsilon^{-5t}\, \delta(t-1)\, dt$

5.31 Simplify the following expressions as far as possible.

(a) $\dfrac{d}{dt}[(2-t) \times u(t-5) \times u(t+5)]$

(b) $\dfrac{d}{dt}[r(t-5)\, u(t-1)]$

o (c) $\dfrac{d}{dt}[\sqrt{t}\, \cos(t-1)\, \delta(t-1)]$

(d) $\dfrac{d}{dt}[\sin 3t\, u(t)]$

o (e) $\dfrac{d}{dt}[u(t+1)\, u(t-1)]$

o (f) $\dfrac{d}{dt}[\sin 2t\, u(t-1)]$

Chapter 6

Simple Circuits with Memory Elements

In this chapter, we shall examine the analysis and the responses of circuits that contain one or two energy-storing elements. As the student will soon see, we will be dealing with differential equations. However, the type of differential equations we will be dealing with is of the simplest type—linear differential equations with constant coefficients. We will describe the procedure for the solution of differential equations associated with circuits of interest. Therefore, a prior knowledge in differential equations, although helpful, is not absolutely necessary.

The solutions that are to be sought here are typically time functions. Hence, we are dealing with the time-domain or dynamic responses of circuits. This type of analysis is sometimes known as *transient analysis*.

6.1 Series RL circuit

We shall first treat in detail a circuit in which we have a resistance and an inductance connected in series and is subjected to various excitations or initial energies.

6.1.1 Step response

The first circuit problem we are going to solve is shown in Fig. 6.1. This circuit could be the singularity-function representation of the

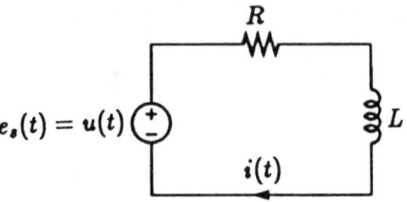

Figure 6.1: An RL series circuit with an unit step excitation.

physical arrangements in Fig. 6.2. In Fig. 6.2(a), the switch is moved from position 1 to position 2 at $t = 0$. Therefore, $e(t)$ is identical to voltage $e_s(t)$ in Fig. 6.1. Also, in Fig. 6.2(b), current $i(t)$ there is identical to that in the circuit in Fig. 6.2(a). This is true because

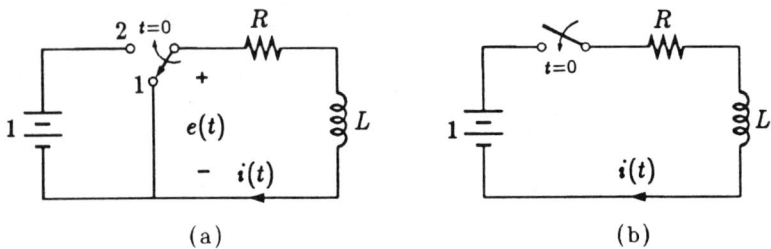

Figure 6.2: Circuit arrangements equivalent to that in Fig. 6.1.

in both circuits, $i(t) = 0$ for $t < 0$. The reasons that this is true in the two circuits are different, however. In Fig. 6.2(a), $e(t) = 0$ for $-\infty < t < 0$. Therefore, the circuit was never energized before $t = 0$. This is also true in the circuit in Fig. 6.1. In Fig. 6.2(b), $i(t) = 0$ before $t = 0$ because the circuit was open-circuited. However, as far as $i(t)$ is concerned, it is identical in all three arrangements.

Just for the record, we first note that

$$i(t) = 0, \quad t < 0 \tag{6.1}$$

To solve for $i(t)$ for $t > 0$, KVL requires

$$L\frac{di}{dt} + Ri = 1 \tag{6.2}$$

There are several techniques for solving Eq. (6.2). We shall utilize the following procedure as it gives the best physical interpretation

SIMPLE CIRCUITS WITH MEMORY ELEMENTS

of the steps taken. The procedure may later be extended to more elaborate circuits.

Step 1: **Forced response.** The first step is to find a function that will satisfy Eq. (6.2). We shall call this function $i_p(t)$. Function $i_p(t)$ needs to satisfy

$$L\frac{di_p}{dt} + Ri_p = 1 \tag{6.3}$$

This step will typically call for some prior knowledge of what type of function will be suitable for each type of excitation. Fortunately, for the types of excitation we are primarily interested in, this step is quite straightforward. That is because we can observe what is taking place in the circuit physically. Through the physical observation, we can anticipate the form of the response and thereby the type of function to use for $i_p(t)$. In this particular case, since the right-hand side of Eq. (6.3) is a constant, making $i_p(t)$ a constant is an obvious choice. We can let

$$i_p(t) = A \tag{6.4}$$

Substituting into Eq. (6.3) gives

$$RA = 1 \quad \text{or} \quad A = \frac{1}{R} \tag{6.5}$$

Hence,

$$i_p(t) = \frac{1}{R} \tag{6.6}$$

The function $i_p(t)$ is known as the *forced response*. It is the response caused entirely by the particular excitation at hand. In differential equations, $i_p(t)$ is known as the *particular integral* or *particular solution*.

Referring to Fig. 6.2, we see that the voltage is dc for $t > 0$. The final current will also be a dc current. It is clear that the value $i(t)$ will finally attain is $1/R$, which is the forced response.

Step 2: **Natural response.** Having found the forced response, we next need to find an additional function, $i_c(t)$, that satisfies

$$L\frac{di_c}{dt} + Ri_c = 0 \tag{6.7}$$

This equation is obtained by setting the right-hand side of the original differential equation to zero. This is known as the *homogeneous equation* of the differential equation. For a linear differential equation with constant coefficients, the solution to its homogeneous equation can always be written in the form

$$i_c(t) = K\epsilon^{st} \tag{6.8}$$

where K is a constant. Substitution of Eq. (6.8) into Eq. (6.7) results in

$$(Ls + R)K\epsilon^{st} = 0 \tag{6.9}$$

Since neither K nor ϵ^{st} is always zero, it is necessary that

$$Ls + R = 0 \tag{6.10}$$

Eq. (6.10) is known as the *characteristic equation* of the circuit. Solving for s gives

$$s = -\frac{R}{L} \tag{6.11}$$

Hence,

$$i_c(t) = K\epsilon^{-\frac{R}{L}t} \tag{6.12}$$

Since Eq. (6.7) is obtained by setting the forcing function (the excitation) to zero, the circuit is left by itself to behave as its nature dictates. This response has nothing to do with the excitation. Function $i_c(t)$ is known as the *natural response* of the circuit. As far as Eq. (6.7) is concerned, K could be any constant. Hence, K is known as the *arbitrary constant*. In differential equations, $i_c(t)$ is known as the *complementary function* or *homogeneous solution*.

Step 3: **Complete response.** Although $i_p(t)$ of Eq. (6.6) satisfies both Eqs. (6.2) and (6.3), it is not complete. This is because $i_c(t)$ of Eq. (6.12) satisfies Eq. (6.7) and, therefore, $i_p(t) + i_c(t)$ also satisfies Eq. (6.2). Hence, the complete solution is

$$i(t) = i_p(t) + i_c(t) = \frac{1}{R} + K\epsilon^{-\frac{R}{L}t} \tag{6.13}$$

SIMPLE CIRCUITS WITH MEMORY ELEMENTS

Step 4: Initial condition. Eq. (6.13) contains an arbitrary constant K. As far as Eq. (6.2) in the range $t > 0$ is concerned, K may assume any value. However, since a certain status already exists before $t = 0$, K must be adjusted so the transition from $t < 0$ to $t > 0$ does not violate any physical constraint present in the circuit.

In this circuit, we have $i(0^-) = 0$.[1] Since $e_s(t)$ is finite at $t = 0$ and an inductance is in series, $i(t)$ cannot change abruptly. (A sudden change in the current through an inductance should be accompanied by a voltage impulse.) Thus, $i(0^+) = 0$. Enforcing this on Eq. (6.13), we obtain

$$\frac{1}{R} + K = 0 \quad \Longrightarrow \quad K = -\frac{1}{R} \tag{6.14}$$

Thus

$$i(t) = \frac{1}{R}\left(1 - \epsilon^{-\frac{R}{L}t}\right), \quad t > 0 \tag{6.15}$$

Combining Eqs. (6.1) and (6.15), we may write

$$g_i(t) = \frac{1}{R}\left(1 - \epsilon^{-\frac{R}{L}t}\right) u(t) \tag{6.16}$$

for all t. The response of a circuit to a unit step excitation is simply known as the *step response* of the circuit. A step response is usually denoted by $g(t)$. In Eq. (6.16), the subscript i is added to indicate that the response is a current, $i(t)$.

The variation of the $g_i(t)$ of Eq. (6.16) is shown in Fig. 6.3. For $t > 0$, this variation is typical of many first-order systems. There is an initial value—in this case, $g_i(0) = 0$. There is a final value—in this case, $g_i(\infty) = 1/R$. The transition from $g_i(0)$ to $g_i(\infty)$ is governed solely by one factor—R/L. This quantity is known as the *damping constant*. The reciprocal of this factor, L/R, is known as the *time constant*, τ, of the circuit. The following are two interpretations of the time constant.

At $t = \tau$,

$$g_i(\tau) = \frac{1}{R}(1 - \epsilon^{-1}) = 0.632\frac{1}{R} \tag{6.17}$$

---

[1] We use $f(0^-)$ to denote $f(t)$ immediately prior to $t = 0$. Also, we shall use $f(0^+)$ to denote $f(t)$ immediately after $t = 0$.

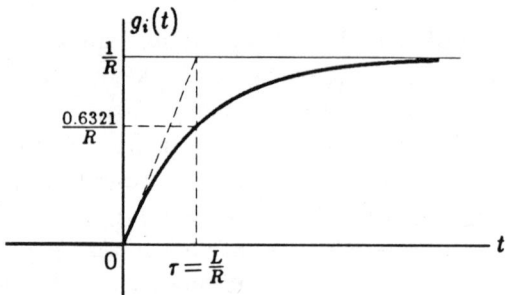

Figure 6.3: Typical step response of an RL series circuit.

In other words, the time constant is the time it takes for $g_i(t)$ to reach 63.2% of its final value.

At $t = 0$,

$$\left.\frac{dg_i}{dt}\right|_{t=0} = \frac{1}{R} \cdot \frac{R}{L} = \frac{1}{\tau} \cdot \frac{1}{R} \tag{6.18}$$

Hence, τ is the time that would take $g_i(t)$ to reach its final value had it been increasing at a constant rate equal to its initial rate. This is indicated in Fig. 6.3.

Both descriptions in the two previous paragraphs lead to the conclusion that the larger the time constant, the longer it will take the current to reach any percentage of its final value. This is illustrated in Fig. 6.4.

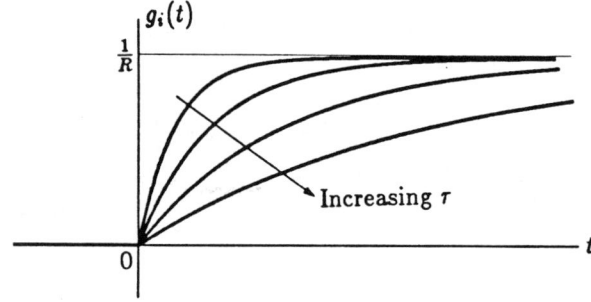

Figure 6.4: Step response of RL circuit for different time constants.

6.1.2 Impulse response

We now change the excitation from a unit step function to a unit impulse function. The arrangement is shown in Fig. 6.5. We first note that

$$i(t) = 0, \quad t < 0 \tag{6.19}$$

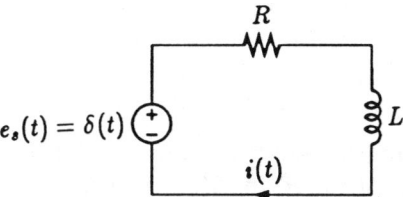

Figure 6.5: An RL series circuit with a unit impulse excitation.

For $t > 0$, the differential equation is

$$L\frac{di}{dt} + Ri = 0 \tag{6.20}$$

Therefore, the forced response is zero. The complete response consists of only the natural response. Since the natural response is the solution of a differential equation with the excitation removed (or set to zero), this part of the response does not change as the excitation changes. Once we have obtained the $i_c(t)$ for a circuit, it can be used for that circuit regardless of what the excitation is. Thus, from Eq. (6.12),

$$i(t) = K\epsilon^{-\frac{R}{L}t} \tag{6.21}$$

For the boundary condition, we examine what takes place at $t = 0$. At $t = 0$, a unit impulse is applied to the RL series combination. If $R = 0$, then the current would be

$$i(t) = \frac{1}{L}\int_{-\infty}^{t} \delta(\lambda)\, d\lambda = \frac{1}{L}u(t) \tag{6.22}$$

In other words, the current would jump from zero to $1/L$ instantaneously. Since this current is finite, even if R is nonzero the voltage drop across this finite R will still be finite. This finite voltage will not affect the voltage across the inductance. Hence, we may conclude that

268   FUNDAMENTALS OF CIRCUIT ANALYSIS

$$i(0^+) = \frac{1}{L} \qquad (6.23)$$

Imposing this initial condition on Eq. (6.21), we have

$$i(t) = \frac{1}{L}\epsilon^{-\frac{R}{L}t} \qquad (6.24)$$

Combining Eqs. (6.19) and (6.24), we can write

$$h_i(t) = \frac{1}{L}\epsilon^{-\frac{R}{L}t}u(t) \qquad (6.25)$$

for all t. This response is known as the *impulse response*. An impulse response is usually denoted by $h(t)$. In Eq. (6.25), a subscript i is added to indicate that the response is a current, $i(t)$.

▲▲ Since a unit impulse is the derivative of a unit step, we can also obtain the impulse response by differentiating the step response. From Eq. (6.16), the impulse response is

$$h_i(t) = \frac{d}{dt}g_i(t) = \frac{d}{dt}\left[\frac{1}{R}\left(1 - \epsilon^{-\frac{R}{L}t}\right)u(t)\right]$$

$$= \frac{1}{R}\left(1 - \epsilon^{-\frac{R}{L}t}\right)\delta(t) - \frac{1}{R}\left(-\frac{R}{L}\right)\epsilon^{-\frac{R}{L}t}u(t) = \frac{1}{L}\epsilon^{-\frac{R}{L}t}u(t) \qquad (6.26)$$

The term $\left(1 - \epsilon^{-\frac{R}{L}t}\right)\delta(t)$ vanishes because $\left(1 - \epsilon^{-\frac{R}{L}t}\right)$ is zero where the impulse occurs—$t = 0$.

6.1.3 Source-free response

We now pose a problem for an RL circuit in which the inductance has an initial current. We wish to know the current after the time at which the initial current is known and without the influence of any source. How the inductance attains the initial current is not an issue here. The arrangement is shown in Fig. 6.6. We shall let initial time be $t = 0$ and $i(0) = I_0$. For $t > 0$, the differential equation is identical to that in Eq. (6.20). Hence the solution is identical to Eq. (6.21). Forcing $i(0) = I_0$ gives

$$i(t) = I_0\,\epsilon^{-\frac{R}{L}t}, \quad t > 0 \qquad (6.27)$$

SIMPLE CIRCUITS WITH MEMORY ELEMENTS

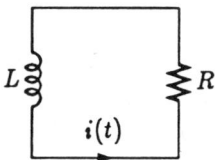

Figure 6.6: An RL circuit with an initial energy.

This is the *source-free response* of an RL circuit.

It is interesting to investigate what happens to the initial energy stored in the inductance. That energy is $\frac{1}{2}LI_0^2$. The power delivered to the resistance R is

$$p(t) = i^2(t) R \qquad (6.28)$$

The total energy delivered to R is

$$\int_0^\infty p(t)\, dt = \int_0^\infty RI_0^2 \epsilon^{-\frac{2R}{L}t}\, dt = \frac{1}{2}LI_0^2 \qquad (6.29)$$

Thus, the energy stored in the inductance L at $t = 0$ is eventually dissipated in the resistance R.

EXAMPLE 1. Obtain $i(t)$ in the circuit in Fig. 6.7.

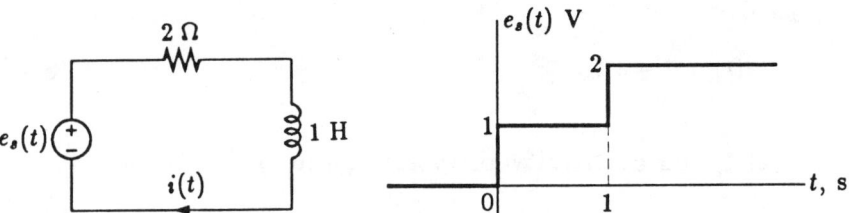

Figure 6.7: Arrangement used in Example 1.

SOLUTION (1) **Solution of the differential equations.** First we note

$$i(t) = 0, \quad t < 0$$

For $0 < t < 1$,

$$\frac{di}{dt} + 2i = 1$$

FUNDAMENTALS OF CIRCUIT ANALYSIS

$$i_p(t) = \frac{1}{2} \quad i_c(t) = K_1\epsilon^{-2t} \quad i(t) = \frac{1}{2} + K_1\epsilon^{-2t}$$

Letting $i(0) = 0$, we get

$$i(t) = \frac{1}{2} - \frac{1}{2}\epsilon^{-2t} \tag{6.30}$$

At $t = 1$,

$$i(1) = \frac{1}{2} - \frac{1}{2}\epsilon^{-2} = 0.4323$$

For $t > 1$,

$$\frac{di}{dt} + 2i = 2$$

$$i_p(t) = 1 \quad i_c(t) = K_2\epsilon^{-2t} \quad i(t) = 1 + K_2\epsilon^{-2t}$$

Forcing $i(1) = 0.4323$, we should have

$$0.4323 = 1 + K_2\epsilon^{-2}$$

Solving, we obtain

$$K_2 = -0.5677\epsilon^2 = -4.1945$$

Hence

$$i(t) = 1 - 4.1945\epsilon^{-2t} \tag{6.31}$$

(2) **Using additivity property.** Since

$$e_s(t) = u(t) + u(t-1)$$

from Eq. (6.16), we can write for our present circuit

$$i(t) = \frac{1}{2}\left(1 - \epsilon^{-2t}\right)u(t) + \frac{1}{2}\left(1 - \epsilon^{-2(t-1)}\right)u(t-1) \tag{6.32}$$

It is obvious that $i(t)$ given by Eq. (6.32) agrees with that given by Eq. (6.30) for $0 < t < 1$. For $t > 1$, $i(t)$ given by Eq. (6.32) reduces to

$$i(t) = \frac{1}{2}\left(1 - \epsilon^{-2t}\right) + \frac{1}{2}\left(1 - \epsilon^{-2(t-1)}\right)$$

$$= 1 - \frac{1}{2}\left(1 + \epsilon^2\right)\epsilon^{-2t} = 1 - 4.1945\epsilon^{-2t}$$

which agrees with Eq. (6.31).

The variation of $i(t)$ in Fig. 6.7 is shown below.

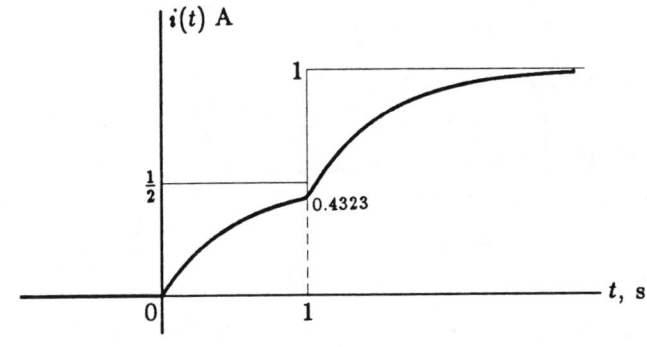

EXAMPLE 2. In the arrangement of Fig. 6.8, the switch has been in position 1 for $t < 0$. At $t = 0$, it is switched to position 2. Find $i(t)$ for all t.

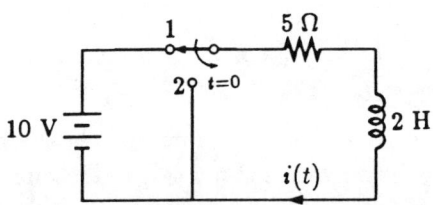

Figure 6.8: The arrangement used in Example 2.

SOLUTION For $t < 0$, $i(t) = \frac{10}{5} = 2$ A. This is the dc value of $i(t)$ after the 10-V source has been applied to the RL series combination since $t = -\infty$.

For $t > 0$, $i(t) = 2\epsilon^{-2.5t}$ A from Eq. (6.27). The complete variation of $i(t)$ is shown below.

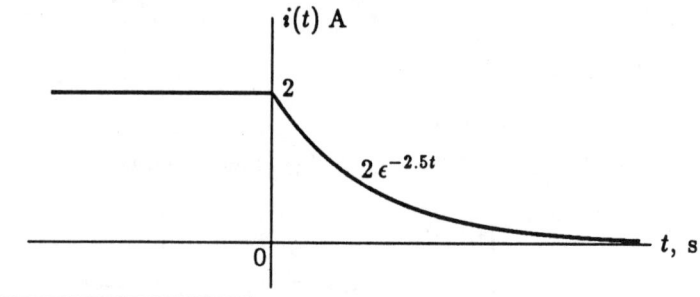

EXAMPLE 3. In the arrangement in Fig. 6.9, switch S_1 has been open and switch S_2 has been closed since $t = -\infty$. At $t = 0$, S_1 is closed. At $t = 2$, S_2 is opened. Determine $i(t)$ for $t > 0$.

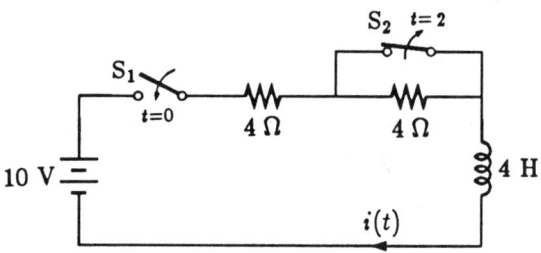

Figure 6.9: Arrangement used in Example 3.

SOLUTION For $0 < t < 2$, the current is the same as a 10-V step applied to a 4-Ω resistance and a 4-H inductance in series. From Eq. (6.16),

$$i(t) = \frac{10}{4}\left(1 - \epsilon^{-\frac{4}{4}t}\right) = 2.5\left(1 - \epsilon^{-t}\right) \text{ A}$$

At $t = 2$, $i(2) = 2.5(1 - \epsilon^{-1}) = 2.1617$.
For $t > 2$, the differential equation is

$$4\frac{di}{dt} + 8i = 10$$

SIMPLE CIRCUITS WITH MEMORY ELEMENTS

$$i_p(t) = \frac{10}{8} = 1.25 \qquad i_c(t) = K\epsilon^{-\frac{8}{4}t} \qquad i(t) = 1.25 + K\epsilon^{-2t}$$

Making

$$i(2) = 1.25 + K\epsilon^{-4} = 2.1617$$

yields $K = 49.777$. Hence,

$$i(t) = 1.25 + 49.777\epsilon^{-2t} \text{ A}, \quad t > 2$$

The variation of $i(t)$ is shown below.

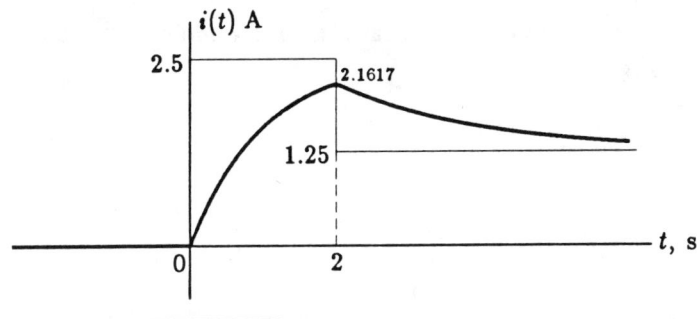

EXAMPLE 4. In Fig. 6.10, $e_s(t) = 10u(t)$ V. Determine $i(t)$.

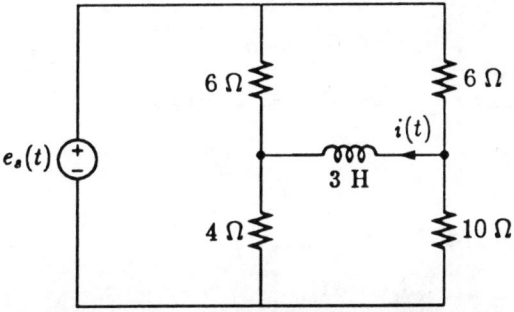

Figure 6.10: Circuit for Example 4.

SOLUTION We first obtain the Thévenin's equivalent of that part of the circuit external to the inductance. The circuits used are shown in Fig. 6.11. From Fig. 6.11(a), we have

$$e_{oc}(t) = -\frac{4}{10}e_s(t) + \frac{10}{16}e_s(t) = \frac{9}{40}e_s(t) = \frac{9}{4}u(t) \text{ V}$$

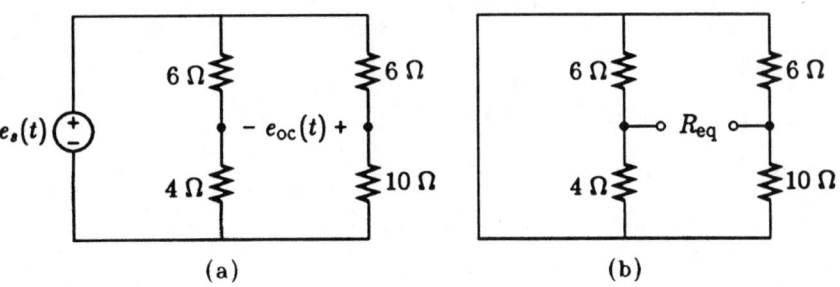

Figure 6.11: Circuits used to obtain the Thevenin's equivalent of the circuit of Fig. 6.10.

From Fig. 6.11(b), we get

$$R_{eq} = 4||6 + 10||6 = \frac{123}{20} \ \Omega$$

From Eq. (6.16), we can write

$$i(t) = \frac{\frac{9}{4}}{R_{eq}} \left(1 - \epsilon^{-\frac{R_{eq}}{3}t}\right) u(t) = \frac{15}{41} \left(1 - \epsilon^{-\frac{41}{20}t}\right) u(t) \ \text{A}$$

EXERCISES

6.1.1 Determine $i(t)$ in the circuit.

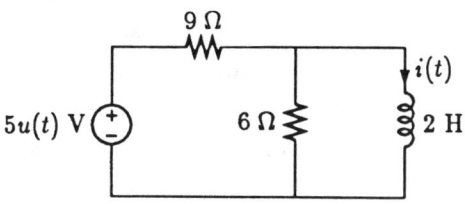

Ans. $\dfrac{5}{9} \left(1 - \epsilon^{-1.8t}\right) u(t)$ A.

6.1.2 Obtain $e_L(t)$ in the arrangement shown below.

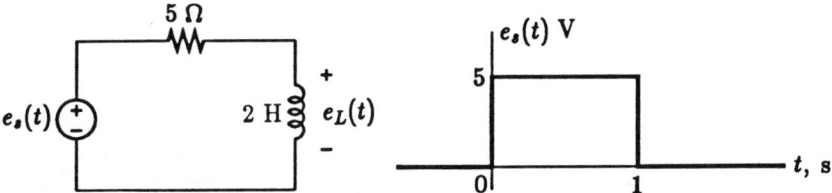

Ans. 0, $t < 0$; $5\epsilon^{-2.5t}$ V, $0 < t < 1$; $5\epsilon^{-2.5t} - 5\epsilon^{-2.5(t-1)}$ V, $1 < t$.

6.1.3 The switch has been in position 1 since $t = -\infty$. At $t = 0$, it is moved to position 2. Find $i(t)$ for $t > 0$.

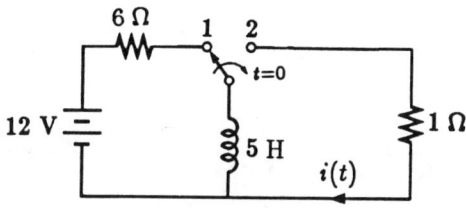

Ans. $-2\epsilon^{-0.2t}$ A.

6.1.4 The switch has been in position 1 since $t = -\infty$. At $t = 0$, it is moved to position 2. Find $i(t)$ for $t > 0$.

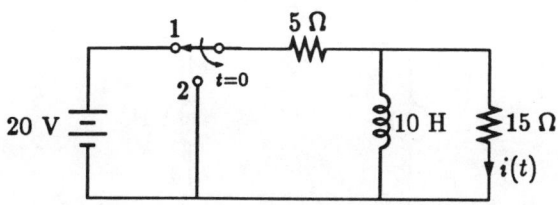

Ans. $-\epsilon^{-0.375t}$ A.

6.1.4 Responses to other excitations

Thus far, we have limited our excitations to constants, at least in the time interval of interest. When the excitation is something other than a constant, the first step of our solution is still to find the forced response or particular solution. The task of finding such a function is not always an easy one. In fact, there is not always a straightforward procedure except for special functions that happen to be the excitation. Fortunately, some of the more easily solvable functions are also the more useful ones in electrical systems. Also, since we are dealing with existing electrical quantities, rather than abstract functions, there are ways of finding what some of these particular functions are for certain types of excitations. The form of some of the better-known particular integrals corresponding to certain excitations are listed in Table 6.1.

Table 6.1 Particular integrals for some excitations

Excitation	Form of particular integral
Constant	Constant
$\epsilon^{\alpha t}$	$A\epsilon^{\alpha t}$
t^n	$At^n + Bt^{n-1} + \cdots$
$\cos \omega t$ or $\sin \omega t$	$A \cos \omega t + B \sin \omega t$
$\epsilon^{\alpha t} \cos \omega t$ or $\epsilon^{\alpha t} \sin \omega t$	$\epsilon^{\alpha t}(A \cos \omega t + B \sin \omega t)$

EXAMPLE 5. For the circuit in Fig. 6.12, $e_s(t) = 5 \cos 4t\, u(t)$ V. Determine $i(t)$.

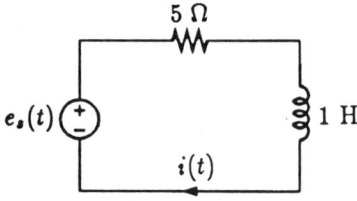

Figure 6.12: Circuit for Example 5.

SOLUTION The circuit is quiescent for $t < 0$. For $t > 0$,
$$\frac{di}{dt} + 5i = 5 \cos 4t$$

We assume

$$i_p(t) = A\cos 4t + B\sin 4t$$

Substitution gives

$$-4A\sin 4t + 4B\cos 4t + 5A\cos 4t + 5B\sin 4t = 5\cos 4t$$

Equating coefficients, we get

$$5A + 4B = 5 \quad \text{and} \quad -4A + 5B = 0$$

Solving for A and B yields

$$A = \frac{25}{41} \quad \text{and} \quad B = \frac{20}{41}$$

The complete solution is

$$i(t) = \frac{25}{41}\cos 4t + \frac{20}{41}\sin 4t + K\,\epsilon^{-5t}$$

At $t = 0$, $i(t) = 0$ because there is a series inductance and there is no voltage impulse in the circuit. Imposing this condition on $i(t)$ results in

$$K = -\frac{25}{41}$$

Hence, the final solution is

$$i(t) = \left[\frac{25}{41}\cos 4t + \frac{20}{41}\sin 4t - \frac{25}{41}\epsilon^{-5t}\right] u(t) \text{ A}$$

EXERCISE

6.1.5 In the circuit, $e_s(t) = \epsilon^{-\alpha t} u(t)$. Determine $i(t)$.

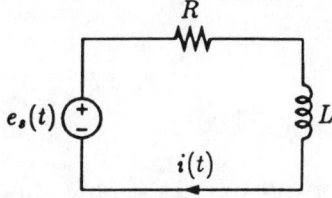

Ans. $\dfrac{1}{R - \alpha L}\left(\epsilon^{-\alpha t} - \epsilon^{-\frac{R}{L}t}\right) u(t)$.

6.1.5 Ramp response

Suppose we have the circuit in Fig. 6.13. This is the same circuit we have been dealing with in this section except the voltage source is now a unit ramp function. The circuit is quiescent for $t < 0$. For $t > 0$, the circuit must satisfy

$$L\frac{di}{dt} + Ri = t \tag{6.33}$$

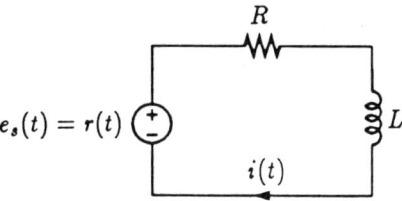

Figure 6.13: An RL series circuit with a unit-ramp excitation.

We assume

$$i_p(t) = At + B \tag{6.34}$$

Substitution gives

$$LA + R(At + B) = t$$

Equating coefficients, we get

$$RA = 1 \quad \text{and} \quad LA + RB = 0$$

Hence,

$$A = \frac{1}{R} \quad \text{and} \quad B = -\frac{L}{R^2}$$

The complete solution is

$$i(t) = \frac{1}{R}t - \frac{L}{R^2} + K\epsilon^{-\frac{R}{L}t} \tag{6.35}$$

Imposing $i(0) = 0$ on Eq. (6.35), we readily obtain

$$i(t) = \frac{1}{R}t - \frac{L}{R^2} + \frac{L}{R^2}\epsilon^{-\frac{R}{L}t} \tag{6.36}$$

We can then write

$$i(t) = \left[\frac{1}{R}t - \frac{L}{R^2}\left(1 - \epsilon^{-\frac{R}{L}t}\right)\right] u(t) \qquad (6.37)$$

for all t.

Since the unit ramp function is the integral of the unit step function, Eq. (6.36) can also be obtained by integrating the step response. From Eq. (6.15), the current due to a unit ramp would be

$$i(t) = \int_0^t \frac{1}{R}\left(1 - \epsilon^{-\frac{R}{L}\lambda}\right) d\lambda$$

$$= \frac{1}{R}\left(\lambda + \frac{L}{R}\epsilon^{-\frac{R}{L}\lambda}\right)\bigg|_0^t = \frac{t}{R} + \frac{L}{R^2}\epsilon^{-\frac{R}{L}t} - \frac{L}{R^2}$$

which is identical to that given in Eq. (6.36).

Conversely, we can also differentiate Eq. (6.36) to get Eq. (6.15), or Eq. (6.37) to get Eq. (6.16).

EXERCISE

6.1.6 Obtain $i(t)$ in the following circuit.

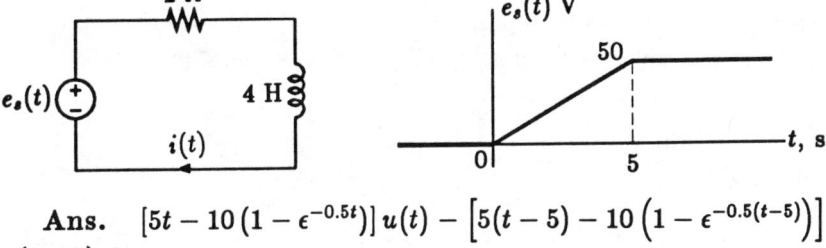

Ans. $[5t - 10(1 - \epsilon^{-0.5t})] u(t) - [5(t-5) - 10(1 - \epsilon^{-0.5(t-5)})] u(t-5)$ A.

6.2 Series RC circuit

Now we turn to another circuit in which a resistance and a capacitance are connected in series. Much of what we develop here is an extension or a variation of what we did for RL series circuits. For that reason, we will not include as much detail as we did in Section 6.1.

6.2.1 Step response

We shall first investigate the response of an RC series circuit subjected to a unit step function voltage excitation. The arrangement is shown in Fig. 6.14. We first note that

$$i(t) = 0, \quad t < 0 \qquad (6.38)$$

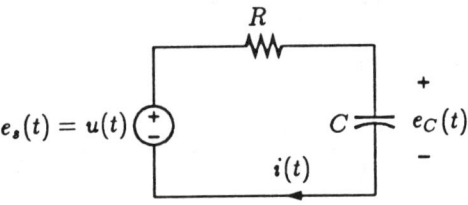

Figure 6.14: An RC circuit with a unit step excitation.

For $t > 0$,

$$Ri + \frac{1}{C}\int_0^t i(\lambda)\, d\lambda = 1 \qquad (6.39)$$

Differentiating Eq. (6.39) with respect to t, we get

$$R\frac{di}{dt} + \frac{1}{C}i = 0 \qquad (6.40)$$

To solve the differential equation of Eq. (6.40), we see that $i_p(t) = 0$ and the characteristic equation of the differential equation is

$$Rs + \frac{1}{C} = 0 \qquad (6.41)$$

which gives

$$s = -\frac{1}{RC} \qquad (6.42)$$

The complete solution is

$$i(t) = i_c(t) = K\epsilon^{-\frac{1}{RC}t} \qquad (6.43)$$

At $t = 0$, the voltage applied to the RC combination is unity. Because of the resistance, $i(0)$ is finite. A finite current cannot deliver any nonzero charge to the capacitance instantaneously. Thus $e_C(0) = 0$ and the unity voltage will appear across the resistance. Therefore, we can say that $i(0) = 1/R$. Imposing this initial condition on the $i(t)$ of Eq. (6.43), we get $K = 1/R$ and

SIMPLE CIRCUITS WITH MEMORY ELEMENTS

$$i(t) = \frac{1}{R}\epsilon^{-\frac{1}{RC}t}, \quad t > 0 \tag{6.44}$$

Hence, the step response is

$$g_i(t) = \frac{1}{R}\epsilon^{-\frac{1}{RC}t} u(t) \tag{6.45}$$

for all t. The variation of this step response is shown in Fig. 6.15 This current is an exponentially decaying function. The initial value

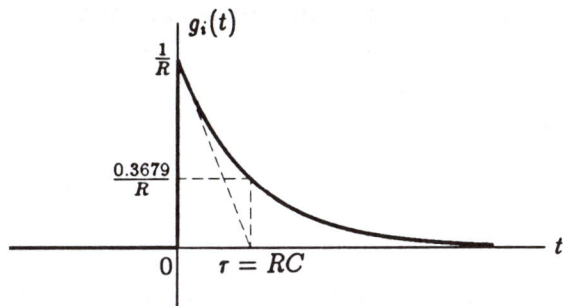

Figure 6.15: The step response of an RC series circuit.

is $1/R$; the final value is zero. The rate of decay depends on the time constant $\tau = RC$, much like the way τ controls the current rise in an RL series circuit in Fig. 6.3.

For the circuit in Fig. 6.14,

$$e_C(t) = u(t) - Ri(t) = \left(1 - \epsilon^{-\frac{1}{RC}t}\right)u(t) \tag{6.46}$$

The variation of $e_C(t)$ is shown in Fig. 6.16.

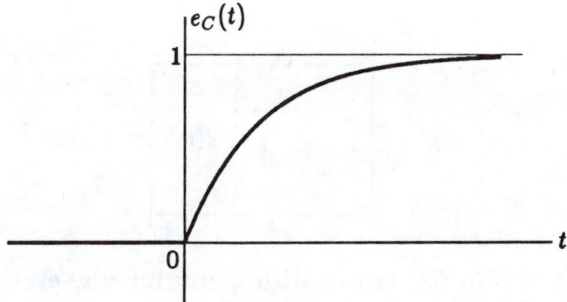

Figure 6.16: Variation of $e_C(t)$ of Fig. 6.14.

EXAMPLE 6. For the arrangement in Fig. 6.17, obtain an expression for $e_C(t)$.

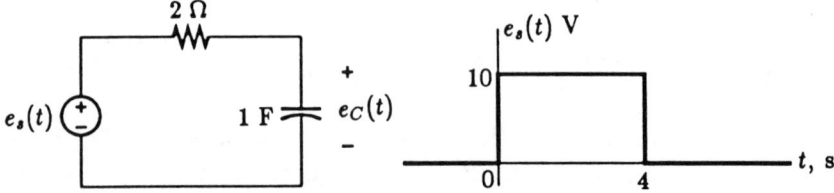

Figure 6.17: Arrangement used in Example 6.

SOLUTION Since

$$e_s(t) = 10\,u(t) - 10\,u(t-4)$$

From Eq. (6.46),

$$i(t) = 5\,\epsilon^{-0.5t}u(t) - 5\,\epsilon^{-0.5(t-4)}u(t-4)$$

$$e_C(t) = e_s(t) - 2i(t)$$

$$= 10\left(1 - \epsilon^{-0.5t}\right)u(t) - 10\left(1 - \epsilon^{-0.5(t-4)}\right)u(t-4)\ \text{V}$$

6.2.2 Impulse response

We now solve the series RC circuit when the excitation is a unit impulse. The arrangement is shown in Fig. 6.18. Again, we have $i(t) = 0$ for $t < 0$.

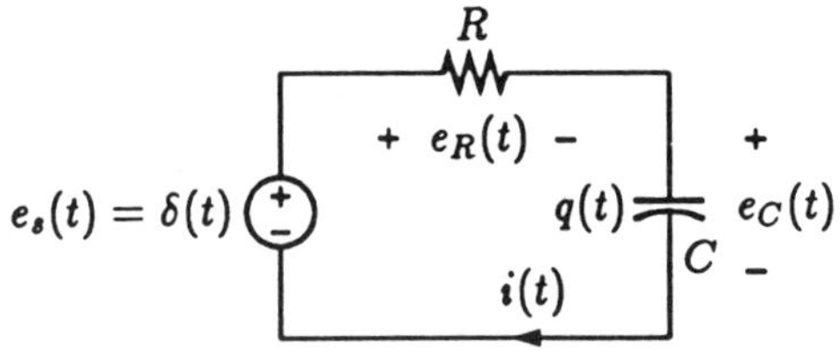

Figure 6.18: An RC circuit with a unit impulse excitation.

SIMPLE CIRCUITS WITH MEMORY ELEMENTS

At $t = 0$, a unit impulse voltage is applied to the RC series combination. This will produce an impulse current of $1/R$ coulombs. This impulse current delivers a charge of $1/R$ coulombs to the capacitance instantaneously. Since $q(0)$ is finite, $e_C(0)$ is also finite. So the presence of this finite $e_C(0)$ does not affect the impulse voltage as far as the resistance is concerned. We can also say that $e_C(0^+) = 1/RC$.

For $t > 0$, the differential equation is

$$Ri + \frac{1}{C}\int_0^t i(\lambda)\,d\lambda = 0 \tag{6.47}$$

$\left(\ q(t) = C \cdot e(t)\ \right)$

Differentiating gives

$$R\frac{di}{dt} + \frac{1}{C}i = 0 \tag{6.48}$$

Eq. (6.48) is identical to Eq. (6.40). Therefore, its complete solution is also

$$i(t) = K\,\epsilon^{-\frac{1}{RC}t} \tag{6.49}$$

For the initial conditions, we have $e_C(0^+) = 1/RC$, $e_s(0^+) = 0$, and $e_R(0^+) = -1/RC$. Hence $i(0^+) = -1/R^2C$. Imposing this condition on the $i(t)$ of Eq. (6.49) leads to

$$i(t) = -\frac{1}{R^2C}\epsilon^{-\frac{1}{RC}t}, \quad t > 0 \tag{6.50}$$

To summarize, the impulse response is

$$h_i(t) = \frac{1}{R}\delta(t) - \frac{1}{R^2C}\epsilon^{-\frac{1}{RC}t}u(t) \tag{6.51}$$

for all t.

This impulse response can also be obtained from the step response of Eq. (6.45).

$$h_i(t) = \frac{d}{dt}g_i(t) = \frac{d}{dt}\left[\frac{1}{R}\epsilon^{-\frac{1}{RC}t}u(t)\right]$$

$$= -\frac{1}{R^2C}\epsilon^{-\frac{1}{RC}t}u(t) + \frac{1}{R}\epsilon^{-\frac{1}{RC}t}\delta(t) = \frac{1}{R}\delta(t) - \frac{1}{R^2C}\epsilon^{-\frac{1}{RC}t}u(t) \tag{6.52}$$

The voltage across the capacitance is

$$e_C(t) = \delta(t) - Ri(t) = \frac{1}{RC}\epsilon^{-\frac{1}{RC}t}u(t) \tag{6.53}$$

The variation of $h_i(t)$ of Eq. (6.51) is shown in Fig. 6.19.

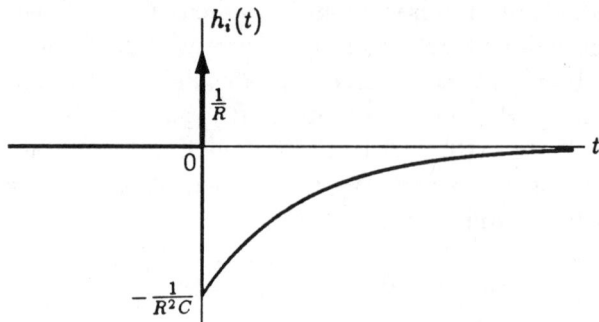

Figure 6.19: Variation of $h_i(t)$ of Eq. (6.51).

6.2.3 Source-free response

Suppose we have a capacitor that has been charged to a voltage of $e(0) = E_0$. At $t = 0$, a resistor is connected across it as shown in Fig. 6.20. We wish to know the current $i(t)$ for $t > 0$.

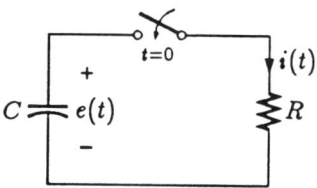

Figure 6.20: The discharge of a capacitor through a resistor.

The differential equation (after differentiating the KVL equation) is

$$R\frac{di}{dt} + \frac{1}{C}i = 0 \tag{6.54}$$

The complete solution is

$$i(t) = K\epsilon^{-\frac{1}{RC}t} \tag{6.55}$$

At $t = 0^+$, $i(0^+) = E_0/R$. Hence,

$$i(t) = \frac{E_0}{R}\epsilon^{-\frac{1}{RC}t} \tag{6.56}$$

L	C
step $i(0^+)=0$	$i(0)=\frac{1}{R}$
Impulse $i(0^+)=\frac{1}{L}$	$i(0^+)=\frac{-1}{R^2C}$

The voltage across both the capacitor and the resistor is

$$e(t) = Ri(t) = E_0 \epsilon^{-\frac{1}{RC}t} \qquad (6.57)$$

The results obtained here are the source-free responses of the RC circuit because no excitation is present in the circuit. The current and voltage of Eqs. (6.56) and (6.57) also describe what happens when an initially charged capacitor is discharged through a resistor.

The results here are exactly the same as those obtained in the previous subsection for $t > 0$. In obtaining the impulse response, we included the entire history of the circuit, including how the capacitor attains an initial charge. In the source-free response, how the capacitor acquires its initial charge is of no concern to us. We are only interested in what happens in the circuit after certain initial charges already exist and the circuit is left to respond without the intervention of any source.

EXAMPLE 7. In Fig. 6.21, the switch has been closed for $t < 0$. At $t = 0$, it is opened. Obtain $i(t)$ for $t > 0$.

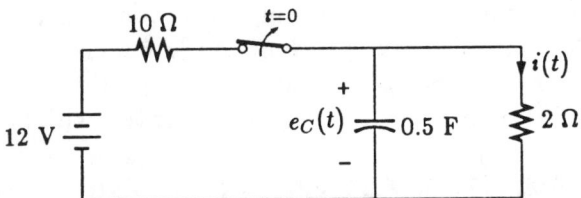

Figure 6.21: Arrangement used in Example 7.

SOLUTION Before $t = 0$, there is no current flowing through the capacitor. Hence, $e_C(0) = 2$ V. From Eq. (6.56), we have

$$i(t) = \frac{2}{2}\epsilon^{-\frac{1}{0.5 \times 2}t} = \epsilon^{-t} \text{ A}$$

EXERCISES

6.2.1 In Fig. 6.20, $C = 0.01$ F, $R = 5$ kΩ, and $e(0) = 20$ V. Find the value of $e(t)$ at $t = 100$ s.

Ans. 2.707 V.

6.2.2 Integrate the power delivered to R in Fig. 6.20 from $t = 0$ to $t = \infty$ to show that the total energy stored in C at $t = 0$ is eventually dissipated in R.

6.2.3 The switch has been closed for $t < 0$. At $t = 0$, it is opened. Give $i(t)$ for $t > 0$.

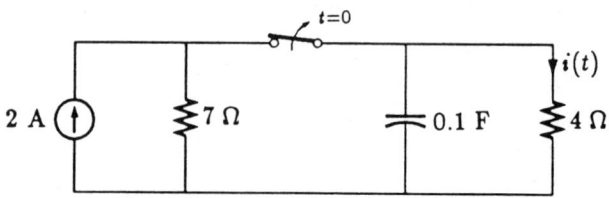

Ans. $\dfrac{14}{11}\epsilon^{-2.5t}$ A.

6.2.4 Response to other excitations

When the excitation is other than singularity functions, the solution procedure is similar to what was done in the RL circuits. Actually, the only difference is the determination of the particular integral for each given forcing function. We shall illustrate this with an example.

EXAMPLE 8. In Fig. 6.22, $e_s(t) = (t^3 + 2t)u(t)$ V. Determine $i(t)$.

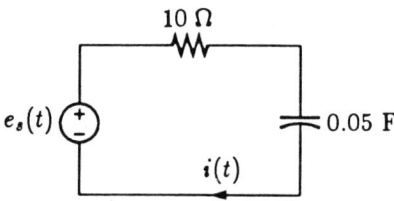

Figure 6.22: Circuit for Example 8.

SOLUTION The circuit is quiescent for $t < 0$. For $t > 0$, KVL requires

$$10i + 20\int_0^\infty i(\lambda)\, d\lambda = t^3 + 2t$$

Differentiating to get
$$10\frac{di}{dt} + 20i = 3t^2 + 2$$
We assume
$$i_p(t) = At^2 + Bt + C$$
Substitution gives
$$20At + 10B + 20At^2 + 20Bt + 20C = 3t^2 + 2$$
Equating coefficients yields
$$20A = 3 \quad 20A + 20B = 0 \quad 20C + 10B = 2$$
which leads to
$$A = \frac{3}{20} \quad B = -\frac{3}{20} \quad C = \frac{7}{40}$$
The complete solution is
$$i(t) = \frac{3}{20}t^2 - \frac{3}{20}t + \frac{7}{40} + K\epsilon^{-2t}$$
Since $e_s(0) = 0$, $i(0) = 0$. Hence $K = -7/40$ and the final answer is
$$i(t) = \left[\frac{3}{20}t^2 - \frac{3}{20}t + \frac{7}{40} - \frac{7}{40}\epsilon^{-2t}\right] u(t) \text{ A}$$

6.2.5 Ramp response

Since the unit ramp is the integral of the unit step, the current in an RC circuit with a unit ramp excitation can be obtained by integrating its step response. From Eq. (6.45), we can write

$$g_i^{(-1)}(t) = \int_{-\infty}^{t} g_i(\lambda)\, d(\lambda) = u(t) \int_0^t \frac{1}{R} \epsilon^{-\frac{1}{RC}\lambda} d\lambda$$

$$= -C\epsilon^{-\frac{1}{RC}\lambda} \bigg|_0^t = C\left(1 - \epsilon^{-\frac{1}{RC}t}\right) u(t) \quad (6.58)$$

EXERCISE

6.2.4 Obtain Eq. (6.58) by solving the differential equation
$$Ri + \frac{1}{C}\int_0^t i(\lambda)\, d\lambda = t$$
for $t > 0$.

288   FUNDAMENTALS OF CIRCUIT ANALYSIS

6.3 Thévenin's and Norton's equivalent of memory elements with initial energies

There is an alternative way of accounting for the initially energized capacitor or inductor. That is to replace the element with its Thévenin's or Norton's equivalent. For convenience, we shall assume that we wish to observe the behavior of the element only for $t > 0$. What happens before $t = 0$ is of no concern to us. If the actual starting time is other than $t = 0$, we merely advance or delay the reference in time.

For a capacitor C with a voltage E_0 at $t = 0$:
(a) The voltage in its Thévenin's equivalent circuit is a constant E_0, since this capacitor will retain all its charge (hence its voltage) when it is open-circuited.
(b) The current in its Norton's equivalent circuit is an impulse function of strength CE_0, since when this capacitor is short- circuited this amount of charge will flow through the short circuit instantaneously.

The Thévenin's or Norton's equivalent circuit of such a capacitor is the appropriate combination of the equivalent source and a capacitor with the same capacitance and no initial energy. This equivalence is shown in Fig. 6.23(a).

For an inductor L with a current I_0 at $t = 0$:
(a) The voltage in its Thévenin's equivalent circuit is an impulse function of strength LI_0, since the opening of a current path abruptly changes the current from I_0 to zero.
(b) The current in its Norton's equivalent circuit is a constant I_0, since a short circuit allows the current in the inductor to flow indefinitely.

The Thévenin's or Norton's equivalent circuit of such an inductor is the appropriate combination of the equivalent source and an inductor of the same inductance with no initial energy. This equivalence is shown in Fig. 6.23(b).

EXAMPLE 9. For the circuit in Fig. 6.24(a), the capacitor has an initial voltage of E_0 at $t = 0^-$. At $t = 0$, the switch is closed.

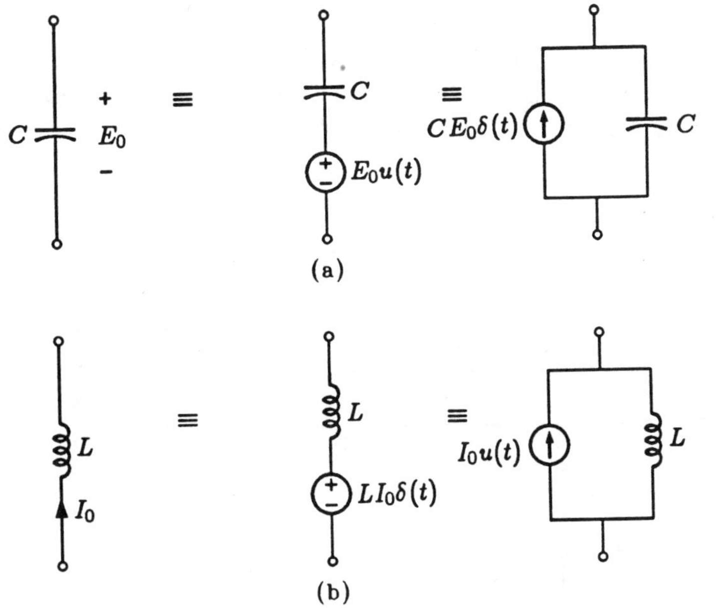

SIMPLE CIRCUITS WITH MEMORY ELEMENTS

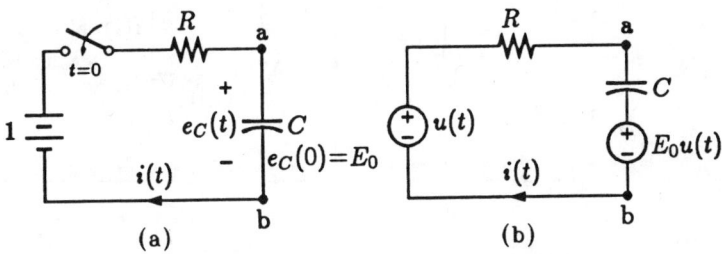

Figure 6.23: Thévenin's and Norton's equivalent of elements with initial energies.

Determine $i(t)$ and $e_C(t)$ for $t > 0$.

Figure 6.24: Circuits for Example 9.

SOLUTION For $t > 0$, the circuit is equivalent to that shown in Fig. 6.24(b). From Eq. (6.45),

$$i(t) = \frac{1 - E_0}{R} \epsilon^{-\frac{1}{RC}t} \qquad (6.59)$$

$$e_C(t) = 1 - Ri(t) = 1 - (1 - E_0)\epsilon^{-\frac{1}{RC}t} \qquad (6.60)$$

A word of caution. The equivalence between the charged C in Fig. 6.24(a) and its Thévenin's equivalent relates only to their behavior at terminals a and b. Hence, if we were to try to obtain $e_C(t)$ from Fig. 6.24(b), we must obtain $e_{ab}(t)$ instead of the voltage across the capacitor there.

EXAMPLE 10. In the circuit in Fig. 6.25(a), the switch has been open for $t < 0$. At $t = 0$, it is closed. Determine $i(t)$ and $e_C(t)$ for $t > 0$.

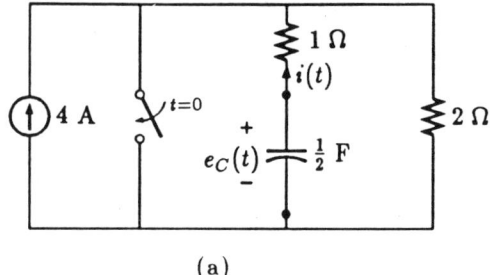

(a)

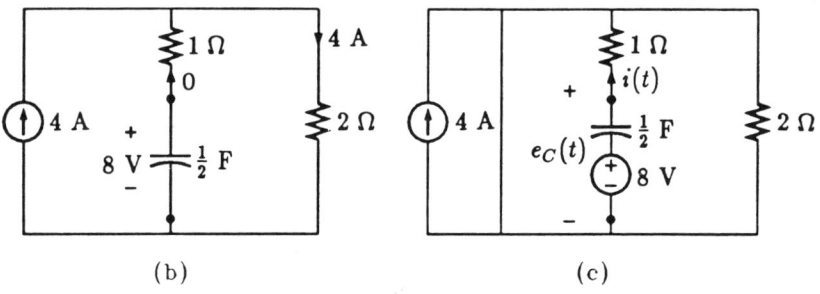

(b)　　　　　　　　　(c)

Figure 6.25: Circuits for Example 10.

SOLUTION The status of the circuit before $t = 0$ is shown in Fig. 6.25(b). The capacitor has an initial voltage of 8 volts. From Fig. 6.23(a), the circuit for $t > 0$ is equivalent to that shown in Fig. 6.25(c). The branch containing the capacitor is short-circuited by the switch and is therefore unaffected by the current source or the 2-Ω resistor. From Eq. (6.45),

$$i(t) = 8\epsilon^{-2t} \text{ A}$$

To obtain $e_C(t)$ in the original circuit, we now take the voltage across the entire series combination of the initially unenergized capacitor and the equivalent voltage source, or

$$e_C(t) = 8 - 2\int_0^t i(\lambda)\,d\lambda = 8 - 8\left(1 - \epsilon^{-2t}\right) = 8\,\epsilon^{-2t} \text{ V}$$

EXERCISE

6.3.1 It is known that $e_C(0) = 5$ V and the switch is open at $t = 0^-$. The switch is closed at $t = 0$. Find $i(t)$ for $t > 0$.

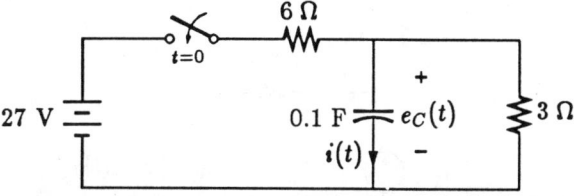

Ans. $2\epsilon^{-5t}$ A.

6.4 RL and RC parallel circuits

Suppose that we now wish to find the voltage $e(t)$ across the parallel RC combination subjected to a unit step current excitation as shown in Fig. 6.26.

Figure 6.26: An RC parallel circuit with a unit step excitation.

We first note that $e(t) = 0$ for $t < 0$. For $t > 0$, KCL requires

$$C\frac{de}{dt} + Ge = 1 \tag{6.61}$$

292 FUNDAMENTALS OF CIRCUIT ANALYSIS

where $G = 1/R$. A comparison of Fig. 6.26 and Fig. 6.1, will show that these two circuits are dual of each other. This is confirmed by the duality between Eqs. (6.2) and (6.61). If we proceed to solve Eq. (6.61), we will merely retrace the steps in solving Eq. (6.2) with every quantity replaced by its dual. Hence, the solution to Eq. (6.61) is simply the dual of Eq. (6.16) or

$$e(t) = \frac{1}{G}\left(1 - \epsilon^{-\frac{G}{C}t}\right)u(t) \tag{6.62}$$

As an alternative to analyzing the circuit in Fig. 6.26 directly, we can apply source transformation to the circuit. As far as the capacitor C is concerned, the circuit of Fig. 6.27 is equivalent to that of Fig. 6.26. What is to the left of the capacitor in Fig. 6.27 is

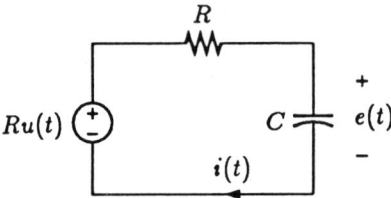

Figure 6.27: Source transformation applied to the circuit of Fig. 6.26.

the Thevenin's equivalent of what is to the left of the capacitor in Fig. 6.26. The circuit in Fig. 6.27 is identical to that in Fig. 6.14 except the step excitation is an R-unit step instead of a unity one. From Eq. (6.45), we can write

$$e(t) = R\left(1 - \epsilon^{-\frac{1}{RC}t}\right)u(t) \tag{6.63}$$

which is the same as Eq. (6.62).

Next, suppose we wish to find the voltage $e(t)$ in the circuit in Fig. 6.28. This circuit is the dual of that in Fig. 6.14. We may, therefore, write

$$g_e(t) = \frac{1}{G}\epsilon^{-\frac{1}{GL}t}u(t) \tag{6.64}$$

and

SIMPLE CIRCUITS WITH MEMORY ELEMENTS

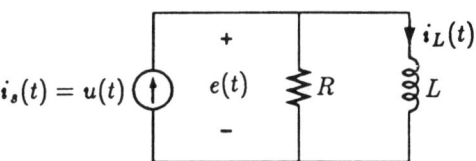

Figure 6.28: An RL parallel circuit subjected to a unit step excitation.

$$i_L(t) = \left(1 - \epsilon^{-\frac{1}{GL}t}\right) u(t) \tag{6.65}$$

which are the dual of Eqs. (6.45) and (6.46), respectively.

By a similar comparison, the impulse responses in RC and RL parallel circuits may also be obtained from those of RL and RC series circuits. The step and impulse responses of the four basic first-order circuits are summarized in Fig. 6.29. We can use these entries to review the principle of duality, the application of source transformation, the initial and final conditions, and to solve further problems that are reducible to one of these forms.

EXAMPLE 11. Find $e_2(t)$ for the circuit in Fig. 6.30(a).

SOLUTION We shall first find the Norton's equivalent for that part of the circuit to the left of terminals a and b. We have

$$i_{sc}(t) = \frac{1}{R_1} u(t) + C_1 \delta(t) \tag{6.66}$$

Hence, as far as R_2 and C_2 are concerned, the circuits in Fig. 6.30(a) and (b) are equivalent. But the circuit in Fig. 6.30(b) can be further reduced to that of Fig. 6.30(c), in which

$$R = \frac{R_1 R_2}{R_1 + R_2} \qquad C = C_1 + C_2 \tag{6.67}$$

We may make use of the additivity property of a linear network since $i_{sc}(t)$ is made up of a step function and an impulse function. The response to each of these is obtained from Fig. 6.29(b). Specifically,

$$e_2(t) = \frac{R}{R_1}\left[1 - \epsilon^{-\frac{1}{RC}t}\right] u(t) + \frac{C_1}{C}\epsilon^{-\frac{1}{RC}t} u(t) \tag{6.68}$$

294 FUNDAMENTALS OF CIRCUIT ANALYSIS

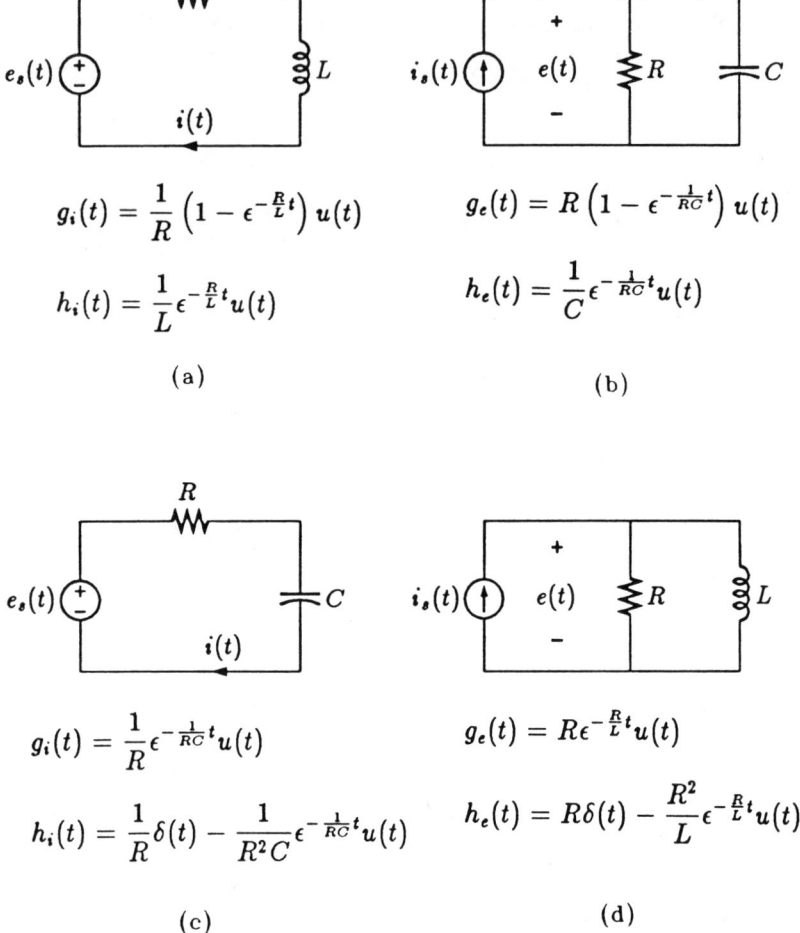

Figure 6.29: Summary of the step and impulse responses of first-order circuits. $g(t)$ is the response when the source is a unit step and $h(t)$ is the response when the source is a unit impulse.

SIMPLE CIRCUITS WITH MEMORY ELEMENTS

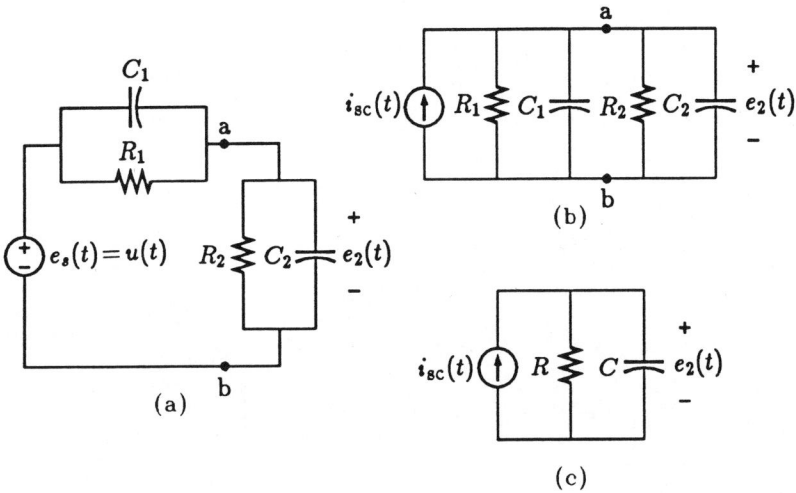

Figure 6.30: Circuits for Example 11.

EXERCISE

6.4.1 Find $e(t)$ in the circuit.

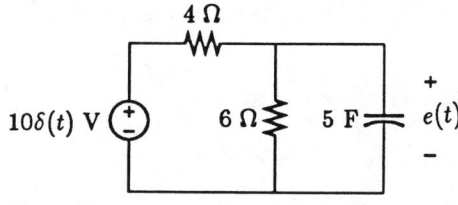

Ans. $0.5\,\epsilon^{-\frac{1}{12}t}\,u(t)$ V.

6.5 First-order op amp circuits

In this section, we will look at several op amp circuits using one memory element. These circuits are used widely in modern signal processing and electronic applications.

6.5.1 The integrator

In the arrangement in Fig. 6.31, the inverting node is virtually grounded. Hence $i_1 = e_1/R$ and $i_2 = C\dfrac{de_2}{dt}$. KCL requires that $i_1 + i_2 = 0$. Thus, we have

$$C\frac{de_2}{dt} + \frac{e_1}{R} = 0 \tag{6.69}$$

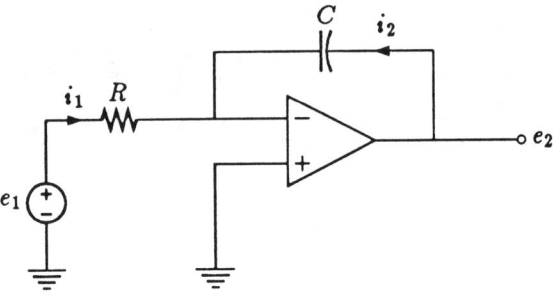

Figure 6.31: An inverting integrator.

or

$$e_2(t) = -\frac{1}{RC}\int_{-\infty}^{t} e_1(\lambda)\,d\lambda \tag{6.70}$$

The circuit in Fig. 6.31 is known as an *inverting integrator*.

6.5.2 The differentiator

In the arrangement in Fig. 6.32, we have

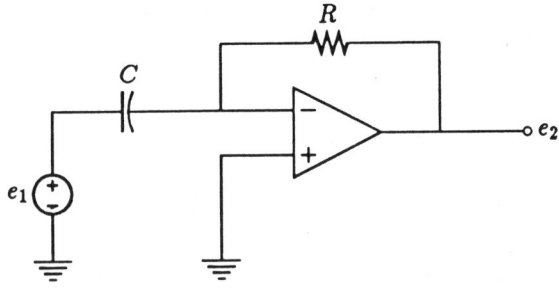

Figure 6.32: An inverting differentiator.

SIMPLE CIRCUITS WITH MEMORY ELEMENTS

$$C\frac{de_1}{dt} + \frac{1}{R}e_2 = 0 \tag{6.71}$$

or

$$e_2(t) = -RC\frac{de_1}{dt} \tag{6.72}$$

Hence, the circuit in Fig. 6.32 is an *inverting differentiator*.

6.5.3 The lossy integrator

In the arrangement in Fig. 6.33, we have

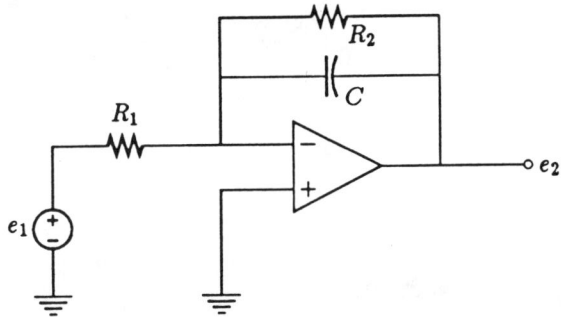

Figure 6.33: A lossy integrator.

$$\frac{e_1}{R_1} + \frac{e_2}{R_2} + C\frac{de_2}{dt} = 0 \tag{6.73}$$

Given an $e_1(t)$, we need to solve the differential equation

$$C\frac{de_2}{dt} + \frac{1}{R_2}e_2 = -\frac{1}{R_1}e_1(t) \tag{6.74}$$

to obtain $e_2(t)$.

To see the effect of R_2 on the output, let's assume $R_1 = 1\ \Omega$ and $C = 1$ F and $e_1(t)$ is as shown in Fig. 6.34(a). If $R_2 = \infty$, the circuit is an inverting (lossless) integrator in Fig. 6.32 and $e_2(t)$ will vary as shown in Fig. 6.34(b). If we let $R_2 = 20\ \Omega$, then for $0 < t < 2$,

$$\frac{de_2}{dt} + \frac{1}{20}e_2 = -1 \tag{6.75}$$

The complete solution to Eq. (6.75) is

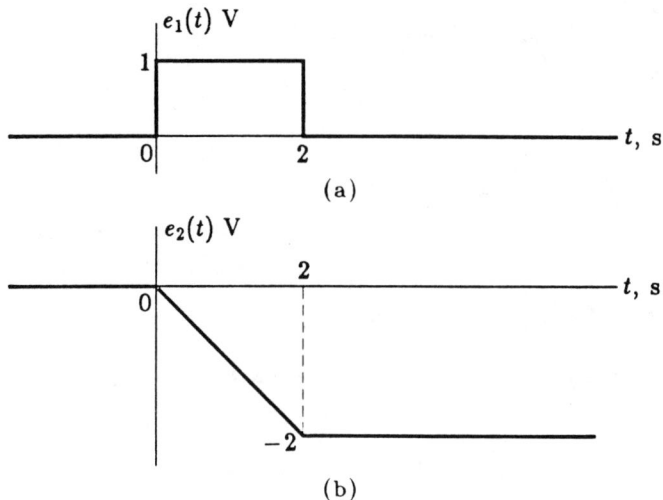

Figure 6.34: Excitation and response of an inverting integrator.

$$e_2(t) = -20 + K\epsilon^{-\frac{1}{20}t} \tag{6.76}$$

Since $e_2(0) = 0$, $K = 20$ and

$$e_2(t) = -20 + 20\epsilon^{-\frac{1}{20}t} \tag{6.77}$$

we obtain $e_2(2) = -1.9033$.

For $t > 2$, the differential equation is

$$\frac{de_2}{dt} + \frac{1}{20}e_2 = 0 \tag{6.78}$$

Hence,

$$e_2(t) = K'\epsilon^{-\frac{1}{20}t} \tag{6.79}$$

Imposing $e_2(2) = -1.9033$, we get

$$e_2(t) = -14.0636\,\epsilon^{-\frac{1}{20}t} = -1.9033\,\epsilon^{-\frac{1}{20}(t-2)} \tag{6.80}$$

The variation of $e_2(t)$ for $R_2 = 20$ Ω is shown in Fig. 6.35. It is seen that, because of the fact that R_2 is finite, the integration is less than perfect. Resistance R_2 represents the leakage of the capacitor C. The larger the value of R_2, the closer the integrator approaches an ideal one.

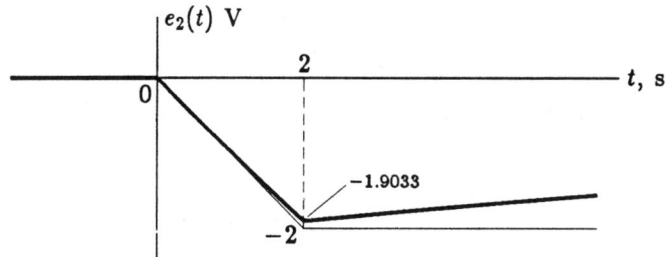

Figure 6.35: Response of a lossy integrator.

6.6 Responses of *RLC* circuits

In this section, we shall deal with circuits containing two memory elements.

6.6.1 Step response of an *RLC* series circuit

We have the circuit of Fig. 6.36. Because the circuit has been idle since $t = -\infty$, we have $i(t) = 0$ for $t < 0$. For $t > 0$, KVL requires

$$L\frac{di}{dt} + Ri + \frac{1}{C}\int_0^t i(\lambda)\, d\lambda = 1 \qquad (6.81)$$

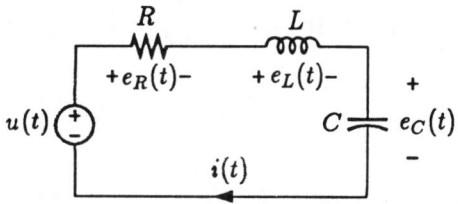

Figure 6.36: An *RLC* series circuit with unit step excitation.

Differentiating gives

$$L\frac{d^2i}{dt^2} + R\frac{di}{dt} + \frac{1}{C}i = 0 \qquad (6.82)$$

The forced response is zero. Hence, the complete solution is made up entirely of the natural response. We let $i_c(t) = K\epsilon^{st}$. Here, s must satisfy the characteristic equation

FUNDAMENTALS OF CIRCUIT ANALYSIS

$$Ls^2 + Rs + \frac{1}{C} = 0 \tag{6.83}$$

which has two roots

$$s_1, s_2 = -\frac{R}{2L} \pm \sqrt{\frac{R^2}{4L^2} - \frac{1}{LC}} \tag{6.84}$$

Depending on the values of the elements, the radical in Eq. (6.84) may be real, imaginary, or zero. We shall investigate these three cases separately.

Case 1. $\frac{R^2}{4L^2} > \frac{1}{LC}$. Under this condition, s_1 and s_2 are both real and negative. Let

$$\alpha = \frac{R}{2L} \quad \lambda = \sqrt{\frac{R^2}{4L^2} - \frac{1}{LC}} \tag{6.85}$$

so

$$s_1, s_2 = -\alpha \pm \lambda \tag{6.86}$$

and

$$i(t) = K_1 \epsilon^{(-\alpha+\lambda)t} + K_2 \epsilon^{(-\alpha-\lambda)t} \tag{6.87}$$

The initial conditions are

$$i(0^+) = 0 \quad e_C(0^+) = 0 \quad e_R(0^+) = 0$$

$$e_L(0^+) = 1 = L\frac{di}{dt}\bigg|_{t=0^+} \implies \frac{di}{dt}\bigg|_{t=0^+} = \frac{1}{L} \tag{6.88}$$

We have

$$K_1 + K_2 = 0 \tag{6.89}$$

$$(-\alpha + \lambda)K_1 + (-\alpha - \lambda)K_2 = \frac{1}{L} \tag{6.90}$$

Solving yields

$$K_1 = \frac{1}{2\lambda L} \quad K_2 = -\frac{1}{2\lambda L} \tag{6.91}$$

Thus,

SIMPLE CIRCUITS WITH MEMORY ELEMENTS

$$i(t) = \frac{1}{2\lambda L} \left[\epsilon^{(-\alpha+\lambda)t} - \epsilon^{(-\alpha-\lambda)t} \right] \qquad (6.92)$$

Using the hyperbolic-function notation,

$$\cosh x = \frac{\epsilon^x + \epsilon^{-x}}{2} \qquad \sinh x = \frac{\epsilon^x - \epsilon^{-x}}{2} \qquad (6.93)$$

we may write Eq. (6.92) as

$$i(t) = \frac{1}{\lambda L} \epsilon^{-\alpha t} \sinh \lambda t \qquad (6.94)$$

Case 1 is known as the *overdamped case* or the *nonoscillatory case*. The step response may be written as

$$g_i(t) = \frac{1}{\lambda L} \epsilon^{-\alpha t} \sinh \lambda t \, u(t) \qquad (6.95)$$

for all t.

Case 2. $\dfrac{R^2}{4L^2} < \dfrac{1}{LC}$. Under this condition, we can let

$$\lambda = \sqrt{\frac{R^2}{4L^2} - \frac{1}{LC}} = j\beta = j\sqrt{\frac{1}{LC} - \frac{R^2}{4L^2}} \qquad (6.96)$$

and

$$s_1, s_2 = -\alpha \pm j\beta \qquad (6.97)$$

Eq. (6.92) becomes

$$i(t) = \frac{1}{2j\beta L} \epsilon^{-\alpha t} \left(\epsilon^{j\beta t} - \epsilon^{-j\beta t} \right) \qquad (6.98)$$

In this case it is more appropriate to make use of Euler's formulas

$$\cos x = \frac{\epsilon^{jx} + \epsilon^{-jx}}{2} \qquad \sin x = \frac{\epsilon^{jx} - \epsilon^{-jx}}{2j} \qquad (6.99)$$

The step response may be written as

$$g_i(t) = \frac{1}{\beta L} \epsilon^{-\alpha t} \sin \beta t \, u(t) \qquad (6.100)$$

for all t. This case is known as the *underdamped case* or the *oscillatory case*.

Case 3. $\dfrac{R^2}{4L^2} = \dfrac{1}{LC}$. Under this condition, the characteristic equation has two repeated roots. They are

$$s_1 = s_2 = -\alpha \tag{6.101}$$

The solution has the form

$$i(t) = (K_1 t + K_2)\epsilon^{-\alpha t} \tag{6.102}$$

Imposing the same initial conditions as we did in the two previous cases, we have

$$i(0^+) = K_2 = 0 \tag{6.103}$$

$$\left.\dfrac{di}{dt}\right|_{t=0^+} = \left[-\alpha(K_1 t + K_2)\epsilon^{-\alpha t} + K_1 \epsilon^{-\alpha t}\right]_{t=0^+}$$

$$= -\alpha K_2 + K_1 = \dfrac{1}{L} \tag{6.104}$$

which leads to $K_1 = 1/L$. The step response is

$$g_i(t) = \dfrac{t}{L}\epsilon^{-\alpha t} u(t) \tag{6.105}$$

for all t. This case is known as the *critically damped case* as it is the borderline situation between the underdamped and the overdamped cases.

Fig. 6.37 shows the step responses for fixed values of L and C for a variety of values of R. Response in (a) is for $R = 0$ and the response is an undamped sinusoid. As the value of R is increased, α increases, which makes the sinusoid damp faster because of the factor $\frac{1}{\beta L}\epsilon^{-\alpha t}$. At the same time, β decreases, making the crossover points of the oscillatory response to occur later along the t axis. These are shown in (b) and (c).

In part (d), the value of R is such that $\beta = 0$. The step response in critically damped. As we further increase R, the response

SIMPLE CIRCUITS WITH MEMORY ELEMENTS

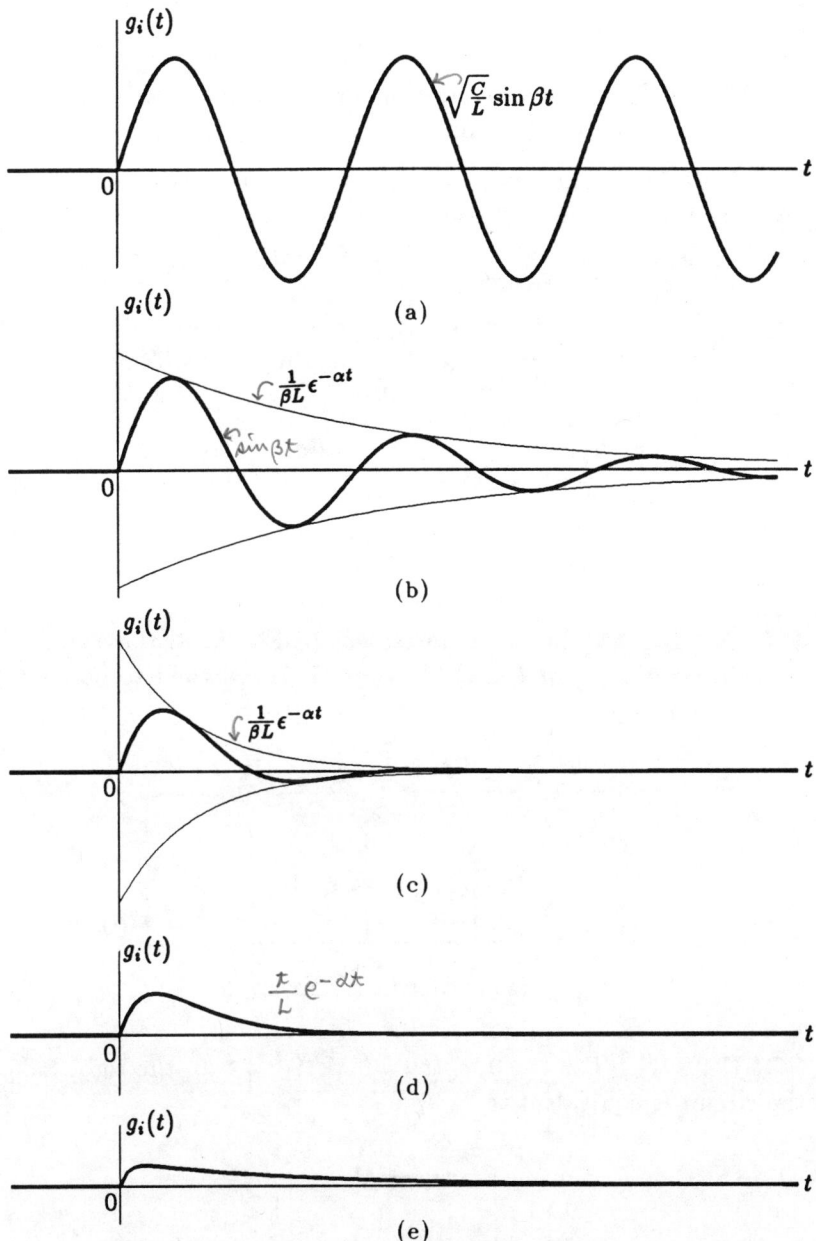

Figure 6.37: Step responses of RLC series circuit as R is varied.

becomes overdamped and the current assumes the general nonoscillatory variation shown in part (e).

EXAMPLE 12. Obtain the response of an RLC series circuit to a unit ramp function. Assume the circuit to be oscillatory.

SOLUTION The ramp response is the integral of the step response. From Eq. (6.100), the ramp response is

$$g_i^{(-1)}(t) = \int_{-\infty}^{t} g_i(\lambda)\, d\lambda = u(t) \int_0^t \frac{1}{\beta L} \epsilon^{-\alpha\lambda} \sin\beta\lambda\, d\lambda$$

$$= -\frac{1}{\beta L(\alpha^2 + \beta^2)} \epsilon^{-\alpha\lambda} \left[\beta\cos\beta\lambda + \alpha\sin\beta\lambda\right]_0^t u(t)$$

$$= C\left[1 - \epsilon^{-\alpha t}\left(\cos\beta t + \frac{\alpha}{\beta}\sin\beta t\right)\right] u(t) \qquad (6.106)$$

since $\alpha^2 + \beta^2 = \frac{1}{LC}$.

EXAMPLE 13. In the arrangement in Fig. 6.38, the switch has been in position 1 for $t < 0$. At $t = 0$ it is switched to position 2. Obtain $i(t)$ for $t > 0$.

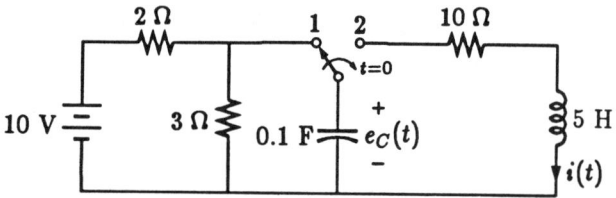

Figure 6.38: Circuit for Example 13.

SOLUTION For $t < 0$, $e_C(t) = 6$ V. For $t > 0$, the right half of the circuit is equivalent to

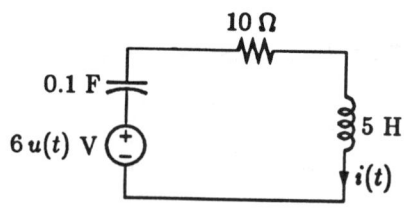

We have

SIMPLE CIRCUITS WITH MEMORY ELEMENTS 305

$$\frac{R^2}{4L^2} = 1 \quad \text{and} \quad \frac{1}{LC} = 2$$

Hence, we have the oscillatory case and

$$\alpha = 1 \quad \beta = 1$$

From Eq. (6.100),

$$i(t) = \frac{6}{5} \epsilon^{-t} \sin t \text{ A}$$

6.6.2 Impulse response of RLC series circuit

To obtain the current response of an RLC series circuit when the excitation is a unit impulse voltage, we can differentiate the step response.

Case 1. Overdamped case. We differentiate Eq. (6.95). The impulse response is therefore

$$h_i(t) = \frac{d}{dt} g_i(t) = \frac{1}{\lambda L} \left[-\alpha \epsilon^{-\alpha t} \sinh \lambda t \, u(t) \right.$$

$$\left. + \lambda \epsilon^{-\alpha t} \cosh \lambda t \, u(t) + \epsilon^{-\alpha t} \sinh \lambda t \, \delta(t) \right]$$

$$= \frac{1}{L} \epsilon^{-\alpha t} \left[\cosh \lambda t - \frac{\alpha}{\lambda} \sinh \lambda t \right] u(t) \quad (6.107)$$

Case 2. Underdamped case. We differentiate Eq. (6.100). The impulse response is

$$h_i(t) = \frac{d}{dt} g_i(t) = \frac{1}{\beta L} \left[-\alpha \epsilon^{-\alpha t} \sin \beta t \, u(t) \right.$$

$$\left. + \beta \epsilon^{-\alpha t} \cos \beta t \, u(t) + \epsilon^{-\alpha t} \sin \beta t \, \delta(t) \right]$$

$$= \frac{1}{L} \epsilon^{-\alpha t} \left[\cos \beta t - \frac{\alpha}{\beta} \sin \beta t \right] u(t) \quad (6.108)$$

Case 3. Critically damped case. We differentiate Eq. (6.105). The impulse response is

$$h_i(t) = \frac{d}{dt} g_i(t) = \frac{1}{L} \left[\epsilon^{-\alpha t} u(t) - \alpha t \epsilon^{-\alpha t} u(t) + t \epsilon^{-\alpha t} \delta(t) \right]$$

$$= \frac{1 - \alpha t}{L} \epsilon^{-\alpha t} u(t) \quad (6.109)$$

6.6.3 Responses of an RLC parallel circuit

The circuit in Fig. 6.39 is the dual of that in Fig. 6.36. Hence, we need not repeat the derivation of responses of this circuit to various singularity-function excitations. We can simply replace every result we have obtained for the RLC series circuit with its dual and obtain the result for the circuit of Fig. 6.39. Here

$$\alpha = \frac{G}{2C} \quad \lambda = \sqrt{\frac{G^2}{4C^2} - \frac{1}{CL}} \quad \beta = \sqrt{\frac{1}{CL} - \frac{G^2}{4C^2}} \quad (6.110)$$

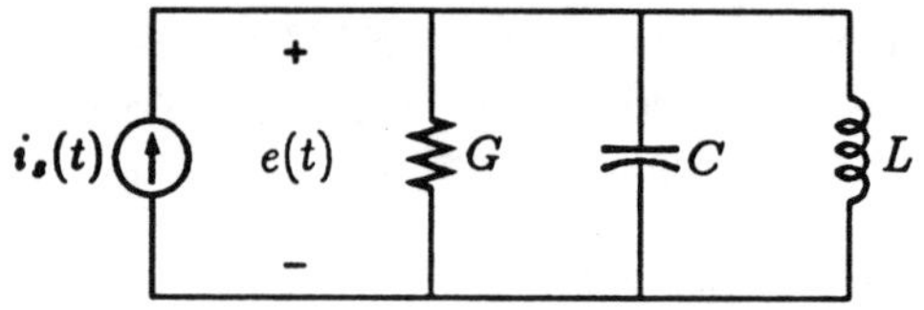

Figure 6.39: An RLC parallel circuit.

For instance, if $i_s(t) = u(t)$, then the response $e(t)$ is simply the dual of Eq. (6.95), or

$$g_e(t) = \frac{1}{\lambda C} \epsilon^{-\alpha t} \sinh \lambda t \, u(t) \qquad (6.111)$$

for the nonoscillatory case.

EXAMPLE 14. Find $e(t)$ for the circuit in Fig. 6.40(a).

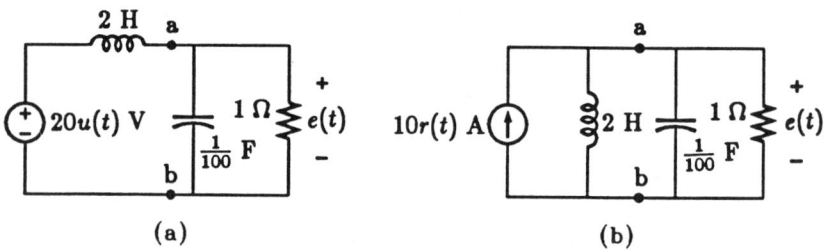

Figure 6.40: Circuits used in Example 14.

SOLUTION We first perform a source transformation for that part of the circuit to the left of terminals a and b. The result is the circuit in Fig. 6.40(b). We have

SIMPLE CIRCUITS WITH MEMORY ELEMENTS

$$\alpha = \frac{G}{2C} = 50 \qquad \lambda = \sqrt{\frac{G^2}{4C^2} - \frac{1}{CL}} = 35\sqrt{2}$$

Taking the dual of the nonoscillatory version of Eq. (6.106), we have

$$e(t) = 20\left[1 - \epsilon^{-50t}\left(\cosh 35\sqrt{2}t + \frac{10}{7\sqrt{2}}\sinh 35\sqrt{2}t\right)\right]u(t) \text{ V}$$

6.6.4 Responses to other excitations

In an RLC series or parallel circuit, if the excitation is other than singularity functions, the forced response needs to be found first. The remainder of the steps of the solution are similar to those taken for singularity-function excitations. To find the forced response due to excitations listed in Table 6.1, we can use the particular functions listed there.

EXAMPLE 15. Find $i(t)$ for $t > 0$ in Fig. 6.41 with $e_s(t) = \cos 8t\, u(t)$ V.

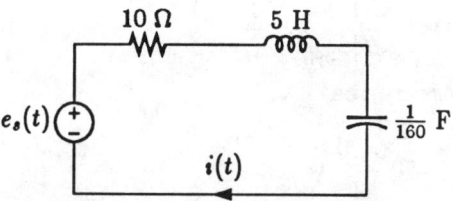

Figure 6.41: Circuit for Example 15.

SOLUTION Here we have

$$\alpha = \frac{R}{2L} = 1 \qquad \beta = \sqrt{\frac{1}{LC} - \frac{R^2}{4L^2}} = \sqrt{31}$$

The differential equation for $t > 0$ is

$$5\frac{d^2 i}{dt^2} + 10\frac{di}{dt} + 160i = -8\sin 8t$$

From Table 6.1, we can assume the forced response to be

$$i_p(t) = A\cos 8t + B\sin 8t$$

Substitution of this $i_p(t)$ into the differential equation gives

$$-320A \cos 8t - 320B \sin 8t - 80A \sin 8t$$
$$+80B \cos 8t + 160A \cos 8t + 160B \sin 8t = -8 \sin 8t$$

Equating coefficients of like terms yields

$$-160A + 80B = 0 \qquad -80A - 160B = -8$$

Solving gives

$$A = \frac{1}{50} \qquad B = \frac{1}{25}$$

or

$$i_p(t) = \frac{1}{50} \cos 8t + \frac{1}{25} \sin 8t$$

Incorporating the natural response, we can write the complete response as

$$i(t) = K_1 \epsilon^{(-1+j\sqrt{31})t} + K_2 \epsilon^{(-1-j\sqrt{31})t} + \frac{1}{50} \cos 8t + \frac{1}{25} \sin 8t$$

Since $e_s(0^+) = 1$, the initial conditions are identical to those for the step response, namely,

$$i(0^+) = 0 \qquad \left.\frac{di}{dt}\right|_{t=0^+} = \frac{1}{L} = \frac{1}{5}$$

Thus, it is necessary that

$$K_1 + K_2 + \frac{1}{50} = 0$$

$$(-1 + j\sqrt{31})K_1 + (-1 - j\sqrt{31})K_2 + \frac{8}{25} = \frac{1}{5}$$

Solving, we get

$$K_1 = -\frac{1}{100} + j\frac{7}{100\sqrt{31}} \qquad K_2 = -\frac{1}{100} - j\frac{7}{100\sqrt{31}}$$

The complete solution is

$$i(t) = \left(-\frac{1}{100} + j\frac{7}{100\sqrt{31}}\right) \epsilon^{(-1+j\sqrt{31})t}$$
$$+ \left(-\frac{1}{100} - j\frac{7}{100\sqrt{31}}\right) \epsilon^{(-1-j\sqrt{31})t} + \frac{1}{50} \cos 8t + \frac{1}{25} \sin 8t$$
$$= -\frac{1}{50} \epsilon^{-t} \cos \sqrt{31}t - \frac{7}{50\sqrt{31}} \epsilon^{-t} \sin \sqrt{31}t$$
$$+ \frac{1}{50} \cos 8t + \frac{1}{25} \sin 8t \text{ A}$$

SIMPLE CIRCUITS WITH MEMORY ELEMENTS

6.7 Other second-order circuits

So far, we have treated only the *RLC* series and parallel circuits. The solution of these circuits—or circuits that are reducible to these forms—involves only one differential equation because they are single-loop and single node-pair circuits. Although these circuits are very important second-order circuits, there are numerous other configurations that also result in second-order differential equations. When a second-order circuit has more complex topology, the solution becomes more laborious and less straightforward. We shall illustrate the solution with two examples.

EXAMPLE 16. Obtain $i_1(t)$ and $i_2(t)$ in Fig. 6.42 for $t > 0$.

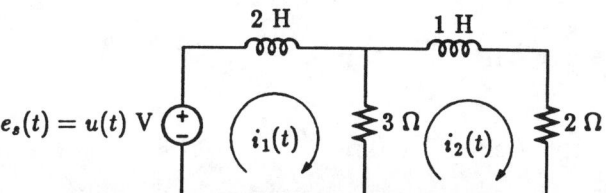

Figure 6.42: Circuit of Example 16.

SOLUTION The mesh equations are

$$2\frac{di_1}{dt} + 3(i_1 - i_2) = 1 \tag{6.112}$$

$$3(i_2 - i_1) + \frac{di_2}{dt} + 2i_2 = 0 \tag{6.113}$$

Since the excitation is a constant, the forced responses are relatively easy to obtain. We can see that $i_1(\infty) = \frac{1}{3} + \frac{1}{2} = \frac{5}{6}$ A and $i_2(\infty) = \frac{1}{2}$ A.

To obtain the homogeneous solution, we shall assume that

$$i_{1c}(t) = K_1 \epsilon^{st} \quad \text{and} \quad i_{2c}(t) = K_2 \epsilon^{st} \tag{6.114}$$

Substitution of Eqs. (6.114) into Eqs. (6.112) and (6.113) with the forcing function removed gives

$$(2s + 3)K_1 - 3K_2 = 0 \tag{6.115}$$

$$-3K_1 + (s+5)K_2 = 0 \tag{6.116}$$

Since K_1 and K_2 are to be nonzero, Eqs. (6.115) and (6.116) require that the characteristic equation

$$\begin{vmatrix} 2s+3 & -3 \\ -3 & s+5 \end{vmatrix} = 2s^2 + 13s + 6 = 0 \tag{6.117}$$

be satisfied. Solving Eq. (6.117), we obtain

$$s_1 = -0.5 \qquad s_2 = -6 \tag{6.118}$$

Since the characteristic equation has two roots, we need two arbitrary constants for each current. Hence, the complete solutions for the two currents are

$$i_1(t) = \frac{5}{6} + K_{11}\epsilon^{-0.5t} + K_{12}\epsilon^{-6t} \tag{6.119}$$

$$i_2(t) = \frac{1}{2} + K_{21}\epsilon^{-0.5t} + K_{22}\epsilon^{-6t} \tag{6.120}$$

At $t = 0^+$, we have

$$i_1 = 0 \qquad i_2 = 0 \qquad 2\frac{di_1}{dt} = 1 \qquad \frac{di_2}{dt} = 0$$

Imposing these conditions on $i_1(t)$ and $i_2(t)$ of Eqs. (6.119) and (6.120), we obtain

$$K_{11} = -\frac{9}{11} \quad K_{12} = -\frac{1}{66} \quad K_{21} = -\frac{6}{11} \quad K_{22} = \frac{1}{22}$$

Hence

$$i_1(t) = \frac{5}{6} - \frac{9}{11}\epsilon^{-0.5t} - \frac{1}{66}\epsilon^{-6t} \text{ A}$$

$$i_2(t) = \frac{1}{2} - \frac{6}{11}\epsilon^{-0.5t} + \frac{1}{22}\epsilon^{-6t} \text{ A}$$

EXAMPLE 17. Determine $i_1(t)$ and $i_2(t)$ in the circuit in Fig. 6.43 for $t > 0$.

SIMPLE CIRCUITS WITH MEMORY ELEMENTS

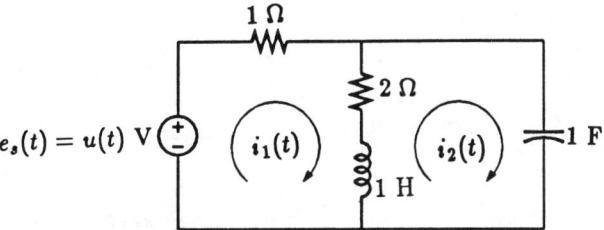

Figure 6.43: Circuit used in Example 17.

SOLUTION The mesh equations are

$$i_1 + 2(i_1 - i_2) + \frac{d}{dt}(i_1 - i_2) = 1 \tag{6.121}$$

$$\frac{d}{dt}(i_2 - i_1) + 2(i_2 - i_1) + \int_0^t i_2(\lambda)\,d\lambda = 0 \tag{6.122}$$

It is easy to see that the forced responses are

$$i_{1p}(t) = \frac{1}{3} \qquad i_{2p}(t) = 0 \tag{6.123}$$

To obtain the homogeneous solution, we assume

$$i_{1c}(t) = K_1 \epsilon^{st} \qquad i_{2c}(t) = K_2 \epsilon^{st} \tag{6.124}$$

Substitution of Eq. (6.124) into Eqs. (6.121) and (6.122) with the right-hand side set to zero gives

$$(s+3)K_1 - (s+2)K_2 = 0 \tag{6.125}$$

$$-(s+2)K_1 + \left(s + 2 + \frac{1}{s}\right)K_2 = 0 \tag{6.126}$$

By the same reasoning we used in Example 16, we find the characteristic equation to be

$$s^2 + 3s + 3 = 0 \tag{6.127}$$

Solving, we get

$$s_1, s_2 = -\frac{3}{2} \pm j\frac{\sqrt{3}}{2} \tag{6.128}$$

Since these roots are complex, we shall write the homogeneous solution in trigonometric form. The complete solutions are

$$i_1(t) = \frac{1}{3} + A_1 \epsilon^{-\frac{3}{2}t} \cos \frac{\sqrt{3}}{2}t + A_2 \epsilon^{-\frac{3}{2}t} \sin \frac{\sqrt{3}}{2}t \qquad (6.129)$$

$$i_2(t) = A_3 \epsilon^{-\frac{3}{2}t} \cos \frac{\sqrt{3}}{2}t + A_4 \epsilon^{-\frac{3}{2}t} \sin \frac{\sqrt{3}}{2}t \qquad (6.130)$$

For the initial conditions, we first see that voltage across the capacitor must be zero at $t = 0^+$ because there is no current impulse in the circuit. Thus, we can say that

$$i_1(0^+) = 1 \qquad i_2(0^+) = 1 \qquad (6.131)$$

The initial voltage across the inductance is also zero. Hence,

$$\left. \frac{d}{dt}(i_1 - i_2) \right|_{t=0^+} = 0 \qquad (6.132)$$

For the fourth condition, we write the KVL equation around the outside mesh to get

$$i_1(t) + \int_0^t i_2(\lambda) \, d\lambda = 1 \qquad (6.133)$$

Differentiating, we get

$$\frac{di_1}{dt} + i_2 = 0 \qquad (6.134)$$

Hence,

$$\left. \frac{di_1}{dt} \right|_{t=0^+} = -i_2(0^+) = -1 \qquad (6.135)$$

Substituting Eq. (6.135) into Eq. (6.132), we obtain

$$\left. \frac{di_2}{dt} \right|_{t=0^+} = -1 \qquad (6.136)$$

Now we shall impose these initial conditions on $i_1(t)$ and $i_2(t)$ of Eqs. (6.129) and (6.130). Eq. (6.131) calls for

$$\frac{1}{3} + A_1 = 1 \quad \implies \quad A_1 = \frac{2}{3}$$

and

$$A_3 = 1$$

SIMPLE CIRCUITS WITH MEMORY ELEMENTS 313

Eq. (6.135) leads to

$$-\frac{3}{2}A_1 + \frac{\sqrt{3}}{2}A_2 = -1 \quad \Longrightarrow \quad A_2 = 0$$

Eq. (6.136) leads to

$$-\frac{3}{2}A_3 + \frac{\sqrt{3}}{2}A_4 = -1 \quad \Longrightarrow \quad A_4 = \frac{1}{\sqrt{3}}$$

The final answers are

$$i_1(t) = \frac{1}{3} + \frac{2}{3} \epsilon^{-\frac{3}{2}t} \cos \frac{\sqrt{3}}{2}t \text{ A}$$

$$i_2(t) = \epsilon^{-\frac{3}{2}t} \cos \frac{\sqrt{3}}{2}t + \frac{1}{\sqrt{3}} \epsilon^{-\frac{3}{2}t} \sin \frac{\sqrt{3}}{2}t \text{ A}$$

6.8 Concluding remarks

The few circuits treated in this chapter are not only important in their own right but also serve to introduce some basic concepts that will be used frequently in more complex circuits.

Both the simple RC and RL circuits are easy to analyze and are used in many actual electrical or electronic circuits. (Or, many circuits have component parts that can be identified as such circuits.) The role of time constant is universal to first-order circuits, or indeed, to all systems that have this type of response.

The RLC series and parallel circuits are also important because they are the next more complex—and therefore more useful—circuit than the first-order circuits. They occur in many applications, some of which will be further studied in Chapter 8.

The principle of duality has good applications in this chapter. It essentially reduces our required treatment to one half of all problems in this chapter. Conversely, it enables us to double our repertoire of treated problems by taking the dual of every solved problem.

The complete solutions of some of the simple circuits in this chapter also serve to point out that there are two distinct types of response in a memory circuit under dynamic excitation—the forced response and the natural response. In some situations, only the forced response is of interest. This will be the subject of most of

Chapters 7 and 8. In other situations, both forced and natural responses are material.

When the forced response of a circuit has a definite pattern, that response is usually referred to as the *steady-state response*. Chief examples of these are when the excitation is a constant (dc), an exponential, or a steady sinusoid (ac).

When a circuit has more than one loop or node pair, the solution of its differential equations becomes very laborious, as was illustrated in Section 6.7. Usually, the characteristic equation is not too difficult to obtain. However, the initial conditions become increasingly difficult and tricky to ascertain. If fact, we seldom attempt to perform transient analysis by differential equation solution on circuits more complicated than what we have studied in this chapter. Fortunately, there is a much more powerful method of analysis—the Laplace transform—that will supplant the need to solve any differential equations.

Problems

6.1 Determine $i(t)$ for all t for $e_s(t) = 10\,r(t) - 10\,r(t-1)$ V. Do this by (a) solving the differential equation in each time interval and (b) using known step responses. Give your answer in both analytical and graphical forms.

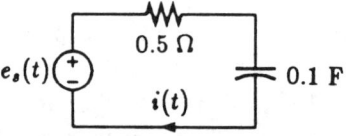

6.2 Determine $i(t)$ for all t by solving the differential equations for successive time intervals. Then verify your answer by considering $e_s(t)$ as the sum of two ramp functions.

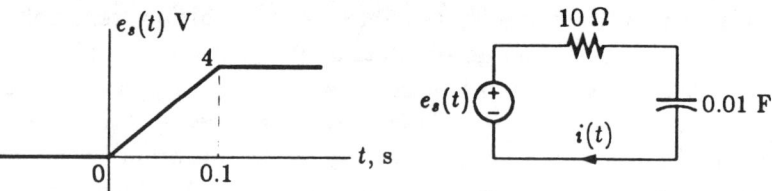

6.3 It is known that $e_C(0) = 6$ V. The switch is closed at $t = 0$.

SIMPLE CIRCUITS WITH MEMORY ELEMENTS

Find $e_C(t)$ for $t > 0$.

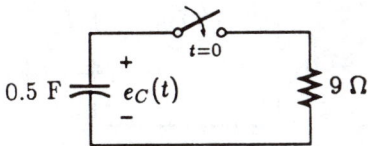

6.4 The switch has been in position B for $t < 0$. At $t = 0$, it is switched to position A for 4 seconds and then returned to position B. Find the equations for $e_C(t)$ and $i(t)$ for all t.

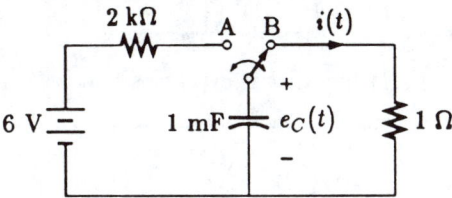

6.5 Determine $e(t)$ for all t.

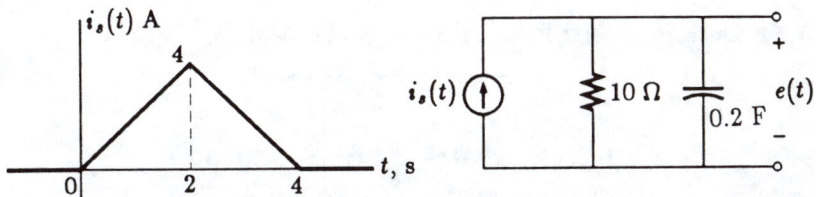

6.6 Obtain the response $i(t)$ for the excitation $e_s(t) = u_{-3}(t)$ by integrating the ramp response of the circuit.

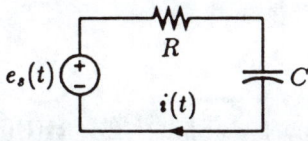

6.7 It is known that $e(0^-) = 10$ V. At $t = 0$, the switch is closed. Obtain $e(t)$ for $t > 0$.

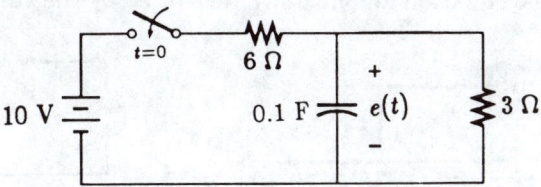

6.8 Determine $i(t)$ for all t if $e_s(t) = 5\epsilon^{-2t}u(t)$ V.

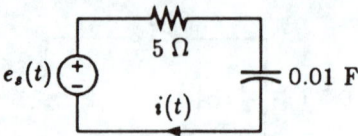

6.9 Determine $e(t)$ for all t if $i_s(t) = 10\epsilon^{-2t}u(t)$ A.

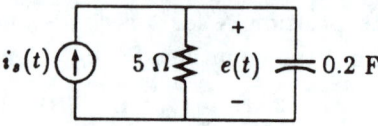

6.10 Determine $i(t)$ for all t if $e_s(t) = \sin 5t\, u(t)$ V.

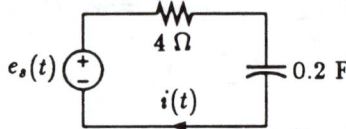

6.11 Determine $e(t)$ for all t if $i_s(t) = 2t^2\, u(t)$ A.

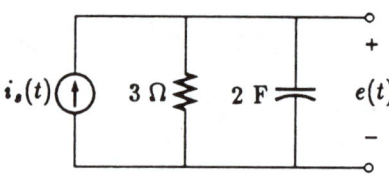

6.12 Determine $i(t)$. Give your answer in both analytical and graphical forms.

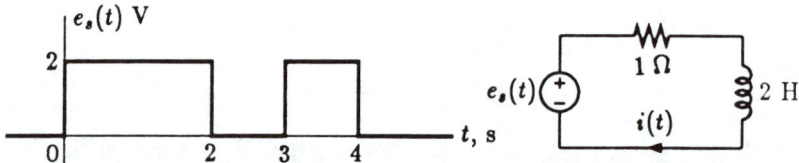

6.13 Find the equation for the power delivered by the voltage source.

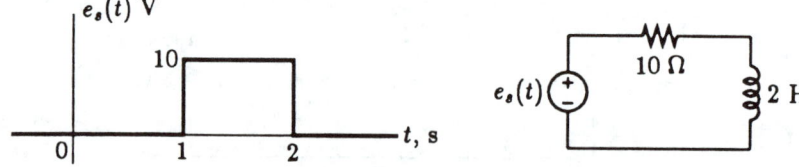

SIMPLE CIRCUITS WITH MEMORY ELEMENTS

6.14 It is known that $e_C(0) = 6$ V. Find $i(t)$ for $t > 0$.

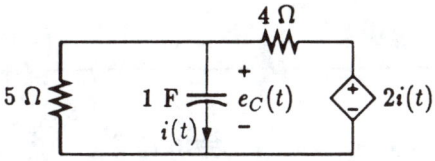

6.15 Determine $i(t)$ if $e_s(t) = (t^2 + 5)\,u(t)$ V.

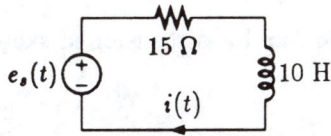

6.16 Determine $i(t)$ if $e_s(t) = 10\sin(10t + 20°)u(t)$ V.

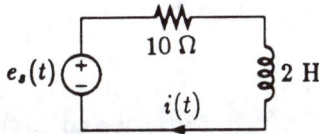

6.17 Determine $i(t)$ if $e_s(t) = (5\cos 3t + 6\sin 3t)\,u(t)$ V.

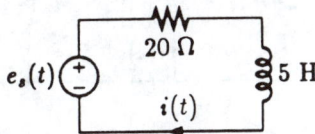

6.18 Give the equation and a dimensioned sketch of $i(t)$.

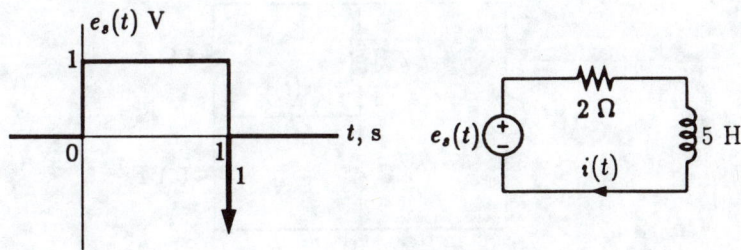

6.19 Determine the amount of charge in the capacitor at $t = 1.5$ s.

$i_s(t) = 2\,u(t)$ A and $e_s(t) = u(t)$ V.

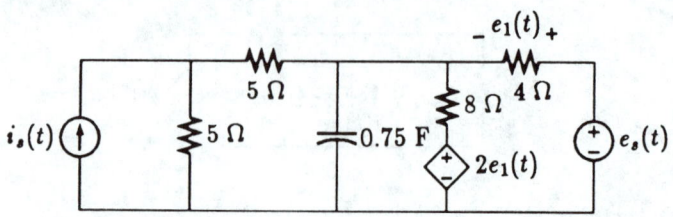

6.20 Give the equation and a dimensioned sketch of $i(t)$.

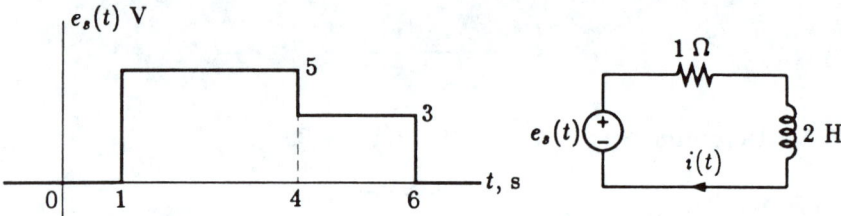

6.21 Determine $i(t)$ if $e_s(t) = 5\,u(t)$ V and $i_s(t) = 2\,u(t)$ A.

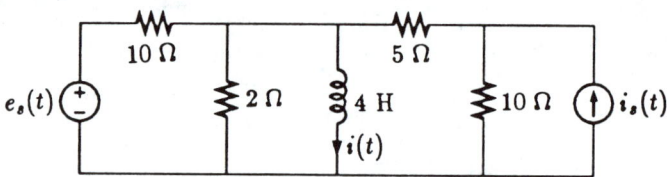

6.22 Determine $i(t)$ if $i_s(t) = u(t)$ A.

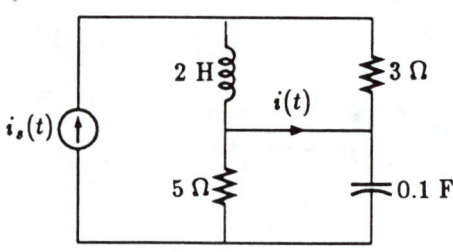

6.23 In the circuit, $e_1(t) = 10\,u(t)$ V, $i_1(t) = 3\,u(t)$ A, and $i_2(t) =$

$5\,r(t)$ A. Determine $i(t)$.

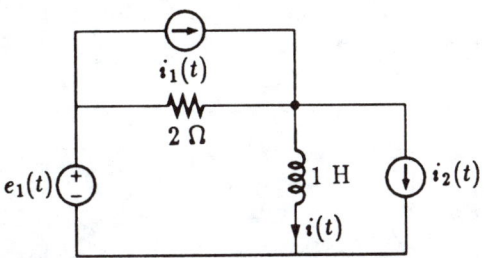

6.24 Determine $e(t)$.

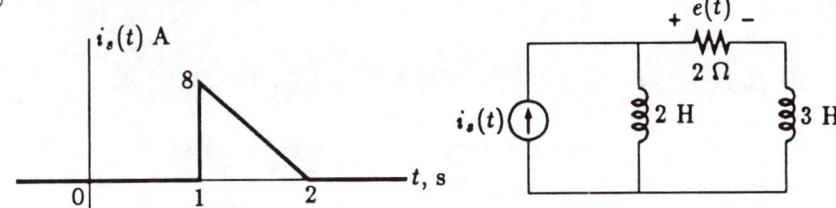

6.25 Determine $i(t)$ if $e_s(t) = 5\,u(t) - 5\,u(t-2)$ V.

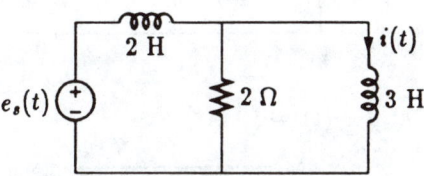

6.26 Determine $i(t)$ if $e_s(t) = 10\,u(t)$ V.

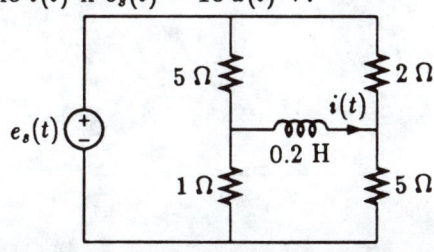

6.27 Determine $e(t)$ if $e_s(t) = u(t) - u(t-2)$ V.

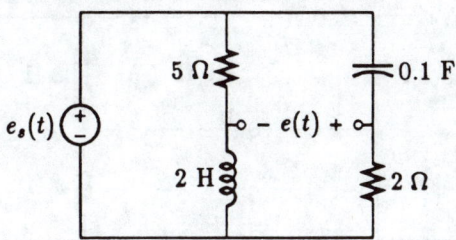

6.28 First, obtain the Norton's equivalent circuit of the network in the dashed box with respect to terminals A and B. Then obtain $i(t)$. $e_s(t) = 6\,u(t)$ V.

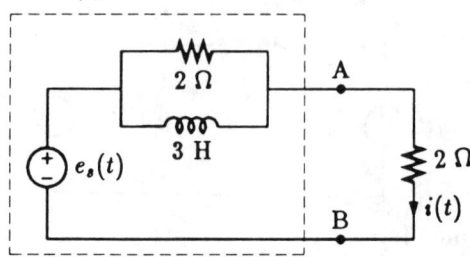

6.29 Determine $e(t)$ if $e_s(t) = 2\,\delta(t)$ V and $i_s(t) = 5\,u(t)$ A.

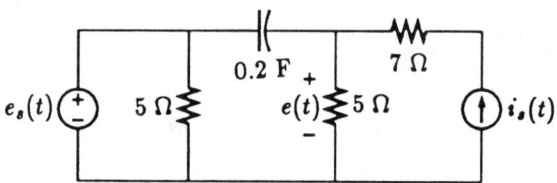

6.30 Determine $i(t)$ if $e_s(t) = 10\,u(t)$ V.

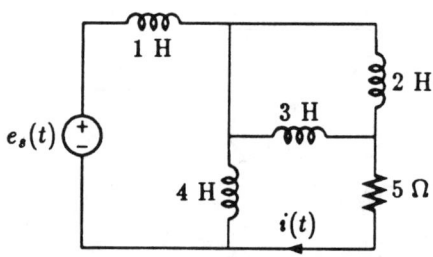

6.31 Determine $i(t)$ if $e_s(t) = 10\,u(t)$ V.

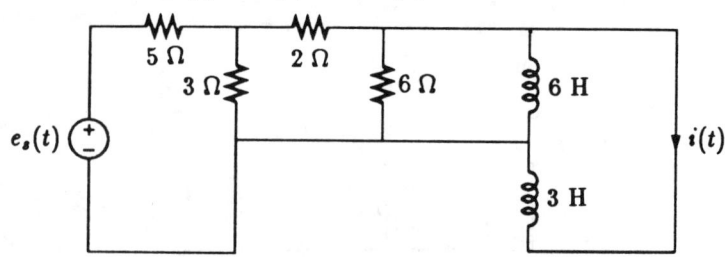

SIMPLE CIRCUITS WITH MEMORY ELEMENTS

6.32 Determine $e_{ab}(t)$ if $e_s(t) = 10\,u(t)$ V.

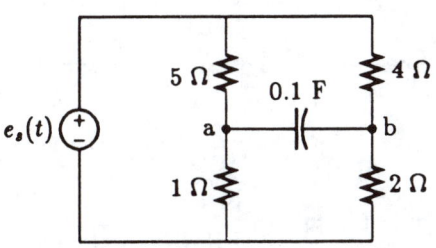

6.33 Obtain the Thévenin's equivalent circuit of the network with respect to terminals A and B. $e_s(t) = u(t)$ V.

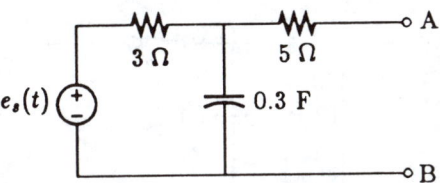

6.34 Determine $e(t)$ if $i_s(t) = u(t)$ A.

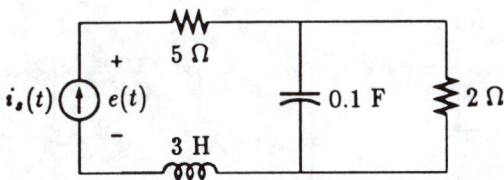

6.35 For the circuit, obtain $i_1(0^+)$, $i(0^+)$, $e_C(0^+)$, $e_L(0^+)$, and $e_R(0^+)$.

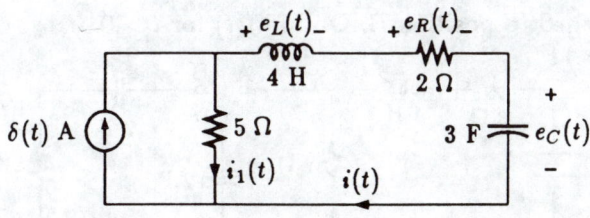

6.36 For the circuit, determine $i_1(t)$, $i_2(t)$, $i_3(t)$, $i_4(t)$, $e_1(t)$, $e_2(t)$,

$\dfrac{di_1}{dt}$, $\dfrac{di_2}{dt}$, and $\dfrac{di_4}{dt}$ at $t = 0^+$.

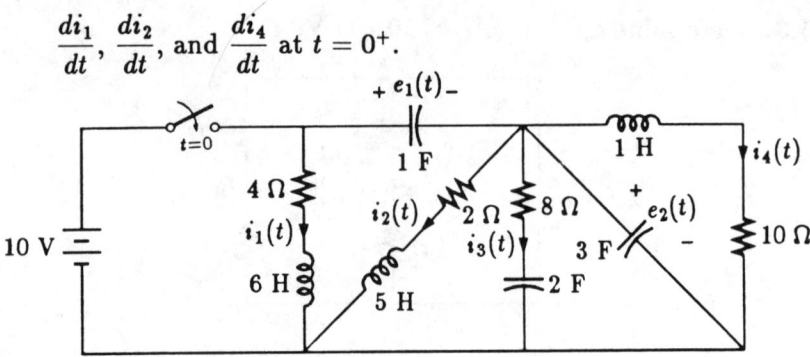

6.37 The switch has been closed for $t < 0$. At $t = 0$, it is opened. Obtain $i(t)$ for $t > 0$.

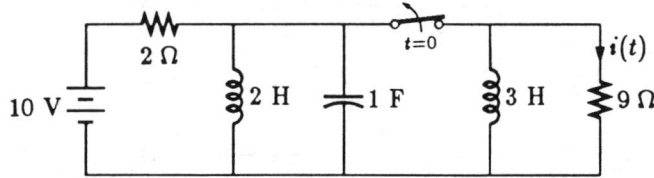

6.38 The switch has been open for $t < 0$. At $t = 0$, it is closed. Give $i(t)$ for $t > 0$.

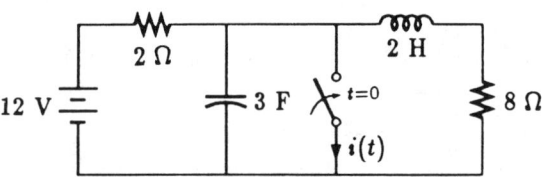

6.39 The switch has been in position A for $t < 0$. At $t = 0$, it is switched to position B. Obtain $i(t)$ for $t > 0$.

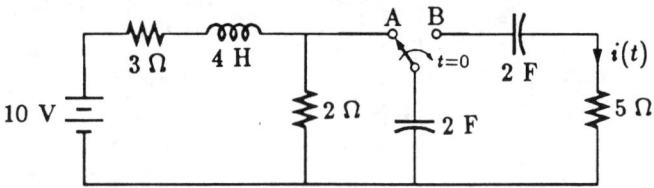

6.40 The switch has been open for $t < 0$. At $t = 0$, it is closed.

SIMPLE CIRCUITS WITH MEMORY ELEMENTS

Determine $i(t)$ for $t > 0$.

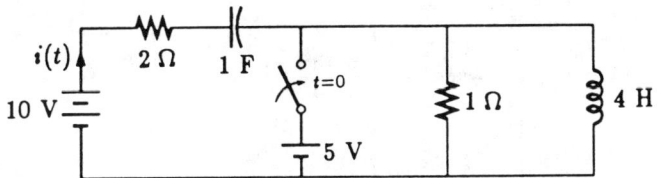

6.41 The switch has been in position A for $t < 0$. At $t = 0$, it is switched to position B. Determine $e(t)$ and $i(t)$ for $t > 0$.

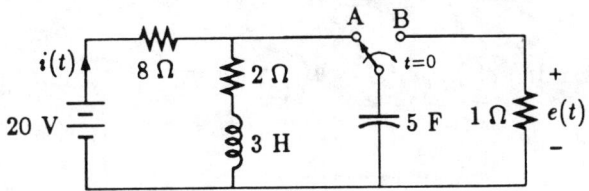

6.42 The switch has been open for $t < 1$. At $t = 1$, it is closed. Determine $i_1(t)$ and $i_2(t)$ for $t > 1$. $e_s(t) = 10u(t) - 10u(t-2)$ V.

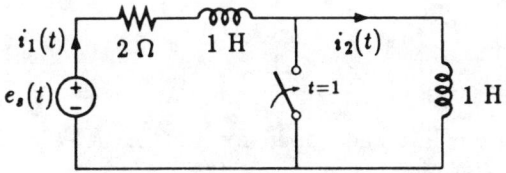

6.43 The switch has been in position A for $t < 0$. At $t = 0$, it is switched to position B. Obtain $i_1(t)$ and $i_2(t)$ for $t > 0$.

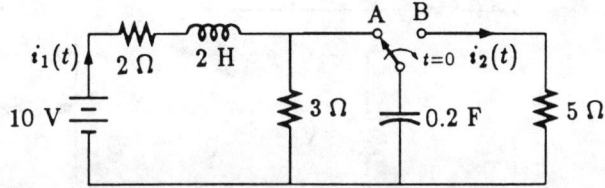

6.44 The switch has been in position A for $t < 0$. At $t = 0$, it is switched to position B. Obtain $i(t)$ for $t > 0$.

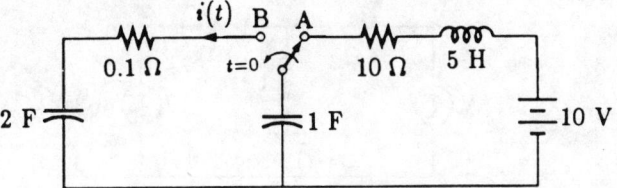

6.45 Determine $e(t)$ if $e_s(t) = 5\epsilon^{-2t}u(t)$ V.

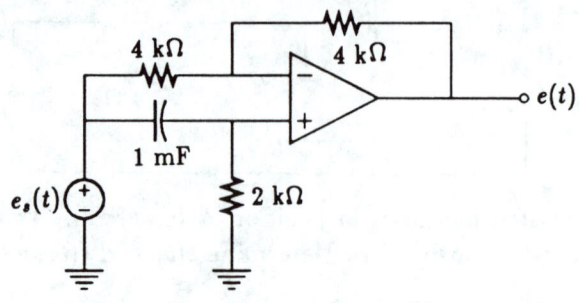

6.46 Determine $e(t)$ if $e_s(t) = u(t)$ V.

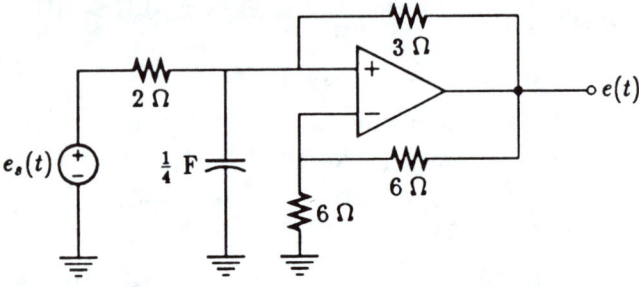

6.47 The switch has been open for $t < 0$. At $t = 0$, it is closed. Obtain $i(t)$ for all t.

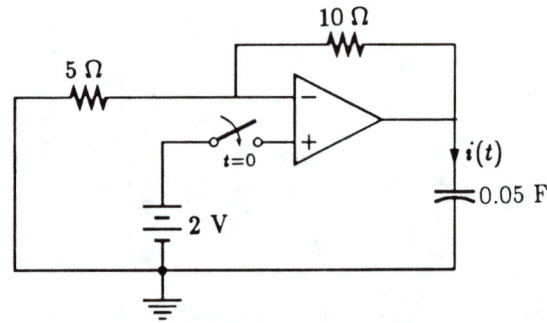

▷ **6.48** Determine $i(t)$.

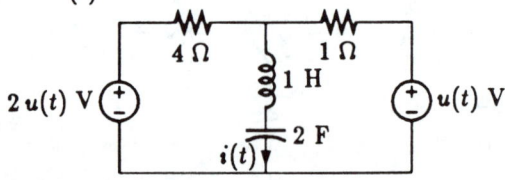

SIMPLE CIRCUITS WITH MEMORY ELEMENTS   325

6.49 Determine $e(t)$ if $e_s(t) = u(t)$ V.

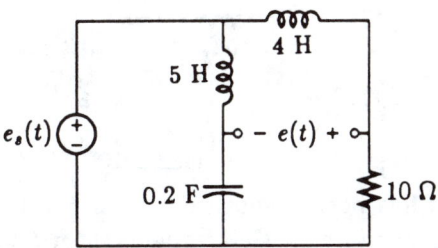

6.50 Determine $i(t)$ if $e_s(t) = 10\,u(t)$ V.

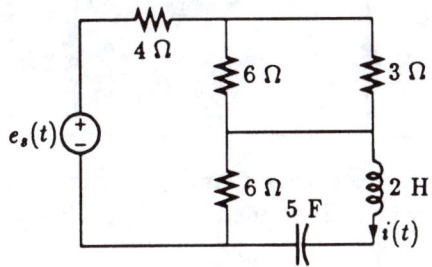

6.51 Determine $e(t)$ if $e_s(t) = 2\,\delta(t)$ V and $i_s(t) = 5\,u(t)$ A.

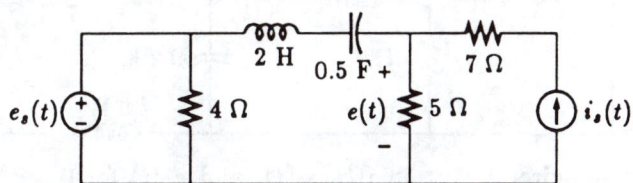

6.52 Determine $i(t)$ if $i_s(t) = 2\,u(t-2)$ A.

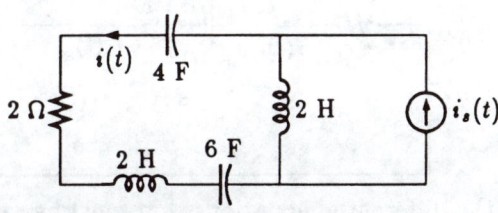

6.53 Determine $i(t)$ if $i_s(t) = 2\,u(t-1)$ A.

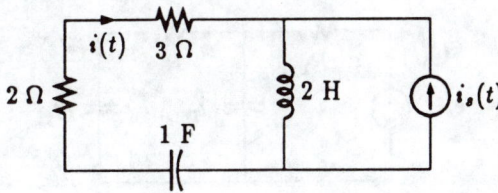

6.54 Determine $i(t)$ if $e_s(t) = 2\sin 6t\, u(t)$ V.

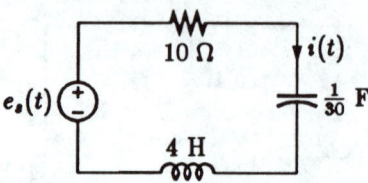

6.55 The switch has been in position A for $t < 0$. At $t = 0$, it is switched to position B. Determine $e(t)$ for $t > 0$.

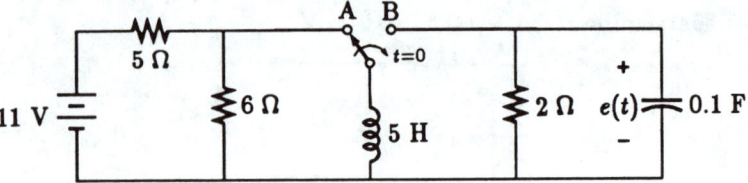

6.56 Switch S_1 has been closed and switch S_2 has been open for $t < 0$. At $t = 0$, S_1 is opened and S_2 is closed. Obtain $i(t)$ for $t > 0$.

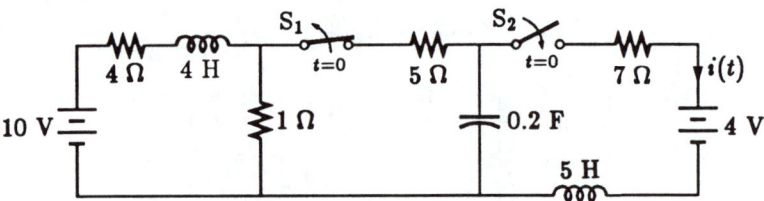

6.57 For the circuit, obtain $i(t)$, $i_2(t)$, and $e_L(t)$ for $0^- < t < 0^+$. Also, obtain $i_1(0^+)$ and $e_C(0^+)$.

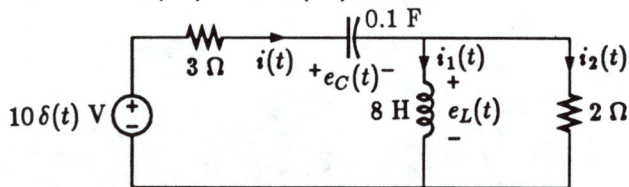

6.58 Obtain the differential equations that must be satisfied by $i_1(t)$ and $i_2(t)$ respectively.

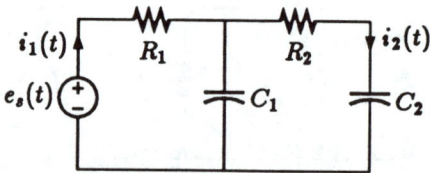

Chapter 7

Steady-State Analysis of Circuits with Exponential and Sinusoidal Excitations

In Chapters 2 through 4, we presented several properties and developed analysis techniques for circuits that include only memoryless elements. In Chapters 5 and 6, we saw the treatment of simple circuits including limited numbers of memory elements—inductors and capacitors. It is seen that the introduction of one or two energy-storing elements greatly complicates the mathematical derivation required to obtain the response to even some of the simplest excitations. When there are several energy-storing elements and excitations of different types, the difficulty and amount of work required can easily become very extensive. Treatment of such circuits usually requires special techniques and is obviously beyond the scope of this text.

However, there is a class of circuit problems that are both practically important and analytically reasonable to manage. These are situations when only the steady-state solutions in the circuits are of interest. These situations arise when the transient part of the response is relatively short in duration. An example of such a situation is a sustained musical note. In other situations, the steady-state quantities last indefinitely. Examples of such situations are the household or industrial ac power applications.

In this chapter, we shall deal with the mathematical foundations useful in handling this class of problems. Although every voltage or current has to commence at some time, our interest is confined to the steady-state response. When each voltage or current began is of no importance to us. Hence we shall regard these quantities to have been initiated at $t = -\infty$ and lasts until $t = \infty$. There is no $u(t)$ to contend with and there is no discontinuity in any of these quantities.

7.1 Complex numbers and arithmetic

Before we formally undertake the development of this class of steady-state analysis, we shall summarize some operations, notations, and algorithms involving complex numbers or variables. It is assumed that the student is already familiar with what a complex number is. However, there are a number of special operations and particular notational forms that are peculiar to circuit analysis—as well as more specialized areas in electrical engineering—that bear some added emphasis here. We shall first introduce several terms and definitions.

7.1.1 Rectangular form

A complex number in rectangular form has two parts: the real and the imaginary. The two parts are distinguished by the quantity $\sqrt{-1} = j$, which multiplies the imaginary part. Let

$$A = a_1 + ja_2 \tag{7.1}$$

in which A denotes a complex number, a_1 its real part, and a_2 its imaginary part. Fig. 7.1 is a geometric representation of a complex number in the Cartesian coordinate system.

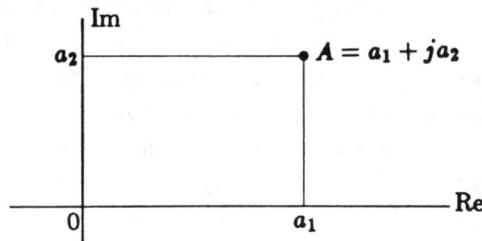

Figure 7.1: A complex number in rectangular form.

Also let

EXPONENTIAL AND SINUSOIDAL EXCITATIONS

$$B = b_1 + jb_2 \tag{7.2}$$

The four rational operations of complex numbers can be performed as if each number is the sum of two numbers. Thus

$$A + B = a_1 + ja_2 + b_1 + jb_2 = (a_1 + b_1) + j(a_2 + b_2) \tag{7.3}$$

$$A - B = a_1 + ja_2 - (b_1 + jb_2) = (a_1 - b_1) + j(a_2 - b_2) \tag{7.4}$$

$$\begin{aligned} A \times B &= (a_1 + ja_2)(b_1 + jb_2) \\ &= a_1b_1 + ja_1b_2 + ja_2b_1 + j^2 a_2b_2 \\ &= (a_1b_1 - a_2b_2) + j(a_1b_2 + a_2b_1) \end{aligned} \tag{7.5}$$

$$\begin{aligned} \frac{A}{B} &= \frac{a_1 + ja_2}{b_1 + jb_2} = \frac{(a_1 + ja_2)(b_1 - jb_2)}{(b_1 + jb_2)(b_1 - jb_2)} \\ &= \frac{a_1b_1 + a_2b_2}{b_1^2 + b_2^2} + j\frac{a_2b_1 - a_1b_2}{b_1^2 + b_2^2} \end{aligned} \tag{7.6}$$

The operation in Eq. (7.6), which deals with the division of two complex numbers, is known as rationalization. This procedure is very useful in many situations—not only in the computation of the ratio of two complex numbers but also in the derivation and simplification of certain expressions.

The rectangular form of complex numbers is used in most calculators and computer programming. It is not very suitable to be used when a complex quantity is a function of time or space.

The following are a few illustrative examples. Let $A = 5 - j8$, $B = -7 + j2$, and $C = j6$. Then

$$A + B + C = -2$$

$$(A + C)(A + B) = (5 - j2)(-2 - j6) = -22 - j26$$

$$(A - B)(A + B)(A - C) = (12 - j10)(-2 - j6)(5 - j14)$$
$$= -1148 + j916$$

$$\frac{A + C}{B - C} = \frac{5 - j2}{-7 - j4} = -\frac{27}{65} + j\frac{34}{65}$$

$$\frac{(2 + j5)A}{(B - A)C} = \frac{(2 + j5)(5 - j8)}{(-12 + j10)(j6)} = \frac{50 + j9}{-60 - j72} = -\frac{76}{183} + j\frac{85}{244}$$

7.1.2 Exponential form

Exponential functions play important roles in many areas in electrical engineering. The exponential function of a complex number is another complex number. Therefore, every complex number may be written in exponential form. The key relationship here is the Euler's formula, which reads

$$\epsilon^{j\theta} = \cos\theta + j\sin\theta \qquad (7.7)$$

Adding a scalar to the equation, we have

$$A = |A|\,\epsilon^{j\theta} = |A|\cos\theta + j\,|A|\sin\theta \qquad (7.8)$$

where $|A|$ is the *magnitude* or *absolute value* of A and θ is the *angle* or *argument* of A.

The following are a few numerical examples.[1]

$$\epsilon^{j45°} = \cos 45° + j\sin 45° = \frac{1}{\sqrt{2}} + j\frac{1}{\sqrt{2}}$$

$$4\,\epsilon^{-j\frac{\pi}{3}} = 4\left[\cos\frac{\pi}{3} - j\sin\frac{\pi}{3}\right] = 2 - j3.464$$

$$2\,\epsilon^{-j\pi} = -2$$

$$\epsilon^{j\frac{\pi}{2}} = \epsilon^{90°} = j$$

$$\epsilon^{j2\pi} = 1$$

A complex number in rectangular form can be converted to exponential form if we set

$$|A|\,\epsilon^{j\theta} = |A|\cos\theta + j|A|\sin\theta = a_1 + ja_2 \qquad (7.9)$$

By equating the real and imaginary parts we will get

$$a_1 = |A|\cos\theta \quad\text{and}\quad a_2 = |A|\sin\theta \qquad (7.10)$$

---

[1] The student should note the two different units used for the angle. When the ° sign is given, the angle is in degrees. When no unit is given, the angle is in radians. An angle in radians is the ratio of the length of a circular arc to the radius and is therefore dimensionless. The degree is an artificial subdivision of a complete revolution—$\frac{1}{360}$.

EXPONENTIAL AND SINUSOIDAL EXCITATIONS

Conversely,

$$\theta = \tan^{-1}\left(\frac{a_2}{a_1}\right) \quad \text{and} \quad |A| = \sqrt{a_1^2 + a_2^2} \qquad (7.11)$$

Thus,

$$\frac{1}{2} + j\frac{\sqrt{3}}{2} = 1\,\epsilon^{j60°} = 1\,\epsilon^{j\frac{\pi}{3}}$$

$$5 + j5 = 5\sqrt{2}\,\epsilon^{j45°}$$

$$10 + j20 = \sqrt{500}\,\epsilon^{j63.43°}$$

$$-j2 = 2\,\epsilon^{j\frac{3\pi}{2}}$$

$$10 = 10\,\epsilon^{j0}$$

The exponential form of complex numbers offers a certain simplification in many derivations and calculations. For example,

$$3\,\epsilon^{j20°} \times 5\,\epsilon^{-j5°} = 15\,\epsilon^{j15°}$$

$$8\,\epsilon^{j\pi}/2\,\epsilon^{-j} = 4\,\epsilon^{j(\pi+1)}$$

$$300\,\epsilon^{j\omega t} \times 2\,\epsilon^{j30°} = 600\,\epsilon^{j(\omega t+30°)}$$

$$\ln(4 + j5) = \ln\left(6.403\,\epsilon^{j0.896}\right) = 1.857 + j0.896$$

$$|A_1|\,\epsilon^{j\omega t} + |A_2|\,\epsilon^{j(\omega t+\theta)} = \left(|A_1| + |A_2|\,\epsilon^{j\theta}\right)\epsilon^{j\omega t}$$

Euler's formula also offers a clear link between an exponential function and a trigonometric function. From Eq. (7.7), we get

$$\epsilon^{-j\theta} = \cos\theta - j\sin\theta \qquad (7.12)$$

Adding Eq. (7.7) and Eq. (7.12), we easily get

$$\cos\theta = \frac{1}{2}\left(\epsilon^{j\theta} + \epsilon^{-j\theta}\right) \qquad (7.13)$$

Subtracting Eq. (7.7) and Eq. (7.12), we get

$$\sin\theta = \frac{1}{2j}\left(\epsilon^{j\theta} - \epsilon^{-j\theta}\right) \qquad (7.14)$$

Thus,

$$5\cos 10t = 2.5\,\epsilon^{j10t} + 2.5\,\epsilon^{-j10t}$$

$$10\sin 377t = -j5\left(\epsilon^{j377t} - \epsilon^{-j377t}\right)$$

7.1.3 Polar form

In Fig. 7.1, a complex number is depicted as a point in the Cartesian coordinate system whose horizontal axis represents the real part and whose vertical axis the imaginary part of the number.

Complex numbers share many properties (but by no means all) with those of vectors. Hence, analogous to the vector, a complex number is frequently represented in polar form as shown in Fig. 7.2, in which

$$A = a_1 + ja_2 = |A|\underline{/\theta} \tag{7.15}$$

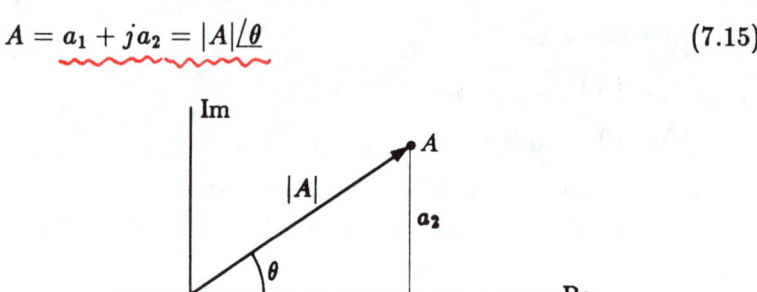

Figure 7.2: A complex number in polar form.

where

$$\theta = \tan^{-1}\left(\frac{a_2}{a_1}\right) \quad \text{and} \quad |A| = \sqrt{a_1^2 + a_2^2} \tag{7.16}$$

Eq. (7.16) is identical to Eq. (7.11). $|A|$ is also the magnitude of A and θ is the angle of A. Thus,

$$|A|\,\epsilon^{j\theta} = |A|\underline{/\theta} \tag{7.17}$$

Although the exponential form and the polar from of a complex number may have different origins and mathematical rationales, for all practical purposes, they are interchangeable. Each includes the same information and each may be preferred by persons working in different areas. The exponential form is more suitable for theoretical developments and derivational purposes. The polar form offers a more visually suggestive perception of the relative magnitudes and angles of a number of complex numbers.

Because of the close link between the exponential form and polar form of a complex number, one can easily appreciate the following relationships:

EXPONENTIAL AND SINUSOIDAL EXCITATIONS 333

$$|A|\underline{/\alpha} \times |B|\underline{/\beta} = |A||B|\underline{/\alpha + \beta}$$

$$\frac{|A|\underline{/\alpha}}{|B|\underline{/\beta}} = \frac{|A|}{|B|}\underline{/\alpha - \beta}$$

$$\ln\left(|A|\underline{/\alpha}\right) = \ln|A| + j\alpha$$

$$5\underline{/j2} = 5\,\epsilon^{j(j2)} = 5\,\epsilon^{-2} = 0.6767$$

$$7\underline{/2 + j3} = 7\,\epsilon^{j(2+j3)} = 7\,\epsilon^{-3}\,\epsilon^{j2} = 0.3485\underline{/2}$$

$$= -0.1450 + j0.3169$$

$$1\underline{/90° + j} = \epsilon^{-1}\,\epsilon^{j90°} = j\epsilon^{-1} = j0.2679$$

$$5\underline{/\theta + 180°} = -5\underline{/\theta}$$

$$5\underline{/\theta + 90°} = j5\underline{/\theta}$$

$$\underline{/90°} = j$$

$$\underline{/180°} = -1$$

$$\underline{/270°} = -j$$

7.1.4 Other terminology

The notation "Re" denotes "the real part of" [the complex number that follows]. Thus

$$\text{Re}[x + jy] = x \qquad (7.18)$$

$$\text{Re}[|A|\,\epsilon^{j\theta}] = |A|\cos\theta \qquad (7.19)$$

$$\text{Re}[A + B] = \text{Re}[A] + \text{Re}[B] \qquad (7.20)$$

The notation "Im" denotes "the imaginary part of" [the complex number that follows]. Thus

$$\text{Im}[x + jy] = y \qquad (7.21)$$

$$\text{Im}[|A| \epsilon^{j\theta}] = |A| \sin\theta \qquad (7.22)$$

$$\text{Im}[A + B] = \text{Im}[A] + \text{Im}[B] \qquad (7.23)$$

Note that the imaginary part of a complex number is itself a real number.

A complex number with the sign of its imaginary part changed is known as the *conjugate* of that number. It is denoted by an asterisk.[2] That is,

$$(x + jy)^* = x - jy \qquad (7.24)$$

In exponential and polar forms, we have

$$\left(|A| e^{j\theta}\right)^* = |A| \epsilon^{-j\theta} \quad \text{and} \quad (|A|\underline{/\theta})^* = |A|\underline{/-\theta} \qquad (7.25)$$

The following relationships are easily shown to be correct:

$$(A^*)^* = A \qquad (7.26)$$

$$AA^* = |A|^2 \qquad (7.27)$$

$$\frac{A}{B} = \frac{AB^*}{|B|^2} \qquad (7.28)$$

$$A + A^* = 2\text{Re}[A] \qquad (7.29)$$

$$A - A^* = j2\text{Im}[A] \qquad (7.30)$$

Eq. (7.28) is the same as Eq. (7.6).

A complex number is multivalued when it is expressed in exponential or polar form. That is,

$$|A| \epsilon^{j\theta} = |A| \epsilon^{j(\theta + 2k\pi)} \quad \text{and} \quad |A|\underline{/\theta} = |A|\underline{/\theta + 2k\pi} \qquad (7.31)$$

where k is any integer. Hence,

$$1 = 1\epsilon^{j2k\pi} = 1\underline{/k \times 360°}$$

$$\ln 5 = 1.609 + j2k\pi$$

---

[2] Some publications use an overbar to denote the conjugate of a complex quantity.

EXPONENTIAL AND SINUSOIDAL EXCITATIONS 335

The n-th root of a complex number has n distinct values.

$$\left(|A|\epsilon^{j\theta}\right)^{\frac{1}{n}} = |A|^{\frac{1}{n}} \epsilon^{j[(\theta + 2k\pi)/n]}, \quad k = 0, 1, 2, \ldots, (n-1) \quad (7.32)$$

For instance,

$$(1)^{1/3} = 1\underline{/0°} \quad \text{or} \quad 1\underline{/120°} \quad \text{or} \quad 1\underline{/240°}$$

$$(j3)^{1/3} = 1.442\underline{/30°} \quad \text{or} \quad 1.442\underline{/150°} \quad \text{or} \quad 1.442\underline{/270°}$$

EXERCISE

7.1.1 Simplify the following expressions. Do this in small steps, one at a time. The purpose of this exercise is not just to obtain the answers. Rather, it is to help you understand the notation and how complex expressions can be manipulated. So in working these problems, use only the very basic relationships, such as rectangular-polar conversion, logarithm of real numbers, exponentiation of real numbers, and rational operations of real numbers. For example,

$$5\epsilon^{j(\pi/6)}(3\underline{/j2}) = 5\underline{/30°} \times 3\epsilon^{-2} = 15 \times \epsilon^{-2}\underline{/30°} = 2.03\underline{/30°}$$

Use software or powerful calculators only to double-check your answers. Give your answers in both rectangular and polar forms.

a. $\dfrac{5\underline{/20°}}{-2 + j14}$

b. $\dfrac{20 - j24}{-2 - j} \times 20\underline{/-25°} \times \epsilon^{-j50°}$

c. $\dfrac{4\underline{/15°} \times 6\epsilon^{j\pi/3} \times 2\epsilon^{-2}}{(-3 - j2) \times 2\underline{/-\pi°}}$

d. $20\epsilon^{-2+j1} + 3\underline{/2 - j}$

e. $\cos(1 + j2)$

f. $(-2 - j3)^{(-1-j5)}$

g. $(-2 + j2)^{(1/3)}$

336   FUNDAMENTALS OF CIRCUIT ANALYSIS

Ans. a. $0.3536 \underline{/-78.13°} = 0.0727 - j0.3460$,
b. $279.4 \underline{/28.24°} = 246.2 + j132.2$,
c. $0.9008 \underline{/224.45°} = -0.6431 - j0.6309$,
d. $9.883 \underline{/101.27°} = -1.931 + j9.693$,
e. $3.667 \underline{/-56.33°} = 2.033 - j3.052$,
f. $5.692 \times 10^{-6} \underline{/-243.71°} = (-2.521 + j5.104) \times 10^{-6}$,
g. $1.414 \underline{/45°} = 1 + j$ or $1.414 \underline{/165°} = -1.366 + j0.366$
or $1.414 \underline{/285°} = 0.366 - j1.366$.

7.2   Analysis of circuits with exponential excitation

We shall now formulate the steady-state analysis of circuits in which all excitations are of the type $A\epsilon^{st}$, where A and s are, in general, complex. Two examples of such arrangements are given in Fig. 7.3. Note that the expressions for all excitations are for the entire duration from $t = -\infty$ to $t = \infty$.

Take the circuit in Fig. 7.3(a), with the four branch currents—i_1, i_2, i_3, and i_4—assumed. KCL and KVL require that

$$R_1 i_1 + \frac{1}{C_1} \int_{-\infty}^{t} i_2(\lambda) d\lambda = e_s$$

$$L_1 \frac{di_3}{dt} + R_2 i_4 - \frac{1}{C_1} \int_{-\infty}^{t} i_2(\lambda) d\lambda = 0 \qquad (7.33)$$

$$i_1 = i_2 + i_3$$

$$i_4 - i_3 = i_s$$

If e_s and i_s are known, we need to solve these four equations simultaneously. That is quite a formidable task if e_s and i_s are arbitrary and a general solution is being sought. However, if e_s and i_s are of the exponential form and we are only interested in the steady-state solution, then the effort involved is relatively modest.

We assume that

$$i_1 = I_1 \epsilon^{st}, \quad i_2 = I_2 \epsilon^{st}, \quad i_3 = I_3 \epsilon^{st}$$

EXPONENTIAL AND SINUSOIDAL EXCITATIONS

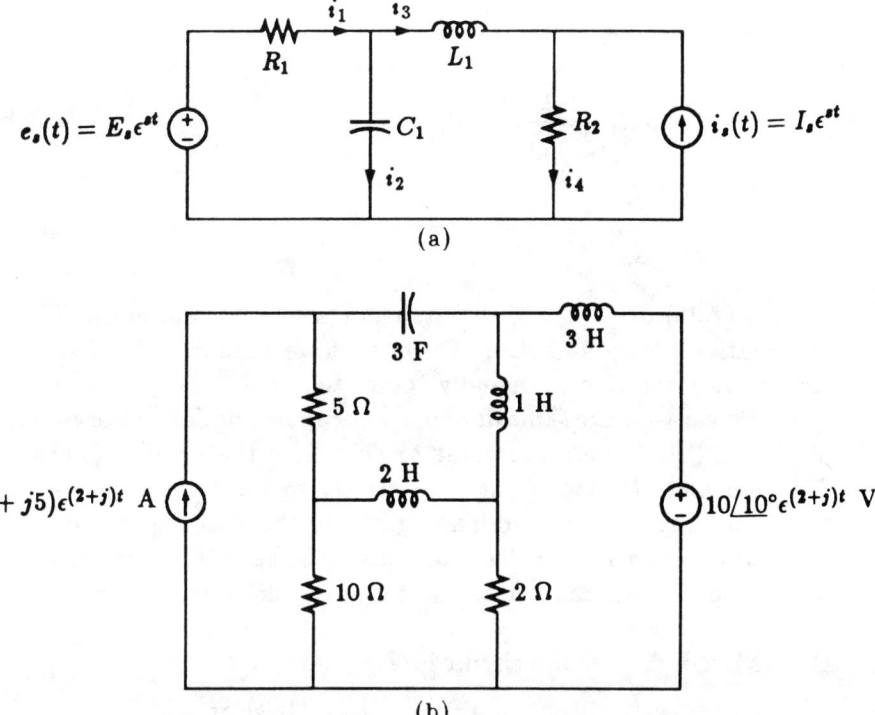

Figure 7.3: Two examples of steady-state analysis problems.

$$i_4 = I_4 \epsilon^{st}, \quad i_s = I_s \epsilon^{st}, \quad e_s = E_s \epsilon^{st}$$

Then Eqs. (7.33) become

$$R_1 I_1 \epsilon^{st} + \frac{1}{C_1} \int_{-\infty}^{t} I_2 \epsilon^{s\lambda} d\lambda = E_s \epsilon^{st}$$

$$L_1 \frac{d}{dt}\left(I_3 \epsilon^{st}\right) + R_2 I_4 \epsilon^{st} - \frac{1}{C_1}\int_{-\infty}^{t} I_2 \epsilon^{s\lambda} d\lambda = 0 \qquad (7.34)$$

$$I_1 \epsilon^{st} = I_2 \epsilon^{st} + I_3 \epsilon^{st}$$

$$I_4 \epsilon^{st} - I_3 \epsilon^{st} = I_s \epsilon^{st}$$

Performing the differentiation and integration and leaving out the common factor ϵ^{st}, we reduce Eqs. (7.34) to

$$R_1 I_1 + \frac{1}{sC_1} I_2 = E_s$$

$$sL_1 I_3 + R_2 I_4 - \frac{1}{sC_1} I_2 = 0 \qquad (7.35)$$

$$I_1 = I_2 + I_3$$

$$I_4 - I_3 = I_s$$

Eqs (7.35) are a set of simultaneous algebraic equations. They are relatively easy to solve. Once we have obtained the I's, the steady-state solution is virtually completed.

It is clear that the same development can be applied to the circuit in Fig. 7.3(b). We are not going to delve into the details of such a development. We include it here merely to illustrate the type of circuit problems we are addressing. Note the absence of the $u(t)$ term in all excitations in Fig. 7.3. This is the key difference between what we do in this chapter as compared to the previous one.

EXAMPLE 1. For the circuit in Fig. 7.4,

$$e_1(t) = 5\,\epsilon^{(2+j3)t} \text{ V} \quad \text{and} \quad e_2(t) = j4\,\epsilon^{(2+j3)t} \text{ V}$$

Determine $i_1(t)$ and $i_2(t)$.

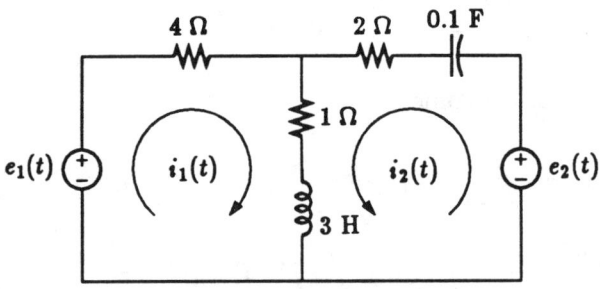

Figure 7.4: Circuit for Example 1.

SOLUTION With the assumption that

$$i_1(t) = I_1\,\epsilon^{(2+j3)t} \quad \text{and} \quad i_2(t) = I_2\,\epsilon^{(2+j3)t}$$

and following the pattern of Eqs. (7.33), (7.34), and (7.35), we may reduce the two mesh equations to

$$4I_1 + \left[1 + (2+j3) \times 3\right](I_1 + I_2) = 5$$

$$\left[2 + \frac{1}{(2+j3) \times 0.1}\right] I_2 + \left[1 + (2+j3) \times 3\right](I_1 + I_2) = j4$$

Solving gives

$$I_1 = 0.8437 - j0.3463 = 0.9120\underline{/-22.13°}$$

$$I_2 = -0.6603 + j0.3083 = 0.7288\underline{/154.97°}$$

Hence,

$$i_1(t) = 0.9120\underline{/-22.13°}\, \epsilon^{(2+j3)t} = 0.9120\epsilon^{[(2+j3)t-j22.13°]}\ \text{A}$$

$$i_2(t) = 0.7288\underline{/154.97°}\, \epsilon^{(2+j3)t} = 0.7288\epsilon^{[(2+j3)t+j154.97°]}\ \text{A}$$

EXERCISES

7.2.1 Determine $i(t)$ in the circuit for (a) $i_s(t) = j5\,\epsilon^{-2t}$ A and (b) $i_s(t) = 10\,\epsilon^{j3t}$ A.

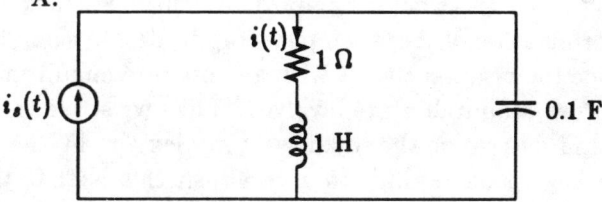

Ans. (a) $j4.167\epsilon^{-2t}$ A, (b) $(10 - j30)\epsilon^{j3t}$ A.

7.2.2 In the circuit, $e_s(t) = (3 + j5)\,\epsilon^{(-2+j4)t}$ V. Determine $i(t)$.

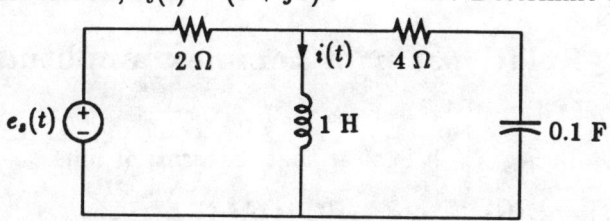

Ans. $\left(\dfrac{8}{13} - j\dfrac{43}{52}\right)\epsilon^{(-2+j4)t}$ A.

340 FUNDAMENTALS OF CIRCUIT ANALYSIS

7.3 Steady-state analysis using complex amplitudes

From the previous section, the student should be able to observe the general pattern by which each steady-state analysis of a circuit with exponential excitations is followed. Here, we assume a suitable number of voltage or current unknowns exactly as we did in the analysis of memoryless circuits with constant excitations—dc circuits. The difference here is that every unknown quantity is of the same exponential form as the excitation. Specifically, every quantity will be of the form $f(t) = C\epsilon^{st}$. We shall call the coefficient C the *complex amplitude* of the quantity $f(t)$.

Then we employ any suitable analysis method to set up the KCL and KVL equations. Since the factor ϵ^{st} will appear with every term in these equations, this factor may be dropped. We are then left with a set of algebraic equations that involve only the complex amplitudes of the assumed voltage or current unknowns. Once these unknown complex amplitudes are found, the steady-state time functions of all unknown voltages or currents are also determined.

Our strategy here is to replace every electrical quantity of the type $C\epsilon^{st}$ by C—its complex amplitude. Then we can concentrate on the determination of these complex amplitudes. We find it expedient to replace the original circuit with a substitute circuit in which only the complex amplitudes are involved. Then we strive to analyze the circuit and determine the unknown complex amplitudes.

One key to our ability to accomplish this is that the voltage-current relationship in each linear element is very simple in terms of its complex amplitudes. We shall now point out these simple relationships.

7.3.1 Relationship of complex amplitudes in a resistance

As shown in Fig. 7.5 it is clear that, in terms of time

$$e(t) = R\,i(t) \quad \text{or} \quad i(t) = G\,e(t) \tag{7.36}$$

which implies that

$$E\epsilon^{st} = R I\epsilon^{st} \quad \text{or} \quad I\epsilon^{st} = G E\epsilon^{st}$$

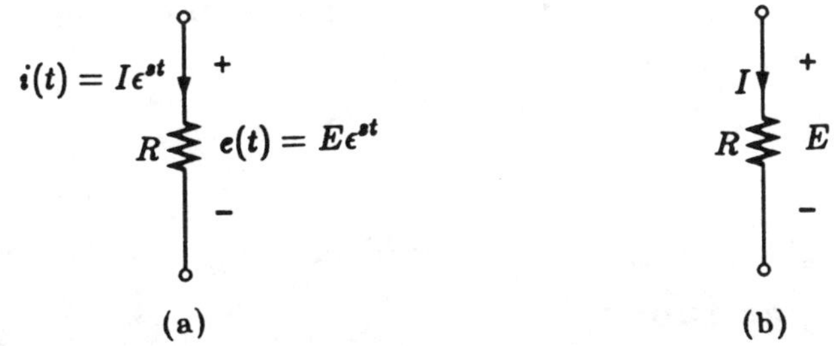

EXPONENTIAL AND SINUSOIDAL EXCITATIONS   341

Figure 7.5: Voltage-current relationship in a resistance.

Thus, in the complex-amplitude domain,

$$E = RI \quad \text{or} \quad I = GE \tag{7.37}$$

Eq. (7.37) is identical in appearance to the Ohm's law relationship in dc circuits. Just as we have stressed in the dc situations, the associated directions and signs must be carefully observed in applying Eq. (7.37).

7.3.2 Relationship of complex amplitudes in an inductance.

Referring to Fig. 7.6, we can write,

$$e(t) = L\frac{di}{dt} \quad \text{or} \quad i(t) = \frac{1}{L}\int_{-\infty}^{t} e(\lambda)\,d\lambda \tag{7.38}$$

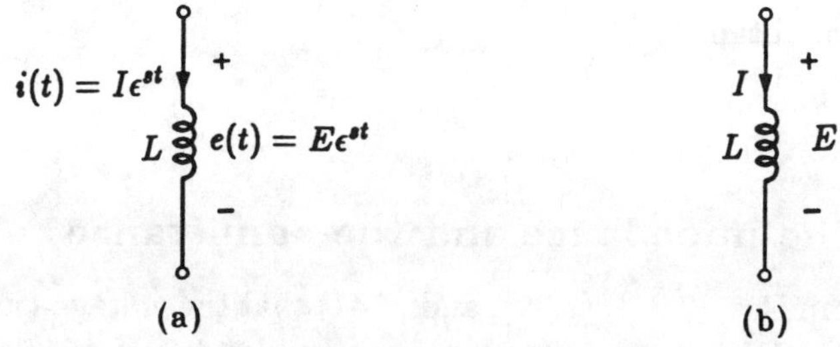

Figure 7.6: Voltage-current relationship in an inductance.

which lead to

$$E\,\epsilon^{st} = sLI\,\epsilon^{st} \quad \text{or} \quad I\,\epsilon^{st} = \frac{1}{sL}E\,\epsilon^{st}$$

Dropping the exponential factor, we obtain in terms of complex amplitudes the relationship

$$E = sL\,I \quad \text{or} \quad I = \frac{1}{sL} E \tag{7.39}$$

7.3.3 Relationship of complex amplitudes in a capacitance

Referring to Fig. 7.7, we have

$$e(t) = \frac{1}{C} \int_{-\infty}^{t} i(\lambda)\,d\lambda \quad \text{or} \quad i(t) = C\frac{de}{dt} \tag{7.40}$$

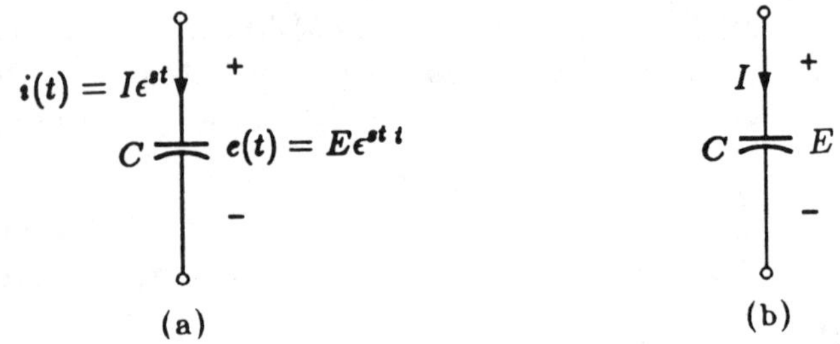

$i(t) = I\epsilon^{st}$ C $e(t) = E\epsilon^{st}$

(a)   (b)

Figure 7.7: voltage-current relationship in a capacitance.

which lead to

$$E\,\epsilon^{st} = \frac{1}{sC} I\,\epsilon^{st} \quad \text{or} \quad I\,\epsilon^{st} = sC\,E\,\epsilon^{st}$$

Similarly, we obtain

$$E = \frac{1}{sC} I \quad \text{or} \quad I = sC\,E \tag{7.41}$$

7.3.4 The impedance and the admittance

It is seen from Eqs. (7.37), (7.39), and (7.41) that the voltage-current relationships of linear elements are very similar to those of resistances in dc analysis. When complex amplitudes are used, those of the voltage and current of each element are proportional to each other. Thus, each element can be quantitatively characterized by a single quantity the same way the resistance or the conductance characterizes a resistor in dc analysis.

EXPONENTIAL AND SINUSOIDAL EXCITATIONS   343

In the steady-state analysis of circuits with linear elements and exponential excitations, we call the ratio of the complex amplitude of the voltage to that of a current of an element the *impedance* of that element and denote it by Z. That is

$$Z = \frac{E}{I} \quad \text{or} \quad E = ZI \tag{7.42}$$

The ratio of the complex amplitude of the current to that of the voltage of an element is called the *admittance* of that element. It is denoted by Y. That is

$$Y = \frac{I}{E} \quad \text{or} \quad I = YE \tag{7.43}$$

Obviously, the impedance and the admittance of an element are reciprocal of each other. We could characterize an element by either its impedance or its admittance.

Thus, for a resistance,

$$Z_R = R \quad \text{and} \quad Y_R = G \tag{7.44}$$

For an inductance,

$$Z_L = sL \quad \text{and} \quad Y_L = \frac{1}{sL} \tag{7.45}$$

For a capacitance,

$$Z_C = \frac{1}{sC} \quad \text{and} \quad Y_C = sC \tag{7.46}$$

Eqs. (7.39) and (7.41) may be called the *Ohm's law* for the complex amplitudes of an inductance and a capacitance, respectively.

Since all instantaneous currents must satisfy KCL, so must all their complex amplitudes. Likewise, since all instantaneous voltages must satisfy KVL, so must all their complex amplitudes. Hence, all analysis techniques and theorems developed in Chapters 2 through 4 are all directly applicable to the steady-state analysis of linear circuits if every current or voltage is replaced by its complex amplitude and every element is replaced by its impedance or admittance.

One major difference in dealing with dc circuits and in dealing with our current type of problems is that now we are generally dealing with complex numbers instead of real numbers. Other than this generalization, everything we learned in Chapters 2 through 4 is directly applicable to the problems in this chapter. Conversely, if a circuit here happens to have only real complex amplitudes and real impedances, the circuit will have exactly the same appearance in its analysis and solutions as those in Chapters 2, 3, and 4.

In applying the concept of the impedance or admittance, the associated direction in each element is again extremely important. Eqs. (7.37), (7.39), (7.41), (7.42), and (7.43) all imply a consistent relative direction of the associated voltage and current complex amplitudes of an element.

EXAMPLE 2. For the circuit in Fig. 7.8(a), $e_s(t) = 5\epsilon^{-2t}$ V. Determine all currents.

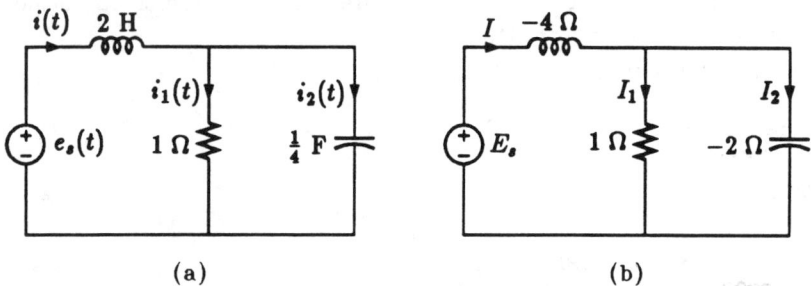

Figure 7.8: Circuits used in Example 2.

SOLUTION We first replace every electrical quantity by its complex amplitude and every element by its impedance with $E_s = 5\underline{/0°}$ V and $s = -2$. The replacement circuit is shown in Fig. 7.8(b). The analysis of the replacement circuit is quite straightforward. We have

$$I = \frac{5}{-4 + \frac{1 \times (-2)}{1-2}} = \frac{5}{-4 + 2} = -2.5 \text{ A}$$

$$I_1 = \frac{-2}{1-2} \times (-2.5) = -5 \text{ A}$$

$$I_2 = \frac{1}{1-2} \times (-2.5) = 2.5 \text{ A}$$

EXPONENTIAL AND SINUSOIDAL EXCITATIONS

Therefore in the circuit in Fig. 7.8(a), we have

$$i(t) = -2.5\,\epsilon^{-2t}\ \text{A}$$

$$i_1(t) = -5\,\epsilon^{-2t}\ \text{A}$$

$$i_2(t) = 2.5\,\epsilon^{-2t}\ \text{A}$$

EXAMPLE 3. In the circuit in Fig. 7.9(a),

$$e_1(t) = 10\,\epsilon^{j(2t-30°)}\ \text{V} \quad \text{and} \quad e_2(t) = 5\,\epsilon^{j(2t+45°)}\ \text{V}$$

Determine all currents in the circuit.

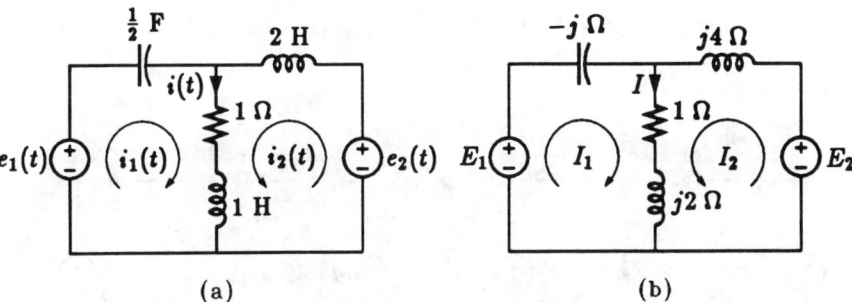

(a) (b)

Figure 7.9: Circuits used in Example 3.

SOLUTION In the complex amplitude-impedance circuit in Fig. 7.9(b), all impedances are indicated in the figure for $s = j2$ and

$$E_1 = 10\underline{/-30°}\ \text{V} \quad \text{and} \quad E_2 = 5\underline{/45°}\ \text{V}$$

Applying mesh analysis, we obtain

$$(1+j)I_1 + (1+j2)I_2 = 10\underline{/-30°}$$

$$(1+j2)I_1 + (1+j6)I_2 = 5\underline{/45°}$$

Solving gives

$$I_1 = 1.898 - j15.331 = 15.448\underline{/-82.94°}$$

FUNDAMENTALS OF CIRCUIT ANALYSIS

$$I_2 = 1.659 + j5.114 = 5.376\underline{/72.02°}$$

$$I = I_1 + I_2 = 3.557 - j10.217 = 10.818\underline{/-70.80°}$$

Hence, for the circuit in Fig. 7.9(a),

$$i_1(t) = 15.448\underline{/-82.94°}\ \epsilon^{j2t} = 15.448\epsilon^{j(2t-82.94°)}\ \text{A}$$

$$i_2(t) = 5.376\underline{/72.02°}\ \epsilon^{j2t} = 5.376\epsilon^{j(2t+72.02°)}\ \text{A}$$

$$i(t) = 10.818\underline{/-70.80°}\ \epsilon^{j2t} = 10.818\ \epsilon^{j(2t-70.80°)}\ \text{A}$$

EXAMPLE 4. In Fig. 7.10(a),

$$i_s(t) = (2 + j5)\epsilon^{(-2+j4)t}\ \text{A} \quad \text{and} \quad e_s(t) = (4 - j2)\epsilon^{(-2+j4)t}\ \text{V}$$

Determine i(t).

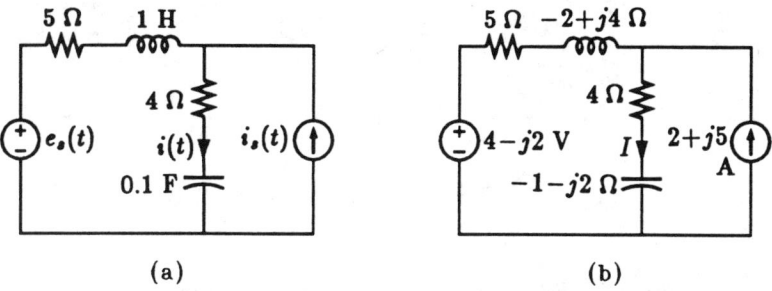

(a)   (b)

Figure 7.10: Circuits used in Example 4.

SOLUTION Here, $s = -2 + j4$. We replace the circuit in Fig. 7.10(a) by that shown in Fig. 7.10(b). Note the impedance of the capacitor is $\frac{1}{(-2+j4)\times 0.1} = -1 - j2\ \Omega$. To determine I, we use the superposition theorem, which gives

$$I = \frac{4 - j2}{5 - 2 + j4 + 4 - 1 - j2}$$

$$+ \frac{5 - 2 + j4}{5 - 2 + j4 + 4 - 1 - j2} \times (2 + j5)$$

EXPONENTIAL AND SINUSOIDAL EXCITATIONS

$$= \frac{1}{2} - j\frac{1}{2} - \frac{19}{20} + j\frac{.83}{20} = -\frac{9}{20} + j\frac{.73}{20}$$

$$= -0.45 + j3.65 = 3.678\underline{/97.03°}$$

Hence

$$i(t) = (-0.45 + j3.65)\epsilon^{(-2+j4)t} = 3.678\,\epsilon^{[(-2+j4)t+j97.03°]}\text{ A}$$

EXERCISE

7.3.1 Determine $i(t)$ in the following circuit for (a) $e_s(t) = 10\epsilon^{-2t}$ V, (b) $e_s(t) = 10\epsilon^{j3t}$ V, and (c) $e_s(t) = (3 + j8)\epsilon^{(-5+j2)t}$ V.

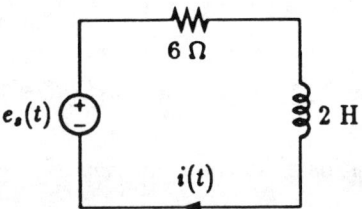

Ans. (a) $5\epsilon^{-2t}$ A, (b) $1.179\epsilon^{j(3t-45°)}$ A, (c) $1.510\epsilon^{[(-5+j2)t-j65.57°]}$ A.

7.4 Interpretations and applications

The development in this chapter deals with quantities that are more philosophical than physical. We have no problem understanding or identifying quantities such as a 5-ampere current, a voltage of $5t$ volts, a current that is a time function described by $5\sin 20t$, or even a $\sqrt{8}$-volt source. But what is a j-ampere current or a voltage that is equal to $5\epsilon^{(3+j5)}$ volts? Now, we shall try to provide a rationale for why we have been dealing with these non-real quantities. We shall also attempt to furnish real-world interpretations of some of the quantities we have been dealing with.

Actually, one could say that what we did in this chapter is purely mathematical and whether or not these quantities correspond to any physical quantities does not matter. That is, we need not justify our effort here on the basis that these quantities have physical counterparts. We are really developing a mathematical foundation and

and methodology that will lead to a number of extremely useful tools later on. But some justification and explanation of why we have been undertaking this development might be useful at this point.

Much of what we have been doing—and what we anticipate taking advantage of in the future—can be explained by the fact that the only property we need to apply to make these conclusions useful is the *linearity* property (which includes proportionality and additivity) of linear networks. Take the simple circuit of Fig. 7.11. Simple analysis will show that if $E = 9$ V then $I = 1$ A.

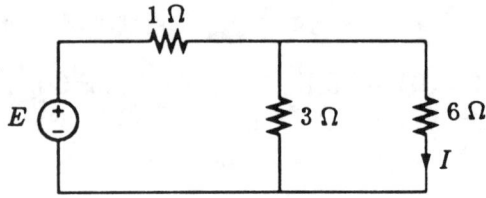

Figure 7.11: A simple resistance circuit.

Linearity assures us that the following E-I pairs would have to hold:

If $E = K$ V, then $I = \dfrac{K}{9}$ A.

If $E = j3$ V, then $I = j\dfrac{1}{3}$ A.

If $E = 3 + j5$ V, then $I = \dfrac{1}{3} + j\dfrac{5}{9}$ A.

If $E = 5\,\epsilon^{jt}$ V, then $I = \dfrac{5}{9}\epsilon^{jt}$ A.

Although we have started out with the simplest E-I pair, actually, all five E-I pairs imply one another. We could have started with any of the five pairs and deduce the other four from it as well.

The point here is that we can assert that each E-I pair is correct, sound, and consistent among themselves. It doesn't matter what a $j3$-volt voltage is. We are assured that if E happens to be $j3$ volts, I must be $j\tfrac{1}{3}$ amperes in order for all rules for linear networks to be satisfied. At this point, what a $j3$-volt voltage and a $j\tfrac{1}{3}$-ampere

EXPONENTIAL AND SINUSOIDAL EXCITATIONS

current are is of no importance. Our only concern is that they are true, accurate, and consistent.

Although the importance of the foundation we have laid down in this chapter will become quite apparent in the next chapter, we will do well to pause and elaborate on some pertinent points associated with our effort so far. Even without this justification, we can cite at least two reasons that what we have undertaken in the chapter has some significance in its own right.

First is that, for certain special situations, the function $C\epsilon^{st}$ is meaningful in terms of real-world problems. The following are three such situations:

1. $K\epsilon^{\sigma t}$, where K and σ are both real. Typical variations of such functions are shown in Fig. 7.12. Example 2 presented earlier is an illustration of this class of functions.

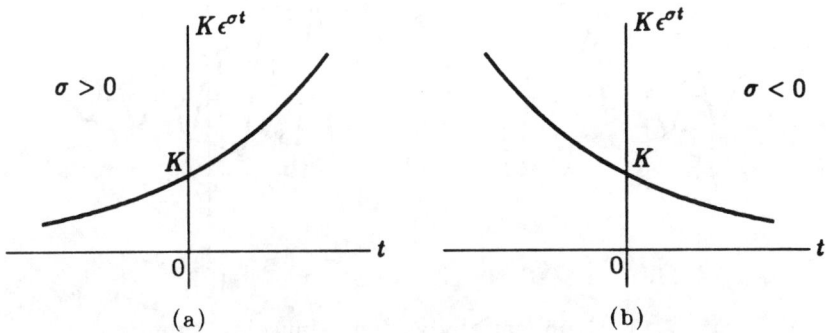

Figure 7.12: Variations of real exponential functions.

2. $K\cos(\omega t + \phi)$, where K, ω, and ϕ are all real. Functions of this type are sustained sinusoidal quantities and are extremely important in electrical engineering. Since

$$K\,\epsilon^{j\phi}\epsilon^{j\omega t} = K\cos(\omega t + \phi) + j\,K\sin(\omega t + \phi) \tag{7.47}$$

it follows that

$$K\cos(\omega t + \phi) = \mathrm{Re}\left[K\,\epsilon^{j\phi}\epsilon^{j\omega t}\right] \tag{7.48}$$

Hence, when we encounter an excitation of the form $K\cos(\omega t + \phi)$, we can, instead, find the answer to another excitation $A\epsilon^{st}$ and let

350 FUNDAMENTALS OF CIRCUIT ANALYSIS

$A = K\underline{/\phi}$ and $s = 0 + j\omega$. After the answer to the excitation $A\epsilon^{st} = K\underline{/\phi}\epsilon^{j\omega t}$ has been found, we simply keep the real part of the answer. This real part is the answer when the excitation is $K\cos(\omega t + \phi)$. In fact, this is the basis of what we are going to expound upon in the next chapter.

3. $K\epsilon^{\sigma t}\cos(\omega t + \phi)$, where K, σ, ω, and ϕ are all real. Functions of this type are exponentially varying sinusoids as illustrated in Fig. 7.13. Since

$$K\,\epsilon^{\sigma t}\cos(\omega t + \phi) = \mathrm{Re}\left[K\,\epsilon^{j\phi}\epsilon^{(\sigma+j\omega)t}\right] \qquad (7.49)$$

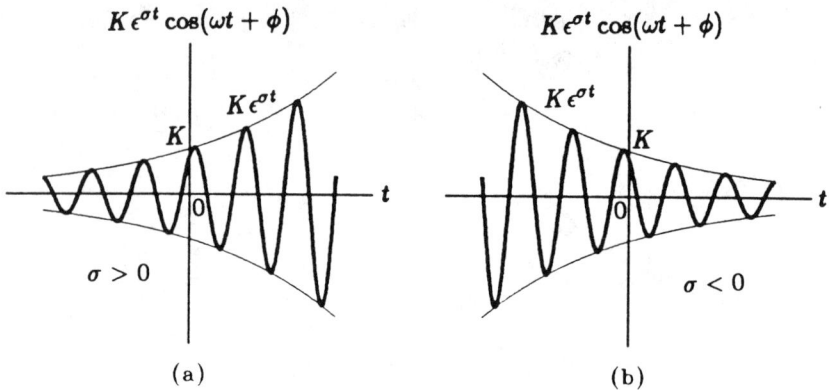

Figure 7.13: Exponentially varying sinusoidal functions.

if we have an excitation $K\epsilon^{\sigma t}\cos(\omega t + \phi)$, we can analyze the circuit subject to another excitation $A\epsilon^{st}$, $A = K\underline{/\phi}$ and $s = \sigma+j\omega$, instead. Then we simply keep the real part of the answer, which must be the answer to the real function.

It is interesting to note that in the two sinusoidal functions discussed here, the two corresponding exponential functions actually contain more than the sinusoidal functions. Each exponential function actually contains two sinusoidal functions—the cosine and the sine. However, the exponential function is notationally more compact. For example, when we need to deal with $K\cos\omega t$, we augment it with an additional part, $j\sin\omega t$. In doing this we get

$$K\cos\omega t + jK\sin\omega t = K\epsilon^{j\omega t}$$

EXPONENTIAL AND SINUSOIDAL EXCITATIONS

The augmented function, which actually contains more, is not only notationally more compact, but also symbolically easier to manipulate. (For instance, it's easier to differentiate and integrate an exponential function than a trigonometric function.) Since the real and imaginary parts can easily be separated, it is more advantageous to deal with the exponential function rather than the trigonometric function. This is one of those happier things in life in which we actually get more for less. This observation can also be viewed as one of the reasons that we prefer to deal with the exponential, albeit non-physical, functions in our current development.

The other application of the development in this chapter is that these analyses are actually applicable when obtaining the forced response or the particular integral of the complete solution of circuits including energy-storing elements, when the excitation can be related to an exponential function. The complex amplitude-impedance method provides a straightforward and systematic procedure for obtaining the particular-integral part of the complete solution. Thus, roughly half of the task of solving the differential equation of the circuit is taken care of. We merely have to add the natural response (homogeneous solution or complementary function) and then impose the initial conditions to complete the solution.

EXAMPLE 5. In the circuit in Fig. 7.14(a),

$$e_s(t) = 10\,\epsilon^{-2t}\cos(6t + 30°)\,u(t)\ \text{V}$$

Determine the steady-state part of $i(t)$.

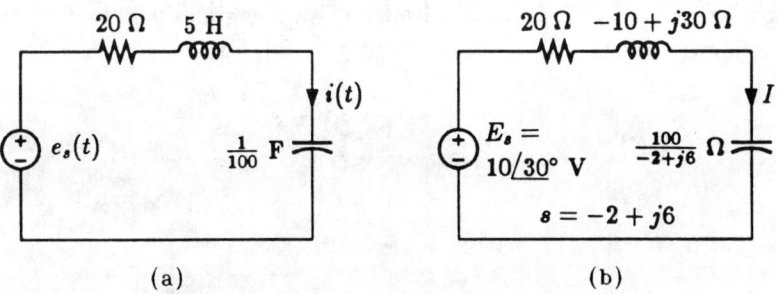

Figure 7.14: Circuits used in Example 5.

SOLUTION Since

$$e_s(t) = \text{Re}\left[10\underline{/30°}\,\epsilon^{(-2+j6)t}\right]$$

to obtain the steady-state part of $i(t)$, we may analyze the circuit in Fig. 7.14(b), in which

$$I = \frac{10\underline{/30°}}{20 - 10 + j30 + \dfrac{100}{-2 + j6}} = \frac{10\underline{/30°}}{5 + j15}$$

$$= 0.4732 - j0.4196 = 0.6325\underline{/-41.57°}$$

Hence, the steady-state part of $i(t)$ is

$$i_p(t) = \text{Re}\left[0.6325\underline{/-41.57°}\,\epsilon^{(-2+j6)t}\right]$$

$$= 0.6235\,\epsilon^{-2t}\cos(6t - 41.57°)\text{ A}$$

EXERCISE

7.4.1 In the circuit, $e_1(t) = \epsilon^{-3t}\,10\cos(2t+50°)$ V. $e_2(t) = \epsilon^{-3t}\,15\sin(2t - 20°)$ V. First, set up the appropriate associated circuit using impedances and complex amplitudes. Use mesh analysis to find the complex amplitudes I_1 and I_2. Then give $i_1(t)$ and $i_2(t)$.

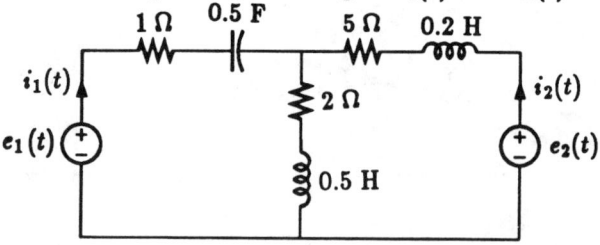

Ans. $i_1(t) = 9.579\epsilon^{-3t}\cos(2t + 42.75°)$ A, $i_2(t) = 4.803\epsilon^{-3t}\cos(2t - 110.97°)$ A.

7.5 Steady-state solution of circuits with sinusoidal excitations

In the last section, we indicated how sinusoidal and exponential functions are closely and simply related. Specifically, we illustrated how the complex amplitude-impedance method of analyzing networks with exponential excitations can be interpreted and applied to the analysis of networks with sinusoidal excitations. Since sinusoids are widely used in electrical engineering, we shall formalize this procedure here.

When we have a sinusoidal excitation[3]

$$e_s(t) = |E_s| \cos(\omega t + \psi) \qquad (7.50)$$

and we wish to determine the response $i(t)$, as symbolically represented in Fig. 7.15(a), we recognize that

$$e_s(t) = \text{Re}\left[|E_s|\, \epsilon^{j\psi}\, \epsilon^{j\omega t}\right] = \text{Re}\left[|E_s|\, \underline{/\psi}\, \epsilon^{j\omega t}\right] \qquad (7.51)$$

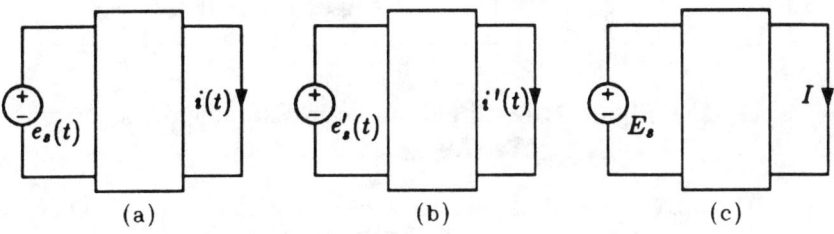

Figure 7.15: Networks with different excitations and responses.

So, we analyze a substitute network with the excitation replaced by

$$e'_s(t) = |E_s|\, \epsilon^{j\psi}\, \epsilon^{j\omega t} = |E_s|\, \underline{/\psi}\, \epsilon^{j\omega t} \qquad (7.52)$$

as represented in Fig. 7.15(b). Once we have determined $i'(t)$, which will have the form

---

[3] Here we use a voltage function just for the sake of being specific. It could just as well have been a current. The student will find that on other occasions we might choose a current to represent a general quantity. These choices are more for expediency rather than restrictive.

354 FUNDAMENTALS OF CIRCUIT ANALYSIS

$$i'(t) = |I|\, e^{j\phi}\, e^{j\omega t} = |I|\,\underline{/\phi}\, e^{j\omega t} \qquad (7.53)$$

we can say that the desired $i(t)$ is[4]

$$i(t) = \text{Re}\big[i'(t)\big] = \text{Re}\left[|I|\, e^{j\phi}\, e^{j\omega t}\right] = |I|\cos(\omega t + \phi) \qquad (7.54)$$

In determining $i'(t)$, we can use another substitute network so as to utilize the complex amplitude-impedance method of analysis. This is indicated in Fig. 7.15(c), in which

$$E_s = |E_s|\,\underline{/\psi} \quad \text{and} \quad I = |I|\,\underline{/\phi} \qquad (7.55)$$

EXAMPLE 6. Determine $i(t)$ in Fig. 7.16(a) for $e_s = 20\,\cos(10^6 t + 30°)$ V.

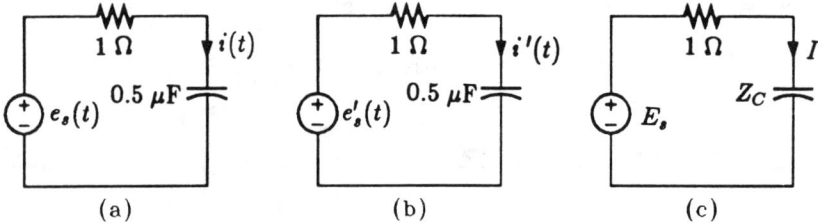

Figure 7.16: Circuits used in Example 6.

SOLUTION The first substitute circuit with exponential excitations is shown in Fig. 7.16(b) in which

$$e'_s(t) = 20\,\underline{/30°}\, e^{j10^6 t}$$

We can note that

$$i(t) = \text{Re}[i'(t)]$$

To determine $i'(t)$, we shall use another substitute circuit, shown in Fig. 7.16(c), in which

$$E_s = 20\,\underline{/30°}\text{ V}$$

---

[4] This is because, in a linear network, the response to a real excitation must be real.

EXPONENTIAL AND SINUSOIDAL EXCITATIONS

$$Z_C = \frac{1}{sC} = \frac{1}{j10^6 \times 0.5 \times 10^{-6}} = -j2 \; \Omega$$

Hence,

$$I = \frac{20\underline{/30°}}{1-j2} = \frac{20\underline{/30°}}{\sqrt{5}\underline{/-63.43°}} = 8.944\underline{/93.43°} \; \text{A}$$

We then have

$$i'(t) = I\,e^{j10^6 t} = 8.944\underline{/93.43°}\,\epsilon^{j10^6 t} = 8.944\epsilon^{j(10^6 t+93.43°)} \; \text{V}$$

and

$$i(t) = \text{Re}[i'(t)] = 8.944\,\cos(10^6 t + 93.43°) \; \text{A}$$

As far as the steady-state solution of a network with sinusoidal excitations is concerned, the intermediate substitute network with exponential excitations can really be bypassed. We can go directly from Fig. 7.15(a) to Fig. 7.15(c) and back without bothering to set up Fig. 7.15(b) every time. Of course, bypassing the circuit of Fig. 7.15(b) does not mean that it doesn't exist. Actually, this step is implicit in the process. However, the substitutions are so routine and perfunctory that they don't have to be spelled out every time. With this understanding, we often simply call $|K|\underline{/\theta}$ the complex amplitude of the sinusoid $|K|\cos(\omega t + \theta)$.

In the future, unless it is otherwise noted, we shall use the same capital letter (with possibly the same subscript) to denote the complex amplitude of a time quantity denoted by a lowercase letter. Thus, E will denote the complex amplitude of $e(t)$, E_2 will denote the complex amplitude of $e_2(t)$, I_a will denote the complex amplitude of $i_a(t)$, etc.

One should not, however, overlook the importance of the exponential-excitation formalism. If it were not for this formalism, going from a sinusoidal quantity to a complex amplitude would seem more like some kind of black magic rather than a step-by-step development. The exponential-function intermediate step provides us with a rigorous and logical development of a mathematical basis as to why the complex amplitude-impedance method works for solving circuits with sinusoidal quantities.[5]

EXAMPLE 7. In the circuit in Fig. 7.17(a), $e_s(t) = 10\cos(10t + 50°)$ V. Determine $i(t)$.

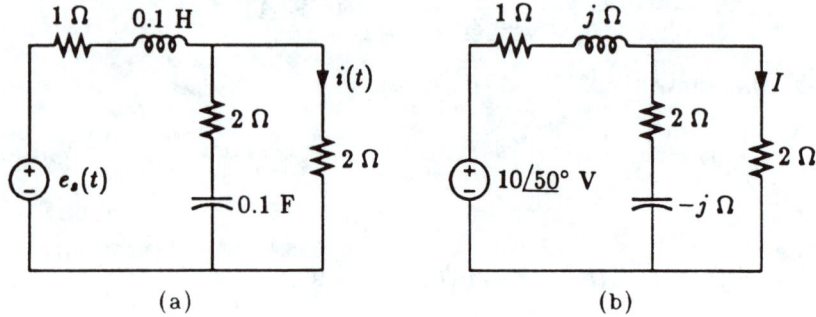

Figure 7.17: Circuits used in Example 7.

SOLUTION The complex amplitude-impedance substitute circuit is shown in Fig. 7.17(b). Using the series-parallel combination and current-division rule, we obtain readily

$$I = \frac{10\underline{/50°}}{1+j+\dfrac{2(2-j)}{4-j}} \times \frac{2-j}{4-j} = 2.469\underline{/17.09°}$$

Hence

$$i(t) = 2.469\cos(10t + 17.09°) \text{ A}$$

---

[5] As a historical note, when Steinmetz first invented this method of analyzing ac circuits, he did not include this intermediate step. What approach he took, we'll probably never know. But his paper left a lot of engineers amazed. They knew that the method worked and could follow the procedure to perform the analysis but did not necessarily understand why. Numerous papers were published to try to explain or rationalize his invention. Our present development is the result of many years of evolution on the delineation on this subject. It is the simplest and neatest, and largely regarded as the standard approach. It not only solves the mystery of Steinmetz's genius, but also offers many other useful insights into and applications on many modern technologies.

EXPONENTIAL AND SINUSOIDAL EXCITATIONS

EXERCISE

7.5.1 $e_1(t) = 10\cos(2t + 50°)$ V. $e_2(t) = 15\sin(2t - 20°)$ V. Obtain $i(t)$. First set up the appropriate associated circuit using impedances and complex amplitudes. Use node analysis to find the complex amplitude I. Then give $i(t)$.

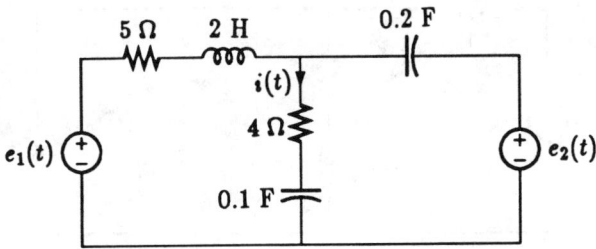

Ans. $2.412\cos(2t - 24.99°)$ A.

7.6 Network relationships as functions of s

There is another important reason that we undertake the development of the analysis of network with exponential excitations. In many situations, the solution of problems with s as a variable can prove to be very useful and informative. This is in part, because how a network response behaves as s is varied will reveal a great deal about the nature and certain characteristics of the network.

Although the student may not yet be able to appreciate the connection between a network function as a function of s and some of its properties and characteristics, one thing we can certainly declare is that if we leave s in literal form and obtain the solution in terms of s, then once an excitation is given, all we have to do is extract the value of that specific s and substitute it into the solution (together with the complex amplitude) and simplify to get the answer. This is particularly true if we have several excitations in a network whose values of s are not all the same. We can then invoke superposition

358   FUNDAMENTALS OF CIRCUIT ANALYSIS

and obtain the response to one excitation at a time. Then we combine the responses to the several excitations to obtain the complete answer.

EXAMPLE 8. For the circuit in Fig. 7.18(a), determine $e(t)$ if

(a) $e_s(t) = 5\epsilon^{-2t}$ V and $i_s(t) = 1\epsilon^{jt}$ A

(b) $e_s(t) = 5\cos 10t$ V and $i_s(t) = 1\cos(10t + 45°)$ A

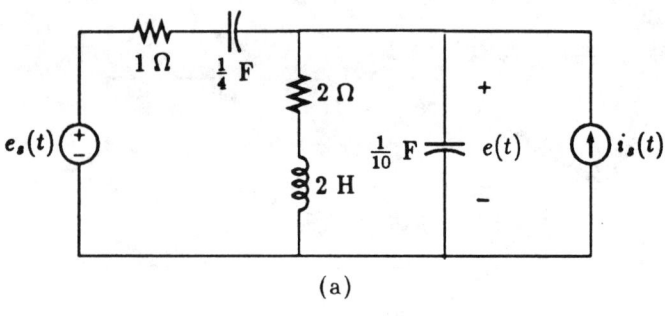

(a)

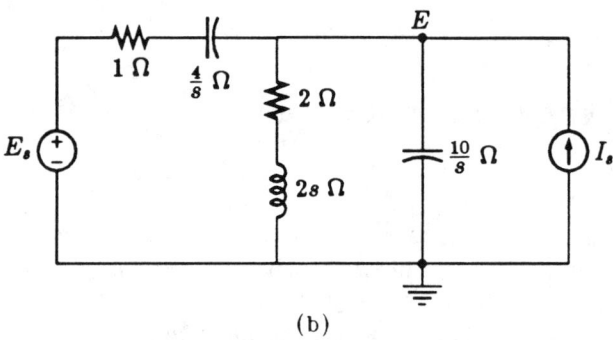

(b)

Figure 7.18: Circuits used in Example 8

SOLUTION The circuit in Fig. 7.18(b) is ready for the complex amplitude-impedance method of analysis of the circuit in Fig. 7.18(a). We shall use node analysis. The node equation is

$$\frac{E - E_s}{1 + \dfrac{4}{s}} + \frac{E}{2 + 2s} + \frac{E}{\dfrac{10}{s}} = I_s$$

Solving for E, we get

EXPONENTIAL AND SINUSOIDAL EXCITATIONS

$$E = \frac{10s(s+1)E_s + 10(s+1)(s+4)I_s}{s^3 + 15s^2 + 19s + 20} \qquad (7.56)$$

(a) For $e_s(t)$ active alone, we make $E_s = 5$, $I_s = 0$, and $s = -2$ in Eq. (7.56). We get

$$E = \frac{50}{7} \implies e(t) = \frac{50}{7}\epsilon^{-2t}$$

For $i_s(t)$ active alone, we make $E_s = 0$, $I_s = 1$, and $s = j$ in Eq. (7.56). We get

$$E = \frac{1050}{349} - j\frac{290}{349} \implies e(t) = \left(\frac{1050}{349} - j\frac{290}{349}\right)\epsilon^{jt}$$

Hence, when both $e_s(t)$ and $i_s(t)$ are active,

$$e(t) = \frac{50}{7}\epsilon^{-2t} + \left(\frac{1050}{349} - j\frac{290}{349}\right)\epsilon^{jt} \text{ V}$$

(b) Here the two values of s are the same. So we let $E_s = 5\underline{/0°}$, $I_s = 1\underline{/45°}$, and $s = j10$ in Eq. (7.56) which gives

$$E = 3.087 - j1.807 = 3.577\underline{/-30.25°}$$

From Section 7.5, we can write[6]

$$e(t) = \text{Re}\left[3.577\underline{/-30.35°}\,\epsilon^{j10t}\right] = 3.577\cos(10t - 30.35°) \text{ V}$$

EXAMPLE 9. Determine the network function $H(s) = E_2/E_1$, in the circuit in Fig. 7.19(a). E_1 is the complex amplitude of the source voltage $e_1(t)$ and E_2 is that of the output voltage $e_2(t)$.

SOLUTION The complex amplitude-impedance analysis circuit is shown in Fig. 7.19(b). We shall use the ladder analysis.

$$E_2 = 1$$

---

[6] Here we go from the circuit with sinusoidal excitation directly to the complex amplitude-impedance circuit and back, bypassing the exponential excitation circuit, as in Example 7.

FUNDAMENTALS OF CIRCUIT ANALYSIS

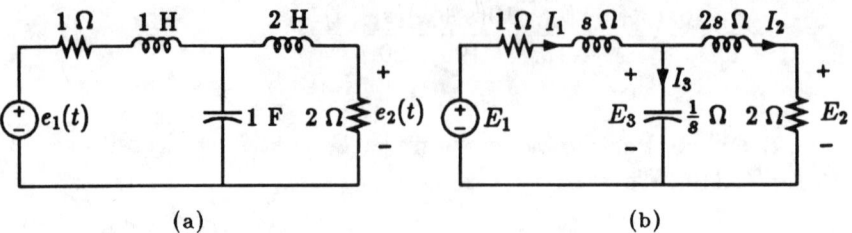

(a)   (b)

Figure 7.19: Calculation of a network function in terms of s.

$$I_2 = \frac{1}{2}$$

$$E_3 = \frac{1}{2} \times (2s + 2) = s + 1$$

$$I_3 = \frac{s+1}{\frac{1}{s}} = s(s+1)$$

$$I_1 = I_2 + I_3 = \frac{1}{2} + s(s+1) = s^2 + s + \frac{1}{2}$$

$$E_1 = E_3 + (s+1)I_1 = s + 1 + (s+1)(s^2 + s + \frac{1}{2})$$

$$= \frac{1}{2}(s+1)(2s^2 + 2s + 3)$$

$$H(s) = \frac{E_2}{E_1} = \frac{2}{(s+1)(2s^2 + 2s + 3)}$$

EXERCISE

7.6.1 In the following circuit, we have $e_1(t) = E_1 \epsilon^{st}$ and $e_2(t) = E_2 \epsilon^{st}$. Determine E_2/E_1 as a function of s.

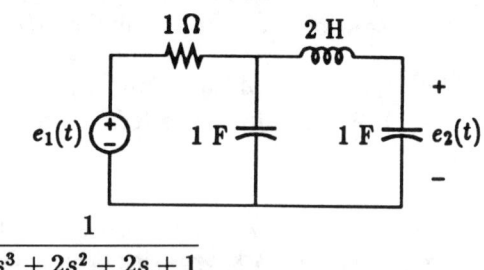

Ans. $\dfrac{1}{2s^3 + 2s^2 + 2s + 1}$

EXPONENTIAL AND SINUSOIDAL EXCITATIONS 361

7.7 An alternative approach to relating networks with sinusoidal and exponential excitations

In Section 7.5, we made use of the fact that the real part of an exponential function is a sinusoidal function. From that fact we showed that the steady-state response of a network subjected to a sinusoidal excitation can be extracted from the response of the same network subjected to an augmented excitation function—the exponential function.

Here we shall show that the same result can be arrived at by a different consideration. Suppose, in the arrangement in Fig. 7.20,

$$e_s(t) = |E_s| \cos(\omega t + \psi) \tag{7.57}$$

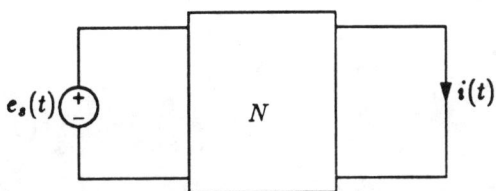

Figure 7.20: The response in an LTI network with excitation $e_s(t)$.

We can write

$$e_s(t) = e_1(t) + e_2(t) \tag{7.58}$$

where

$$e_1(t) = \frac{1}{2}|E_s|\epsilon^{j(\omega t+\psi)} \quad \text{and} \quad e_2(t) = \frac{1}{2}|E_s|\epsilon^{-j(\omega t+\psi)} \tag{7.59}$$

We denote the response due to $e(t)$ by

$$i(t) = i_1(t) + i_2(t) \tag{7.60}$$

in which $i_1(t)$ is the response due to $e_1(t)$ and $i_2(t)$ due to $e_2(t)$. We can then use the complex amplitude-impedance method to find the responses due to $e_1(t)$ and $e_2(t)$ each active alone and then combine the two responses.

With $e_1(t)$ active alone, we let

$$E_1 = \frac{1}{2}|E_s|\underline{/\psi} \tag{7.61}$$

We call the response due to E_1

$$I_1 = \frac{1}{2}|I|\underline{/\phi} \tag{7.62}$$

which implies, in terms of time,

$$i_1(t) = I_1\, \epsilon^{j\omega t} = \frac{1}{2}|I|\underline{/\phi}\, \epsilon^{j\omega t} \tag{7.63}$$

It is not necessary to repeat the analysis to obtain the response $i_2(t)$. We observe that

$$e_2(t) = [e_1(t)]^* \tag{7.64}$$

Hence,[7]

$$i_2(t) = [i_1(t)]^* = \frac{1}{2}|I|\underline{/-\phi}\, \epsilon^{-j\omega t} \tag{7.65}$$

and

$$i(t) = i_1(t) + i_2(t)$$
$$= \frac{1}{2}|I|\, \epsilon^{j(\omega t+\phi)} + \frac{1}{2}|I|\, \epsilon^{-j(\omega t+\phi)} = |I|\cos(\omega t + \phi) \tag{7.66}$$

The relationship between E_1 of Eq. (7.61) and I_1 of Eq. (7.62) is exactly the same as the relationship between E_s and I in Eq. (7.55). The factor $\frac{1}{2}$ appears in both Eq. (7.61) and Eq. (7.62) and may be regarded simply as a scaling factor as it does not affect the relationship between E_1 and I_1. Thus, both the development in Section 7.5 and that in this section lead to the same conclusion: the steady-state analysis of a network with sinusoidal excitation may be performed using the complex amplitude-impedance method with the same interpretation of the results obtained in the complex amplitude domain.

---

[7] This is because the response to a real excitation must be real. It follows that the response to an imaginary excitation must be imaginary. Hence, if $f_1(t)+jf_2(t)$ produces $g_1(t)+jg_2(t)$, then $f_1(t)-jf_2(t)$ produces $g_1(t)-jg_2(t)$.

EXPONENTIAL AND SINUSOIDAL EXCITATIONS

7.8 Summary

In this chapter, we developed a procedure for obtaining the steady-state response of a network when the excitation is either exponential or sinusoidal. This procedure—the complex amplitude-impedance method of analysis—is useful both for obtaining the forced response of a network and when only the steady-state response is of interest.

We are now in a position to apply this procedure to a class of important electrical circuits—the ac circuits—in which all excitations are sinusoidal and only steady-state responses are of interest. This specialized area of circuit analysis will be the topic of the next chapter.

Problems

7.1 Simplify the following expressions. Give your answers in rectangular form.

(a) $\dfrac{(5+j2)(10\,\epsilon^{j30°})(-2-j8)}{5\underline{/8°} \times \text{Re}[3\underline{/-10°}]}$

(b) $\epsilon^{j2\underline{/40°}}$

(c) $\ln[(8-j20) \times 3\underline{/-50°}]$

(d) $8^{\frac{1}{3}}$

(e) $(2+j5)^{(2-j)}$

(f) $10\underline{/j2}$

(g) $\sin^{-1}(2)$

(h) $\cos^{-1}(2+j5)$

(i) $\epsilon^{j[\ln(5-j3)]}$

(j) $(5+j2)^{\frac{1}{2}}$

(k) $\dfrac{(10+j8) \times 5\epsilon^j \times \mathrm{Re}[2\underline{/-j30°}}{\mathrm{Im}[2+j4] \times 2^j}$

7.2 Determine $e_2(t)$.

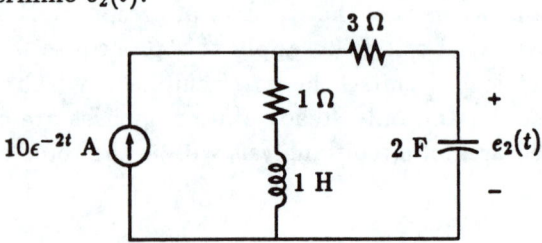

7.3 Determine $i_1(t)$, $i_2(t)$, and $i(t)$.

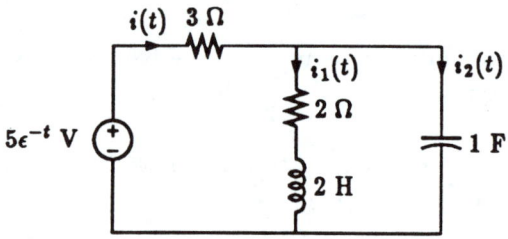

7.4 Determine $i(t)$ and $e(t)$.

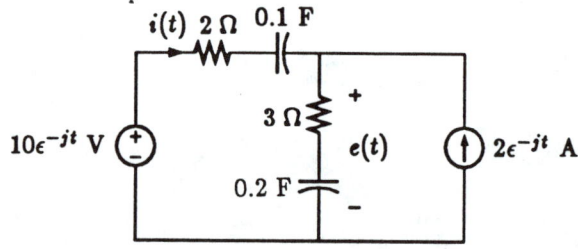

7.5 Determine $i(t)$ and $e(t)$.

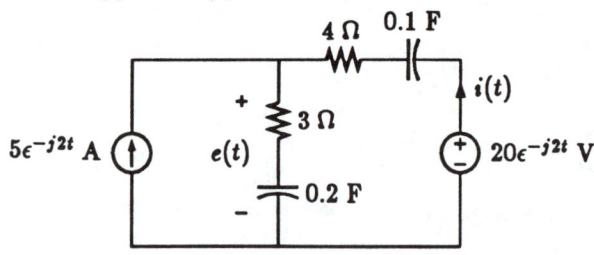

7.6 Determine $i(t)$.

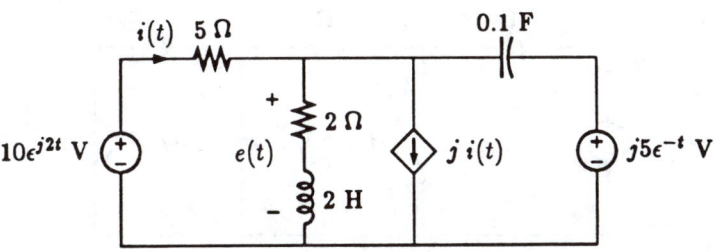

7.7 Determine $e(t)$.

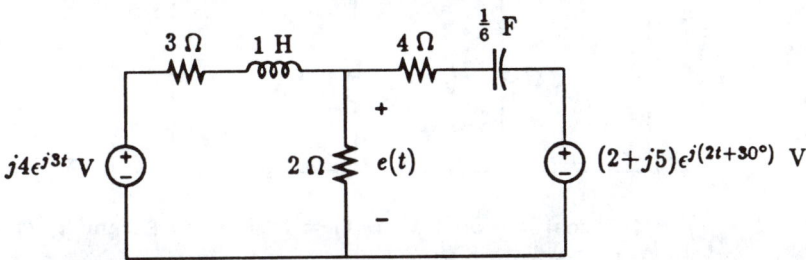

7.8 Determine $i(t)$ and $e(t)$.

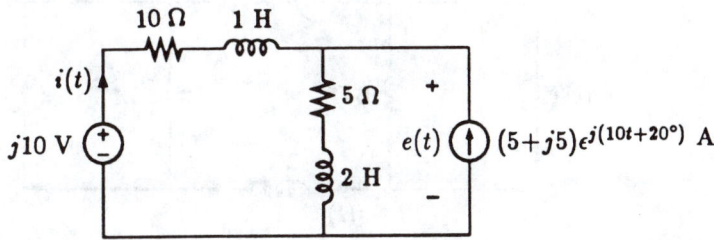

7.9 Determine $i_1(t)$ and $i_2(t)$.

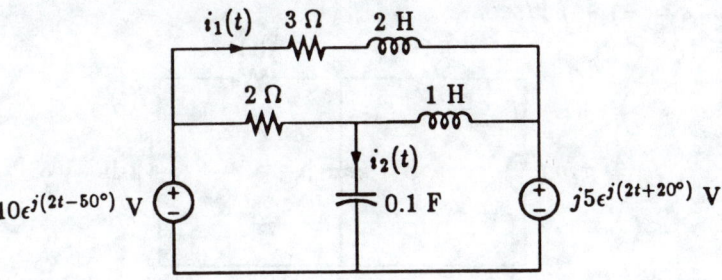

7.10 Determine $e_1(t)$ and $e_2(t)$.

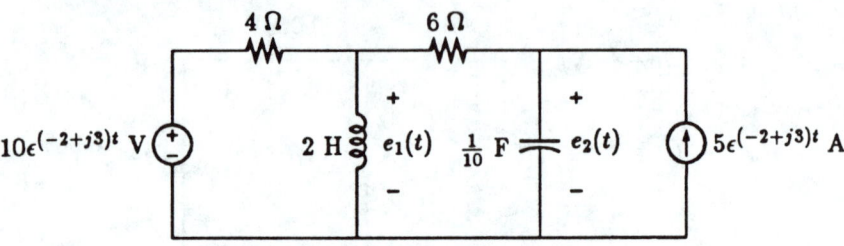

7.11 $i_s(t) = (10 + j5)\epsilon^{-2t}\cos(10t + 30°)$ A. Determine $i(t)$ and $e(t)$.

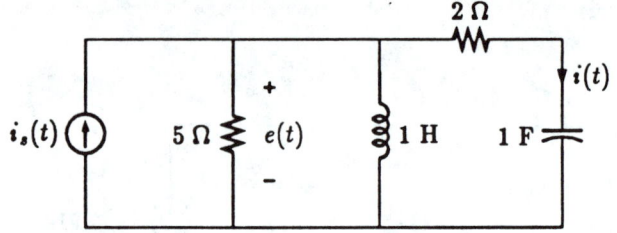

7.12 $e_s(t) = 4\epsilon^{-t}\cos(3t - 50°)$ V, $i_1(t) = 5\epsilon^{-t}\cos 3t$ A, and $i_2(t) = \epsilon^{-t}\sin 3t$ A. Determine $i(t)$.

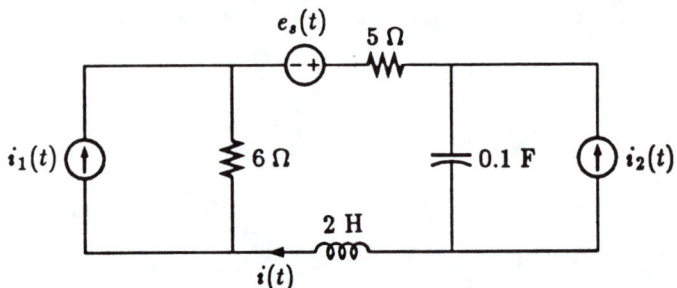

7.13 Determine $i(t)$ for (a) $E = 5$ V and $s = -0.5$ 1/sec, (b) $E = 3 + j8$ V and $s = -j2$ 1/sec, and (c) $E = 10\underline{/15°}$ V and $s = -1 + j$ 1/sec.

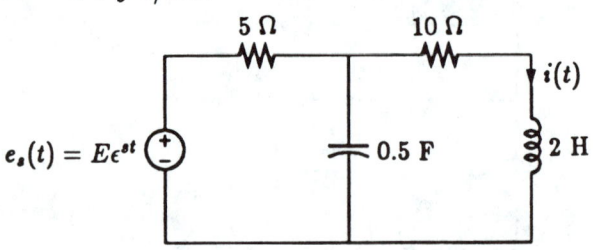

7.14 Evaluate Z_{eq} for $s = 3 - j5$.

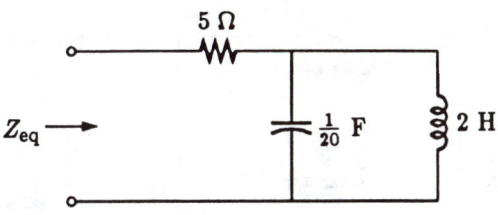

7.15 Evaluate Y_{eq} for $s = -2 + j3$.

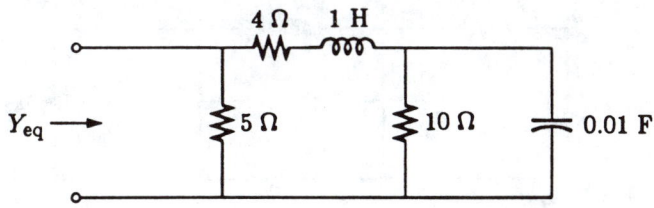

7.16 $i_s(t) = 4\cos 3t$ A, $e_1(t) = 5\cos 3t$ V, and $e_2(t) = 2\cos 2t$ V. Determine $e(t)$.

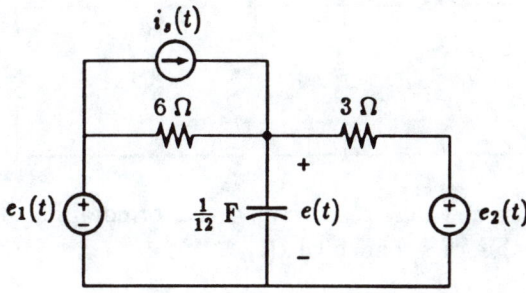

△**7.17** $i_1(t) = 10\cos 5t$ A and $i_2(t) = j10\cos 5t$ A. Determine $e_1(t)$ and $e_2(t)$.

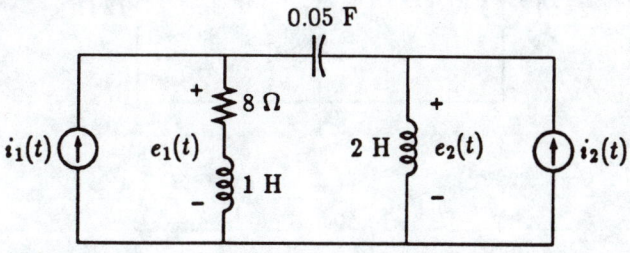

7.18 Determine $Y(s)$.

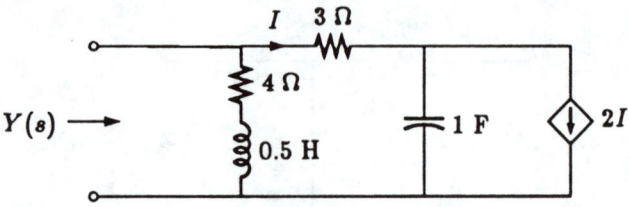

7.19 In the circuit, $e_1(t) = E_1 \epsilon^{st}$ and $e_2(t) = E_2 \epsilon^{st}$. Obtain the ratio E_2/E_1 as a function of s.

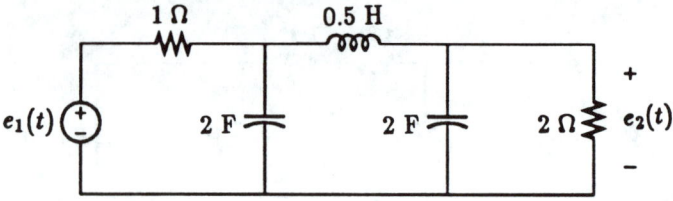

7.20 In the circuit, $e_1(t) = E_1 \epsilon^{st}$ and $i_2(t) = I_2 \epsilon^{st}$. Obtain the ratio I_2/E_1 as a function of s.

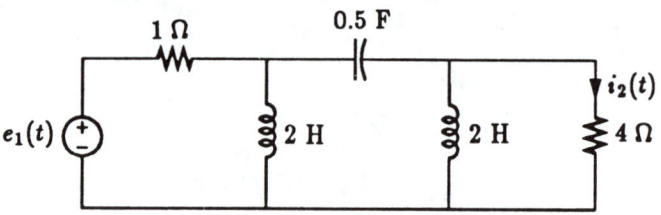

7.21 The value of s may assume any (real or complex) value. What is the value of s for which $i(t) \equiv 0$?

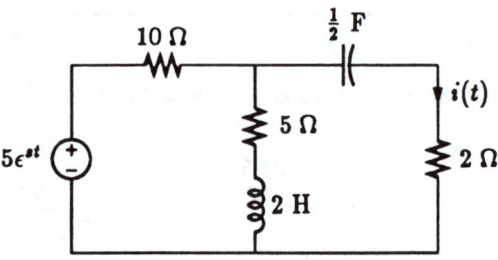

Chapter 8

Analysis of AC Circuits

In this chapter, we shall deal in detail with the many aspects of linear circuits subjected to sinusoidal excitations. Circuits of this class are known as the alternating-current (ac) circuits. Actually, the methodology and the theoretical bases for this class of circuit problems have already been established in the previous chapter, and we will draw heavily on the development and the conclusions reached there. Here, we will concentrate on the applications of those concepts and procedures to ac circuits. We will also introduce the student to the general manner in which these problems are typically handled, as well as to some of their jargon.

Because sinusoidal quantities play such important roles in electrical engineering, the study of circuits with these quantities is extremely important. In the early part of this century, ac circuit analysis was important because ac power became the standard way of furnishing consumers with electrical power. Until high-temperature superconductors become practical, ac will continue to be the form of electrical power delivered to every household or plant.

Sinusoids also play important roles in telecommunication. Carriers are largely sinusoidal. Even if they are not, other periodic waveforms can be expressed in terms of, or related to, sinusoids. Hence, ac circuit analysis is indispensible even for circuits whose excitations are not purely sinusoidal.

FUNDAMENTALS OF CIRCUIT ANALYSIS

8.1 The sinusoids

Sinusoidal quantities occur widely in nature—the swing of a pendulum, the vibration of a musical instrument, the response of an LC circuit with no loss. In this section, we shall describe some elementary properties and mathematical algorithms that are useful in dealing with sinusoidal quantities.

8.1.1 Terminology of sinusoids

In dealing with sinusoidal quantities, we can use either the sine function or the cosine function to represent a sinusoid. We shall choose to use the cosine function.[1] A sinusoidal voltage will have the form

$$e(t) = E_{\mathrm{m}} \cos(\omega t + \psi) \qquad (8.1)$$

where E_{m} is real and represents the maximum or peak value of the sinusoid as shown in Fig. 8.1. A sinusoid repeats itself when

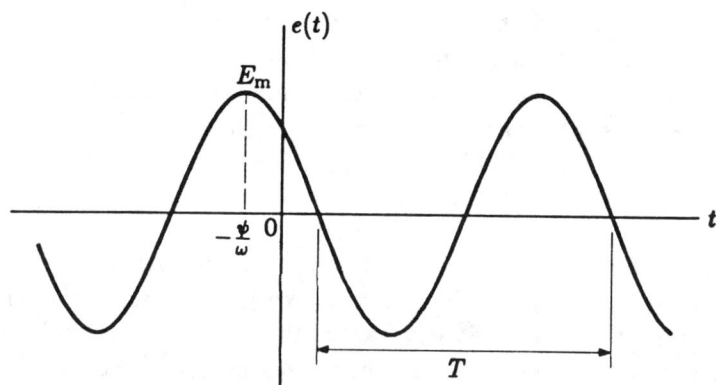

Figure 8.1: The variation of a sinusoidal voltage.

its angle is increased by 2π. The minimum duration for which a sinusoid repeats itself is known as its *period* and is denoted by T. Therefore,

---

[1]This is somewhat motivated by the fact that $\cos \omega t = \mathrm{Re}\left[\epsilon^{j\omega t}\right]$. But there is no reason why we could not use the sine function and make use of the relationship $\sin \omega t = \mathrm{Im}\left[\epsilon^{j\omega t}\right]$. The choice of using the cosine function is, therefore, largely arbitrary.

ANALYSIS OF AC CIRCUITS

$$T = \frac{2\pi}{\omega} \tag{8.2}$$

The reciprocal of T is known as the *frequency* of the sinusoid. It represents the number of repetitions the sinusoid goes through in one unit of time. We denote the frequency by f and

$$f = \frac{1}{T} = \frac{\omega}{2\pi} \tag{8.3}$$

When T is in seconds, f has the unit *hertz* (abbreviated Hz).[2] The coefficient

$$\omega = 2\pi f = \frac{2\pi}{T} \tag{8.4}$$

in radians per second is known as the *angular frequency* or *radian frequency*.[3]

The quantity ψ in Eq. (8.1) is the *phase*, *angle*, or *phase angle* of the voltage. One of the positive maxima of $e(t)$ of Eq. (8.1) occurs at $\omega t + \psi = 0$ or

$$t = -\frac{\psi}{\omega} \tag{8.5}$$

8.1.2 Conversion of sinusoids

It is our choice here to use the cosine function as the standard representation of sinusoids. Sometimes quantities do occur as sine functions. Hence, we may wish to convert a sine function into a cosine function or vice versa. The following identities are useful:

$$\cos x = \sin(x + 90°) \tag{8.6}$$

$$\sin x = \cos(x - 90°) \tag{8.7}$$

Thus,

$$e_1(t) = 5\sin(10t + 10°) = 5\cos(10t - 80°)$$

---

[2] The unit cps (for cycles per second) was used in earlier days. It is no longer being used.

[3] Sometimes ω is simply referred to as the frequency or the frequency variable for brevity when the context of its usage clearly indicates that it signifies the angular frequency.

$$i_2(t) = 100 \cos(2t + 25°) = 100 \sin(2t + 115°)$$

Also,

$$\cos(x \pm 180°) = -\cos x \qquad (8.8)$$

$$\sin(x \pm 180°) = -\sin x \qquad (8.9)$$

We can also write

$$e_1(t) = 5 \sin(10t + 10°)$$

$$= -5 \sin(10t - 170°) = -5 \cos(10t + 100°)$$

$$i_2(t) = 100 \cos(2t + 25°)$$

$$= -100 \cos(2t + 205°) = -100 \sin(2t + 295°)$$

Since sinusoids are periodic, in terms of t,

$$\cos(\omega t + \psi) = \cos[\omega(t + kT) + \psi] \qquad (8.10)$$

In terms of the angle,

$$\cos(\omega t + \psi) = \cos(\omega t + \psi + 2k\pi) \qquad (8.11)$$

where k is any integer. Thus,

$$\sin(3t + 25°) = \sin(3t - 335°) = \sin(3t + 385°)$$

$$= \cos(3t + 295°) = \cos(3t - 65°)$$

EXERCISE

8.1.1 Convert each of the following functions into a cosine function with a positive coefficient.

(a) $-20 \cos(\omega t + 10°)$ (b) $30 \sin(10t - 50°)$
(c) $10 \sin(10t + \frac{\pi}{3})$ (d) $6 \sin(10t + \frac{\pi}{2})$

Ans. (a) $20 \cos(\omega t - 170°)$, (b) $30 \cos(10t - 140°)$, (c) $10 \cos(10t - \frac{\pi}{6})$, (d) $6 \cos 10t$.

ANALYSIS OF AC CIRCUITS

There are occasions on which we wish to express a sinusoid as combinations of cosine and sine functions with zero angles. The following identities are useful:

$$\cos(x \pm y) = \cos x \cos y \mp \sin x \sin y \quad (8.12)$$

$$\sin(x \pm y) = \sin x \cos y \pm \cos x \sin y \quad (8.13)$$

Thus,

$$\cos(\omega t + 30°) = \cos \omega t \cos 30° - \sin \omega t \sin 30°$$

$$= 0.866 \cos \omega t - 0.5 \sin \omega t$$

$$\sin(\omega t - 60°) = \sin \omega t \cos 60° - \cos \omega t \sin 60°$$

$$= 0.5 \sin \omega t - 0.866 \cos \omega t$$

Conversely, if we have both a cosine function and a sine function (both with zero angle), we may wish to combine them into a single sinusoid. It is easy to show that

$$A \cos x + B \sin x = C \cos(x - \theta) \quad (8.14)$$

where

$$C = \sqrt{A^2 + B^2} \quad \text{and} \quad \theta = \tan^{-1} \frac{B}{A} \quad (8.15)$$

Thus,

$$3 \cos \omega t + 4 \sin \omega t = 5 \cos(\omega t - 53.13°)$$

$$10 \cos \omega t - 8 \sin \omega t = \sqrt{164} \cos(\omega t + 38.66°)$$

$$-4 \cos 2t + 2 \sin 2t = \sqrt{20} \cos(2t + 206.57°)$$

$$-2 \cos 10t - 5 \sin 10t = \sqrt{29} \cos(2t + 111.80°)$$

Note that the value of θ depends on the signs of both A and B, not the sign of B/A.

8.1.3 Combination of sinusoids

We frequently have the need to combine several sinusoids into a single cosine function. This can be accomplished by first applying Eqs. (8.12) and (8.13) to expand all terms into sine and cosine functions with zero angles. Then we can group all sine and cosine terms and apply Eq. (8.14) to combine them into a single term. For example,

$$5 \cos(\omega t + 10°) + 4 \sin(\omega t - 50°)$$

$$= 5 \cos 10° \cos \omega t - 5 \sin 10° \sin \omega t$$

$$+ 4 \cos 50° \sin \omega t - 4 \sin 50° \cos \omega t$$

$$= 1.860 \cos \omega t + 1.703 \sin \omega t$$

$$= 2.522 \cos(\omega t - 42.48°) \qquad (8.16)$$

There is an alternative to this combination process. We can make use of the relationship between the sinusoidal and exponential functions explained in Chapter 7. For the same example, we can write

$$5 \cos(\omega t + 10°) + 4 \sin(\omega t - 50°)$$

$$= 5 \cos(\omega t + 10°) + 4 \cos(\omega t - 140°)$$

$$= \text{Re}\left[(5\underline{/10°} + 4\underline{/-140°})\epsilon^{j\omega t}\right]$$

$$= \text{Re}\left[(2.522\underline{/-42.48°})\epsilon^{j\omega t}\right]$$

$$= 2.522 \cos(\omega t - 42.48°) \qquad (8.17)$$

This version is, of course, equivalent to the previous version—Eq. (8.16)—which is totally elementary and is somewhat lengthier to write. The advantage of using the exponential function is again apparent here. When we perform the process in Eq. (8.17), we are making use of the complex amplitude notation. We can concentrate on the magnitude and phase of each sinusoid and dispense with the

ANALYSIS OF AC CIRCUITS

recurring quantities in the process. The essential parts of Eq. (8.17) are:

The complex amplitude of $5\cos(\omega t + 10°)$ is $5\underline{/10°}$.

The complex amplitude of $4\sin(\omega t - 50°)$ is $4\underline{/-140°}$.

We then add the complex amplitudes, viz.

$$5\underline{/10°} + 4\underline{/-140°} = 2.522\underline{/-42.48°}$$

The final sum implies the sinusoid

$$2.522\cos(\omega t - 42.48°)$$

We have reduced the combination of sinusoids to the combination of complex numbers. In doing so, we can focus our attention on the essential parts of each sinusoid and perform only the addition or subtraction of complex numbers. This greatly simplifies the amount of computation. Further, as we shall see shortly, this idea and procedure enable us to simplify the method of analyzing many physical problems in which all physical quantities are sinusoids.

EXERCISE

8.1.2 Combine the following expressions into a single cosine term with positive coefficients.

(a) $2\sin(\omega t + 50°) + 10\sin(\omega t - 20°)$

(b) $10\cos(\omega t + 20°) + 15\sin(\omega t - 30°)$

(c) $8\cos\omega t + 6\sin\omega t + 10\sin(\omega t + 20°)$

Ans. (a) $10.85\cos(\omega t - 100.02°)$, (b) $9.756\cos(\omega t - 78.79°)$, (c) $19.17\cos(\omega t - 53.43°)$.

8.2 The effective value of a periodic function

A quantity is periodic with a period T if

$$f(t) = f(t + kT)$$

where k is any integer. A pertinent question about such a quantity is exactly how much such a quantity is worth or, just how strong such a quantity is? The answer to such a question will, of course, depend on the criterion we use to assess the strength of such a quantity. For example, in an electroplating process, if $f(t)$ is a periodic current, the strength of such a current should be measured by its average value, because this value is an indication of the rate by which the metal is deposited on the object.

We shall now describe the criterion that is most commonly used in the electrical engineering profession to describe the strength of a periodic electrical quantity—the effective value.

8.2.1 The root-mean-square (rms) value of a periodic function

The most commonly used measure to assess the strength of a periodic quantity is its capability of producing power. Let's say that we have a periodic current $i_p(t)$. Using this criterion, we should compare the periodic current with a dc current I_{dc}. If these two currents produce the same power when each flows through the same resistance, then they have the same power-producing capability. Then we shall say that the periodic current has an *effective value* equal to that constant current I_{dc}. Mathematically, if

$$\text{the average value of } [i_p^2(t)R] = I_{dc}^2 R \qquad (8.18)$$

then

$$\text{the effective value of } i_p(t) = I_{eff} = I_{dc} \qquad (8.19)$$

Leaving R out of both sides of Eq. (8.18), Eq. (8.19) gives

$$I_{eff} = \sqrt{\text{the average of } [i_p^2(t)]} \qquad (8.20)$$

ANALYSIS OF AC CIRCUITS

Since $i_p(t)$ is periodic, the average of $[i_p^2(t)]$ may be obtained by averaging $i_p^2(t)$ over an arbitrary period. Thus,

$$I_{\text{eff}} = \sqrt{\frac{1}{T} \int_{t_0}^{t_0+T} i_p^2(t)\, dt} \qquad (8.21)$$

in which t_0 is any arbitrary point along the t axis. Eq. (8.21) states that the "effective value" of $i_p(t)$ is the "square root of the average of $i_p(t)$ squared." Hence it is also known as the root-mean-square (rms) value of $i_p(t)$. The effective value of a periodic quantity is independent of the location of the origin since the averaging is over either a sample period or a long duration.

EXAMPLE 1. Find the effective value of the currents shown in Fig. 8.2.

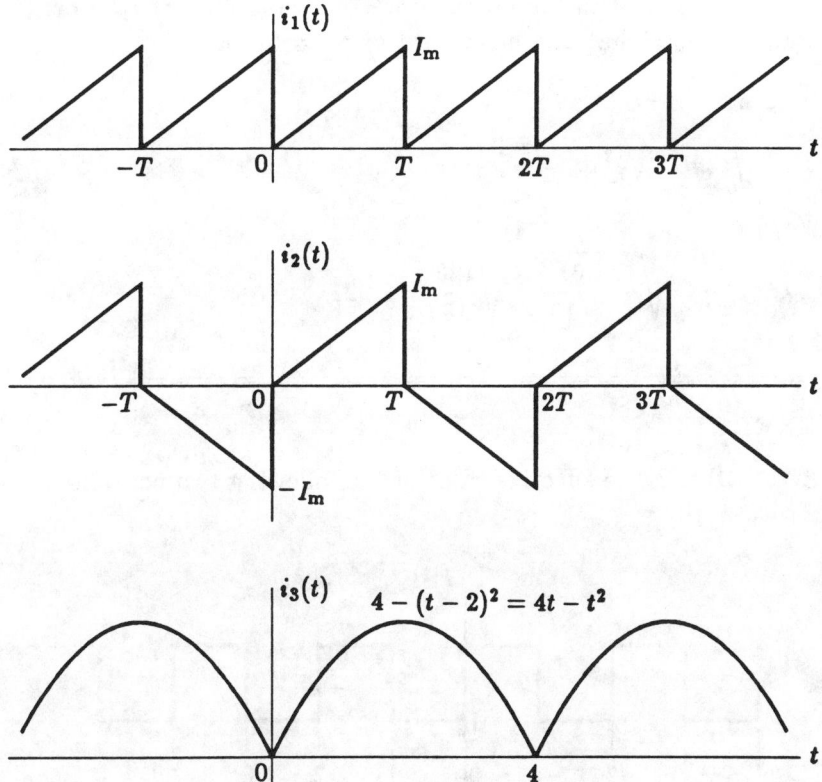

Figure 8.2: Waveforms for Example 1.

SOLUTION (a) For $0 < t < T$,

$$i_1(t) = \frac{I_m}{T} t$$

$$\int_0^T i_1^2(t)\, dt = \int_0^T \frac{I_m^2}{T^2} t^2\, dt = \frac{I_m^2}{T^2} \times \frac{T^3}{3} = \frac{I_m^2 T}{3}$$

$$I_{1\text{eff}} = \sqrt{\frac{1}{T} \times \frac{I_m^2 T}{3}} = \frac{I_m}{\sqrt{3}}$$

(b) Since $i_2^2(t) = i_1^2(t)$ for all t,

$$I_{2\text{eff}} = \frac{I_m}{\sqrt{3}}$$

If $I_m = 100$ A, then the power-producing capability of $i_1(t)$ or $i_2(t)$ is the same as that of a dc current of $\frac{100}{\sqrt{3}} = 57.74$ A.

(c) Since

$$\int_0^4 (4t - t^2)^2\, dt = \frac{512}{15}$$

we have

$$I_{3\text{eff}} = \sqrt{\frac{1}{4} \times \frac{512}{15}} = \sqrt{\frac{128}{15}} = 2.921$$

EXERCISE

8.2.1 Obtain the effective value of the following two periodic waveforms.

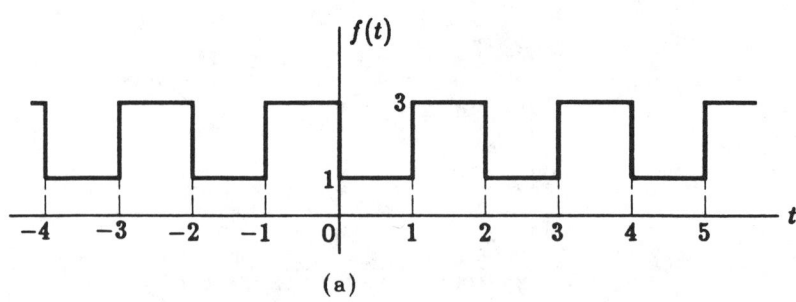

(a)

ANALYSIS OF AC CIRCUITS

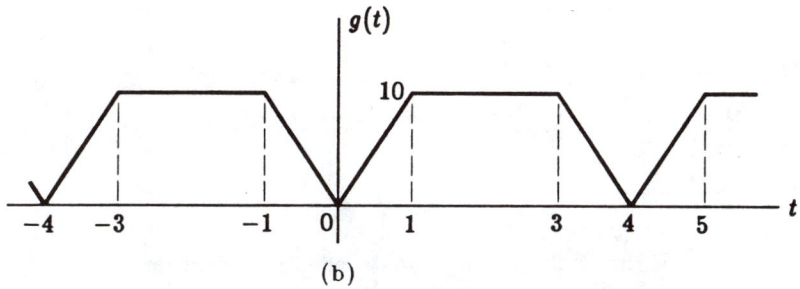

(b)

Ans. 2.236, 8.165.

8.2.2 The effective value of sinusoids

We shall now apply the definition of the rms value of a periodic function to sinusoids. If we have a voltage

$$e_s(t) = E_m \cos \omega t \tag{8.22}$$

its effective value is

$$E_{\text{eff}} = \sqrt{\frac{1}{T} \int_0^T E_m^2 \cos^2 \omega t \, dt} \tag{8.23}$$

But

$$\int_0^T \cos^2 \omega t \, dt = \frac{T}{2}$$

Hence,

$$E_{\text{eff}} = \frac{E_m}{\sqrt{2}} \tag{8.24}$$

The effective value of a sinusoid is its maximum value divided by $\sqrt{2}$.

EXAMPLE 2. In Fig. 8.3, calculate the average power dissipated in each resistor if $e_1(t) = 20 \cos(5t + 10°)$ V and $i_2(t) = 5 \sin(10t - 20°)$ A.

SOLUTION (a)

$$E_{1\text{eff}} = \frac{20}{\sqrt{2}} = 14.14 \text{ V}$$

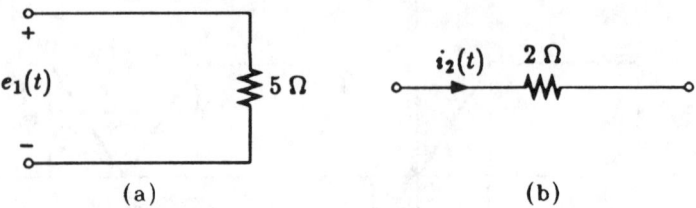

Figure 8.3: Arrangements used in Example 3.

$$P_{\text{avg}} = \frac{E_{1\text{eff}}^2}{R} = \frac{14.14^2}{5} = 40 \text{ W}$$

(b)

$$I_{2\text{eff}} = \frac{5}{\sqrt{2}} = 3.536 \text{ A}$$

$$P_{\text{avg}} = I_{2\text{eff}}^2 \times 2 = 3.236^2 \times 2 = 25\text{W}$$

[Note: the value of ω and ϕ or ψ does not figure in the calculation of the average power. Why?]

Because the effective value of a sinusoidal voltage or current is much more meaningful from practical points of view, the strength of a sinusoid is usually described in terms of its effective value rather than its magnitude or maximum value. When we describe a household voltage to be 115 volts, its instantaneous voltage is really

$$e(t) = \sqrt{2} \times 115 \cos(\omega t + \psi) = 162.63 \cos(\omega t + \psi) \text{ V}$$

It makes so much more practical sense to describe an ac quantity in terms of its power-producing capability—its effective value—than its maximum value. All ac voltmeters and ammeters give effective values, rather than the maximum values, of a voltage or current. In the calculation of powers in an ac circuit, effective values are much more convenient to use.

ANALYSIS OF AC CIRCUITS

8.3 Phasors

In Chapter 7, we showed that, when linear networks are subjected to sinusoidal excitations, the analysis can be carried out in the complex-number domain using the complex amplitude-impedance method. Suppose the excitation is the voltage

$$e_s(t) = E_m \cos(\omega t + \psi) \qquad (8.25)$$

We define its complex amplitude to be

$$E_s = E_m \underline{/\psi} \qquad (8.26)$$

If the response is the current I in the complex amplitude domain and we find it to be

$$I = I_m \underline{/\phi} \qquad (8.27)$$

we then interpret this result to imply that the current response in the time domain is

$$i(t) = I_m \cos(\omega t + \phi) \qquad (8.28)$$

As we work more and more of these problems, it becomes natural to simply dwell in the complex amplitude domain as all the essential information of all electrical quantities are contained in their complex amplitudes. We can actually stay in the complex amplitude domain exclusively and let the implied time functions of all the complex amplitudes stay in the background. We will only spell out these time-domain functions when we have a good reason to display their explicit functions. Indeed, this practice is so common that every electrical engineer accepts it as standard.

There is an additional refinement that is also common and standard in ac circuit analysis. Instead of using the maximum values of sinusoids to be the magnitudes of their complex amplitudes, we use their effective values. This makes the complex amplitudes more meaningful, especially in the calculation of powers. The only thing that we need to do in using this convention is to scale down all complex amplitudes by the constant $\sqrt{2}$. Because of the proportionality property of linear networks, the scaling of all electrical quantities by a constant still leaves all quantities consistent among them.

Actually, when we use the complex amplitude-impedance method of ac circuit analysis, it doesn't matter whether the complex amplitudes are the maximum values or the effective values of the sinusoids, as long as all electrical quantities are interpreted consistently. For example, in the circuit in Fig. 8.4, if the maximum values are used, then the proper interpretation is

$$e_s(t) = 50 \cos(\omega t + 60°) \text{ V} \tag{8.29}$$

produces

$$i(t) = 8.577 \cos(\omega t + 29.04°) \text{ A} \tag{8.30}$$

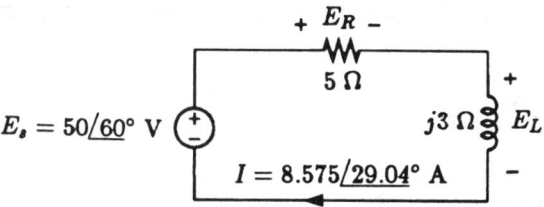

Figure 8.4: A simple ac circuit.

On the other hand, if the effective values are used, then the interpretation is

$$e_s(t) = 70.71 \cos(\omega t + 60°) \text{ V} \tag{8.31}$$

produces

$$i(t) = 12.127 \cos(\omega t + 29.04°) \text{ A} \tag{8.32}$$

Hence, as long as we stay in the complex domain, we do not have to be concerned with whether the complex amplitudes represent the maximum values or the effective values.

Primarily because of the power calculations, it is more convenient and sensible to use the effective values. *Henceforth the complex amplitude of a sinusoid shall be taken to mean its effective value instead of its maximum value.* From now on, the factor $\sqrt{2}$ that relates the maximum value and the effective value of sinusoids shall be built in when we state the complex amplitude of a sinusoid as a matter of standard practice. The complex amplitude of the voltage $e_s(t)$ of Eq. (8.25) is

ANALYSIS OF AC CIRCUITS 383

$$\frac{E_m}{\sqrt{2}}\underline{/\psi}$$

instead of that given in Eq. (8.26) and the complex amplitude of the current $i(t)$ of Eq. (8.28) is

$$\frac{I_m}{\sqrt{2}}\underline{/\phi}$$

instead of that given in Eq. (8.27).

In the earlier development of ac circuits, the complex amplitudes were called "vectors" because complex amplitudes add and subtract exactly like vectors. Beyond these two operations, complex amplitudes can no longer be regarded as vectors. For example, we can multiply complex numbers, but we cannot multiply vectors. The term "phasor" was then coined to replace the term "vector." Ostensibly, a phasor is a quantity that has not only a magnitude, but also a phase [angle]. This term is now widely accepted, and we shall use it as a synonym for "complex amplitude."

8.4 Circuit analysis in the phasor domain

We shall now illustrate the steady-state analysis of linear networks with sinusoidal excitations. We use the word "illustrate" advisedly. This is because the basic laws that all electrical quantities must satisfy are still Ohm's law, KVL, and KCL. Since these laws must be satisfied by all instantaneous voltages and currents, they must be satisfied by all phasors. Thus, the analysis of ac circuits is simply a generalization of the various methods of analysis used on dc resistive circuits (Chapters 3 and 4). The only difference is that all electrical quantities are phasors and the voltage-current relationship of each element is represented by the impedance or admittance of that element. Controlled sources usually remain the same, except that sometimes the proportionality quantity can be complex.

8.4.1 The impedance and the admittance

In ac circuit analysis, the impedance of an element or a branch plays a role similar to that of the resistance in dc resistive circuit analysis. Likewise, the admittance plays a role similar to that of the conductance.

The impedance of an R-Ω resistor is simply R Ω (or $R + j0 = R\underline{/0}$ Ω). The admittance of an R-Ω resistor is $1/R = G$ ℧ (or $G + j0 = G\underline{/0}$ ℧).

The impedance of an L-H inductor is $j\omega L$ Ω (or $0 + j\omega L = \omega L\underline{/90°}$ Ω). The admittance of an L-H inductor is $\frac{1}{j\omega L} = -j\frac{1}{\omega L}$ ℧ (or $0 - j\frac{1}{\omega L} = \frac{1}{\omega L}\underline{/-90°}$ ℧).

The impedance of a C-F capacitor is $\frac{1}{j\omega C} = -j\frac{1}{\omega C}$ Ω (or $0 - j\frac{1}{\omega C} = \frac{1}{\omega C}\underline{/-90°}$ Ω). The admittance of a C-F capacitor is $j\omega C$ ℧ (or $0 + j\omega C = \omega C\underline{/90°}$ ℧).

These impedances and admittances serve to relate the voltage phasor and the current phasor associated with an element. Again, Ohm's law applied to these elements implies an associated direction relationship between each pair of voltage and current phasors. This cannot be over-stressed.

Generally, the impedance or admittance of a branch that consists of more than one element is complex. When a *complex impedance*, Z, is expressed in rectangular form

$$Z = R + jX \tag{8.33}$$

its real part, R, is known as the *resistance* and its imaginary part, X, the *reactance* of the impedance.

When an admittance, Y, is expressed in rectangular form

$$Y = G + jB \tag{8.34}$$

its real part, G, is known as the *conductance* and its imaginary part, B, the *susceptance* of the admittance.

It is important to note that although the impedance and the admittance of the same branch are reciprocal of each other, their real parts and imaginary parts individually are not necessarily reciprocal of each other. For example, if

$$Z = 10 + j8 = R + jX$$

then

ANALYSIS OF AC CIRCUITS

$$Y = \frac{1}{10+j8} = \frac{5}{82} - j\frac{2}{41} = G + jB$$

Clearly, $G \neq 1/R$ and $B \neq 1/X$.

Another point should be clarified here. In general, impedances and admittances are complex numbers. They can be operated, treated, and manipulated as complex numbers in all respects. However, they are not complex amplitudes. Rather, they represent ratios of complex amplitudes. They differ from complex amplitudes in that they do not implicitly represent any time functions. Their utility is only to relate pairs of complex amplitudes or phasors. In this consideration, impedances and admittances are very different from complex amplitudes or phasors.

8.4.2 Relationship among phasors, phasor diagrams

If we have two phasors

$$A = 5\underline{/70°} \quad \text{and} \quad B = 10\underline{/20°}$$

we may describe their phase relationship with any of the following:

(1) A leads B by 50°.

(2) B lags A by 50°.

(3) A lags B by $-50°$.

(4) B leads A by $-50°$.

(5) A leads B by 410°.

(6) A leads B by $-310°$.

Etc.

Statements (5) and (6) are the consequence of the fact that

$$B = 10\underline{/20°} = 10\underline{/380°} = 10\underline{/-340°}$$

Although all six statements above (as well as numerous others) are correct, we usually prefer to describe the phase angle relationship by positive angles of less than 180°. Thus, statements (1) and (2) will usually be the preferred way to describe phase relationships, although the remainder of the statements are also correct and acceptable.

If two phasors have the same phase angle, they are said to be *in phase*.

If the phase angles of two phasors differ by 90°, they are said to be in *quadrature phase* with each other.

If the phase angles of two phasors differ by 180°, they are actually the negative of each other.

We shall now elaborate briefly on some of the fine points of how an impedance or admittance relates the voltage and current phasors in a branch. Fig. 8.5 shows the symbols associated with a branch. With the indicated relative directions of I and E, we can write

$$E = |E|\underline{/\psi} = ZI = |Z|\underline{/\theta}|I|\underline{/\phi} = |Z||I|\underline{/\phi + \theta} \qquad (8.35)$$

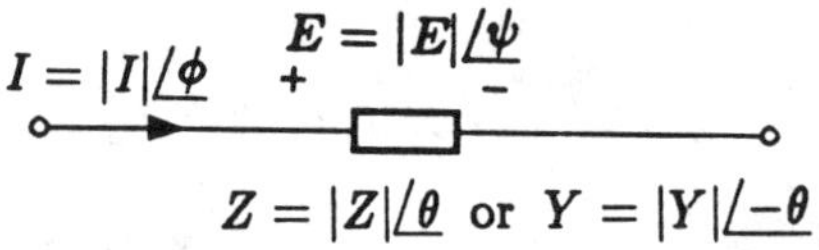

$$I = |I|\underline{/\phi} \quad E = |E|\underline{/\psi}$$
$$Z = |Z|\underline{/\theta} \text{ or } Y = |Y|\underline{/-\theta}$$

Figure 8.5: Quantities associated with a branch in ac circuits.

Hence,

$$|E| = |Z||I| \quad \text{and} \quad \psi = \phi + \theta$$

Also,

$$I = |I|\underline{/\phi} = YE = |Y|\underline{/-\theta}|E|\underline{/\psi} = |Y||E|\underline{/\psi - \theta} \qquad (8.36)$$

Thus,

$$|I| = |Y||E| \quad \text{and} \quad \phi = \psi - \theta$$

The impedance or admittance of a branch relates the magnitudes as well as the phases of their phasors. The magnitude parts are related exactly like the voltage and the current are related by the resistance or conductance of a branch in a dc resistive network. The phase-angle relationship did not exist in dc resistive networks and will be described briefly here.

The phase angle of Z is the angle by which E leads I (or the angle by which I lags E) since

ANALYSIS OF AC CIRCUITS

$$Z = |Z|\underline{/\theta} = \frac{E}{I} = \frac{|E|}{|I|}\underline{/\psi - \phi} \qquad (8.37)$$

The angle of Y is the angle by which I leads E (or the angle by which E lags I) since

$$Y = |Y|\underline{/-\theta} = \frac{I}{E} = \frac{|I|}{|E|}\underline{/\phi - \psi} \qquad (8.38)$$

We can easily see the following phase relationships in the three pure elements. These observations are frequently helpful in writing circuit equations.

The voltage and current in a resistance are in phase.

The voltage in an inductance leads the current by 90°. Or, the current in an inductance lags the voltage by 90°.

The current in a capacitance leads the voltage by 90°. Or, the voltage in a capacitance lags the current by 90°.

As mentioned earlier, phasors do combine—add or subtract—like vectors. Phasors in an ac circuit are frequently plotted to scale to show how they combine as well as how they are situated relative to one another. For example, in the circuit in Fig. 8.4, we have, in addition to the phasors with their values shown,

$$E_L = j3 \times I = 25.73\underline{/119.04°} \text{ V}$$

$$E_R = 5 \times I = 42.88\underline{/29.04°} \text{ V}$$

The three voltage "vectors" are shown in Fig. 8.6. As $E = E_R + E_L$, it is seen that E_R and E_L add according to the parallelogram rule of vector addition. Their addition may also be carried out in a head-to-tail fashion as shown in the same figure. Also, the vector of the current, I, is included.[4]

A diagram of the type shown in Fig. 8.6 is known as a *phasor diagram*. It is frequently helpful in displaying the relative magnitudes and phases of the complex amplitudes in a network. The diagram

---

[4] Here we use solid arrowheads to represent voltage phasors and hollow ones to represent current phasors. This distinction is visually helpful in using a phasor diagram.

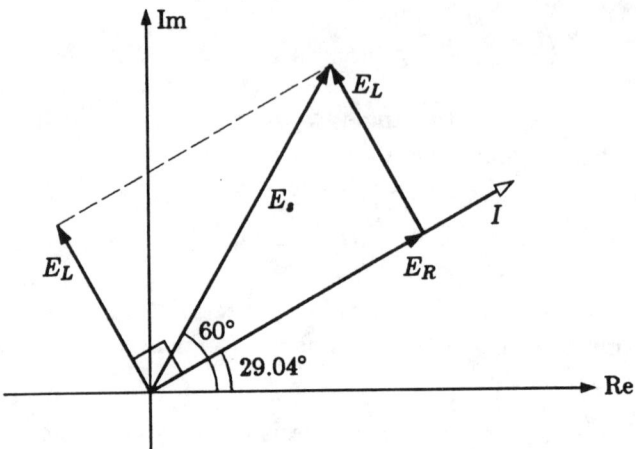

Figure 8.6: The phasor diagram for the circuit in Fig. 8.4.

in Fig. 8.6, for instance, shows that E_R and I are in phase and that E_L leads I and E_R by 90°.

In analyzing an ac circuit, we often prefer to choose a quantity as the *reference*. By this we simply mean that we arbitrarily assign an angle (usually, but not necessarily, zero) to one of the phasors. We can do this because a change of all phasors in a circuit by the same angle θ is really a multiplication of all phasors by the constant $\epsilon^{j\theta} = 1\underline{/\theta}$. Because of the proportionality property of linear networks, all phasors remain consistent with one another by this change of angles.

One reason for doing this might be to simplify the angle values to make the phasors easier to visualize. Another reason might be that in a problem, none of the phase angles are known in advance. After we have analyzed the circuit using this reference, we can always change all angles to fit the problem at hand.

For example, in the circuit in Fig. 8.4, if we choose E_s as the reference or we let $E_s = 50\underline{/0°}$ V, then $I = 8.575\underline{/-30.96°}$ A, $E_R = 42.88\underline{/-30.96°}$ V, and $E_L = 25.73\underline{/59.04°}$ V.

If, instead, we choose I as the reference or we let $I = 8.575\underline{/0°}$ A, then $E_R = 42.88\underline{/0°}$ V, $E_L = 25.73\underline{/90°}$ V, and $E_s = 50\underline{/30.96°}$ V.

When we choose a different quantity as the reference, it is equivalent to rotate the entire phasor diagram by some convenient angle. Doing this amounts to locating the origin at different locations along the time axis with regard to all time functions implied by all phasors.

ANALYSIS OF AC CIRCUITS

8.4.3 Series-parallel combinations

The total impedance of a number of impedances connected in series is the sum of all impedances. Hence, in Fig. 8.7,

$$Z = Z_1 + Z_2 + \cdots + Z_n \qquad (8.39)$$

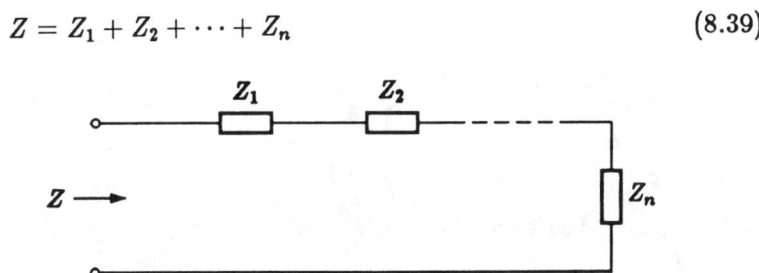

Figure 8.7: Impedances connected in series.

The total admittance of a number of admittances connected in parallel is the sum of all admittances. Therefore, in Fig. 8.8,

$$Y = Y_1 + Y_2 + \cdots + Y_n \qquad (8.40)$$

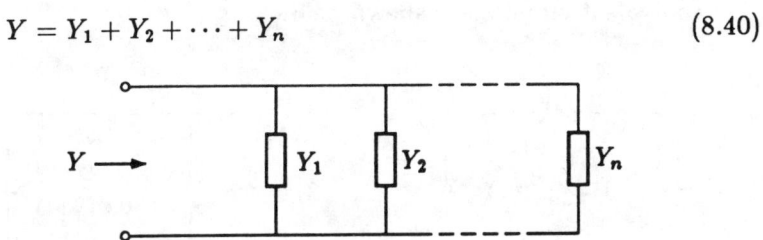

Figure 8.8: Admittances connected in parallel.

EXAMPLE 3. Find the equivalent impedances, Z_1 and Z_2, for the circuits in Fig. 8.9. For each impedance find both a series and a parallel equivalent R-L or R-C circuits.

SOLUTION.

$$Z_1 = 5 - j2 + \frac{(6 - j3)(5 + j6)}{6 - j3 + 5 + j6} = 9.546 - j1.331 \; \Omega$$

$$Y_1 = \frac{1}{Z_1} = 0.1028 + j0.0143 \; \mho$$

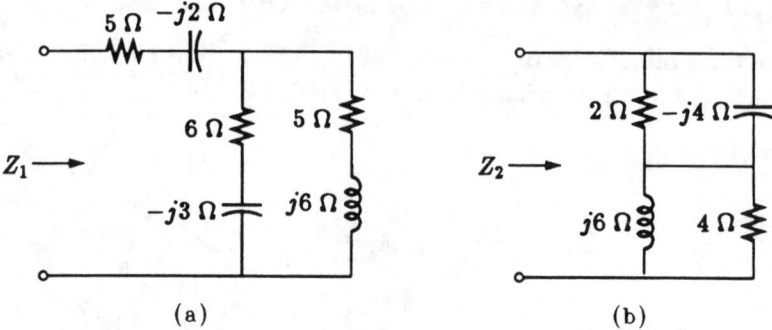

Figure 8.9: Circuits used in Example 3.

$$Z_2 = \frac{2(-j4)}{2-j4} + \frac{4(j6)}{4+j6} = \frac{284}{65} + j\frac{68}{65}\ \Omega$$

$$Y_2 = \frac{1}{Z_2} = \frac{71}{328} - j\frac{17}{328}\ \mho$$

The equivalent circuits are shown below.

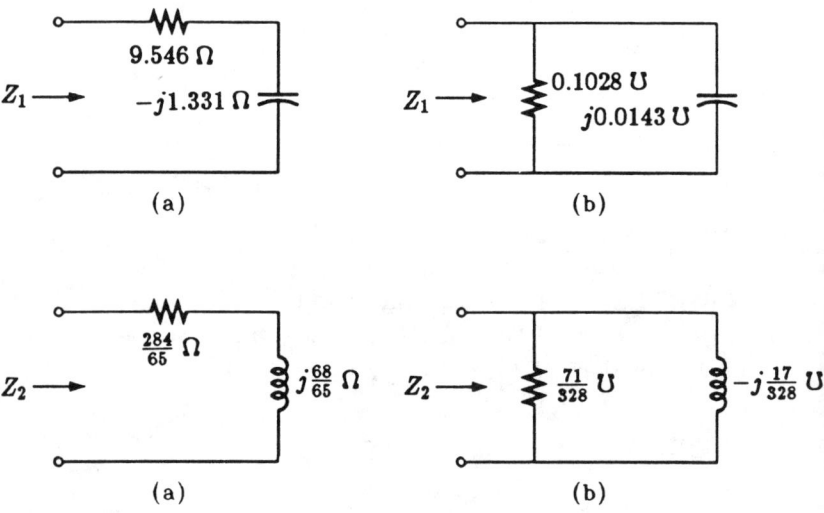

ANALYSIS OF AC CIRCUITS

EXERCISE

8.4.1 Calculate the impedances, Z_a and Z_b, in the circuits.

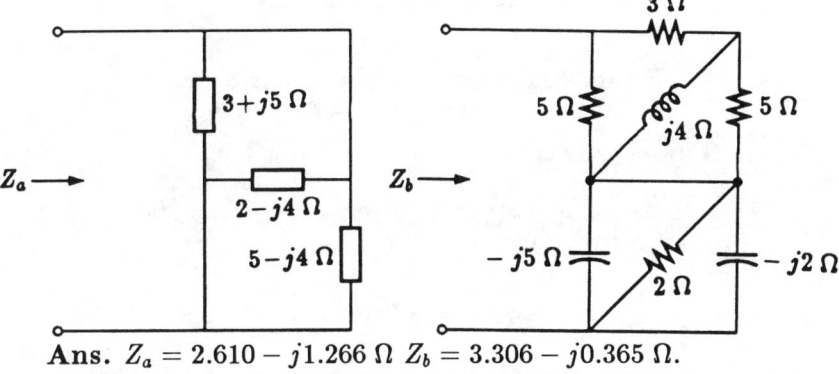

Ans. $Z_a = 2.610 - j1.266 \ \Omega \quad Z_b = 3.306 - j0.365 \ \Omega$.

8.4.4 Voltage-division and current-division rules

In a number of series-connected branches, the voltage across each branch compared to the total voltage is the same as the impedance of that individual branch compared to the total impedance. Therefore, for the arrangement in Fig. 8.10,

$$E_i = \frac{Z_i}{Z_1 + Z_2 + \cdots + Z_n} \times E, \quad i = 1, 2, \ldots, n \quad (8.41)$$

Figure 8.10: The voltage division rule.

EXAMPLE 4. Determine all voltages in the circuit in Fig. 8.11. Then construct a phasor diagram to show how the component voltages add up to equal the total voltage.

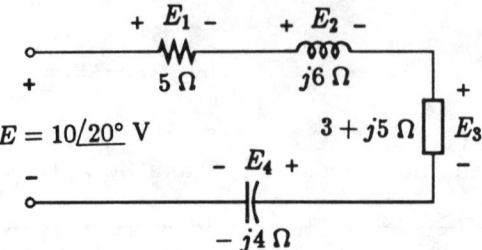

Figure 8.11: Circuit used in Example 4.

SOLUTION

$$E_1 = \frac{5}{8+j7} \times E = 4.386 - j1.700 = 4.704\underline{/-21.18°} \text{ V}$$

$$E_2 = \frac{j6}{8+j7} \times E = 2.040 + j5.263 = 5.644\underline{/68.81°} \text{ V}$$

$$E_3 = \frac{3+j5}{8+j7} \times E = 4.331 + j3.366 = 5.485\underline{/37.85°} \text{ V}$$

$$E_4 = \frac{-j4}{8+j7} \times E = -1.360 - j3.509 = 3.763\underline{/-111.19°} \text{ V}$$

The phasor diagram for these voltages is shown in Fig. 8.12.

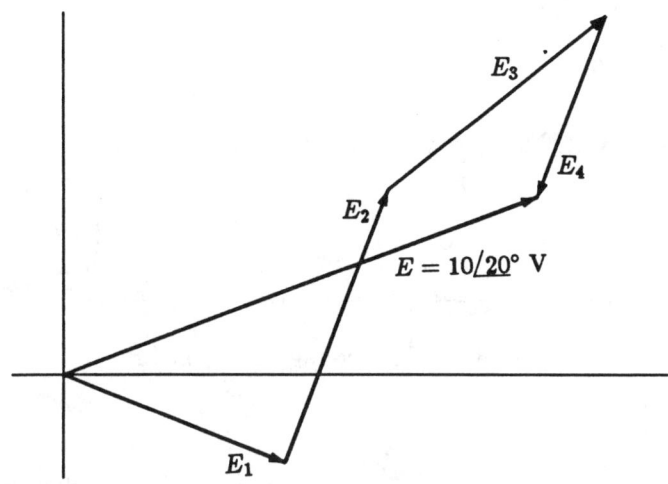

Figure 8.12: Phasor diagram for voltages in Example 4.

ANALYSIS OF AC CIRCUITS

In a number of parallel-connected branches, the component current compared to the total current is the same as the individual admittance compared to the total admittance. Therefore, in the arrangement in Fig. 8.13,

$$I_i = \frac{Y_i}{Y_1 + Y_2 + \cdots + Y_n} \times I, \quad i = 1, 2, \ldots, n \qquad (8.42)$$

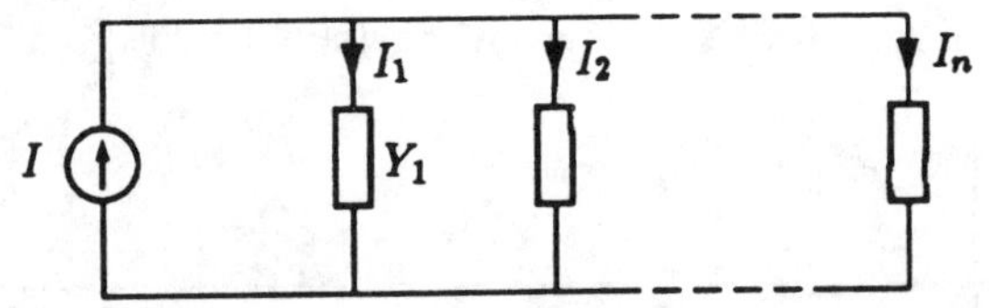

Figure 8.13: The current-division rule.

EXAMPLE 5. Determine all currents for the circuit in Fig. 8.14. Then construct a phasor diagram to show how the component currents add up to equal the total current.

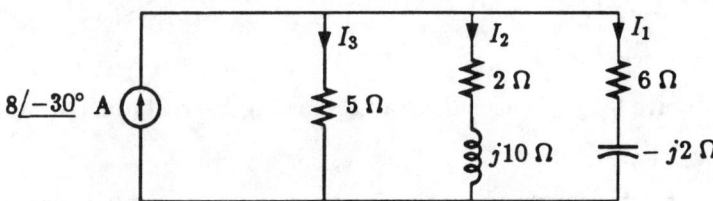

Figure 8.14: Circuit used in Example 5.

$$I_1 = \frac{\frac{1}{6-j2}}{\frac{1}{5} + \frac{1}{2+j10} + \frac{1}{6-j2}} \times 8\underline{/-30°} = 3.389 - j0.263$$

$$= 3.399\underline{/-4.44°} \text{ A}$$

$$I_2 = \frac{\frac{1}{2+j10}}{\frac{1}{5} + \frac{1}{2+j10} + \frac{1}{6-j2}} \times 8\underline{/-30°} = -0.423 - j2.065$$

$$= 2.108\underline{/-101.57°} \text{ A}$$

$$I_3 = \cfrac{\cfrac{1}{5}}{\cfrac{1}{5} + \cfrac{1}{2+j10} + \cfrac{1}{6-j2}} \times 8\underline{/-30°} = 3.962 - j1.672$$

$$= 4.300\underline{/-22.88°} \text{ A}$$

The phasor diagram of the currents is shown in Fig. 8.15.

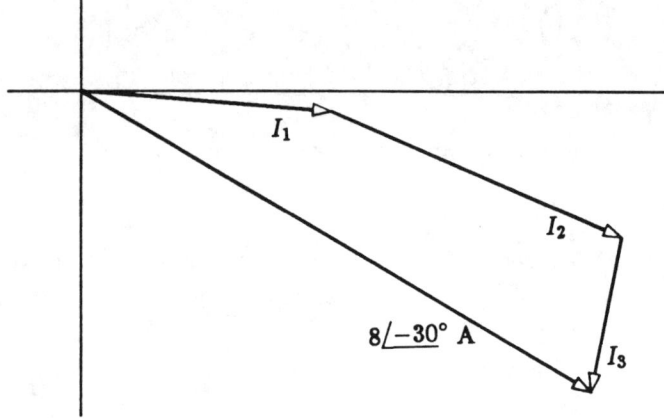

Figure 8.15: Phasor diagrams showing currents in Example 5.

EXERCISES

8.4.2 Calculate currents I_1, I_2, and I_3 in the circuit.

Ans. $I_1 = 5.172 + j0.431$ A, $I_2 = 1.293 - j3.017$ A, $I_3 = 2.155 - j0.862$ A.

ANALYSIS OF AC CIRCUITS

8.4.3 Find the value of current I in the following circuit.

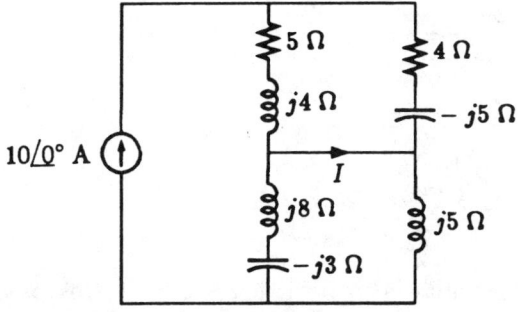

Ans. $-j5$ A.

8.4.5 Delta-wye and wye-delta equivalence

A simple extension of the equivalence of delta-connected and wye-connected resistors will lead to similar equivalence of delta-connected and wye-connected impedances. The two networks in Fig. 8.16 are equivalent if

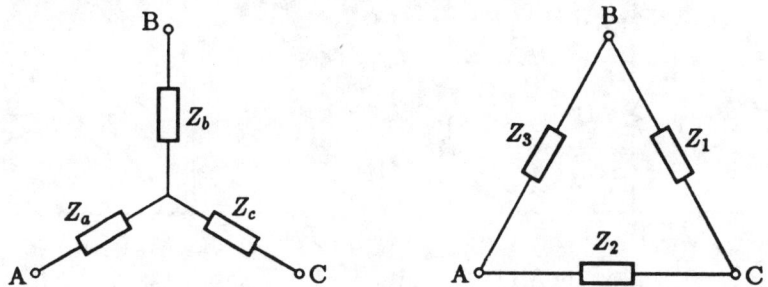

Figure 8.16: Equivalence of delta- and wye-connected impedances.

$$Z_a = \frac{Z_2 Z_3}{Z_1 + Z_2 + Z_3}$$

$$Z_b = \frac{Z_3 Z_1}{Z_1 + Z_2 + Z_3} \tag{8.43}$$

$$Z_c = \frac{Z_2 Z_1}{Z_1 + Z_2 + Z_3}$$

or

$$Z_1 = \frac{Z_a Z_b + Z_b Z_c + Z_c Z_a}{Z_a}$$

$$Z_2 = \frac{Z_a Z_b + Z_b Z_c + Z_c Z_a}{Z_b} \quad (8.44)$$

$$Z_3 = \frac{Z_a Z_b + Z_b Z_c + Z_c Z_a}{Z_c}$$

This equivalence may also be expressed in terms of admittances. The conversion formulas are similar to those in Chapter 3 using conductances.

EXAMPLE 6. Find the equivalent impedances, Z_{eq}, between terminals A and B for the circuits in Fig. 8.17.

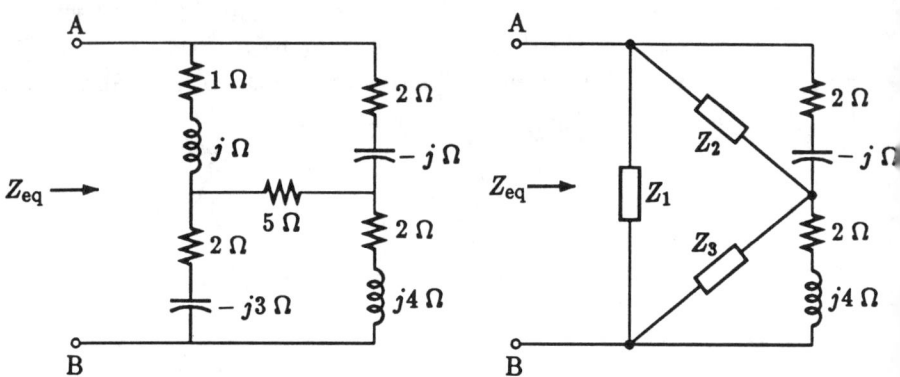

Figure 8.17: Circuits used in Example 6.

$$Z_1 = \frac{5(1+j) + 5(2-j3) + (1+j)(2-j3)}{5} = 4 - j2.2$$

$$Z_2 = \frac{5(1+j) + 5(2-j3) + (1+j)(2-j3)}{2-j3} = \frac{73}{13} + j\frac{38}{13}$$

$$Z_3 = \frac{5(1+j) + 5(2-j3) + (1+j)(2-j3)}{1+j} = 4.5 - j15.5$$

$$Z_4 = Z_2 \| (2-j) = \frac{1407}{802} - j\frac{331}{802}$$

ANALYSIS OF AC CIRCUITS

$$Z_5 = Z_3 \| (2 + j4) = \frac{1222}{349} + j\frac{1464}{349}$$

$$Z_4 + Z_5 = 5.256 + j3.782$$

$$Z_{eq} = Z_1 \| (Z_4 + Z_5) = 3.144 - j0.152 = 3.148\underline{/-2.77°} \; \Omega$$

EXERCISE

8.4.4 Find the equivalent wye-connected circuit for the delta-connected network shown below.

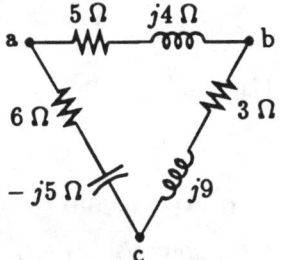

Ans. $Z_a = 2.662 - j1.592 \; \Omega$, $Z_b = 0.623 + j3.715 \; \Omega$, and $Z_c = 4.592 + j0.162 \; \Omega$.

8.4.6 General methods of analysis

The general analysis techniques developed in Chapter 3 also apply to networks consisting of impedances, controlled sources, and phasors. The present subsection simply provides the reader with some examples to illustrate these techniques as they are applied to ac circuits. The general statements and the descriptions of these techniques are identical to those made for memoryless networks in Chapter 3.

EXAMPLE 7. [Ladder analysis] In the circuit of Fig. 8.18, $E_s = 50\underline{/0°}$ V. Calculate E.

SOLUTION. We first assume $E = 1$ V. Then we obtain successively

$$I_5 = 0.1$$

FUNDAMENTALS OF CIRCUIT ANALYSIS

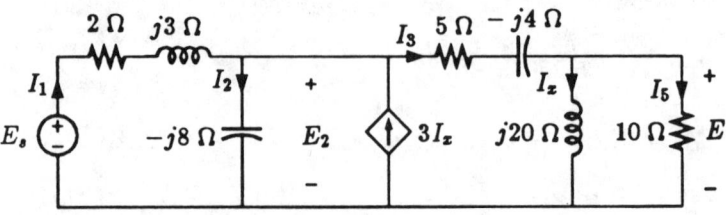

Figure 8.18: An example of ladder analysis.

$$I_x = \frac{1}{j20} = -j0.05$$

$$I_3 = I_x + I_5 = 0.1 - j0.05$$

$$E_2 = E + (5 - j4)I_3 = 1.3 - j0.65$$

$$I_2 = \frac{E_2}{-j8} = 0.0813 + j0.163$$

$$I_1 = I_2 - 3I_x + I_3 = 0.181 + 0.263$$

$$E_1 = (2 + j3)I_1 + E_2 = 0.875 + j0.419$$

By proportionality, the actual E should satisfy

$$\frac{E}{50} = \frac{1}{0.875 + j0.419}$$

Thus

$$E = 46.49 - j22.25 = 51.54\underline{/-25.57°} \text{ V}$$

EXERCISE

8.4.5 Use ladder analysis to determine I in the circuit.

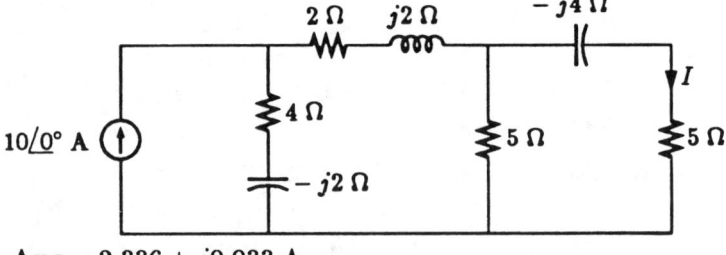

Ans. $2.336 + j0.033$ A.

ANALYSIS OF AC CIRCUITS

EXAMPLE 8. [Node analysis] Use node analysis to determine E_x in Fig. 8.19.

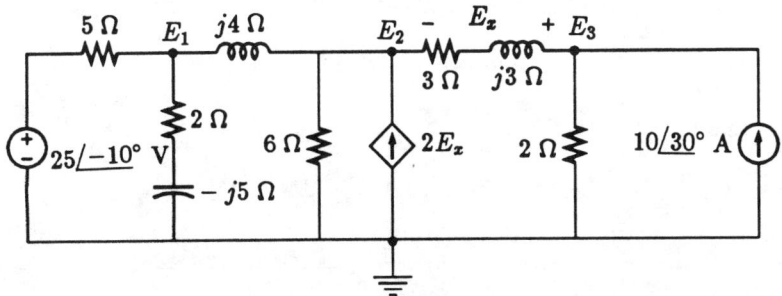

Figure 8.19: An example of node analysis.

SOLUTION. The node equations are

$$\frac{E_1 - 25\underline{/-10°}}{5} + \frac{E_1}{2-j5} + \frac{E_1 - E_2}{j4} = 0$$

$$\frac{E_2 - E_1}{j4} + \frac{E_2}{6} - 2E_x + \frac{E_2 - E_3}{3+j3} = 0$$

$$\frac{E_3 - E_2}{3+j3} + \frac{E_3}{2} = 10\underline{/30°}$$

The controlling quantity is

$$E_x = E_3 - E_2$$

Substitute, rationalize all complex fractions, and collect like terms to get

$$\left(\frac{39}{145} - j\frac{9}{116}\right) E_1 + j\frac{1}{4} E_2 = 5\underline{/-10°}$$

$$j\frac{1}{4} E_1 + \left(\frac{7}{3} - j\frac{5}{12}\right) E_2 + \left(-\frac{13}{6} + j\frac{1}{6}\right) E_3 = 0$$

$$\left(-\frac{1}{6} + j\frac{1}{6}\right) E_2 + \left(\frac{2}{3} - j\frac{1}{6}\right) E_3 = 10\underline{/30°}$$

Solving yields

FUNDAMENTALS OF CIRCUIT ANALYSIS

$$E_1 = 27.518 - j7.464 = 28.51\underline{/-15.18°}$$

$$E_2 = 13.097 + j7.593 = 15.14\underline{/30.10°}$$

$$E_3 = 15.653 + j10.037 = 18.59\underline{/32.67°}$$

The answer is

$$E_x = E_3 - E_2 = 2.557 + j2.445 = 3.537\underline{/43.72°} \text{ V}$$

EXERCISE

8.4.6 Use node analysis to determine E_1 and E_2 in the following circuit.

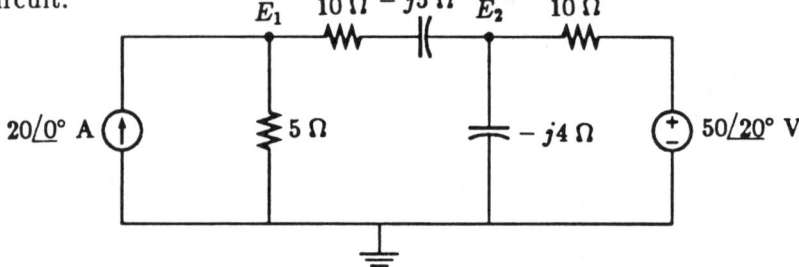

Ans. $E_1 = 80.59 - j14.23$ V, $E_2 = 27.55 - j23.30$ V.

EXAMPLE 9. [Mesh analysis] Use mesh analysis to determine I_1 and I_2 in the circuit in Fig. 8.20. Then calculate the voltage across the current source, E_i.

SOLUTION The mesh equations are

$$2I_1 + (3 + j4)(I_1 + I_3) + (1 + j3)(I_1 - I_2) = 10\underline{/0°}$$

$$(1 + j3)(I_2 - I_1) + (2 - j2)(I_2 + I_3) + j3I_2 + 2E_x = 0$$

We also have

$$I_3 = 2\underline{/-20°}$$

$$E_x = -3(I_1 + I_3)$$

ANALYSIS OF AC CIRCUITS

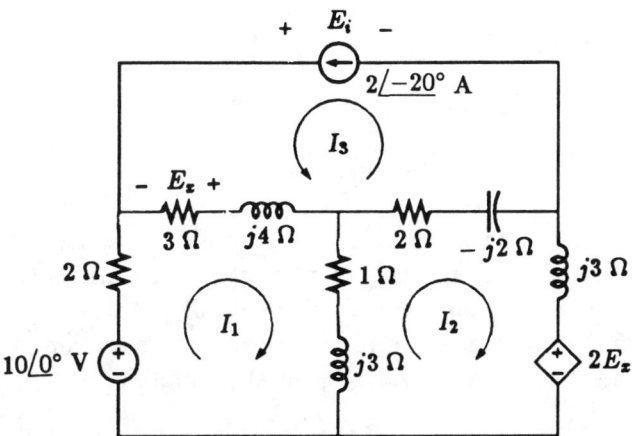

Figure 8.20: An example of mesh analysis.

Substituting and solving, we get

$$I_1 = 0.224 - j1.636 \quad \text{and} \quad I_2 = 0.282 - j3.628$$

Hence,

$$E_i = (2 - j2)(I_2 + I_3) + (3 + j4)(I_1 + I_3)$$

$$= 11.29 - j11.49 = 16.11\underline{/-45.52°} \text{ V}$$

EXERCISE

8.4.7 Use mesh analysis to determine I_1, I_2, and I_3 in the circuit.

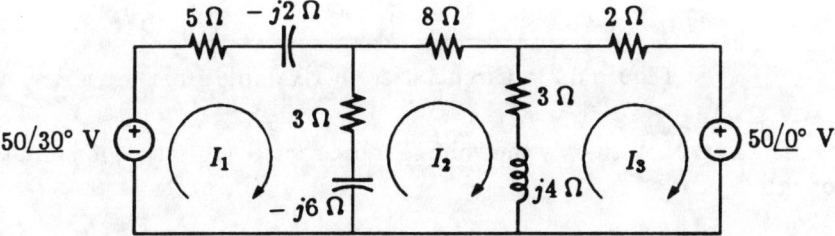

Ans. $I_1 = 0.687 + j4.163$ A, $I_2 = -0.675 - j0.414$ A, $I_3 = -6.527 + j4.434$ A.

8.5 Network theorems in the phasor domain

Because impedance networks and resistive networks are governed by the same network laws—Ohm's law, KVL, and KCL—all network theorems developed in Chapter 4 can be applied to ac circuits. The only modification necessary is that all electrical quantities (voltages and currents) are phasors and all elements or branches are represented by their impedances or admittances. In this section, we shall illustrate this extension by way of several examples.

8.5.1 Superposition

The superposition property of linear networks simply states that the total response is the sum of all responses when excitations are either activated one at a time or subdivided into a number of groups. The following two examples should illustrate how superposition is used in ac circuits.

EXAMPLE 10. Determine I in Fig. 8.21 by evaluating it with one source active at a time.

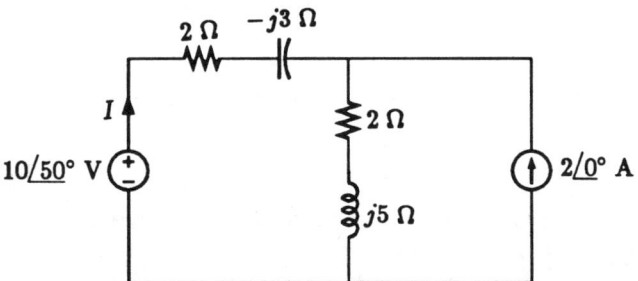

Figure 8.21: Circuit used in Example 10.

SOLUTION With only the voltage source active (the current source open),

$$I_a = \frac{10/\underline{50°}}{2 - j3 + 2 + j5} = 2.052 + j0.889$$

With only the current source active (the voltage source shorted),

ANALYSIS OF AC CIRCUITS

$$I_b = -\frac{2+j5}{2-j3+2+j5} \times 2 = -1.8 - j1.6$$

Hence, with both sources active

$$I = I_a + I_b = 0.252 - j0.711 = 0.754\underline{/-70.50°} \text{ A}$$

EXAMPLE 11. Determine $e(t)$ for the circuit in Fig. 8.22. $e_1(t) = 20 \cos 3t$ V and $e_2(t) = 10 \cos(2t + 30°)$ V.

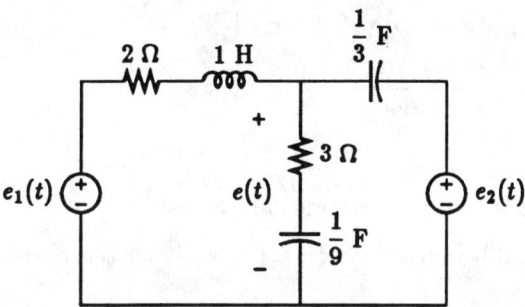

Figure 8.22: Circuit with two sources with different frequencies.

SOLUTION Since the two voltages have different values of ω, it is necessary to obtain $e(t)$ due to these two sources separately.

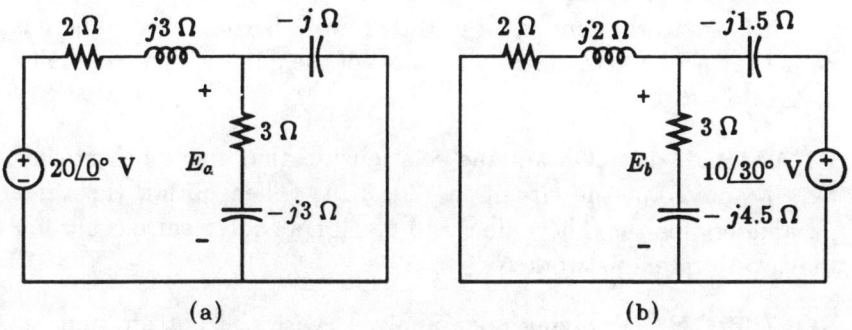

Figure 8.23: Phasor circuits for Example 11.

With only $e_1(t)$ active, the phasor circuit is shown in Fig. 8.23(a), in which[5]

$$(-j)||(3 - j3) = \frac{(-j)(3 - j3)}{3 - j4} = 0.12 - j0.84$$

$$E_a = \frac{0.12 - j0.84}{2 + j3 + 0.12 - j0.84} \times 20$$

$$= -3.406 - j4.454 = 5.607\underline{/-127.41°}$$

With only $e_2(t)$ active, the phasor circuit is shown in Fig. 8.23(b), in which

$$(2 + j2)||(3 - j4.5) = 2.64 + j0.72$$

$$E_b = \frac{2.64 + j0.72}{-j1.5 + 2.64 + j0.72} \times 10\underline{/30°}$$

$$= 4.710 + j8.754 = 9.940\underline{/61.72°}$$

Thus, when both voltage sources are active

$$e(t) = 5.607\cos(3t - 127.41°) + 9.940\cos(2t + 61.72°) \text{ V}$$

8.5.2 Equivalent circuits of two-terminal networks—Thévenin's and Norton's theorems

The following example illustrates how a two-terminal network can be characterized by any two of the three related parameters, as far as how it looks to the remainder of the network is concerned. There is no new principle here except that most of the quantities are complex instead of real.

EXAMPLE 12. Obtain the relationship that must be satisfied by E_L and I_L in the circuit in Fig. 8.24. Then obtain the three parameters for the Thévenin's and Norton's equivalent circuits for the two-terminal network N.

SOLUTION Applying node analysis, with the bottom node as the datum node, we have the node equations

---
[5] Here, we did not use the effective values of the sinusoids because our answer is sought in the time domain. Hence, it doesn't matter whether the maximum values or the effective values are used in the complex-amplitude domain.

ANALYSIS OF AC CIRCUITS

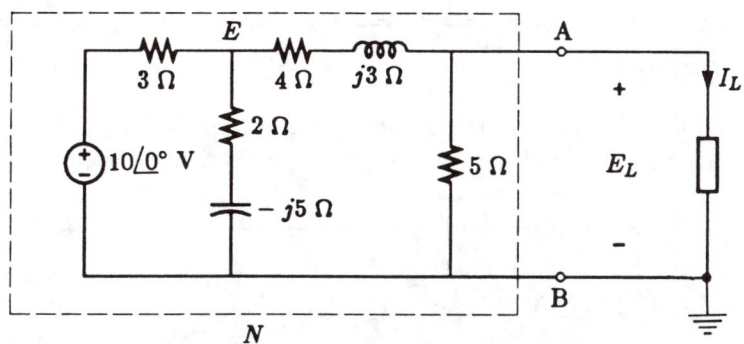

Figure 8.24: Circuit used in Example 12.

$$\frac{E-10}{3} + \frac{E}{2-j5} + \frac{E-E_L}{4+j3} = 0$$

$$\frac{E_L - E}{4+j3} + \frac{E_L}{5} + I_L = 0$$

Eliminating E between the two equations, we get either

$$E_L = 2.797 - j1.881 - (2.826 + j0.411)I_L$$

or

$$I_L = 0.875 - j0.793 - \frac{E_L}{2.826 + j0.411}$$

By comparing these expressions with Eqs. (4.16) and (4.19), we can identify the following quantities

$$E_{oc} = 2.797 - j1.881 = 3.371\underline{/-33.91°} \text{ V}$$

$$I_{sc} = 0.875 - j0.793 = 1.180\underline{/-42.20°} \text{ A}$$

$$R_{eq} = 2.826 + j0.411 = 2.855\underline{/8.28°} \text{ } \Omega$$

[The student is encouraged to obtain these results by other methods as exercises.]

EXERCISE

8.5.1 Obtain the parameters for the Thévenin's equivalent circuit for the two-terminal network N with respect to terminals A and B.

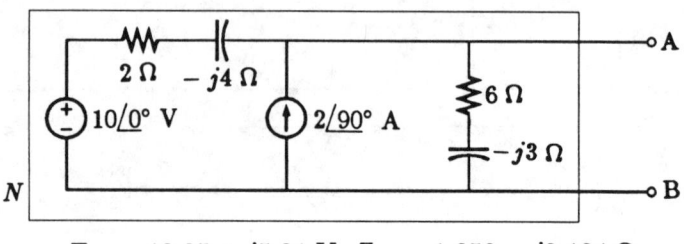

Ans. $E_{oc} = 10.35 + j5.31$ V, $Z_{eq} = 1.858 - j2.124$ Ω.

8.5.3 Other theorems and techniques

In Chapter 4, we established and described several other theorems and techniques for modifying networks while preserving certain equivalences. They are; Tellegen's theorem, reciprocal theorem, source transformation, and source shifting. Tellegen's theorem will be used in the next chapter with regard to certain aspects of power calculation. Reciprocal theorem can be considered as a special case of the Tellegen's theorem—Tellegen's corollary. We shall not elaborate on these, except to state that they all work in exactly the same manner as they do in resistive networks.

We shall now show an example of the source transformation as applied to ac circuits. We start with the circuit in Fig. 8.24 and successively transform the circuit, using the Thévenin-Norton equivalence, until the final Norton's equivalent circuit is reached. This sequence of steps is shown in Fig. 8.25.

8.6 Special aspects of ac circuits

As students work through several network problems with phasors and impedances, it must have occurred to them that the basic principles and techniques of ac circuit analysis are simply an extension of dc resistive circuit analysis. However, because phasors and

ANALYSIS OF AC CIRCUITS

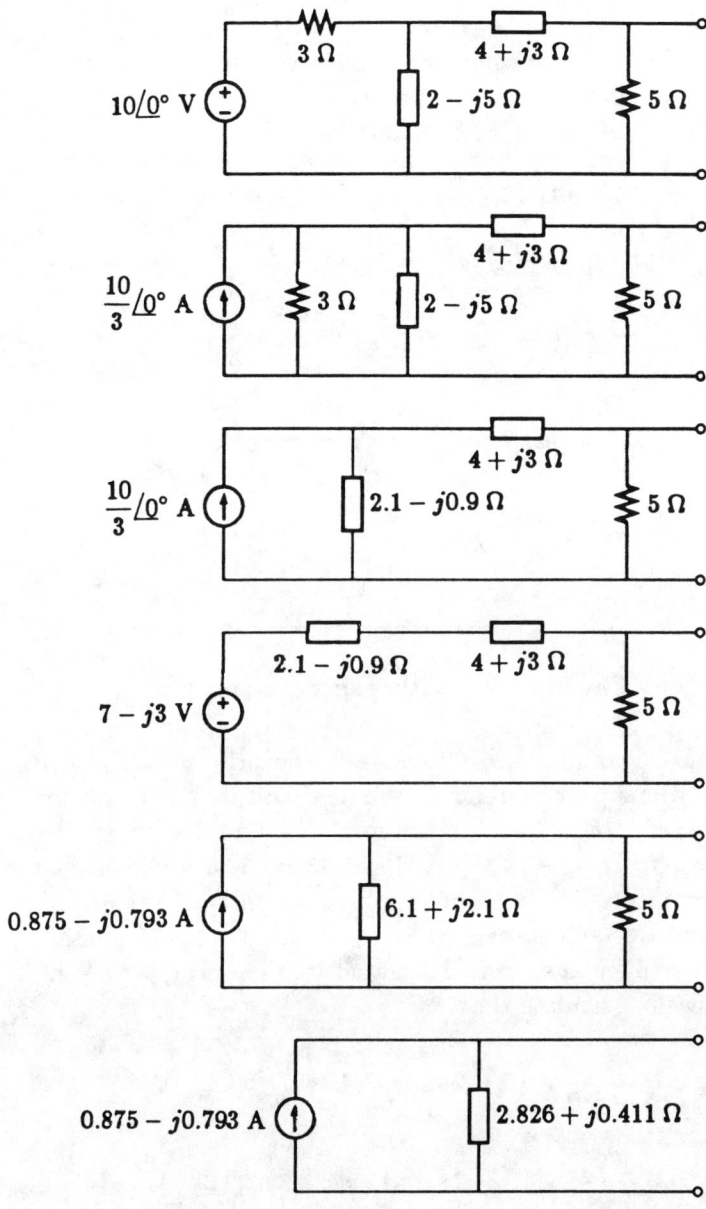

Figure 8.25: An example of source transformation.

impedances are complex, the outcomes of the analysis of ac circuits frequently give results that are somewhat surprising if one anticipates the general results expected with dc circuits. The following examples demonstrate some of the potential pitfalls that should be avoided when dealing with ac circuits.

In the circuit in Fig. 8.26, we easily obtain

$$I = 8.575\underline{/-59.04°} \text{ A}$$

$$E_R = 51.45\underline{/-59.04°} \text{ V}$$

$$E_L = 85.75\underline{/30.96°} \text{ V}$$

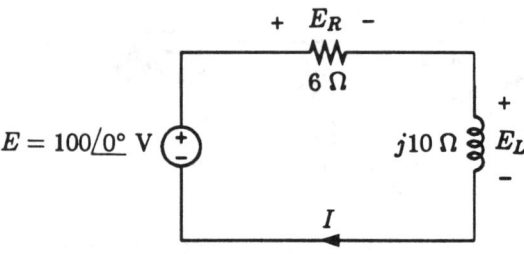

Figure 8.26: Another simple ac circuit.

When thinking of phasors, it comes as no surprise that $|E_R|+|E_L| \neq |E|$. But if one were working in the field and took a voltmeter and measured, say, $|E|$ and $|E_R|$, one might be tempted to erroneously infer that $|E_L| = |E|-|E_R| = 100-51.45 = 48.55$ V which would, of course, be incorrect. *We should not add or subtract the magnitudes of ac quantities* unless they are in phase or 180° out of phase.

Another phenomenon is illustrated by the example in Fig. 8.27. We can easily calculate that

$$E_R = \frac{5}{5+j10} \times 10 = 2 - j4 = 4.472\underline{/-63.43°} \text{ V}$$

$$E_L = \frac{j30}{5+j10} \times 10 = 24 + j12 = 26.83\underline{/26.57°} \text{ V}$$

$$E_C = \frac{-j20}{5+j10} \times 10 = -16 - j8 = 17.89\underline{/-153.43°} \text{ V}$$

ANALYSIS OF AC CIRCUITS

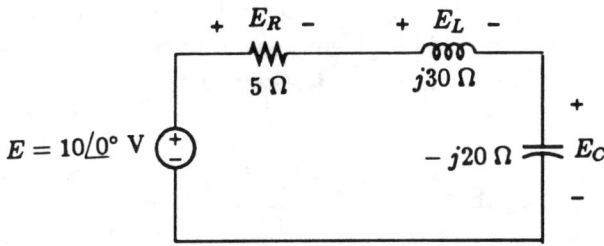

Figure 8.27: Another ac circuit.

Here we see that some component voltages can be larger than the total voltage. That both $|E_L|$ and $|E_C|$ are greater than $|E|$ in this example is the consequence of the fact that E_L and E_C actually are negative of each other and their magnitudes cancel. Broadly speaking, *in an ac circuit, some responses can be greater than the applied excitations.*

Another aspect of ac circuit problems that may arise that does not exist in dc circuits is that each ac quantity has two components—magnitude and phase, or real and imaginary parts. Occasions may arise when either one or the other of the two components of a phasor is not known. For example, the reading of a voltmeter or ammeter gives the magnitude of a voltage or current; no phase information is known from the meter readings.

We shall illustrate this type of problem by the circuit in Fig. 8.28, in which we know that

$$|E| = 100 \text{ V} \quad \text{and} \quad |I| = 10 \text{ A}$$

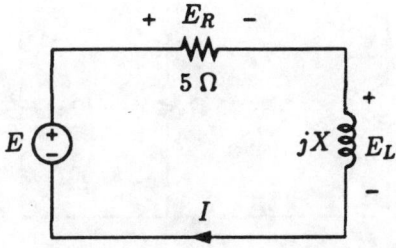

Figure 8.28: Another simple ac circuit.

and we wish to determine the value of X and $|E_L|$. The proper approach in solving this problem is as follows.

We choose, arbitrarily, I as the reference, or

$$I = 10\underline{/0°}$$

Then we can say that

$$E_R = 50\underline{/0°} \quad \text{and} \quad E_L = j10X$$

$$E = 50 + j10X \tag{8.45}$$

Equating the magnitudes of both sides of Eq. (8.45), we get

$$|E| = 100 = |50 + j10X| = \sqrt{50^2 + (10X)^2} \tag{8.46}$$

Solving for X, we get

$$X = 8.66 \ \Omega \quad \text{and} \quad |E_L| = 86.6 \text{ V}$$

The key point of this example is that Eq. (8.45) is a complex equation. When we have two complex quantities that are equal to each other, their magnitudes, phase angles, and real and imaginary parts must all be equal. Eq. (8.46) is the consequence of equating the magnitudes of both sides of Eq. (8.45).

EXAMPLE 13. For the circuit in Fig. 8.29, it is known that $|E| = 100$ V and $|E_1| = 50$ V. What is the value of R?

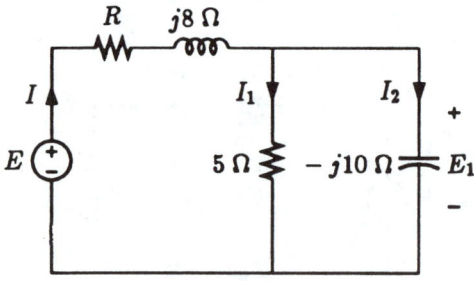

Figure 8.29: Circuit used in Example 13

ANALYSIS OF AC CIRCUITS

SOLUTION We let

$$E_1 = 50\underline{/0°}$$

Then we have

$$I_1 = \frac{50}{5} = 10\underline{/0°}$$

$$I_2 = \frac{50}{-j10} = j5$$

$$I = 10 + j5$$

$$E = 50 + (10 + j5)(R + j8) = 10 + 10R + j(80 + 5R)$$

$$100^2 = (10 + 10R)^2 + (80 + 5R)^2$$

Solving and keeping only the positive solution, we have

$$R = 2.633 \, \Omega$$

EXAMPLE 14. In the circuit in Fig. 8.30, it is known that I_s leads I_2 by 30°. Determine the value of X.

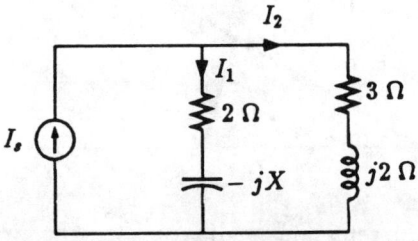

Figure 8.30: Circuit used in Example 14.

SOLUTION From the circuit,

$$I_1 = \frac{(3 + j2)I_2}{2 - jX}$$

FUNDAMENTALS OF CIRCUIT ANALYSIS

$$I_s = I_1 + I_2 = \left(\frac{3+j2}{2-jX} + 1\right)I_2 = \frac{5+j(2-X)}{2-jX} \times I_2$$

Rationalize to get

$$I_s = \frac{[5+j(2-X)](2+jX)}{4+X^2}I_2 = \frac{10 - 2X + X^2 + j(4+3X)}{4+X^2}I_2$$

Since the denominator is real, it necessary that

$$\frac{4+3X}{10 - 2X + X^2} = \tan 30° = \frac{1}{\sqrt{3}}$$

Solving, we get

$$X = 0.456 \quad \text{or} \quad 6.740 \; \Omega$$

Both answers are valid. If $X = 0.456\;\Omega$, $I_1 = (2.209 + j1.276)I_2$. If $X = 6.740$, $I_1 = (-0.151 + j0.490)I_2$. Fig. 8.31 is a phasor diagram that shows the two possible I_1's and the two resultant I's that satisfy the phase requirement.

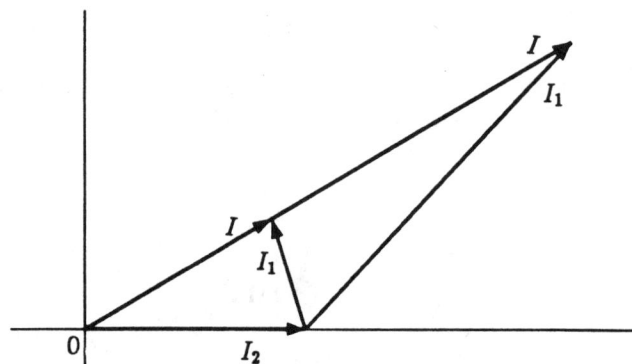

Figure 8.31: Current phasor diagram for the circuit of Fig. 8.30.

These examples illustrate that, in working with ac circuits, one should be aware that all quantities are phasors and must be treated with care in that regard. Generally, magnitudes are but parts of these quantities; so are the phase angles. One should never draw oversimplified conclusions based solely on magnitudes or angles without sound justifications.

EXERCISES

8.6.1 In the circuit, $|E| = 20$ V and $|I| = 2.5$ A. Determine $|E_C|$ and X_C.

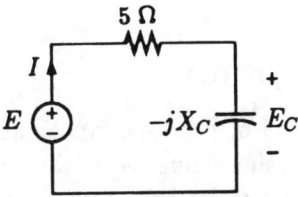

Ans. 15.61 V, 6.245 Ω.

8.6.2 In the circuit, $|E| = 100$ V and $E_L = 60$ V. Determine the value of R.

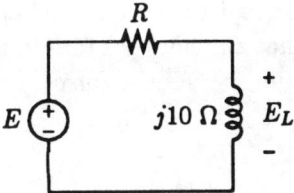

Ans. 13.33 Ω.

8.6.3 In the circuit, $|E| = 15$ V and $|E_2| = 10$ V. Determine the value of R.

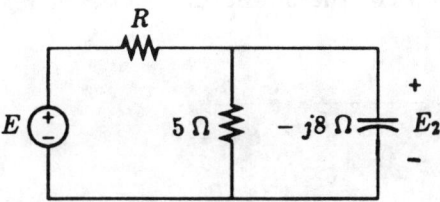

Ans. 2.354 Ω.

8.7 The phasor circle diagrams

The locus of a network function (such as the impedance or gain) when a certain parameter (such as a circuit element, ω, or s) is varied is a very useful indication of certain properties of the network function. Here we shall deal with certain particularly simple loci of impedances and admittances.

8.7.1 General theory

We shall discuss the loci of network functions in the complex plane as the value of *one* of the elements in the network is varied. The network function may be the impedance, the admittance, or the transfer function (such as the voltage ratio). The element value may be a resistance (or conductance), a reactance (or susceptance), or the strength of a controlled source. In this discussion, it is important to stress that, when one of the values is varied, *all* other element values in the network are assumed to remain unchanged.

Bode[6] has shown that any network function, z, expressed in terms of a single element value, x, has the form

$$z = \frac{A + Bx}{C + Dx} \tag{8.47}$$

where A, B, C, and D are functions of $j\omega$ (or more generally s). Hence, for a given value of $j\omega$ (or s), A, B, C, and D are complex constants.

For example, in the circuit in Fig. 8.32, if the value of the inductance L is varied, the admittance between terminals A and B is

$$Y = \frac{1}{20} + \frac{1}{10 + j\omega L} + \frac{1}{16 + \dfrac{8}{j\omega}} \tag{8.48}$$

The corresponding impedance is

$$Z = \frac{1}{Y} = \frac{(1600 + j3200\omega) + j\omega(320j\omega + 160)L}{(240 + j680\omega) + j\omega(36j\omega + 8)L} \tag{8.49}$$

---

[6]H. W. Bode, *Network Analysis and Feedback Amplifier Design*, Van Nostrand Company, Inc., New York, p. 10, 1945.

ANALYSIS OF AC CIRCUITS

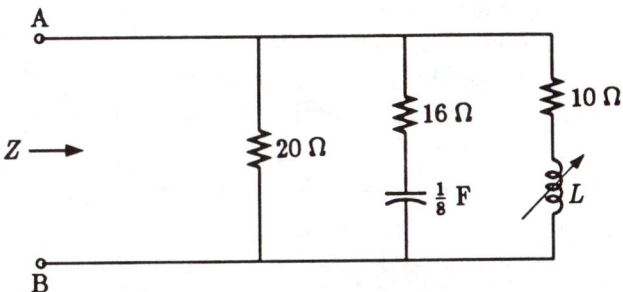

Figure 8.32: A circuit with one variable element.

When A, B, C, and D are complex constants, the relationship in Eq. (8.47) is said to be *bilinear*. Solving for x in terms of z from Eq. (8.47), we obtain

$$x = -\frac{A - Cz}{B - Dz} \qquad (8.50)$$

which is also bilinear.

When x and z are complex variables, Eq. (8.47) can be regarded as a *complex transformation*—specifically, the bilinear transformation. For each value of x, there is a corresponding value of z. Thus, each point in the x plane is *mapped* onto another point in the z plane. As x moves along a certain locus in the x plane, the corresponding z also moves along another locus in the z plane. The locus of z is known as the *image*, the *map*, or the *transformation* of the locus of x.

Since Eq. (8.50) is equivalent to Eq. (8.47), Eq. (8.50) may be regarded as the transformation from z to x—or the inverse transformation. It maps a locus in the z plane onto a locus in the x plane.

The bilinear transformation has several interesting properties:

(1) The mapping is *unique*. For each value of x, there is only one corresponding z; and vice versa. Thus mapping a locus from one plane to another, and back, always returns to the original locus.

(2) A *circular* locus in one plane is mapped onto another circular locus in the other plane.[7] In this regard, it should be noted that straight lines are special cases of circles—circles with infinite radii.

---

[7] E. A. Guillemin, *The Mathematics of Circuit Analysis*, John Wiley & Sons, Inc., New York, pp.363-367, 1949.

(3) From properties (1) and (2), it follows that a circular arc in one plane maps onto another circular arc in the other plane. Again, a segment of a straight line is a special case of a circular arc.

(4) The following transformations are all special cases of the bilinear transformation. Therefore properties (1), (2), and (3) also apply to these transformations.

$$z = A + Bx \qquad (8.51)$$

$$z = \frac{A}{x} \qquad (8.52)$$

$$z = \frac{A}{B + Cx} \qquad (8.53)$$

$$z = A + \frac{B}{Cx} \qquad (8.54)$$

(5) Since a circle is fixed if we know three points on this circle, the transformation of a circle can be determined by locating the transformations of three points on the circle.

(6) Property (5) becomes especially useful when the transformation of an arc is being sought. The two end points of an arc map onto the ends of its transformation. Thus, in addition to the two end points, we need to locate only the transformation of one intermediate point to determine the entire transformation.

For example, if we wish to determine the locus of Z for the circuit in Fig. 8.32 as L is varied from 0 to ∞, we are really trying to locate the Z-plane transformation of the positive half of the real axis of the L plane. The bilinear transformation between L and Z is Eq. (8.49). The two end points of the Z locus can be found for $L = 0$ and $L = \infty$. These two points, together with a third point, say $L = 10$ H, enable us to locate the complete locus. Specifically, if $\omega = 1$ rad/sec, we have

$$L = 0 \qquad Z = \frac{64}{13} - j\frac{8}{13} \; \Omega$$

$$L = 10 \qquad Z = \frac{240}{33} + j\frac{40}{37} \; \Omega$$

$$L = \infty \qquad Z = \frac{160}{17} - j\frac{40}{17} \; \Omega$$

The locus of Z is shown in Fig. 8.33.

ANALYSIS OF AC CIRCUITS 417

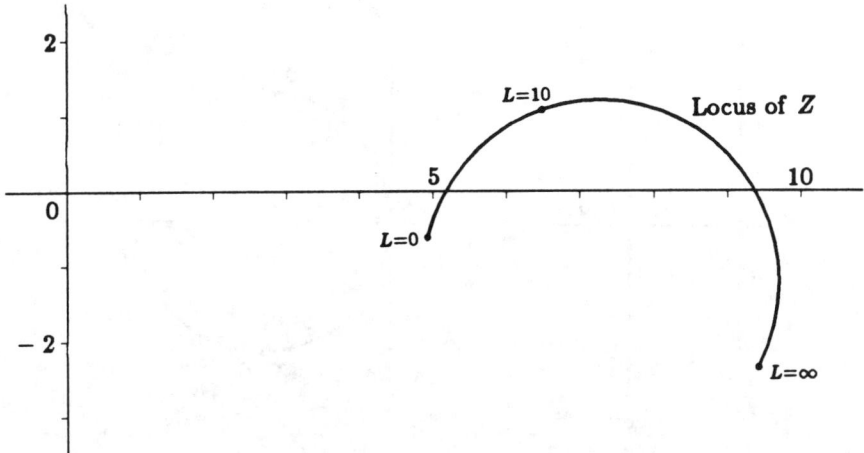

Figure 8.33: Locus of Z for the circuit in Fig. 8.32.

8.7.2 Impedance and admittance loci of simple branches

When one of the elements of a simple R-C or R-L branch is varied, the locus of its impedance or admittance is quite simple. Let us take a series branch consisting of a resistance R_0 and a reactance X, as shown in Fig. 8.34(a). We shall consider the case when X may

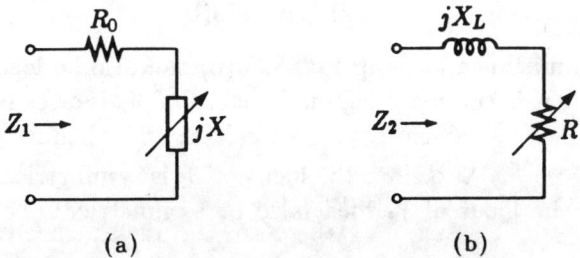

(a)   (b)

Figure 8.34: Two simple two-element circuits.

vary from $-\infty$ to $+\infty$. (X may be either inductive or capacitive.) Clearly, the locus of Z_1 is a vertical line passing through $Z = R_0 + j0$, as shown in Fig. 8.35(a).

The admittance $Y_1 = \frac{1}{Z_1}$ is

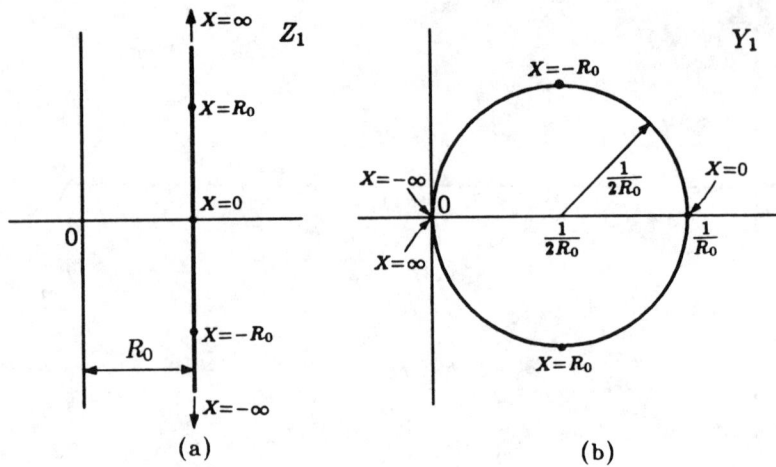

Figure 8.35: Impedance and admittance loci of the circuit in Fig. 8.34(a).

$$Y_1 = G_1 + jB_1 = \frac{R_0}{R_0^2 + X^2} - j\frac{X}{R_0^2 + X^2} \qquad (8.55)$$

It can be shown that G_1 and B_1 satisfy the equation

$$\left(G_1 - \frac{1}{2R_0}\right)^2 + B_1^2 = \left(\frac{1}{2R_0}\right)^2 \qquad (8.56)$$

which is the equation of a circle centered at $\frac{1}{2R_0} + j0$ with a radius of $\frac{1}{2R_0}$. This locus is shown in Fig. 8.35(b).

The admittance locus in Fig. 8.35(b) can also be located graphically since it is the mapping of the locus of Z_1. For example, when $Z_1 = R_0$, $Y_1 = \frac{1}{R_0}$; when $Z_1 = \pm\infty$, $Y_1 = 0$; and when $Z_1 = R_0 \pm jR_0$; $Y_1 = \frac{1}{2R_0} \mp j\frac{1}{2R_0}$. Also since the locus of Z_1 is symmetrical about the real axis, the locus of Y_1 must also be symmetrical about the real axis.

We have let the value of X vary from $-\infty$ to ∞. This is actually more convenient so we can discuss the complete loci of both Z_1 and Y_1 as they are both complete circles. If X is the reactance of an inductance, then X must be positive and the locus of Z_1 is the upper half of the straight line and that of Y_1 the lower semicircle. If X is the reactance of a capacitance, then the Z_1 locus is the lower half of the straight line and that of Y_1 is the upper semicircle.

ANALYSIS OF AC CIRCUITS

Next, let us examine an R-X series branch and let X be fixed and R be variable as shown in Fig. 8.34(b). To be specific, let X be positive (and call it X_L). Also, let R vary from $-\infty$ to ∞. The locus of $Z_2 = R + jX_L$ is obviously a horizontal line passing through $0 + jX_L$ as shown in Fig. 8.36(a). By a similar reasoning by which Fig. 8.35(b) was obtained from Fig. 8.35(a), the locus of $Y_2 = \frac{1}{Z_2}$ is a circle of radius $\frac{1}{2X_L}$ centered at $0 - j\frac{1}{2X_L}$. The locus of Y_2 is shown in Fig. 8.36(b).

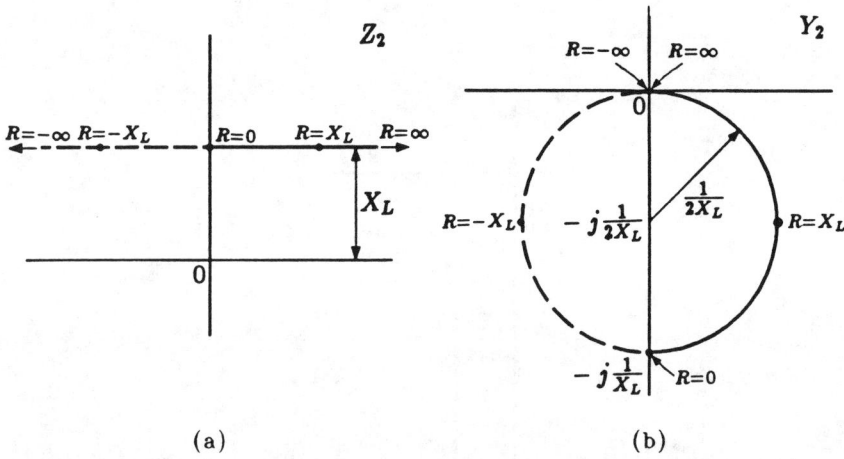

Figure 8.36: Impedance and admittance loci of the circuit in Fig. 8.34(b).

Here, we have allowed the value of R to be both positive and negative. This is done primarily to take advantage of the mathematical symmetry. If R is restricted to be positive, it is clear that both loci of Fig. 8.36 will be restricted to the right half-plane halves of each locus. We have shown those parts of the loci corresponding to negative R dashed.

The above reasoning and development are typical for two-element branches. They can be readily extended to all R-C and R-L combinations. A summary of the loci of all series combinations is shown in Fig. 8.37. A dual summary of the loci of all parallel branches is shown in Fig. 8.38. It should be remarked here that the variation of X can be the result of either a varying ω or a varying circuit element value—L or C.

420 FUNDAMENTALS OF CIRCUIT ANALYSIS

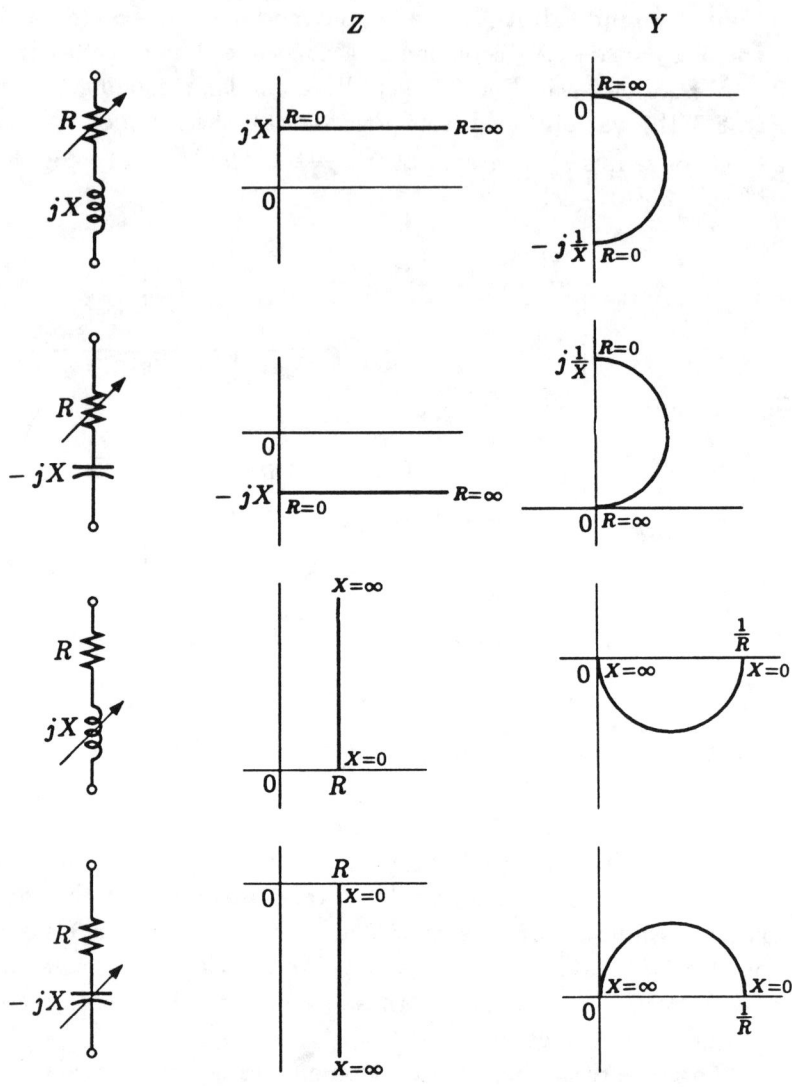

Figure 8.37: Impedance and admittance loci of two-element series branches.

ANALYSIS OF AC CIRCUITS

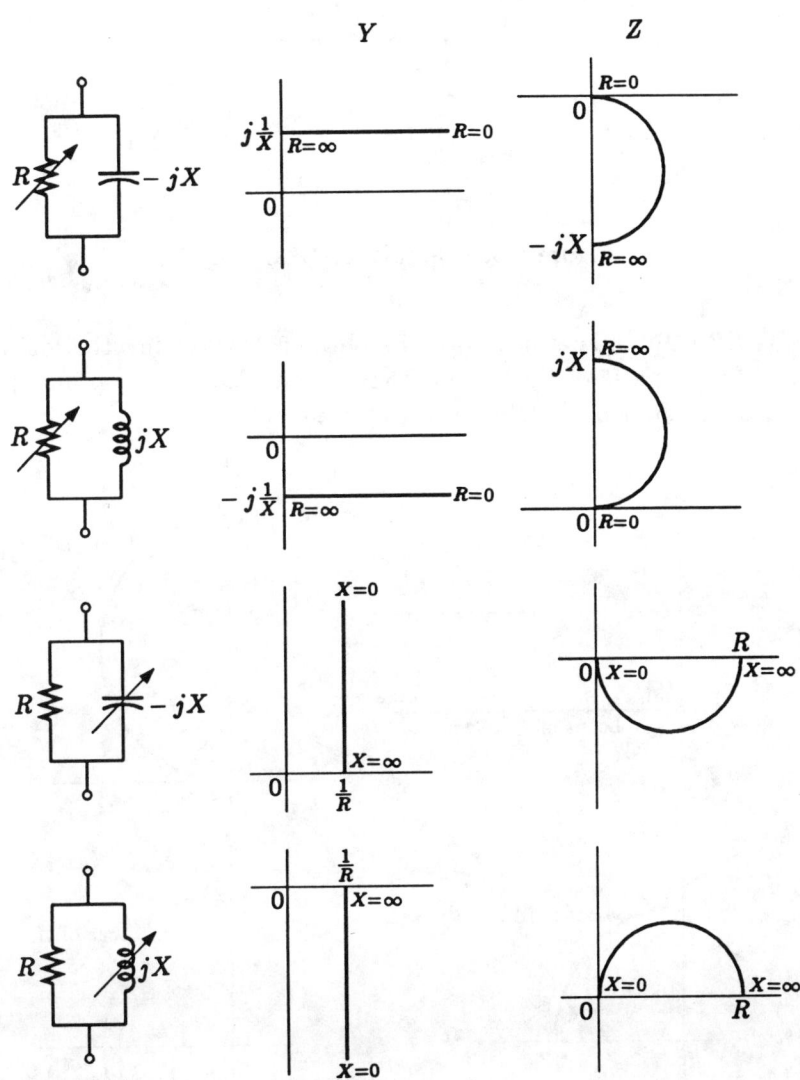

Figure 8.38: Impedance and admittance loci of two-element parallel branches.

EXAMPLE 15. Obtain the locus of impedance Z in Fig. 8.39 as R is varied from 0 to ∞.

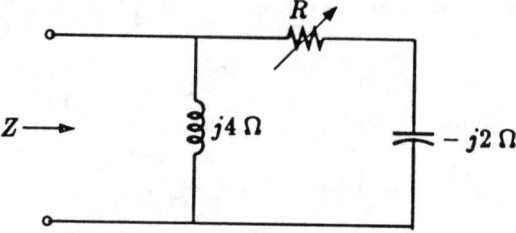

Figure 8.39: Circuit for Example 15

SOLUTION The locus may be obtained by getting the loci of the various parts of the circuit from right to left. The various loci and the final locus are shown in Fig. 8.40.

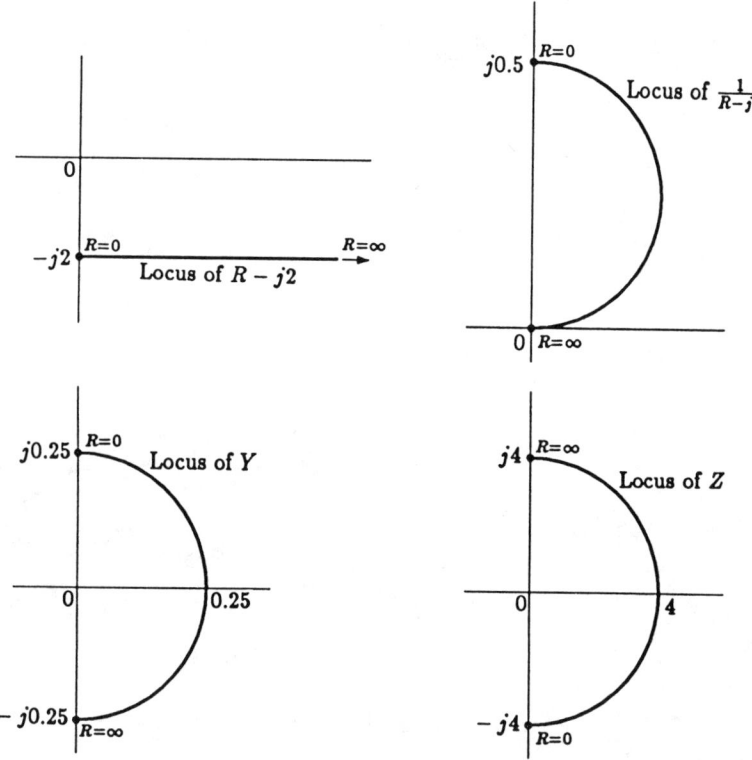

Figure 8.40: Steps for the solution of Example 15.

ANALYSIS OF AC CIRCUITS

EXAMPLE 16. For the circuit in Fig. 8.41, obtain the semicircular locus of impedance Z as X is varied from 0 to ∞. From this locus, determine the maximum and minimum values of $|Z|$ and their corresponding value of X.

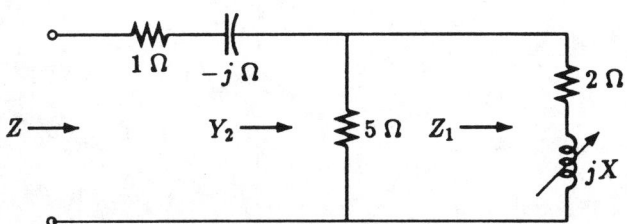

Figure 8.41: Circuit for Example 16.

SOLUTION The locus of Z can be obtained by successively obtaining the loci of Z_1, Y_1, Y_2, Z_2, and finally Z. These loci are shown in Fig. 8.42.

From the locus of Z, it is obvious that $|Z|_{\max}$ occurs at $X = \infty$. For $X = \infty$, $Z = 6 - j$. Hence, $|Z|_{\max} = \sqrt{37} = 6.083$ Ω.

$|Z|_{\min}$ is the point on the locus that is closest to the origin. That point is the intersect of a line drawn between the origin and the center of the circular locus with the locus. From the last diagram in Fig. 8.42

$$|Z|_{\min} = \sqrt{\left(\frac{59}{14}\right)^2 + 1^2} - \frac{25}{14} = 2.547 \text{ Ω}$$

Under this condition,

$$Z = 2.546 \underline{/-13.35°}$$

$$\frac{1}{Y_2} = Z - (1 - j) = 1.477 + j0.412$$

$$Y_2 = \frac{1}{1.477 + j0.412} = 0.628 - j0.175$$

$$\frac{1}{Z_1} = Y_2 - 0.2 = 0.428 - j0.175$$

$$Z_1 = \frac{1}{0.428 - j0.175} = 2 + j0.819$$

Hence, $|Z|_{\min}$ occurs when $X = 0.819$ Ω.

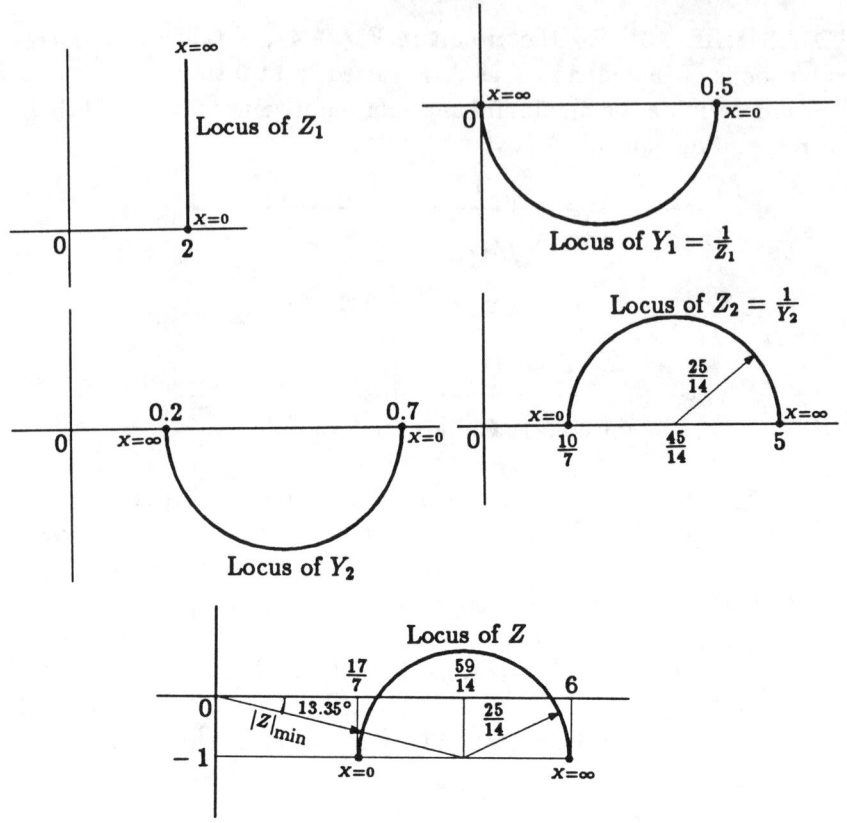

Figure 8.42: Steps for the solution of Example 16.

EXAMPLE 17. [Phase-shifting network] Fig. 8.43 is an example of arrangements that will provide a scheme for deriving, from a given source, an output that has a constant magnitude and a variable phase (typically over a range of 180°) by varying one of the circuit elements.

From Fig. 8.37, as X is varied from 0 to ∞, the locus of the admittance of the R-X branch is shown in Fig. 8.44(a). If we let $E = |E|\underline{/0°}$, the locus of I_2 is shown in Fig. 8.44(b). Since $E_R = I_2 R$, the locus of E_R is shown in Fig. 8.44(c). Since $E_o = \frac{1}{2}|E| - E_R$, the locus of E_o is shown in Fig. 8.44(d). Thus, by varying X (ω or C), we can obtain a voltage—E_o—that is one half the magnitude of E with any angle from 0 to 180° lagging that of E.

ANALYSIS OF AC CIRCUITS

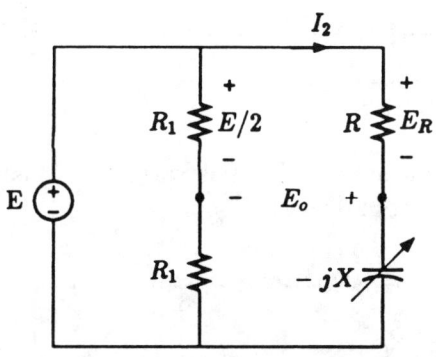

Figure 8.43: A phase-shifting circuit.

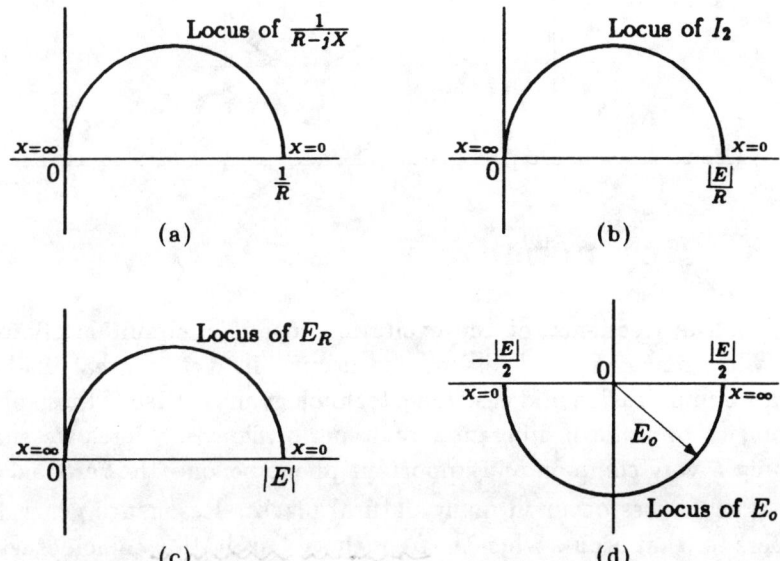

Figure 8.44: Various loci of the phase-shifting circuit.

FUNDAMENTALS OF CIRCUIT ANALYSIS

EXERCISE

8.7.1 In the following circuit, X is variable. Construct the locus of Z as X is varied from 0 to ∞.

Ans.

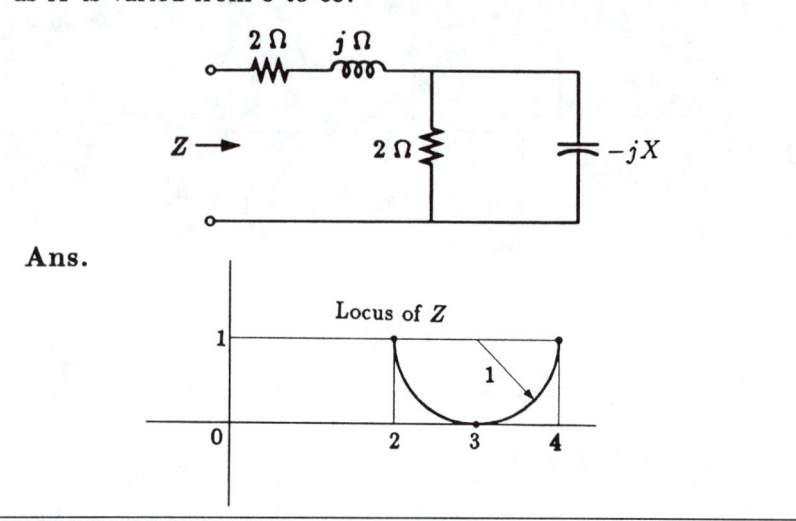

8.8 Resonance

When the frequency of the excitation in an ac circuit is allowed to vary, many unusual phenomena occur. In fact, a great deal of telecommunication and electronic technology makes use of these phenomena to achieve different circuit performances. Here, we shall study a very common and important phenomenon—the resonance.

Resonances occur in many natural places. Resonance generally refers to what occurs when the frequency of excitation coincides with or is very nearly equal to the natural frequency of the system. An example of resonance is when the tires of an automobile are not completely round or are out of balance. At a certain speed, the vibration caused by the irregularity coincides with the natural frequency of the vehicle. This can cause the vehicle to vibrate violently or even destructively.

ANALYSIS OF AC CIRCUITS

8.8.1 Series resonance circuit

The basic series resonance circuit is shown in Fig. 8.45. We assume that E is fixed in magnitude but its ω is variable. We are interested in how $|I|$ varies as ω is varied. We have

$$I = \frac{E}{R + j\left(\omega L - \dfrac{1}{\omega C}\right)} \tag{8.57}$$

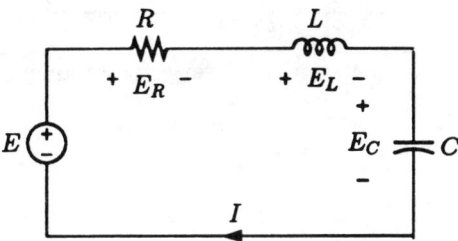

Figure 8.45: A series resonance circuit.

Hence,

$$|I| = \frac{|E|}{\sqrt{R^2 + \left(\omega L - \dfrac{1}{\omega C}\right)^2}} \tag{8.58}$$

It is obvious that $|I|$ is zero at $\omega = 0$ and $\omega = \infty$. Since ω appears only in the parentheses in the denominator, $|I|$ is maximum when the quantity in the parentheses is zero, or

$$\omega L - \frac{1}{\omega C} = 0 \tag{8.59}$$

Hence, $|I|$ is maximum when

$$\omega = \frac{1}{\sqrt{LC}} = \omega_0 \tag{8.60}$$

The quantity ω_0 is known as the resonance (angular) frequency and when, $\omega = \omega_0$, the series circuit is said to be at resonance. Thus, the $|I|$-versus-ω curve must have the shape shown in Fig. 8.46.

At resonance, $\omega = \omega_0$, the following statements may be made:

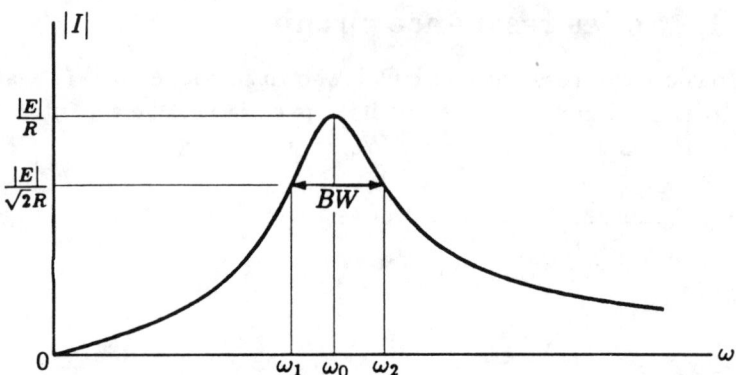

Figure 8.46: Current in a series resonance circuit.

1. $|I|$ is maximum.
2. $X_L = X_C$.
3. $Z = R + j0$.
4. $|E_L| = |E_C| = \frac{\omega_0 L}{R}|E|$. For small R, $|E_L|$ and $|E_C|$ can be much larger than $|E|$.
5. $E_L + E_C = 0$.
6. $E_R = IR = E$.

So far, we have fixed the two end points ($\omega = 0$ and ∞) and the peak point ($\omega = \omega_0$, $|I| = |E|/R$) of the resonance curve. There is another degree of freedom. This third degree of freedom can be used as a measure of the sharpness of the $|I|$-versus-ω curve. This is usually done by locating the two points on the curve whose values are $1/\sqrt{2}$ of the peak value. They are known as the half-power points because the power delivered to R at these points is exactly one-half of that delivered at $\omega = \omega_0$. We shall denote these points by ω_1 and ω_2. Their values are the solution to the equation

$$\sqrt{R^2 + \left(\omega L - \frac{1}{\omega C}\right)^2} = \sqrt{2}R \tag{8.61}$$

or

$$\omega_1, \omega_2 = \sqrt{\frac{R^2}{4L^2} + \frac{1}{LC}} \pm \frac{R}{2L} \tag{8.62}$$

ANALYSIS OF AC CIRCUITS

The half-power bandwidth is

$$\text{BW} = \omega_2 - \omega_1 = \frac{R}{L} \tag{8.63}$$

These quantities are indicated in Fig. 8.46. The measure of sharpness is the ratio of ω_0 to the half-power bandwidth, commonly known as the Q factor (Q connotes "quality").

$$Q = \frac{\omega_0}{\text{BW}} = \frac{\omega_0 L}{R} \tag{8.64}$$

Thus, the higher Q is, the sharper (narrower) the resonance curve will be.

The value of ω_2 and ω_1 are such that

$$\omega_1 \omega_2 = \omega_0^2 \quad \text{or} \quad \frac{\omega_0}{\omega_1} = \frac{\omega_2}{\omega_0} \tag{8.65}$$

Thus, ω_2 and ω_1 are *geometrically symmetric* about ω_0.[8] If we plot the resonance curve along a logarithmic scale of ω, the curve will indeed appear symmetric.

Let us return to Fig. 8.45. If the current in the time domain is

$$i(t) = I_m \cos \omega_0 t \tag{8.66}$$

then the energy stored in the inductor at any time is

$$w_L(t) = \frac{1}{2} L i^2 = \frac{1}{2} L I_m^2 \cos^2 \omega_0 t \tag{8.67}$$

The energy stored in the capacitor is

$$w_C(t) = \frac{1}{2} C e_C^2(t) = \frac{1}{2} \cdot \frac{I_m^2}{\omega^2 C} \sin^2 \omega_0 t = \frac{1}{2} L I_m^2 \sin^2 \omega_0 t \tag{8.68}$$

The total energy stored in both the inductor and the capacitor is

$$\text{Total energy stored} = w_L(t) + w_C(t) = \frac{1}{2} L I_m^2 \tag{8.69}$$

---

[8] The student should not confuse this symmetry with the arithmetic symmetry which would mean $\omega_0 = \frac{1}{2}(\omega_2 + \omega_1)$.

FUNDAMENTALS OF CIRCUIT ANALYSIS

Hence, at resonance, the total energy stored in the inductor and the capacitor is a constant. The energy is continuously being transferred back and forth between the capacitor and the inductor while the total energy remains constant.

As the energy is being transferred between the inductor and the capacitor, some current must flow. As this current flows through the resistor R, some energy is being dissipated in R. The source must replenish this energy loss to maintain the total energy stored. The energy lost during each period is

$$\text{Energy loss per period} = \left(\frac{I_m}{\sqrt{2}}\right)^2 \times R \times T = \frac{I_m^2}{2} \times R \times \frac{2\pi}{\omega_0} \quad (8.70)$$

Taking the ratio of Eq. (8.70) and Eq. (8.69), we get

$$\frac{\text{Total energy stored}}{\text{Energy lost per period}} = \frac{1}{2\pi} \cdot \frac{\omega_0 L}{R} = \frac{1}{2\pi} Q \quad (8.71)$$

Hence, an alternative definition of Q is

$$Q = 2\pi \times \frac{\text{Total energy stored}}{\text{Energy lost per period}} \quad (8.72)$$

This definition is identical to that given in Eq. (8.64) for the simple RLC series circuit. But the definition of Eq. (8.72) is more general and is applicable to other systems to which Eq. (8.64) may not be applicable.

EXAMPLE 17 For the circuit in Fig. 8.45, $R = 100\ \Omega$, $L = 10$ mH, and $C = 5$ nF. Calculate ω_0, $f_0(= \omega_0/2\pi)$, Q, the half-power points, ω_2 and ω_1, and BW.

SOLUTION

$$\omega_0 = \frac{1}{\sqrt{LC}} = 1.414 \times 10^5 \text{ rad/sec}$$

$$f_0 = \frac{\omega_0}{2\pi} = 22{,}508 \text{ Hz}$$

$$Q = \frac{\omega_0 L}{R} = 14.14$$

$$\omega_2 = \sqrt{\frac{R^2}{4L^2} + \frac{1}{LC}} + \frac{R}{2L} = 1.4651 \times 10^5 \text{ rad/sec}$$

ANALYSIS OF AC CIRCUITS

$$\omega_1 = \sqrt{\frac{R^2}{4L^2} + \frac{1}{LC}} - \frac{R}{2L} = 1.3651 \times 10^5 \text{ rad/sec}.$$

$$\text{BW} = \omega_2 - \omega_1 = 10,000 \text{ rad/sec}.$$

EXERCISES

8.8.1 The frequency of the voltage source is variable. What is the value of ω at which $|I|$ is maximum? What is this maximum $|I|$? What is the value of $|E_C|$ for this ω?

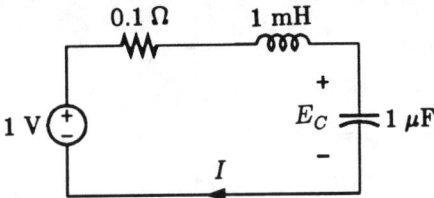

Ans. $\omega_0 = 31.62 \times 10^3$ rad/sec, $|I|_{max} = 10$ A $|E_C| = 316.2$ V

8.8.2 In an RLC series circuit, $R = 1$ Ω, $L = 2$ mH, and $Q = 10$. Determine C, ω_0, ω_2. and ω_1.

Ans. $C = 20$ μF, $\omega_0 = 5,000$ rad/sec, $\omega_2 = 5,256.25$ rad/sec, $\omega_1 = 4,756.25$ rad/sec.

8.8.2 Parallel resonance circuit

The basic GCL parallel circuit shown in Fig. 8.47 is the dual of the series RLC circuit in Fig. 8.45. If we hold I to have a constant magnitude and vary its ω, the response E will behave exactly like the way I behaves in the circuit in Fig. 8.45. In fact, if we apply the principle of duality, everything we said about the series resonance has a dual in parallel resonance. For example, the following statements are correct:

1. $|E|$ is maximum when $\omega = 1/\sqrt{CL}$.

2. At resonance, $B_L = B_C$ and $Y = G$.

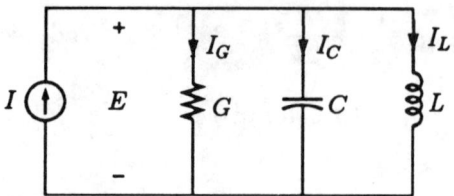

Figure 8.47: A parallel resonance circuit.

3. At resonance, $|I_L| = |I_C| = \frac{\omega_0 C}{G}|I|$.

4. At resonance, the total energy stored in the capacitor and the inductance is a constant and is equal to $\frac{1}{2}CE_m^2$, where E_m is the maximum value of $e(t)$.

Etc.

EXERCISE

8.8.3 For the parallel resonance circuit shown, determine ω_0, Q, ω_2, ω_1, and BW.

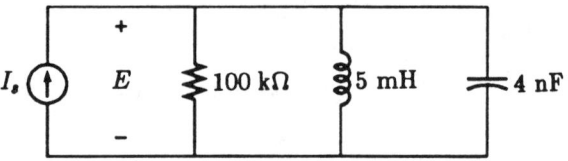

Ans. $\omega_0 = 2.236 \times 10^5$ rad/sec, $Q = 89.44$, $\omega_2 = 2.2486 \times 10^5$ rad/sec, $\omega_1 = 2.2236 \times 10^5$ rad/sec, BW = 2,500 rad/sec.

8.9 Summary

In this chapter, we have shown the methods for analysis of ac circuits. The basic methods of analysis are exactly like those used for memoryless circuits. Once an ac circuit has been formulated in terms of phasors and impedances, its analysis is entirely parallel to dc circuit analysis. However, since a phasor represents a sinusoidal quantity, one should not forget that there is a time function implied by every phasor.

ANALYSIS OF AC CIRCUITS

Because ac circuits involve phasors, there are phenomena and aspects of circuit analysis that do not have comparable situations in dc circuits. Problems discussed in Section 8.6, 8.7, and 8.8 do not occur in dc circuits.

Another aspect of ac circuits that is much more involved than that of dc circuits is the power consideration. In fact, we will devote the whole of Chapter 9 to this subject.

Problems

8.1 Simplifying the following expressions into single cosine terms with positive coefficients.

≤(a) $10 \cos \omega t + 5 \sin \omega t + 4 \sin(\omega t + 20°)$

(b) $4 \cos(\omega t + 120°) + 5 \sin(\omega t - 20°)$

≤(c) $15 \cos \omega t + 10 \sin \omega t + 7 \sin(\omega t + 0.85)$

(d) $12 \cos(\omega t - 120°) - 20 \sin(\omega t + 120°)$
        $+ 8 \sin(\omega t - 0.5)$

≤(e) $25 \cos(\omega t + \pi) - 20 \sin(\omega t - \dfrac{\pi}{2})$
        $+ \cos(\omega t + 30°) - \sin(\omega t - 1)$

≤**8.2** Simplify the following expressions into single terms each consisting of a positive coefficient, an exponential term, and a cosine term.

(a) $15\epsilon^{-3t} \cos \omega t + 20\epsilon^{-3t} \sin \omega t + 8\epsilon^{-3t} \sin(\omega t + 30°)$

(b) $12\epsilon^{-6t} \cos(\omega t - 120°) + 20\epsilon^{-6t} \cos(\omega t + \dfrac{\pi}{2})$
        $- 5\epsilon^{-6t} \sin(\omega t - 30°)$

(c) $10\epsilon^{-\alpha t} \cos \omega t + 20\epsilon^{-\alpha t} \sin \omega t + 7\epsilon^{-\alpha t} \cos(\omega t + 20°)$
        $+ 10\epsilon^{-\alpha t} \sin(\omega t + 60°)$

8.3 Find the effective values of the following periodic currents.

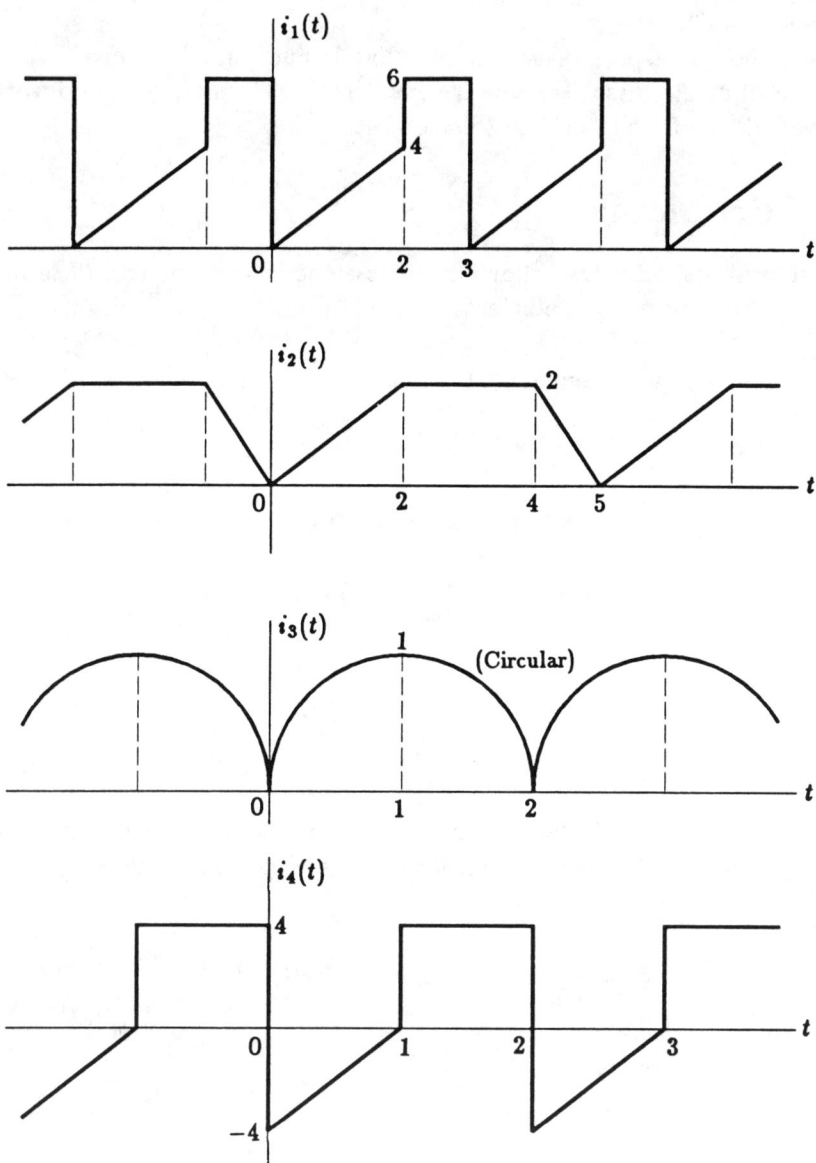

ANALYSIS OF AC CIRCUITS

8.4 $e_s(t) = 50 \cos(100t - 20°)$ V. Determine $i(t)$.

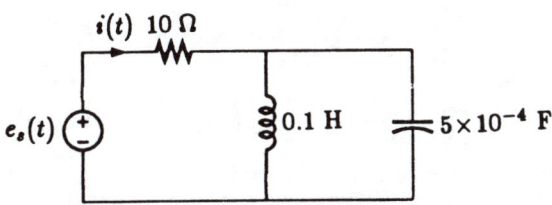

8.5 $i_s(t) = 0.2 \cos(100t - 20°)$ A. Determine $i(t)$.

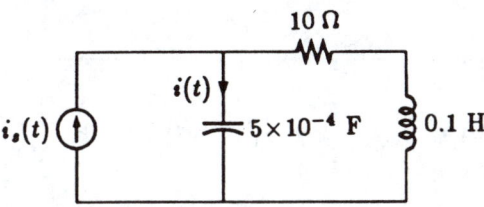

△ **8.6** $e_s(t) = 2 \cos(3t + 30°)$ V. Determine $i(t)$.

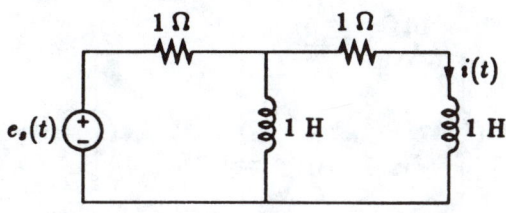

8.7 $e_s(t) = 5 \sin(5t - 20°)$ V. Determine $i(t)$.

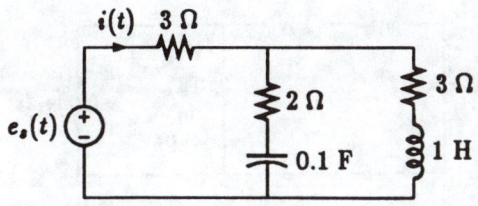

8.8 $e_s(t) = 100 \sin 10t$ V. Determine $i(t)$.

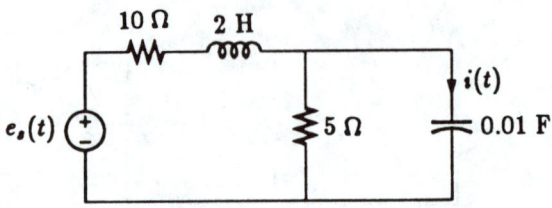

8.9 $i_1(t) = 5 \cos 100t$ A and $i_2(t) = 10 \sin 100t$ A. Determine $e_1(t)$ and $e_2(t)$.

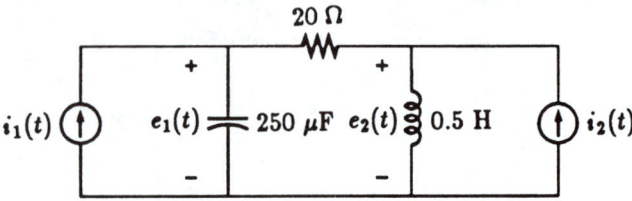

8.10 $e_1(t) = 10 \cos 10t$ V and $i_2(t) = 2 \cos 5t$ A. Determine $e(t)$.

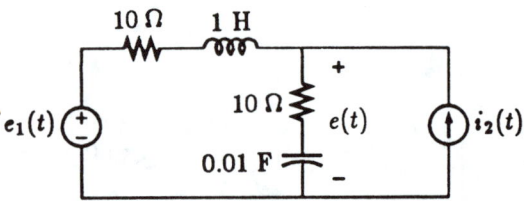

8.11 $e_s(t) = 10 \cos(2t + 20°)$ V and $i_s(t) = 5 \sin t$ A. Determine $i(t)$.

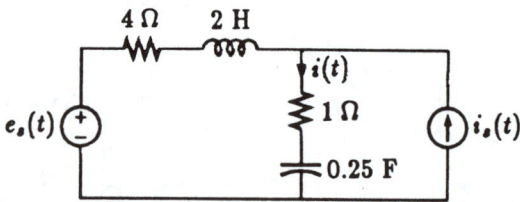

ANALYSIS OF AC CIRCUITS

8.12 Determine $i(t)$.

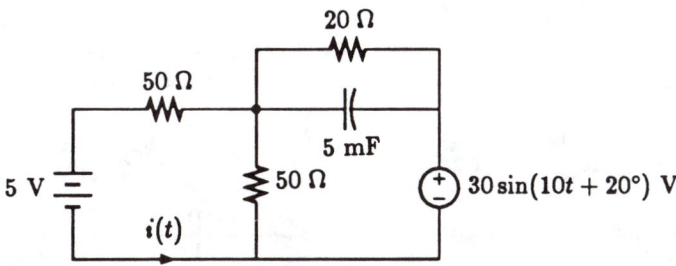

8.13 $e_1(t) = 20\cos 3t$ V and $e_2(t) = 10\cos(2t + 30°)$ V. Determine $e(t)$.

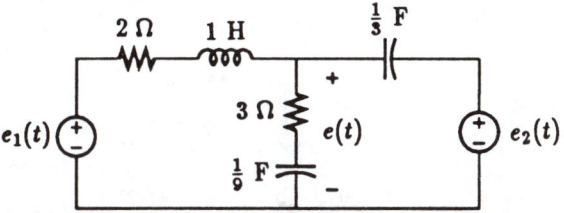

8.14 Calculate the equivalent impedance Z.

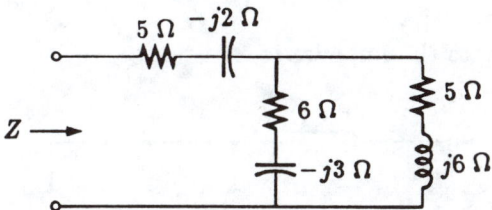

△ **8.15** Calculate the values of R and C such that $Z_1 = Z_2$. The frequency is 50 Hz.

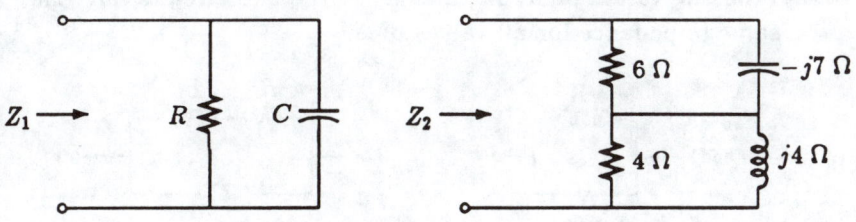

8.16 Calculate the values of R and L such that $Z_1 = Z_2$. The frequency is 50 Hz.

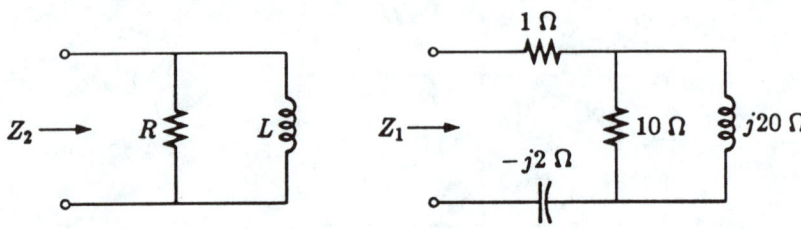

8.17 Calculate the impedance Z. Then find a simple R-L series branch and a simple R-L parallel branch with the same impedance.

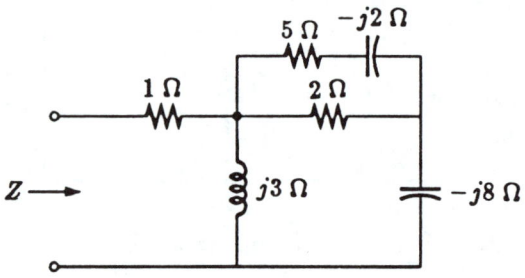

8.18 Calculate the impedance Z.

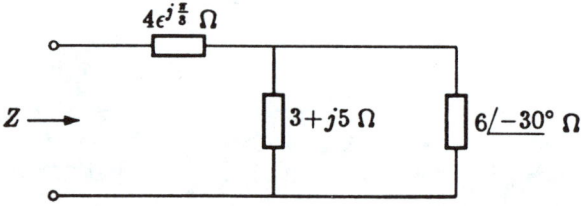

8.19 Find the values of R, L_1, and L_2 so the two circuits have the same impedance for all values of ω.

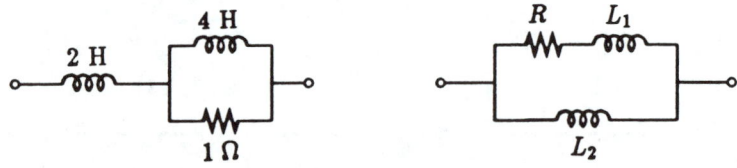

ANALYSIS OF AC CIRCUITS

8.20 Calculate the impedance Z.

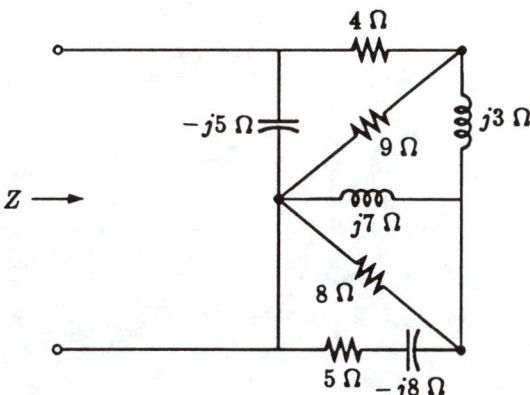

8.21 Determine the values of E.

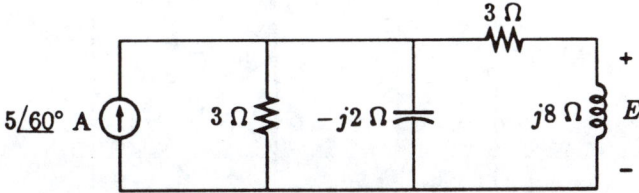

8.22 Determine all currents in the circuit. Then construct a phasor diagram to show how the component currents add up to the total current.

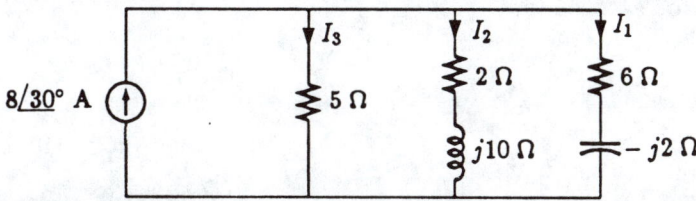

8.23 Determine all currents in the circuit. Then construct a phasor diagram to show $I = I_1 + I_2 = I_3 + I_4$.

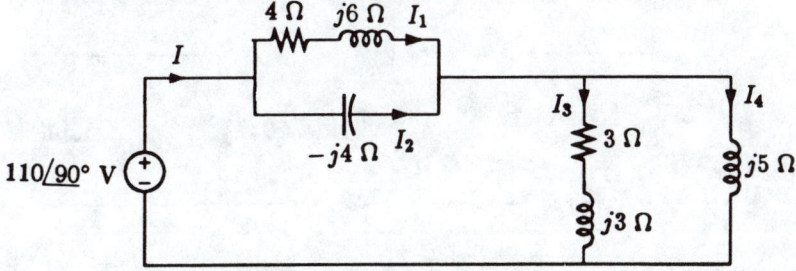

8.24 Determine all currents in the circuit. Then construct a phasor diagram to show $I_2 = I_3 + I_4$ and $I_1 + I_2 = 5$ A.

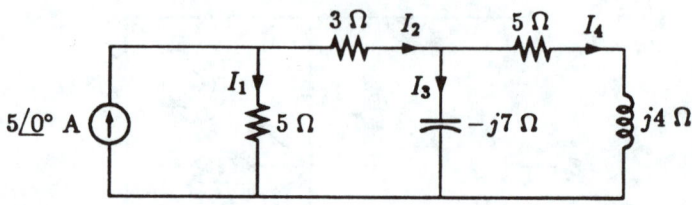

▲ **8.25** Determine I_1 and I_2.

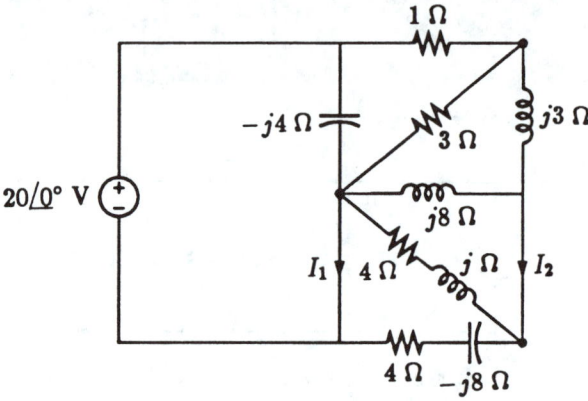

8.26 Find the equivalent delta for each wye, and the equivalent wye for each delta.

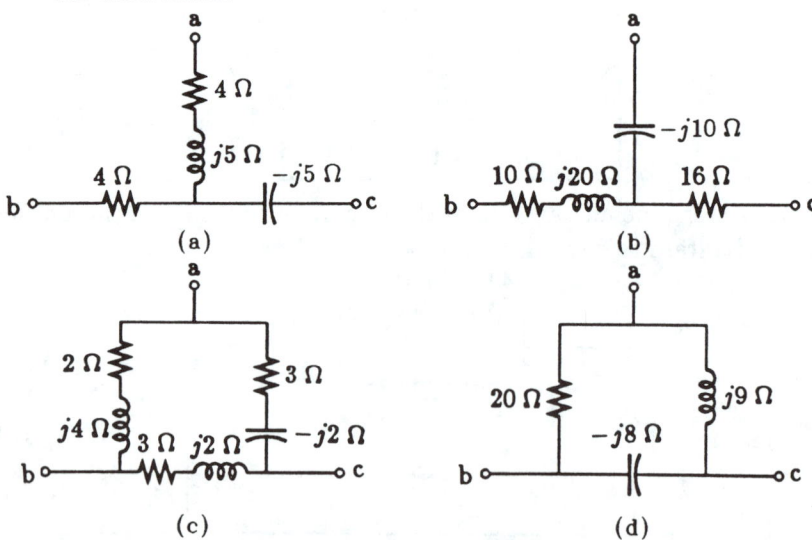

ANALYSIS OF AC CIRCUITS

8.27 First replace each tee by its equivalent pi. Then determine E_2.

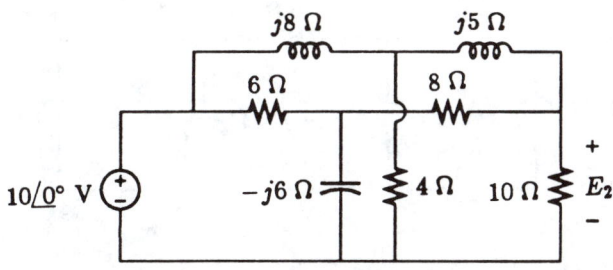

8.28 Determine I.

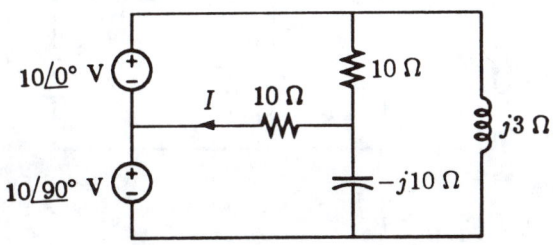

8.29 Determine E.

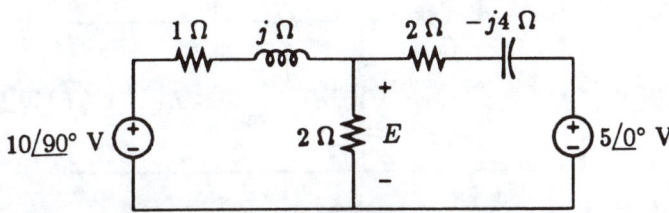

8.30 Determine E.

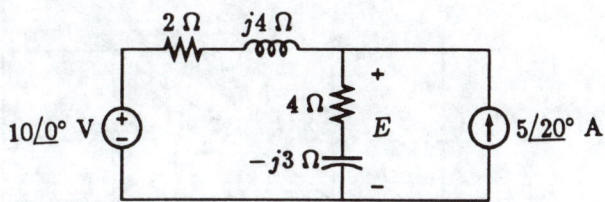

442　　　　　FUNDAMENTALS OF CIRCUIT ANALYSIS

8.31 Determine I.

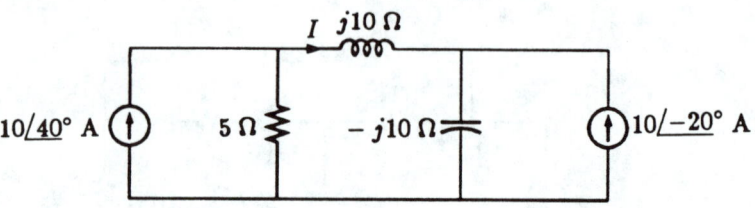

8.32 $I_s = 2/\underline{10°}$ A and $E_s = 20/\underline{-20°}$ V. Determine I.

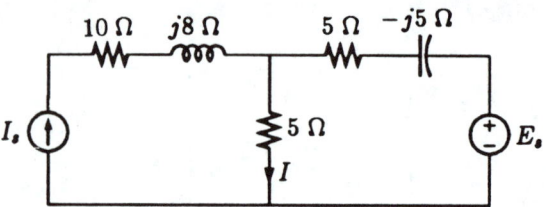

8.33 Determine E_1 and E_2.

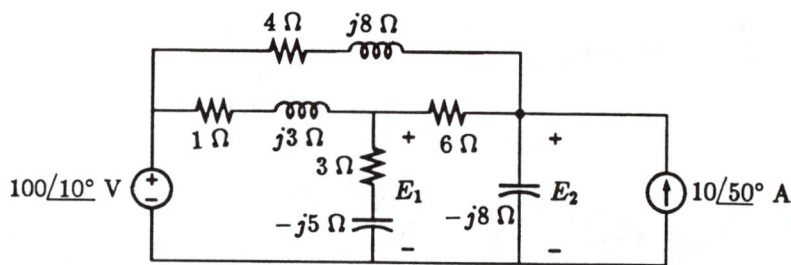

8.34 $E_s = 10/\underline{90°}$ V, $I_1 = 5/\underline{0°}$ A, and $I_2 = 5/\underline{180°}$ A. Determine E.

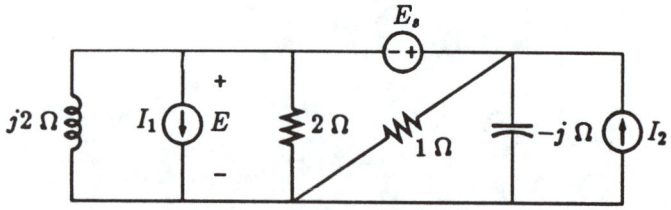

8.35 It is known that $I = 2\underline{/-20°}$ A. Determine E.

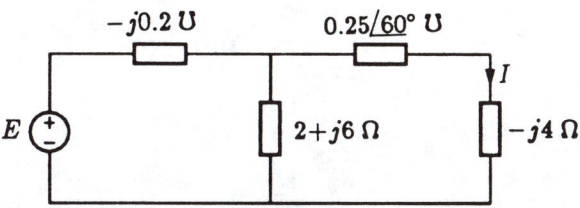

8.36 Determine E, (a) by superposition, (b) by node analysis, and (c) by mesh analysis.

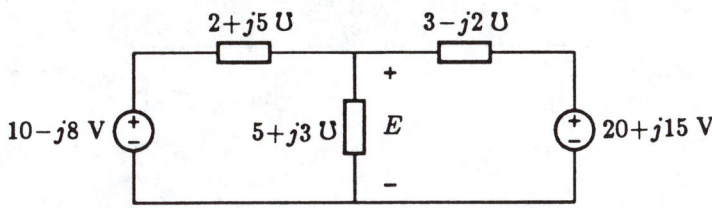

8.37 It is known that $E = 25\underline{/-15°}$ V. Determine Z.

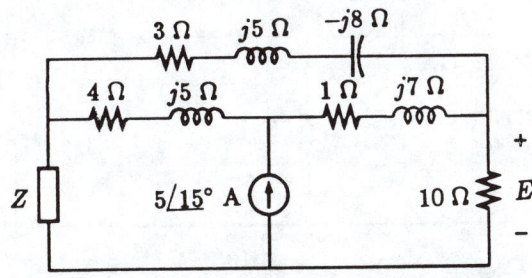

8.38 It is known that $E_2 = 60\underline{/50°}$ V. Determine Z.

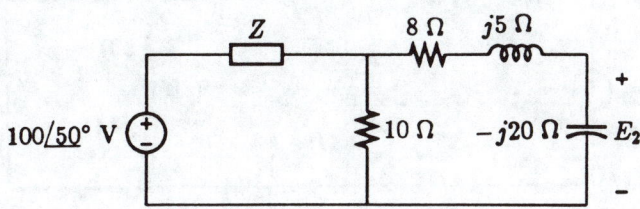

8.39 Determine E_2.

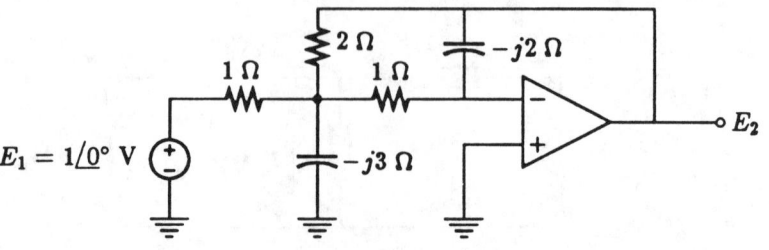

8.40 Determine E_2.

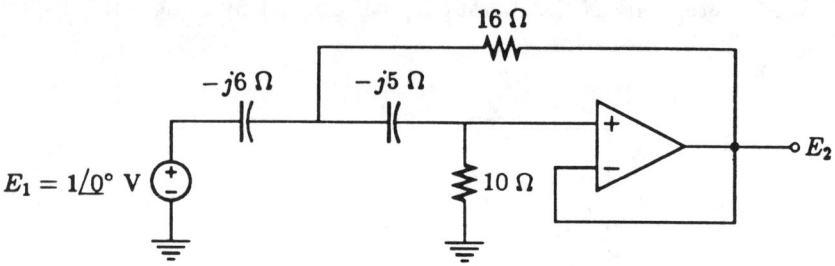

8.41 $E_s = 100\underline{/0°}$ V and $\omega = 2$ rad/sec. Determine all voltages in the circuit. Then construct to scale a phasor diagram to show $E_1 = E_2 + E_3 + E_4$ and $E_s = E_1 + E_5$.

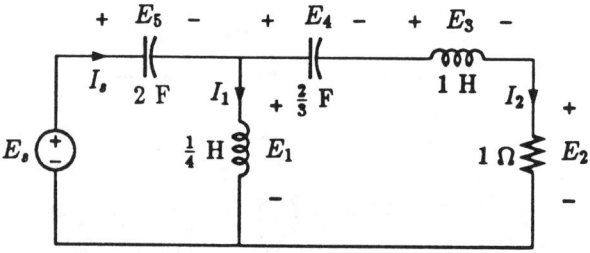

8.42 $E_1 = 100\underline{/30°}$ V and $I_5 = 10\underline{/60°}$ A. Determine all currents in the circuit.

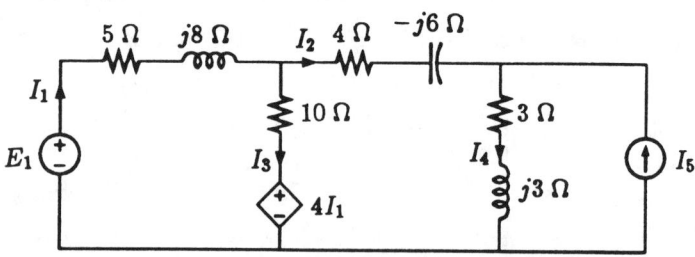

ANALYSIS OF AC CIRCUITS

8.43 Determine the Thevenin's equivalent circuit of the network with respect to terminals A and B.

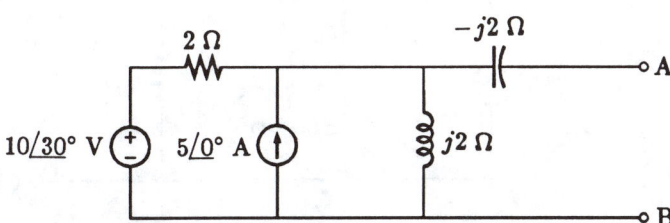

8.44 Determine the Thevenin's equivalent circuit of the network with respect to terminals A and B.

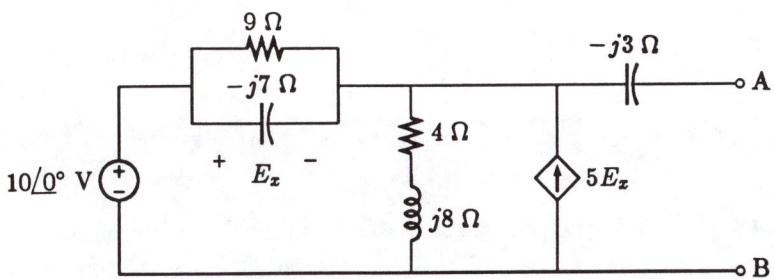

8.45 Determine the Norton's equivalent circuit of the network with respect to terminals A and B.

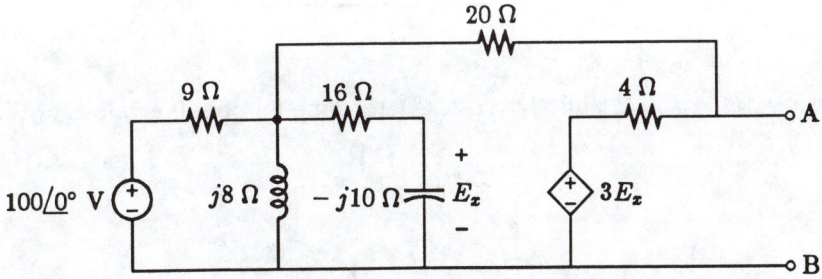

8.46 Determine the Norton's equivalent circuit of the network with respect to terminals A and B.

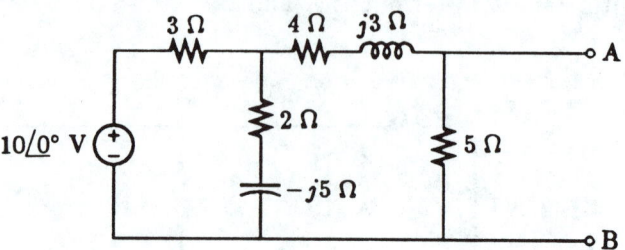

8.47 $|I| = 2$ A. Determine $|E_s|$.

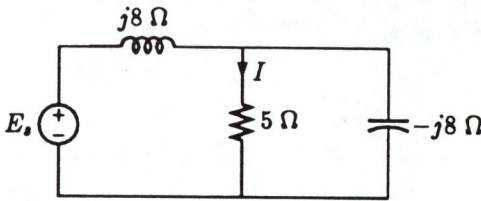

8.48 $|I| = 5$ A and $|E_s| = 100$ V. Determine the value of R.

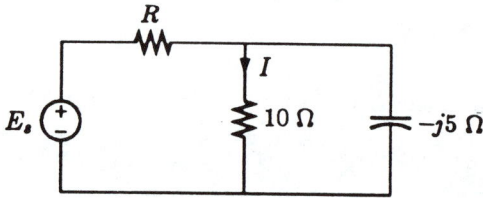

8.49 $|E_s| = 50$ V and $|I| = 5$ A. Determine all possible values of X.

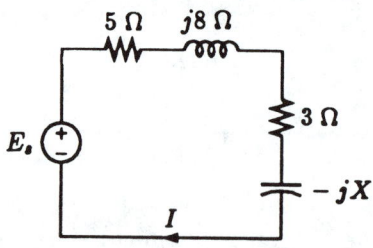

ANALYSIS OF AC CIRCUITS

8.50 $|I_s| = 4$ A and $|I| = 2$ A. Determine R.

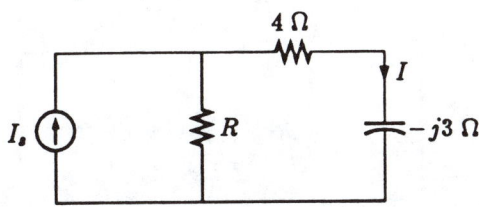

8.51 $|E_s| = 100$ V and $|I| = 7$ A. Determine X.

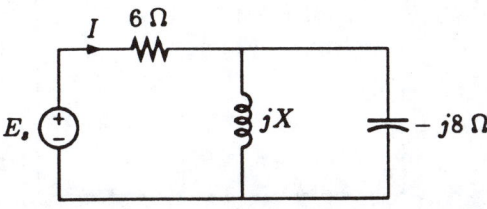

8.52 $|E_s| = 20$ V and $|E| = 10$ V. Determine R.

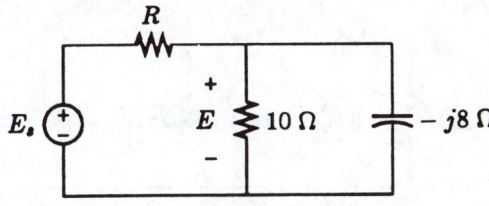

8.53 $|E_s| = 9$ V and $|I| = 4$ A. Determine X.

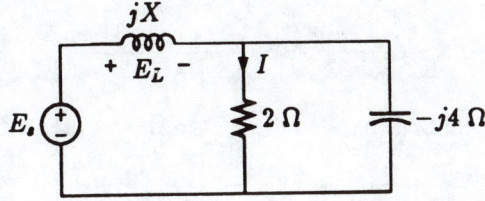

8.54 $|I_s| = 40$ A and $|I| = 10$ A. Determine R.

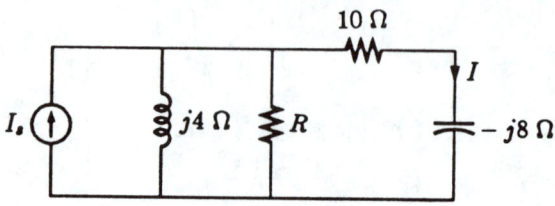

8.55 $|E_s| = 20$ V and $|E| = 10$ V. Determine R.

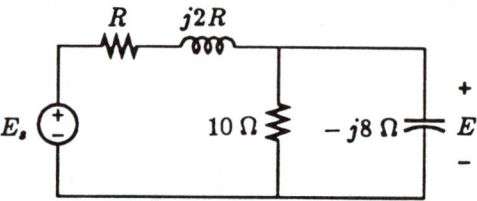

8.56 $|E_s| = 20$ V and $|I| = 3$ A. Determine R.

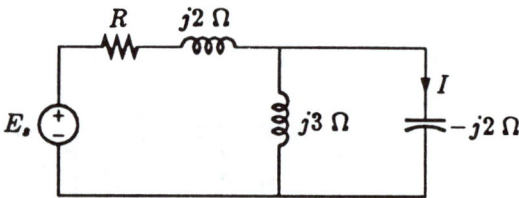

8.57 It is known that I leads E_s by 35°. Determine X.

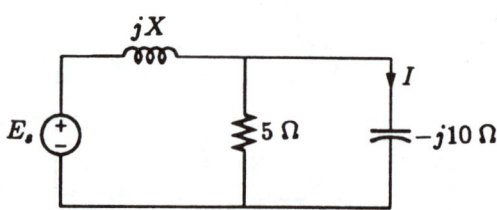

ANALYSIS OF AC CIRCUITS

8.58 $E_s = 20\underline{/20°}$ V and $I = 5\underline{/\theta}$. Determine X and θ.

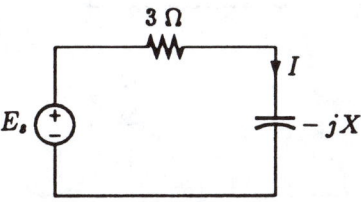

8.59 $|E_s| = 100$ V and $|I| = 6$ A. Determine X.

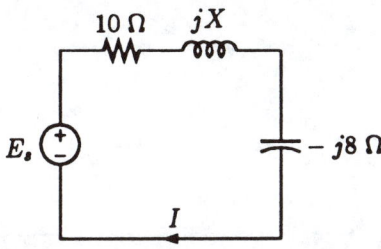

8.60 It is known that $I = 6\underline{/0°}$ A and $E_s = |E_s|\underline{/-60°}$ V. Determine X.

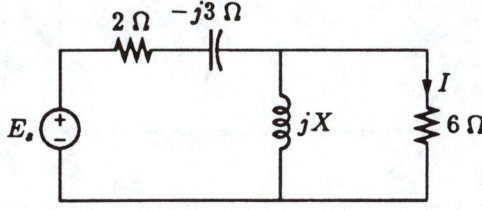

8.61 Give the locus of Y as R is varied from 0 to ∞.

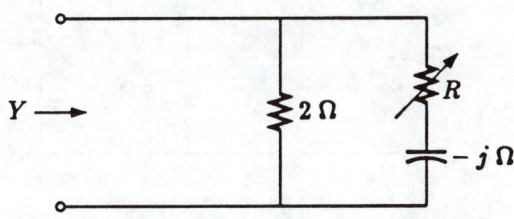

8.62 Give the locus of Y as R is varied from 0 to ∞.

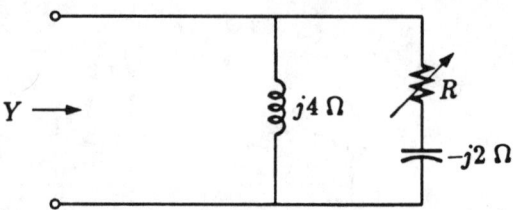

8.63 Give the locus of Z as R is varied from 0 to ∞. From this locus, determine the maximum and minimum values of $|Z|$. Also, determine the values of R at which these extrema occur.

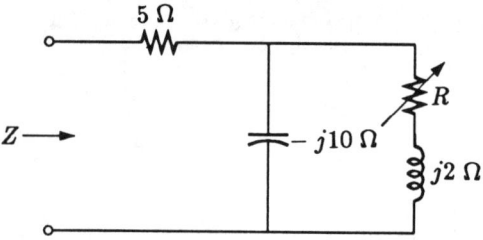

8.64 Give the locus of Z as X is varied from 0 to ∞. What is the minimum value of $|Z|$?

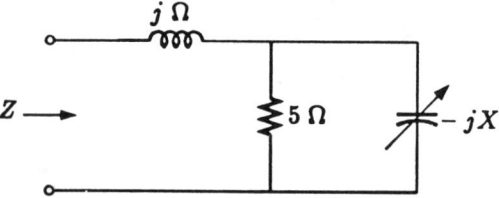

8.65 Give the locus of Z as R is varied from 0 to ∞. What is the minimum value of $|Z|$?

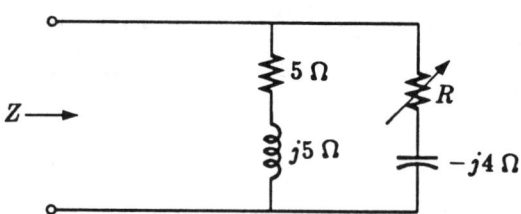

8.66 Give the locus of Z as X is varied from 0 to ∞.

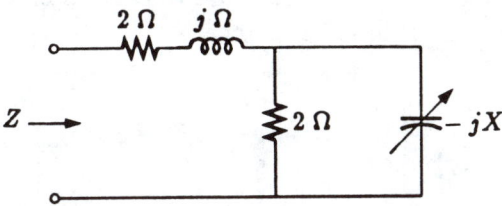

8.67 Give the locus of Z as R is varied from 0 to ∞.

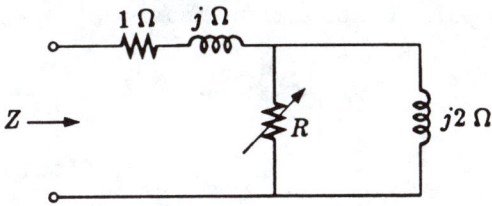

8.68 In the circuit, ω is variable. Calculate (a) the resonance frequency, (b) Q, (c) the half-power frequencies, and (d) the quarter-power frequencies.

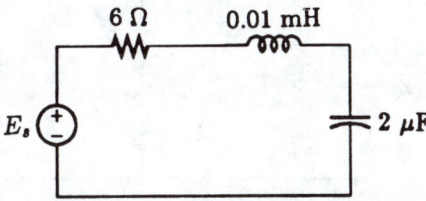

8.69 In the circuit, $\omega = 1$ rad/sec and C is variable. (a) What are the values of C and $|E_C|$ at resonance? (b) What is the value of C at which $|E_C|$ is maximum and what is the maximum $|E_C|$?

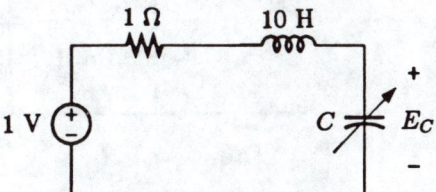

8.70 In the circuit, ω is variable. (a) what are the values of ω and $|I_C|$ at resonance? (b) What are the values of ω at which $|I_C|$ is 60% of its value at resonance?

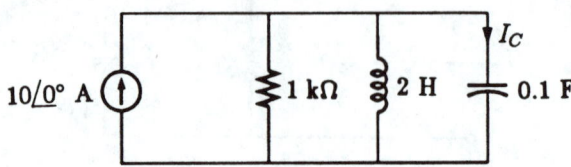

8.71 In the circuit, ω is variable. Determine the value of ω at which $|E_C|$ is maximum and the maximum $|E_C|$.

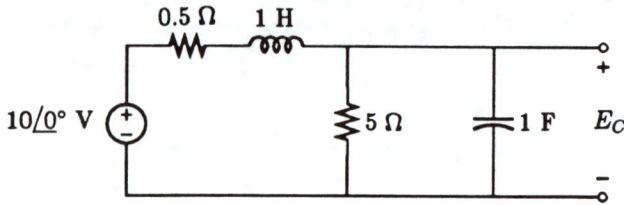

8.72 Derive the expression for the finite value of ω at which Z is pure resistive.

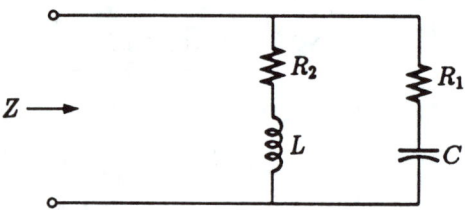

8.73 Determine the value of ω at which Z is purely resistive.

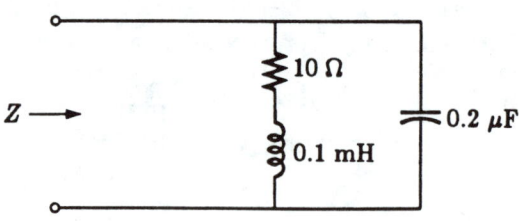

Chapter 9

Alternating-Current Powers

We have discussed the power delivered to a part of a network before. In Fig. 9.1, the power *delivered to* or *consumed by* a part of a network, represented by a rectangular box N_L, at any instant is $p(t) = e(t)i(t)$. It is important to pay particular attention to the associated directions of the voltage and the current when we describe this power. If the direction of one of the quantities is reversed, the $e(t)i(t)$ product represents the power *delivered by* or *supplied by* N_L.

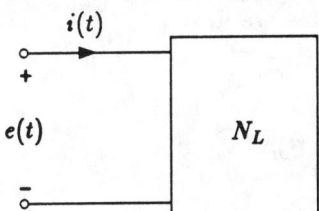

Figure 9.1: The instantaneous power.

In a dc circuit, all quantities are constant. The power delivered to any part of a network, as shown in Fig. 9.2, is

$$P = IE = I^2 R = \frac{E^2}{R} \tag{9.1}$$

In the last two expressions of Eq. (9.1), the load (or N_L) is represented by the equivalent resistance R.

We shall now discuss the details of power description in circuits in which all voltages and currents are sinusoidal and of the same frequency—ac circuits. In ac circuits, all electrical quantities are

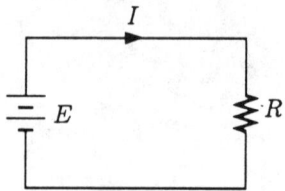

Figure 9.2: A dc circuit.

time functions and, therefore, all powers are also functions of time. Because of the relative phase angles between voltages and currents, there are several types of power in an element, branch, or a part of a network. We shall first examine the power variations in a pure resistance, a pure inductance, and a pure capacitance.

9.1 The power in a resistance

If we have an ac current, I, flowing in a resistance, as shown in Fig. 9.3, the current $I = |I|\underline{/\phi}$ and the voltage $E = |E|\underline{/\phi}$ are in phase. In terms of time, we have

$$i(t) = \sqrt{2}|I| \cos(\omega t + \phi)$$

$$e(t) = \sqrt{2}|E| \cos(\omega t + \phi)$$

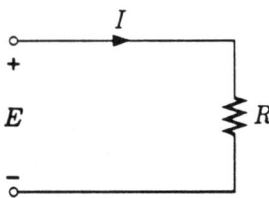

Figure 9.3: A resistance with an ac current.

The instantaneous power delivered to the resistance is

$$p(t) = e(t)i(t) = 2|E||I| \cos^2(\omega t + \phi)$$

$$= |E||I| \{1 + \cos[2(\omega t + \phi)]\} \tag{9.2}$$

ALTERNATING-CURRENT POWERS

The variations of the three quantities—$i(t)$, $e(t)$, and $p(t)$—are depicted in Fig. 9.4. From both Eq. (9.2) and Fig. 9.4, it is clear that the *average power* is

$$P = |E||I| \tag{9.3}$$

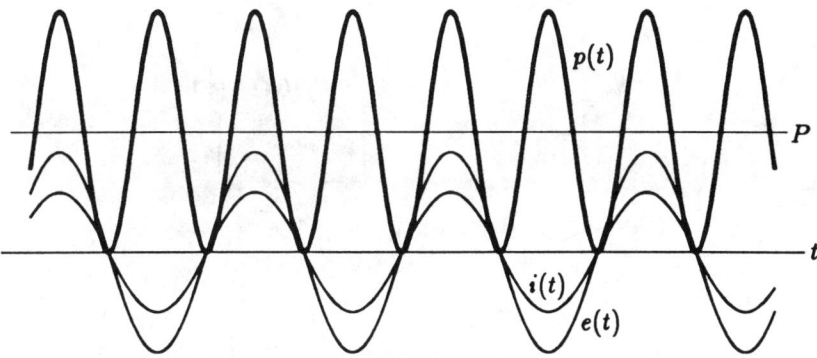

Figure 9.4: Instantaneous power in a resistance.

Thus, the average power delivered to a resistance is simply the product of the effective values of the current and the voltage. It is also quite clear that

$$P = |I|^2 R = \frac{|E|^2}{R} \tag{9.4}$$

Eqs. (9.3) and (9.4) are just like Eq. (9.1). This is precisely the reason we choose to represent the magnitude of an ac quantity by its effective (or rms) value instead of its peak value.

In Fig. 9.4, we have intentionally left out the location of the origin. Over a long period, which is when the steady-state evaluation of circuits becomes meaningful, the average power is independent of where the origin happens to be. In terms of phasors I and E, this simply means that the average power does not depend on the absolute phase angle of the current and the voltage, ϕ. Rather, it's the relative angle between the current and the voltage—in this case, zero—that matters.

To identify the type of power associated with an ac current and an ac voltage that are in phase, we shall call this type of power the *real power*, the *average power*, the *in-phase power*, the *wattage*, or simply *the power*.

EXAMPLE 1. Calculate the real power delivered to the resistor in Fig. 9.5.

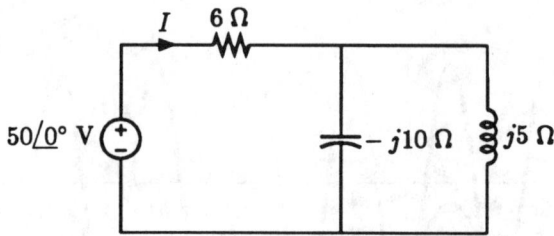

Figure 9.5: Circuit used in Example 1.

SOLUTION

$$I = \frac{50}{6 + \frac{(j5)(-j10)}{j5-j10}} = 2.206 - j3.676 = 4.287\underline{/-59.04°} \text{ A}$$

$$P = |I|^2 \times 6 = 110.29 \text{ W}$$

EXERCISES

9.1.1 Calculate the real power consumed by each resistor.

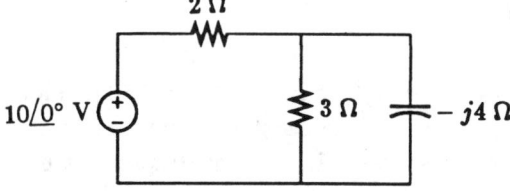

Ans. $P_{2\Omega} = 11.47$ W, $P_{3\Omega} = 11.01$ W.

9.1.2 Calculate the real power delivered to each resistor.

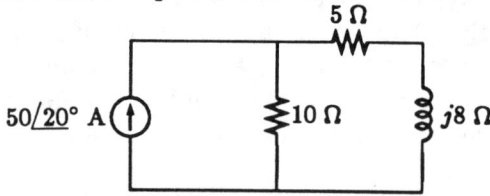

Ans. $P_{5\Omega} = 4.325$ kW, $P_{10\Omega} = 7.699$ kW.

9.2 The power in an inductance

If we have an ac current, I, flowing in an inductance, as shown in Fig. 9.6, the current $I = |I|\underline{/\phi}$ lags the voltage $E = |E|\underline{/\psi}$ by 90°. In terms of time, we have

$$i(t) = \sqrt{2}\,|I|\cos(\omega t + \psi - 90°) = \sqrt{2}\,|I|\sin(\omega t + \psi)$$

$$e(t) = \sqrt{2}\,|E|\cos(\omega t + \psi)$$

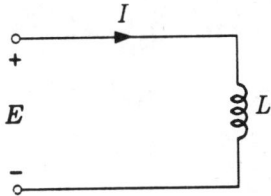

Figure 9.6: An inductance with an ac current.

The instantaneous power delivered to the inductance is

$$p_L(t) = e(t)i(t) = 2\,|E|\,|I|\sin(\omega t + \psi)\cos(\omega t + \psi)$$

$$= |E|\,|I|\sin[2(\omega t + \psi)] \tag{9.5}$$

The variations of the three quantities—$i(t)$, $e(t)$, and $p_L(t)$—are depicted in Fig. 9.7. From both Eq. (9.5) and Fig. 9.7, it is clear that the average power is zero over a long period of time.

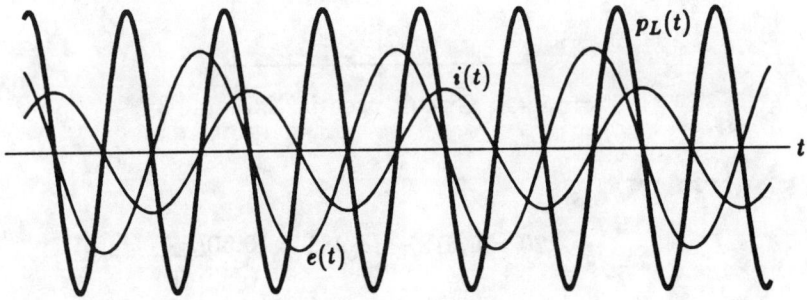

Figure 9.7: Instantaneous power in an inductance.

Although the average power consumed by an inductance is zero, the time function of $p_L(t)$ of Eq. (9.5) or Fig. 9.7 does represent an interchange of energy between the inductance and the rest of the network. The energy absorbed by the inductance during one quarter period in which $p_L(t)$ is positive is stored in the electromagnetic field. This stored energy is returned to the remainder of the circuit during the subsequent quarter period during which $p_L(t)$ is negative. This phenomenon of energy interchange has a definite consequence in a circuit. We call this power variation a *reactive power*. Specifically, when a voltage, E, and a current I, are in quadrature phase, their associated reactive power is defined as

$$Q_L = |E||I| \tag{9.6}$$

Analogous to the real power, the reactive power is also equal to

$$Q_L = |I|^2 X_L = \frac{|E|^2}{X_L} \tag{9.7}$$

where $X_L = \omega L$ is the reactance of the inductance. The dimension of Q_L is the same as that of the real power. However, since the net wattage in a pure inductance is zero, it would not be appropriate to use the watt as the unit of a reactive power. Instead, the unit for a reactive power is VAR (for volt-ampere, reactive).

EXAMPLE 2. Calculate the reactive power delivered to the inductor in Fig. 9.8.

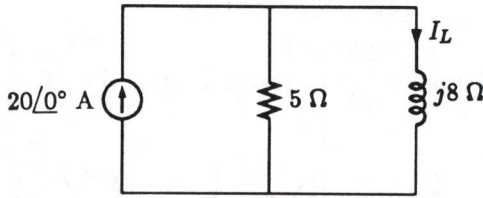

Figure 9.8: Circuit used in Example 2.

SOLUTION

$$I_L = \frac{5}{5 + j8} \times 20 = 5.618 - j8.989 = 10.60\underline{/-57.99°} \text{ A}$$

$$Q = |I_L|^2 \times 8 = 898.9 \text{ VAR}$$

ALTERNATING-CURRENT POWERS

EXERCISES

9.2.1 Calculate the reactive power delivered to the inductor.

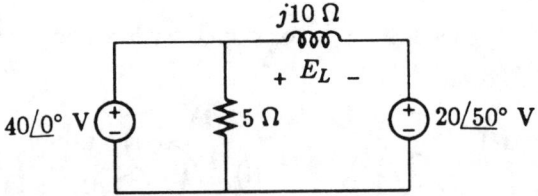

Ans. 97.15 VAR.

9.2.2 Calculate the reactive power delivered to the inductor.

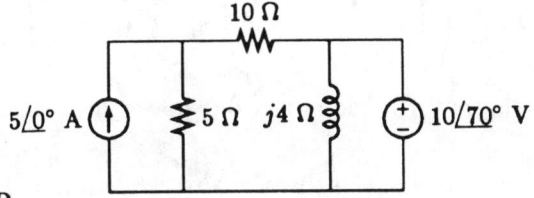

Ans. 25 VAR.

9.3 The power in a capacitance

If we have an ac current, I, flowing in a capacitance, as shown in Fig. 9.9, the current $I = |I|\underline{/\phi}$ leads the voltage $E = |E|\underline{/\psi}$ by 90°. In terms of time, we have

$$i(t) = \sqrt{2}\,|I|\cos(\omega t + \phi)$$

$$e(t) = \sqrt{2}\,|E|\cos(\omega t + \phi - 90°) = \sqrt{2}\,|E|\sin(\omega t + \phi)$$

The instantaneous power delivered to the capacitance is

$$p_C(t) = e(t)i(t) = 2\,|E|\,|I|\cos(\omega t + \phi)\sin(\omega t + \phi)$$

$$= |E|\,|I|\sin[2(\omega t + \phi)] \qquad (9.8)$$

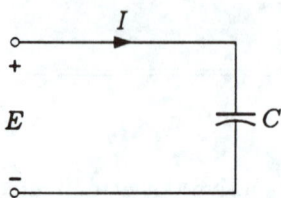

Figure 9.9: A capacitance with an ac current.

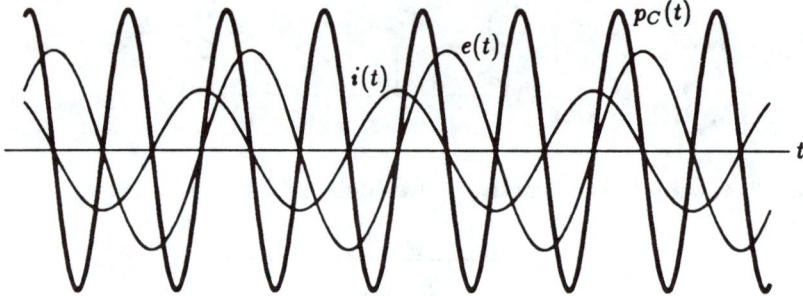

Figure 9.10: Instantaneous power in a capacitance.

The variations of the three quantities—$i(t)$, $e(t)$, and $p_C(t)$—are depicted in Fig. 9.10. From both Eq. (9.8) and Fig. 9.10, it is clear that the average power is zero over a long period of time.

Again, although the average power consumed by a capacitance is zero, the time function $p_C(t)$ of Eq. (9.9) or Fig. 9.10 does represent an interchange of energy between the capacitance and the rest of the network. The energy absorbed by the capacitance during one quarter period in which $p_C(t)$ is positive is stored in the electrostatic field. This stored energy is returned to the remainder of the circuit during the subsequent quarter period during which $p_C(t)$ is negative. This phenomenon of energy interchange is like the energy exchange between inductance and the rest of the network and has a definite consequence in a circuit. This power variation is also a _reactive power_. The reactive power associated with a capacitance is given by

$$Q_C = |E|\,|I| = |I|^2 X_C = \frac{|E|^2}{X_C} \qquad (9.9)$$

where $X_C = 1/\omega C$ is the reactance of the capacitance. The unit for Q_C is also VAR.

ALTERNATING-CURRENT POWERS

EXAMPLE 3. Calculate the reactive power delivered to the capacitor in Fig. 9.11.

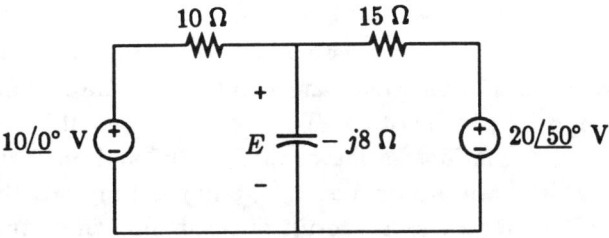

Figure 9.11: Circuit used in Example 3

SOLUTION We use node analysis with E as the voltage unknown. The node equation is

$$\frac{E-10}{10} + \frac{E}{-j8} + \frac{E - 20/\underline{50°}}{15} = 0$$

Solving gives

$$E = 10.073 - j1.426 = 10.173/\underline{-8.06°} \text{ V}$$

$$Q = \frac{|E|^2}{X_C} = \frac{10.173^2}{8} = 12.94 \text{ VAR}$$

EXERCISE

9.3.1 What is the reactive power delivered to the capacitor?

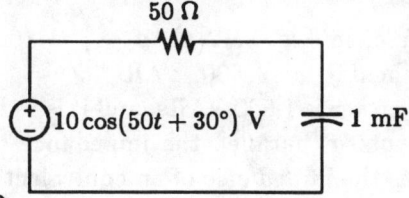

Ans. 0.3448 VAR.

9.4 The inductive and capacitive reactive powers

Although the reactive power associated with an inductance and that associated with a capacitance are similar in nature, they are also directly opposite of each other. This becomes evident if we carefully examine their variations as shown in Figs. 9.7 and 9.10. If we align the voltages in those two figures, $p_L(t)$ and $p_C(t)$ are exactly 180° out of phase. Therefore, the net result of combining these two reactive powers is that they cancel the effects of each other. Similarly, if the currents in those two figures are aligned, $p_L(t)$ and $p_C(t)$ are also 180° out of phase. Therefore, when the inductive reactive power and the capacitive reactive power are combined, they subtract from each other.

Another way to view the relative effects of the two types of reactive power is to consider them to be associated with two equivalent circuit elements. If we have a capacitor and an inductor connected in parallel as shown in Fig. 9.12(a), the reactive power associated with the inductance is

$$Q_L = \frac{|E|^2}{X_L} = |E_L|^2 B_L$$

and that associated with the capacitance

$$Q_C = \frac{|E|^2}{X_C} = |E_L|^2 B_C$$

where $X_L = \omega L$, $B_L = 1/X_L$, $X_C = 1/\omega C$, and $B_C = 1/X_C$. The total reactive power associated with the parallel combination is

$$Q = |E|^2 B = |E|^2 |B_L - B_C| = |Q_L - Q_C|$$

The total reactive power will be of the same type as the one that is greater.

For example, in Fig. 9.12(a), if $|E| = 100$ V, $X_L = 4$ Ω, and $X_C = 10$ Ω; then $Q_L = 2,500$ VARs; $Q_C = 1,000$ VARs; and the total reactive power is 1,500 VARs, and it is inductive. If we combine the two elements in parallel, the impedance of the combination is $j\frac{40}{6}$ Ω, which is the impedance of an equivalent inductance. If a 100-volt voltage is applied across this equivalent inductance, the reactive power will be 1,500 VARs.

ALTERNATING-CURRENT POWERS

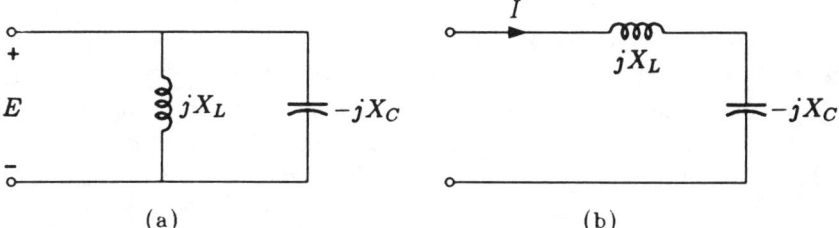

Figure 9.12: Parallel and series combinations of a capacitance and an inductance.

Alternatively, if we have a capacitance and an inductance connected in series as shown in Fig. 9.12(b), the reactive power associated with the inductance is

$$Q_L = |I|^2 X_L = \frac{|I|^2}{B_L}$$

and that associated with the capacitance is

$$Q_C = |I|^2 X_C = \frac{|I|^2}{B_C}$$

The total reactive power associated with the series combination is

$$Q = |I|^2 |X| = |I|^2 |X_L - X_C| = |Q_L - Q_C|$$

The total power will be of the same type as the one that is greater in magnitude.

For example, in Fig. 9.12(b), if $|I| = 5$ A, $X_L = 20$ Ω, and $X_C = 50$ Ω; then $Q_L = 500$ VARs, $Q_C = 1,250$ VARs; and the total reactive power is $Q = 750$ VARs, and it is capacitive. If we combine the two elements in series, the total impedance is $-j30$ Ω, which is the impedance of an equivalent capacitance. If a 5-ampere current is flowing through this equivalent capacitive impedance, the reactive power is 750 VARs.

Since the combined effect of an inductive reactive power and a capacitive reactive power is the difference of the two, it would be appropriate to call one of the reactive powers "positive" and the other "negative." Doing so not only takes care of the opposite effects of the two types of reactive powers, but also shows the type of the resultant reactive power of a number of reactive powers by the algebraic sign of the sum.

Unfortunately, this sign convention is not universal. Not only does it vary from country to country, but it also varies from company to company. Even within one industry, it may change from one chief engineer to another. To circumvent this ambiguity, we shall differentiate these two types of reactive power by attaching a notation to their units. Thus, a capacitive reactive power will have a unit of VAR(C) while an inductive reactive power will have a unit of VAR(L). When several reactive powers are combined, the net reactive power is the difference of the sum of all reactive powers of one type and the sum of all reactive powers of the other type. The type of the combined reactive will be of the type whose sum is greater.

The *reactive power* is also known as the *quadrature power* and *wattless power*.

EXERCISE

9.4.1 Calculate the power delivered to each of the three passive elements.

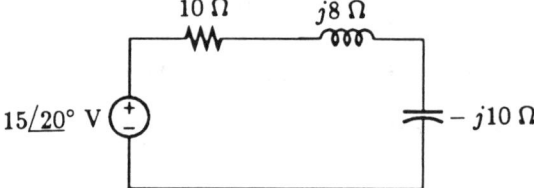

Ans. $P_R = 21.63$ W, $Q_L = 17.31$ VAR(L), $Q_C = 21.63$ VAR(C).

9.5 General ac powers

When a voltage and a current in a circuit are neither in phase nor in quadrature phase with each other, both real and reactive powers are present. Fig. 9.13 shows a general subnetwork N_L for which

$$E = |E|\underline{/\psi} \quad \text{and} \quad I = |I|\underline{/\phi}$$

To be specific, we shall assume that E leads I. We may replace the subnetwork N_L by an equivalent impedance Z where

$$Z = \frac{E}{I} \tag{9.10}$$

ALTERNATING-CURRENT POWERS

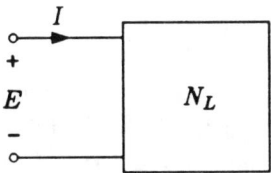

Figure 9.13: Quantities associated with an ac subnetwork.

Thus

$$|Z| = \frac{|E|}{|I|} \quad \text{and} \quad \theta = \psi - \phi$$

and θ is positive. The impedance Z, in turn, may be replaced by either a series or parallel combination of a resistance and an inductance. Let us first use the series combination as shown in Fig. 9.14(a). We have

$$R = |Z|\cos\theta \quad \text{and} \quad X_L = |Z|\sin\theta$$

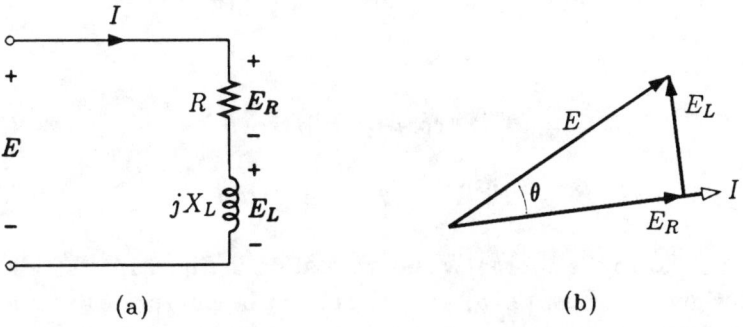

Figure 9.14: The series RL equivalent of an impedance.

Hence,

$$P = |I|^2 R = |I|^2 |Z|\cos\theta = |E||I|\cos\theta \tag{9.11}$$

$$Q = |I|^2 X_L = |I|^2 |Z|\sin\theta = |E||I|\sin\theta \tag{9.12}$$

There is another way to interpret the relationship given in Eqs. (9.11) and (9.12). Using the notation shown in Fig. 9.14(a), we have

$$|E_R| = |E|\cos\theta \quad \text{and} \quad |E_L| = |E|\sin\theta$$

It is obvious that E_R is in phase with I and E_L leads I by 90°. As shown in Fig. 9.14(b), this resolves E into two components—one in phase with I and the other in quadrature phase with I. Eqs. (9.11) and (9.12) may be rewritten as

$$P = |I||E_R| = |I||E_{\text{in-phase}}| \tag{9.13}$$
$$Q = |I||E_L| = |I||E_{\text{quadrature-phase}}| \tag{9.14}$$

Thus, the real power is the magnitude of that part of E that is in phase with I times $|I|$, while the reactive power is the magnitude of that part of E that is in quadrature phase with I times $|I|$.

Next, we shall use the parallel combination. In this case, it is easier to use the reciprocal of Z—the admittance. In other words,

$$Y = \frac{1}{Z} = \frac{I}{E} = G - jB_L$$

and

$$G = |Y|\cos\theta \quad \text{and} \quad B_L = |Y|\sin\theta$$

Hence,

$$P = |E|^2 G = |E|^2|Y|\cos\theta = |E||I|\cos\theta \tag{9.15}$$

$$Q = |E|^2 B_L = |E|^2|Y|\sin\theta = |E||I|\sin\theta \tag{9.16}$$

Similar to the alternative interpretation in Fig. 9.14, the relationships given in Eqs. (9.15) and (9.16) can be interpreted in another way. Using the notation shown in Fig. 9.15(a), we have

$$|I_G| = |I|\cos\theta \quad \text{and} \quad |I_L| = |E|\sin\theta$$

It is obvious that I_G is in phase with E and I_L lags E by 90°. As shown in Fig. 9.15(b), this resolves I into two components—one in phase with E and the other in quadrature phase with E. Eqs. (9.15) and (9.16) may be rewritten as

$$P = |E||I_G| = |E||I_{\text{in-phase}}| \tag{9.17}$$
$$Q = |E||I_L| = |E||I_{\text{quadrature-phase}}| \tag{9.18}$$

ALTERNATING-CURRENT POWERS

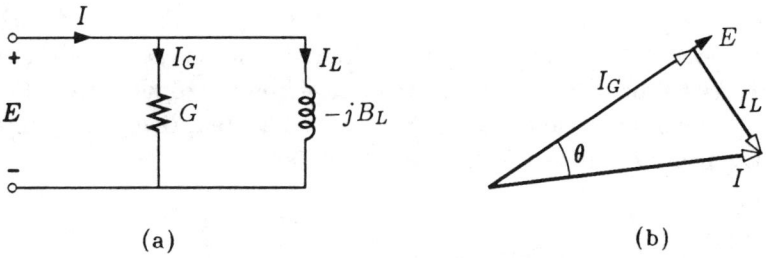

(a)   (b)

Figure 9.15: The parallel RL equivalent of an admittance.

Thus, the real power is the magnitude of that part of I that is in phase with E times $|E|$, while the reactive power is the magnitude of that part of I that is in quadrature phase with E times $|E|$.

EXERCISE

9.5.1 Calculate the real or reactive power delivered to each element in the circuit.

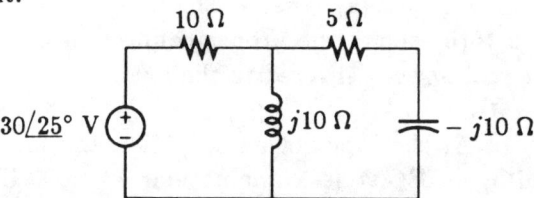

Ans. $P_{10\Omega} = 9$ W, $P_{5\Omega} = 18$ W, $Q_L = 45$ VAR(L), $Q_C = 36$ VAR(C).

9.6 The wattmeter

An ac wattmeter is an instrument that measures the (real) power delivered to a load. A rudimentary wattmeter contains two coils—a current coil and a voltage coil. Therefore, it has four terminals. The instantaneous torque between these two coils is proportional to the product of the two currents in the two coils at that instant. Since the coils are relatively slow-responding compared with the rapidly changing currents, the coil will only sense the average torque. If we hold one coil stationary, the torque affecting the other coil is the average torque. Fig. 9.16(a) gives a representation of such a wattmeter. The instantaneous current in the current coil is $i(t)$. The instantaneous current in the voltage coil is proportional to $e(t)$. Hence, such an arrangement, properly calibrated, reads the average of the product $e(t)i(t)$—the average power.

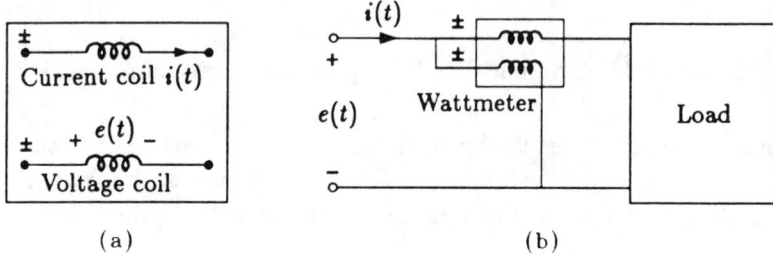

Figure 9.16: A two-coil wattmeter.

The polarities of the two coils are all-important. Usually, one terminal of each coil is labeled with a ± sign. With the terminal labels shown in Fig. 9.16(a), the wattmeter gives the average of $i(t)e(t)$. Fig. 9.16(b) shows the proper connection of a wattmeter that reads the real power delivered to the load.

EXAMPLE 4. At the input of an amplifier, the impedance can be represented by an RC series combination. At 1,000 Hz, the voltmeter reads 25 V, the ammeter reads 1.2 A, and the wattmeter reads 20 W. Calculate R and C.

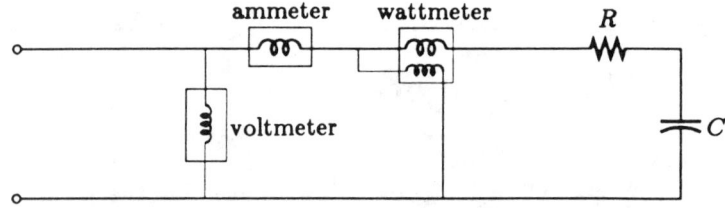

Figure 9.17: Circuit of Example 4.

SOLUTION The equivalent circuit is shown in Fig. 9.17. We have

$$R = \frac{P}{|I|^2} = \frac{20}{1.2^2} = 13.89 \; \Omega$$

ALTERNATING-CURRENT POWERS

$$|Z| = |R - jX_C| = \frac{|E|}{|I|} = \frac{25}{1.2} = 20.85 \;\Omega$$

$$X_C = \sqrt{|Z|^2 - R^2} = \sqrt{20.83^2 - 13.89^2} = 15.52 \;\Omega$$

$$C = \frac{1}{\omega X_C} = \frac{1}{2\pi \times 10^3 \times 15.52} = 10.25 \;\mu F$$

9.7 The apparent power and the power factor

Referring to Fig. 9.13, we see that there are several quantities associated with a subnetwork N when its ac powers are being considered. They include the voltage E (in volts), the current I (in amperes), the *real power* $P = |E||I|\cos\theta$ (in watts), and the *reactive power* $Q = |E||I|\sin\theta$ (in VARs). There are several other alternative terms that are used in place of or in addition to these four quantities. These other terms are used chiefly in the power industry.

The *apparent power*, P_a, is the product of the magnitudes of the voltage and the current.

$$P_a = |E||I| \tag{9.19}$$

It is simply the product of the voltmeter reading and the ammeter reading at a load.

The *power factor*, PF, is the ratio of the real power to the apparent power.

$$\text{PF} = \frac{P}{P_a} \tag{9.20}$$

The *reactive factor*, RF, is the ratio of the reactive power to the apparent power.

$$\text{RF} = \frac{Q}{P_a} \tag{9.21}$$

The power factor angle, θ, is the angle whose cosine is the power factor. It is also the angle difference between the voltage phasor and the current phasor.

Because $P_a^2 = P^2 + Q^2$, the relationship among several of these terms can be simply related by a right triangle known as the *power triangle*, as shown in Fig. 9.18. The two perpendicular sides of the right triangle are P and Q, and the hypotenuse is P_a. The power factor angle is the angle between the base—the P side—and the hypotenuse.

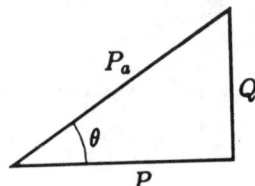

Figure 9.18: The power triangle.

This simple triangle is quite handy in relating the several terms associated with ac powers. It is obvious that if any of the two quantities in a triangle are known, the other two can easily be calculated. Besides, PF and RF are simply related to the θ. However, these quantities, by themselves, furnish no information on whether the reactive power is inductive or capacitive.

The type of reactive power associated with an ac load can be ascertained by investigating the relative angular positions of the E and I phasors. If we always use the angle between these quantities that is less than 180°, then the reactive power is inductive if I lags E and capacitive if I leads E.

In the power industry, the terms *leading power factor* and *lagging power factor* are frequently used. The words "leading" and "lagging" are used to indicate whether I is leading or lagging E. This jargon probably stemmed from the fact that in power generation, the customer is supplied with a voltage bus from which the customer draws power. Hence, the voltage is more naturally viewed as the reference. Different powers are drawn from this voltage bus by drawing different currents, both in terms of magnitude and phase.

The phrases "leading power factor" and "lagging power factor" should be viewed as contractions of more complete descriptions of an ac power. For instance, "a load consumes 500 watts at a *leading* power factor of 0.8" is a contraction of the more proper description "a load consumes 500 watts with a power factor of 0.8, and the current *leads* the voltage."

ALTERNATING-CURRENT POWERS

EXERCISE

9.7.1 In the following, two of the five quantities—P, Q, PF, P_a, and the power-factor angle θ—are given. Obtain the other three quantities. In this exercise, the type of reactive power is no concern.

(a) $P = 500$ W, PF$=0.5$.

(b) $Q = 250$ VAR, $P_a = 300$ VA.

(c) $\theta = 20°$, $Q = 50$ VAR.

(d) $Q = 50$ VAR, PF$= 0.7$.

(e) $P_a = 600$ VA, $P = 500$ W.

Ans. (a) $\theta = 60°$, $P_a = 1000$ VA, $Q = 866.03$ VAR. (b) $\theta = 56.44°$, $P = 165.8$ W, PF$= 0.553$. (c) $P_a = 146.2$ VA, $P = 137.4$ W, PF$= 0.940$. (d) $\theta = 45.57°$, $P_a = 70.01$ VA, $P = 49.01$ W. (e) PF$= 0.833$, $\theta = 33.56°$, $Q = 331.7$ VAR.

9.8 The complex power

We have described several procedures for calculating the different quantities associated with the ac powers of a load. Here we shall describe another procedure that is particularly simple to use in some situations. Let us define the quantity S as

$$S = EI^* \tag{9.22}$$

where I^* denotes the complex conjugate of I. Using the notation associated with Fig. 9.13, we have

$$E = |E|\underline{/\psi} \quad \text{and} \quad I = |I|\underline{/\phi} \quad \text{and} \quad I^* = |I|\underline{/-\phi}$$

and

$$S = |E||I|\cos(\psi - \phi) + j|E||I|\sin(\psi - \phi) = P + jQ \tag{9.23}$$

The quantity S gives us both the real and reactive powers in one operation. Its real part is the real power and its imaginary part is the reactive power. Further, if Q is positive, E leads I and, therefore, Q is inductive. If Q is negative, it must be capacitive. This becomes very convenient when E and I are in rectangular form, as we don't have to ascertain which leads or lags which. The sign of the imaginary part of EI^* automatically reveals the type of reactive power. Also,

$$|S| = |E||I| = P_a$$

The quantity S defined in Eq. (9.22) is known as the *complex power*. The student should not try to treat it as if it were a complex amplitude. It is not a phasor. It is simply a complex number whose real part is equal to the real power, and its imaginary part is equal to the reactive power. Eq. (9.22) should be regarded as a mathematical artifice or a "trick" by which both real and reactive powers can be obtained in one step.

The complex power can alternatively be defined as $E^*I = S^*$. This alternative definition is just as valid and useful. The only difference is in the sign of its imaginary part. If one recognizes this difference and interprets the sign of the imaginary part exactly opposite to the definition of Eq. (9.22), this definition is just as informative. For no special reason, we shall choose to adhere to the definition of Eq. (9.22) from now on.

9.9 Examples of various methods of ac power calculation

We shall now use a numerical example to illustrate the various methods of computing the real and reactive powers. Referring to Fig. 9.13, we let $E = 100\underline{/-20°}$ V and $I = 25\underline{/10°}$ A.

(1) Using Eqs. (9.11) and (9.12), we have

$$P = 100 \times 25 \times \cos 30° = 2{,}165 \text{ W}$$

$$Q = 100 \times 25 \times \sin 30° = 1{,}250 \text{ VAR(C)}$$

Q is capacitive because I leads E by $30°$.

(2) Using the equivalent RC series combination, we have

$$Z = R - jX_C = \frac{E}{I} = \frac{100\underline{/-20°}}{25\underline{/10°}}$$

$$= 4\underline{/-30°} = 3.464 - j2 \; \Omega$$

$$P = |I|^2 R = 25^2 \times 3.464 = 2{,}165 \; \text{W}$$

$$Q = |I|^2 X_C = 25^2 \times 2 = 1{,}250 \; \text{VAR(C)}$$

(3) Using the equivalent RC parallel combination, we have

$$Y = G + jB_C = \frac{I}{E} = \frac{25\underline{/10°}}{100\underline{/-20°}}$$

$$= 0.25\underline{/30°} = 0.2165 + j0.125\mho$$

$$P = |E|^2 G = 100^2 \times 0.2165 = 2{,}165 \; \text{W}$$

$$Q = |E|^2 B_C = 100^2 \times 0.125 = 1{,}250 \; \text{VAR(C)}$$

(4) Resolving E into an in-phase and a quadrature-phase component, we get

$$E_{\text{in-phase}} = 100 \times \cos 30° \; \underline{/10°} = 86.6\underline{/10°} \; \text{V}$$

$$E_{\text{quadrature-phase}} = 100 \times \sin 30° \; \underline{/-80°} = 50\underline{/-80°} \; \text{V}$$

From Eqs. (9.13) and (9.14), we obtain

$$P = 86.6 \times 25 = 2{,}165 \; \text{W}$$

$$Q = 50 \times 25 = 1{,}250 \; \text{VAR(C)}$$

(5) Resolving I into an in-phase and a quadrature-phase component, we get

$$I_{\text{in-phase}} = 25 \times \cos 30° \,/\!-20° = 21.65/\!-20°\text{ A}$$

$$I_{\text{quadrature-phase}} = 25 \times \sin 30° \,/70° = 12.5/70°\text{ A}$$

From Eqs. (9.15) and (9.16), we obtain

$$P = 100 \times 21.65 = 2,165\text{ W}$$

$$Q = 100 \times 12.5 = 1,250\text{ VAR(C)}$$

(6) Using Eq. (9.22), we have

$$S = EI^* = 100/\!-20° \times 25/\!-10°$$

$$= 2500/\!-30° = 2,165 - j1,250\text{ VA}$$

Hence,

$$P = 2,165\text{ W} \quad\text{and}\quad Q = 1,250\text{ VAR(C)}$$

Here, Q is capacitive because the imaginary part of S is negative.

EXERCISE

9.9.1 Calculate the real and reactive powers delivered to the load.

(a) $E = 10/50°\text{ V}, I = 20/20°\text{ A}$.
(b) $E = 200/0°\text{ V}, I = 5/25°\text{ A}$.
(c) $E = 15/90°\text{ V}, I = 20/0°\text{ A}$.
(d) $E = 50/120°\text{ V}, I = 5/10°\text{ A}$.
(e) $E = 20/\!-30°\text{ V}, I = 10/90°\text{ A}$.

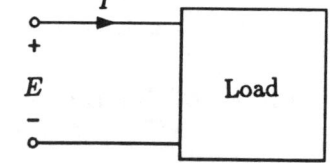

Ans. (a) 173.2 W, 100 VAR(L). (b) 906.3 W, 422.6 VAR(C). (c) 0, 300 VAR(L). (d) −85.5 W, 234.9 VAR(L). (e) −100 W, 173.2 VAR(C).

9.10 Conservation of real and reactive powers

Given a network with B branches. Let us denote the current in the kth branch by I_k and the voltage across the same branch by E_k. Further, let the associated directions of I_k and E_k be consistent through the entire network. If we choose the associated directions as shown in Fig. 9.19, then

$$E_k I_k^* = P_k + jQ_k$$

represents the complex power *delivered to* the kth branch. Also, P_k is the real power delivered to the kth branch and Q_k is the reactive power delivered to the kth branch—inductive if positive and capacitive if negative.

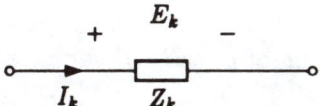

Figure 9.19: Voltage and current directions in a branch.

The E's must satisfy KVL and the I's must satisfy KCL. It follows that the conjugate of all I's—I_k^*, $k = 1, 2, \cdots, B$—must also satisfy KCL. Tellegen's theorem requires that

$$\sum_{k=1}^{B} E_k I_k^* = 0 \qquad (9.24)$$

Separating the real and imaginary part of Eq. (9.24), we get

$$\sum_{k=1}^{B} P_k = 0 \quad \text{and} \quad \sum_{k=1}^{B} Q_k = 0 \qquad (9.25)$$

Eq. (9.24) states that the sum of all complex powers delivered to all branches in a network must be zero. Eq. (9.25) states that the algebraic sum of all real powers delivered to all branches of a network must be equal to zero, and the algebraic sum of all reactive powers delivered to all branches in a network must be zero. Thus, the complex powers, the real powers, and the reactive powers of any complete network must be conserved.

The conservation of these three types of powers is sometimes more conveniently stated in an alternative way. That is, the power (complex, real, or reactive) *consumed by* one part of a network must be equal to the power *supplied by* the rest of the network. This might be preferable in the case when we have a number of sources and a number of passive elements.[1] The total power (of each of the three types) consumed by all the passive elements must be equal to the total power (of the same type) supplied by all the sources.

The student should be cautioned that this conservation does not apply to the apparent powers. In general, the apparent powers are not very meaningful when it comes to the combination of powers. To combine apparent powers would be analogous to combining the magnitudes of phasors. It would usually lead to erroneous or meaningless results.

EXAMPLE 5. We shall use the circuit in Fig. 9.20 to illustrate the principle of conservation of powers. Using mesh analysis, we have

$$(9 - j3)I_1 + (4 - j6)I_2 = 100\underline{/0°}$$

$$(4 - j6)I_1 + (6 - j)I_2 = 50\underline{/20°}$$

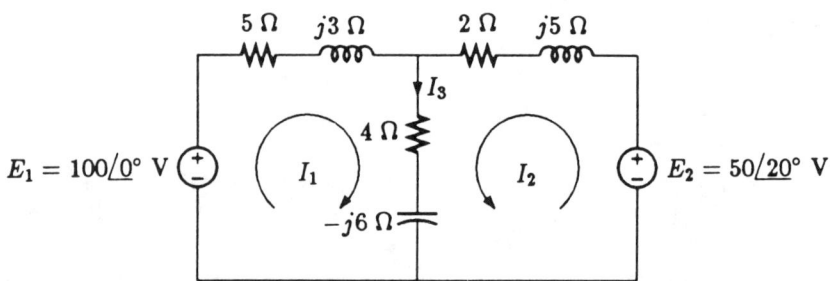

Figure 9.20: An example of power conservation.

Solving, we get

$$I_1 = 4.443 + j0.284 = 4.452\underline{/3.66°} \text{ A}$$

---

[1] Resistor, inductors, and capacitors are passive elements.

ALTERNATING-CURRENT POWERS

$I_2 = 3.309 + j7.655 = 8.339\underline{/66.62°}$ A

$I_3 = 7.751 + j7.939 = 11.096\underline{/45.69°}$ A

Power consumed by the 5-Ω resistor = $|I_1|^2 \times 5 = 99.09$ W

Power consumed by the 2-Ω resistor = $|I_2|^2 \times 2 = 139.08$ W

Power consumed by the 4-Ω resistor = $|I_3|^2 \times 4 = 492.45$ W

Total power consumed by the three resistors = 730.62 W

Reactive power consumed by the 3-Ω inductor

$= |I_1|^2 \times 3 = 59.46$ VAR(L)

Reactive power consumed by the 5-Ω inductor

$= |I_2|^2 \times 5 = 347.69$ VAR(L)

Reactive power consumed by the 6-Ω capacitor

$= |I_3|^2 \times 6 = 738.67$ VAR(C)

Total reactive power consumed by the passive elements

$= 738.67 - 347.69 - 59.46 = 313.52$ VAR(C)

Total complex power supplied by the two sources

$= E_1 I_1^* + E_2 I_2^* = 444.27 - j28.46 + 286.35 - j303.07$

$= 730.62 - j313.52$ VA

Clearly, the total real and reactive powers consumed by the passive elements are equal to those supplied by the sources.

EXAMPLE 6. In Fig. 9.21, Box A consumes 200 VA at 0.9 power factor lagging, Box B consumes 500 W at 0.5 power factor leading, Box C consumes 200 VAR at 0.4 power factor leading, and Box D consumes 50 W at unity power factor. $E_s = 230\underline{/0°}$. Calculate I_s.

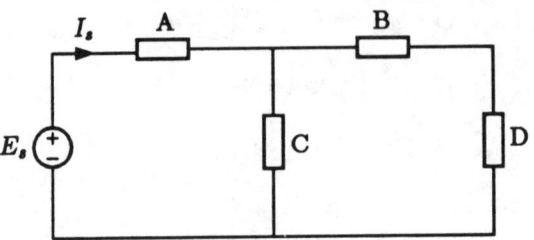

Figure 9.21: The circuit for Example 6.

SOLUTION We construct the power triangles for Boxes A, B, and C. The numbers are shown in Fig. 9.22 (not to scale). From these numbers we have

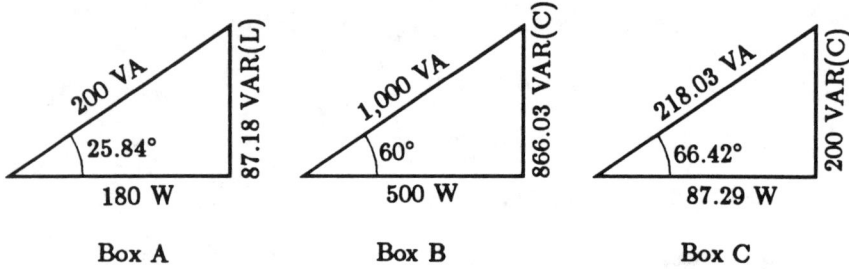

Figure 9.22: Power triangles for Example 6.

$$P = 180 + 500 + 87.29 + 50 = 817.29 \text{W}$$

$$Q = 866.03 + 200 - 87.18 = 978.85 \text{VAR(C)}$$

Hence,

$$E_s I_s^* = 817.29 - j978.85 = 1,275.18 \underline{/-50.14°}$$

$$I_s^* = 3.553 - j4.256 = 5.544 \underline{/-50.14°}$$

$$I_s = 3.553 + j4.256 = 5.544 \underline{/50.14°} \text{ A}$$

Note that the way the four boxes are connected has no bearing on the power calculation. This is, of course, because the powers must be conserved. Also, note that the apparent powers do not enter into the picture of how the loads are combined.

ALTERNATING-CURRENT POWERS

EXERCISE

9.10.1 Calculate the real or reactive power delivered to each element. Then calculate the real and the reactive supplied by the voltage source.

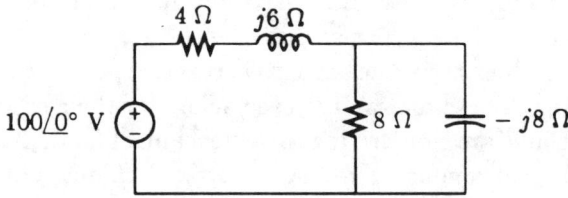

Ans. $P_{4\Omega} = 588.23$ W, $P_{8\Omega} = 588.234$;W, $Q_L = 882.34$ VAR(L), $Q_C = 588.23$ VAR(C), $P_s = 1,176.5$ W, $Q_s = 294.1$ VAR(L).

9.11 Power factor correction

As mentioned earlier, although a reactive power does not consume any average power, it does play a role in the real power transmission. This is particularly important in the power industry where a large amount of power is involved. Here is why:

Suppose a power plant is delivering a power of 1 MW at unity power factor with the generating voltage of 2 kV. The current will be 500 A and in phase with the voltage.

If the power factor is 0.8, the apparent power is 1.25 MVA and the current will have to be 625 A, and it is $\cos^{-1} 0.8 = 36.87°$ out of phase with the voltage. This means that to deliver the same amount of power, the current will have to be 25% higher. The higher current means that the generator will have to be constructed with larger wires. The transformers, which are usually rated in terms of kVA capacity, will have to be larger. The higher the current, the higher the $|I|^2 R$ loss (usually referred to as the copper loss) and as a consequence, an overall lower efficiency. Furthermore, a larger current will mean a larger electromagnetic force and stress to all the apparatus used, and thus shorter useful lives.

With extremely rare exceptions, most industrial machinery as well as household appliances are inductive and consume power at a

480 FUNDAMENTALS OF CIRCUIT ANALYSIS

less-than-unity lagging power factor. The power company would like for all power factors to be as close to unity as possible. The reader may have observed some rectangular boxes mounted on the utility poles with bushings coming out of them. Those devices are oil-filled capacitors. The purpose of those capacitors is to compensate for the inductive reactive powers inherent in the existing loads. Usually, those capacitors are not sufficient or of exactly the right size to bring the power factor to unity. But they do improve the power factor.

Some substations or power plants employ over-excited synchronous motors, which consume capacitive reactive power, to improve the power factor. These motors are used solely for their capacitive reactive power and are not driving any mechanical load. These devices are called synchronous or rotary capacitors. Their advantages are that they occupy less space than the stationary capacitors, and they are easily adjustable continuously by varying their field currents.

The principle of power factor correction is computationally very simple because both the real and the reactive powers must be conserved as described in the previous section. We shall illustrate this procedure with a few examples.

EXAMPLE 7. A substation has four feeder lines. Line 1 delivers 150 kW at a power factor of 0.8 leading. Line 2 delivers 200 kVA at a power factor of 0.9 lagging. Line 3 delivers 300 kW at 0.95 power factor lagging. Line 4 delivers 250 kW at 0.7 power factor lagging. (a) What are the total real and reactive powers delivered by the substation? (b) What additional reactive power is required to bring the power factor to unity?

SOLUTION We first construct a power triangle for each of the four lines, as shown in Fig. 9.23 (not to scale). From these power triangles, we have

(a) Total real power $= 150 + 180 + 300 + 250 = 880$ kW

Total reactive power

$$= 255.05 + 98.61 + 87.18 - 112.5 = 328.34 \text{ kVAR(L)}$$

Total apparent power $= \sqrt{880^2 + 328.34^2} = 939.26$ kVA

Power factor at the substation $= \dfrac{880}{939.26} = 0.937$ lagging

ALTERNATING-CURRENT POWERS

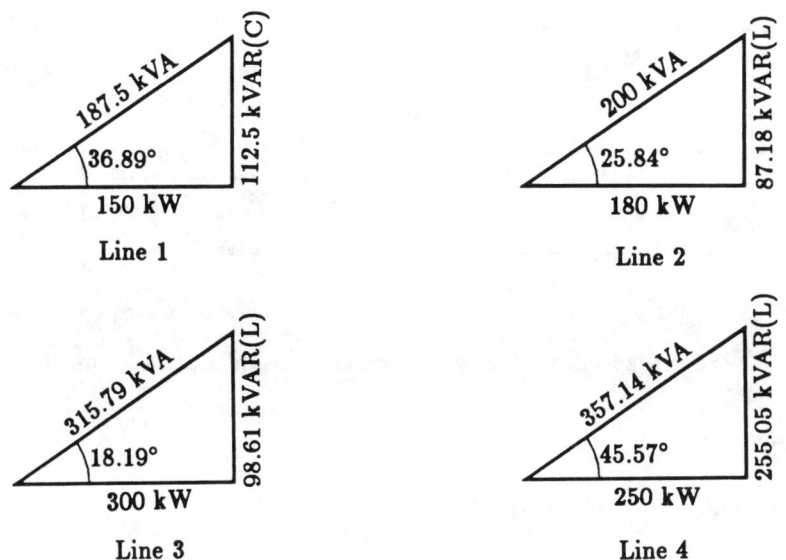

Figure 9.23: Power triangles for four feeder lines.

(b) Reactive power required to bring the power factor to unity

$$= 328.34 \text{ kVAR(C)}$$

EXAMPLE 8. In Fig. 9.24, the induction motor M consumes 5 kW at a lagging power factor of 0.7. What are the values of C that will make the power factor of the combination 0.9?

SOLUTION The reactive power consumed by M is

$$Q_M = 5 \times \tan(\cos^{-1} 0.7) = 5 \times \tan 45.57° = 5.101 \text{ kVAR(L)}$$

For the M-C combination to have a power factor of 0.9, the total reactive power must be

$$Q = 5 \times \tan(\cos^{-1} 0.9) = 2.422 \text{ kVAR(C) or kVAR(L)}$$

If $Q = 2.422$ kVAR(L), then

$$Q_C = 5.101 - 2.422 = 2.679 \text{ kVAR(C)}$$

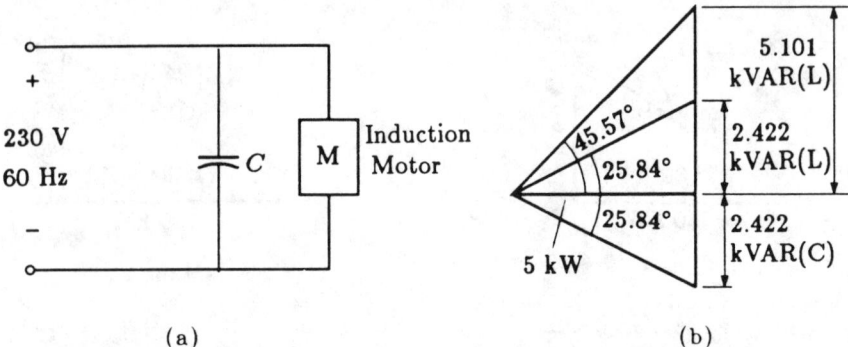

Figure 9.24: Power factor correction of an induction motor.

But

$$Q_C = \frac{|E|^2}{X_C} = \omega C |E|^2$$

$$C = \frac{Q_C}{\omega |E|^2} = \frac{2,697}{2\pi \times 60 \times 230^2} = 134.33 \ \mu\text{F}$$

If we make $Q = 2.422$ kVAR(C), then

$$Q_C = 5.101 + 2.422 = 7.523 \text{ kVAR(C)}$$

and

$$C = \frac{7,523}{2\pi \times 60 \times 230^2} = 377.2 \ \mu\text{F}$$

The power triangle for the motor and its power factor correction are shown in Fig. 9.24(b).

9.12 Maximum power transfer

In the power industry, the main consideration, as far as the power is concerned, is efficiency. In electronic applications, however, the objective is to deliver as much power as possible to a load. A typical situation is a radio transmitter feeding an antenna. There, the optimum condition is to deliver the maximum power to the antenna, thereby broadcasting with the maximum signal strength. Another example is the audio amplifier driving a loudspeaker. The power

ALTERNATING-CURRENT POWERS

consumed by the amplifier is of no significant concern. The key is to deliver power to the speaker. In such situations, the question is how much power can be extracted from a two-terminal network, rather than efficiency.

In the case of the voltage bus of an ac source, we can assume the voltage bus to be nearly an ideal voltage source—constant voltage with no internal impedance. Therefore, the power drawn from the voltage bus is virtually unlimited. This is not the case with most electronic equipment.

The question then is what is the available power from a two-terminal network. To answer this question we shall formulate the problem as shown in Fig. 9.25(a). Network N is a two-terminal network containing linear elements, controlled sources, and independent sources. A load is connected to this two-terminal network. We assume that the load is a variable impedance and the impedance can have any value. By varying this load impedance, what is the maximum power delivered to the load?

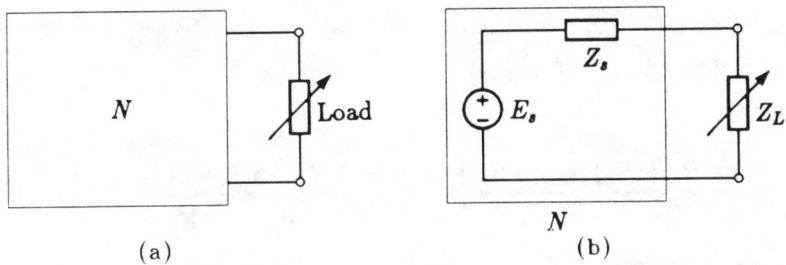

Figure 9.25: Maximum power transfer from a two-terminal network.

To address this question, we first replace the two-terminal network by its Thévenin's equivalent circuit. Further, we let the load be represented by its impedance Z_L. This is shown in Fig. 9.25(b), in which E_s and Z_s are fixed and Z_L is allowed to vary.

We now have

$$I = \frac{E_s}{Z_s + Z_L} \tag{9.26}$$

We let $Z_s = R_s + jX_s$ and $Z_L = R_L + jX_L$, where R_s and R_L are both positive and X_s and X_L may each be positive or negative. Eq. (9.26) becomes

$$I = \frac{E_s}{R_s + R_L + j(X_s + X_L)} \tag{9.27}$$

The power delivered to the load is

$$P_L = |I|^2 R_L = \frac{|E_s|^2 R_L}{(R_s + R_L)^2 + (X_s + X_L)^2} \tag{9.28}$$

Since X_L appears only with X_s in a squared term in the denominator, the best that can be done to maximize P_L is for

$$X_s + X_L = 0 \quad \text{or} \quad X_L = -X_s \tag{9.29}$$

Under this condition, we can now maximize the quantity

$$\frac{|E_s|^2 R_L}{(R_s + R_L)^2}$$

This is done by setting

$$\frac{d}{dR_L}\left[\frac{R_L}{(R_s + R_L)^2}\right] = 0 \tag{9.30}$$

or

$$\frac{2(R_s + R_L)R_L - (R_s + R_L)^2}{(R_s + R_L)^4} = \frac{2R_L - (R_s + R_L)}{(R_s + R_L)^3} = 0$$

which gives

$$R_L = R_s \tag{9.31}$$

Eqs. (9.29) and (9.31) may be combined to read

$$Z_L = Z_s^* \tag{9.32}$$

Eq. (9.32) states that the power delivered by a two-terminal network N to Z_L is maximum when Z_L is the complex conjugate of the Thévenin's equivalent impedance of the two-terminal network. Under this condition, the power delivered by N is

$$P_{\max} = \frac{|E_s|^2}{4R_s} \tag{9.33}$$

ALTERNATING-CURRENT POWERS

This *maximum power* is sometimes referred to as the *available power* of N.

When Eq. (9.32) is satisfied, we describe the condition as a *conjugate impedance match* or *maximum power match*.[2]

EXAMPLE 9. In Fig. 9.26, what is the maximum power deliverable to the impedance Z_L when it is allowed to assume any value?

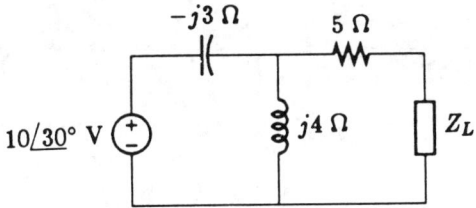

Figure 9.26: Circuit of Example 9.

SOLUTION For the Thévenin's equivalent circuit to the left of Z,

$$E_{oc} = \frac{j4}{j4 - j3} \times 10\underline{/30°} = 40\underline{/30°} \text{ V}$$

$$Z_{eq} = 5 + \frac{-j3 \times j4}{j4 - j3} = 5 - j12 \text{ }\Omega$$

Maximum power is delivered to Z_L if $Z_L = 5 + j12 \text{ }\Omega$, and

$$P_{max} = \frac{40^2}{4 \times 5} = 80 \text{ W}$$

EXAMPLE 10. In this example we examine the maximum power deliverable to Z_L when it is restricted to be a resistance, R_L.

---

[2] In other situations, a different type of matching is appropriate. This other type of matching calls for $Z_L = Z_s$ and is simply referred to as the *impedance matching*.

FUNDAMENTALS OF CIRCUIT ANALYSIS

SOLUTION Refer to Fig. 9.25(b) with $X_L = 0$. Now

$$I = \frac{E_s}{R_s + R_L + jX_s}$$

$$P_L = \frac{|E_s|^2 R_L}{(R_s + R_L)^2 + X_s^2}$$

Setting

$$\frac{d}{dR_L}\left[\frac{R_L}{(R_s + R_L)^2 + X_s^2}\right] = \frac{R_s^2 + X_s^2 - R_L^2}{((R_s + R_L)^2 + X_s^2)^2} = 0$$

we get

$$R_L^2 = R_s^2 + X_s^2 \quad \text{or} \quad R_L = |Z_s| \tag{9.34}$$

Hence, under this restriction, maximum power is delivered to R_L when its value is equal to the magnitude of the source impedance.

EXERCISE

9.12.1 What is the power available to Z_L?

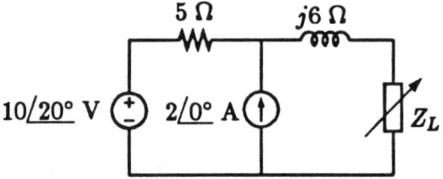

Ans. 19.397 W

9.13 Summary

Power calculation is an important part of the steady-state analysis of ac circuits. There are three types of power. The real power represents the actual transmission of energy from one origin to a destination. The reactive power, which frequently accompanies a real power, does not involve any energy transfer over a long period of time. The apparent power represents the capacity of the apparatus required to generate and transmit the combination of real and reactive powers. Several methods of computing ac powers are given in this chapter.

The complex amplitude or phasor of an ac quantity is customarily represented by its effective (rms) value. This makes the computation of ac powers much more convenient than if maximum value was used. The strength of an ac voltage or current is usually measured by its effective value.

In applications that involve large amounts of power, efficiency is usually the major concern. In situations in which the efficiency is not the main consideration, we may be interested in extracting the maximum power from a two-terminal network. Conditions for attaining maximum power are given in this chapter.

Problems

9.1 Calculate the average power delivered by the voltage source.

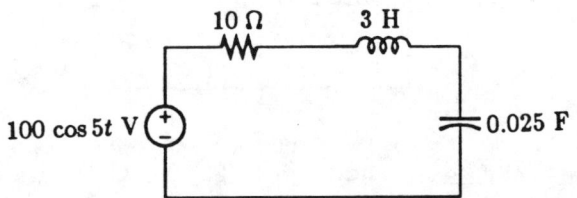

9.2 Current $i(t)$ is flowing through a 5-Ω resistor. Calculate the average power delivered to the resistor.

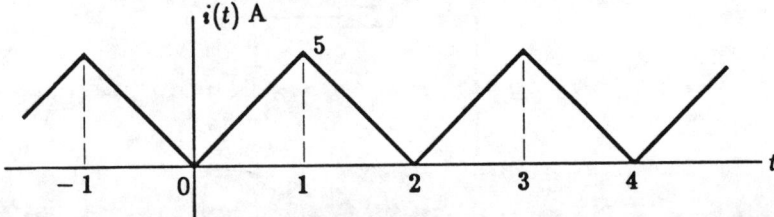

9.3 $|E_s| = 20$ V. Calculate the real or reactive power delivered to each passive element. Also, calculate the real and reactive powers delivered by the voltage source.

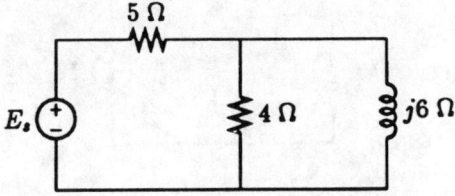

9.4 Calculate the real or reactive power delivered to the each passive element. Also calculate the real and reactive powers delivered by each source.

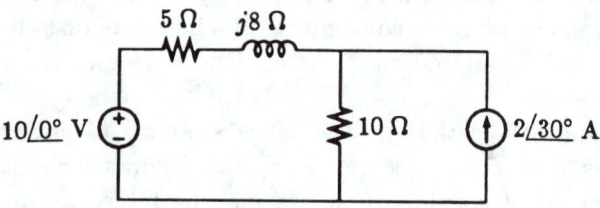

9.5 Calculate the real and reactive powers delivered by the voltage source.

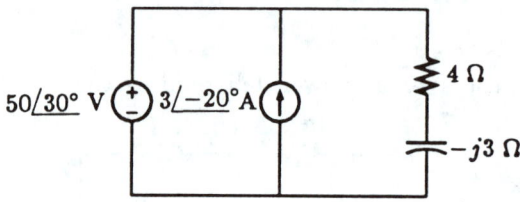

9.6 $E_s = 50\underline{/-20°}$ V and $E = 35\underline{/20°}$ V. Calculate the real and reactive powers delivered by the voltage source.

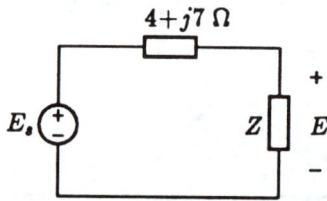

9.7 Calculate the real and reactive power delivered by the current source.

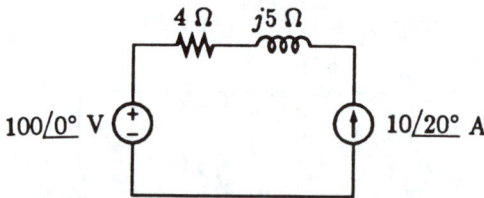

ALTERNATING-CURRENT POWERS

9.8 The powers supplied by the voltage source is 1.2 kW and 200 VAR(C). Calculate I_s.

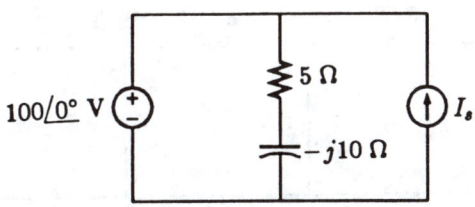

9.9 The powers supplied by the current source is 500 W and 400 VAR(L). Calculate I_e.

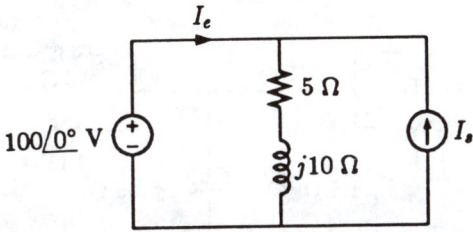

9.10 Determine the real and reactive powers delivered to Box A.

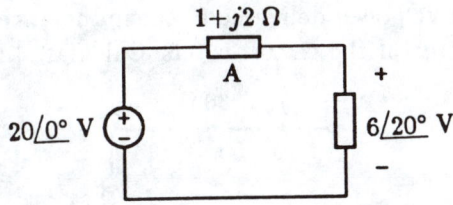

9.11 Calculate the real or reactive power delivered to each passive element. Also, calculate the real and reactive powers supplied by each source.

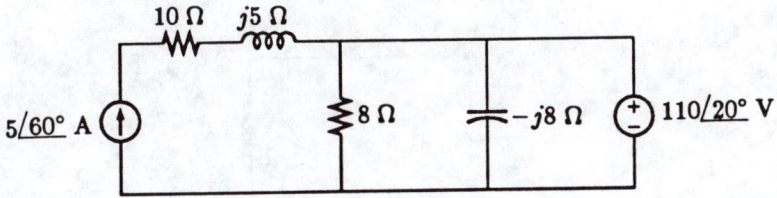

9.12 Calculate the real and reactive powers supplied by the controlled source.

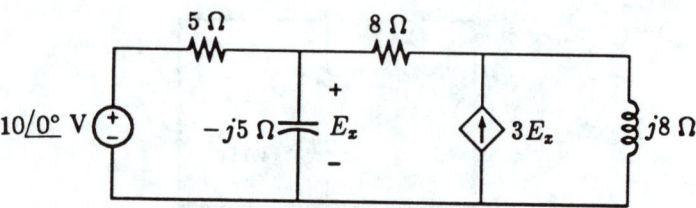

9.13 Calculate the real or reactive power delivered to each passive element. Also, calculate the real and reactive powers supplied by the voltage source. $E_s = 15\underline{/30°}$ V.

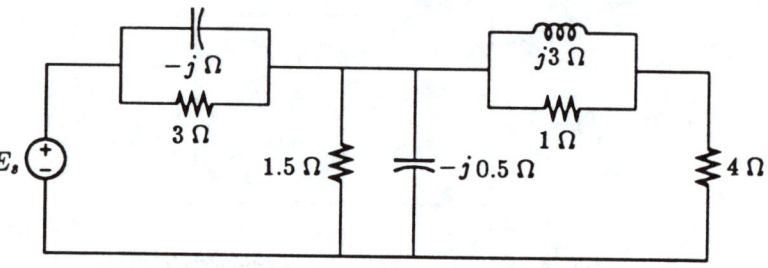

9.14 The reactive power delivered to the inductance is 90 VAR. The power factor of the circuit is 0.6. Calculate $|E|$ and $|I|$.

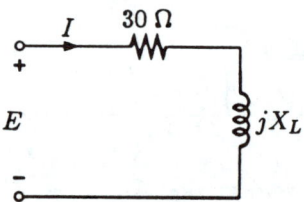

9.15 It is known that $|I| = 5$ A and the reactive power delivered to the inductance is 125 VAR. Determine $|E|$.

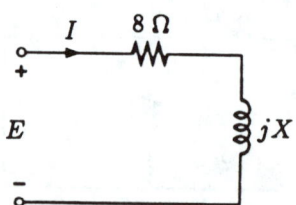

∡ 9.16 Box A consumes 3 kW at 0.6 PF lagging. Box B consumes 1 kVA at 0.5 PF leading. Box C consumes 1 kW at 0.2 PF lagging. Determine I.

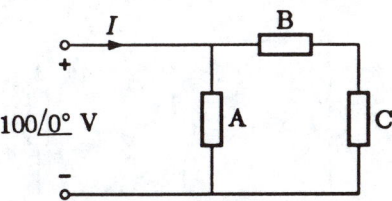

9.17 Box A consumes 500 W at 0.8 PF lagging. Box B consumes 200 VAR at 0.5 PF leading. Box C consumes 600 VA at 0.7 PF lagging. $E = 100\underline{/20°}$ V. Calculate I.

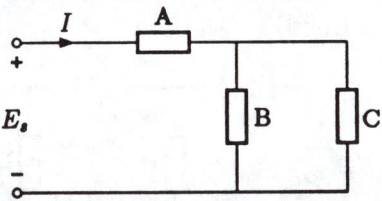

∡ 9.18 Box A consumes 1 kW at 0.8 PF lagging. Box B consumes 1kVAR at 0.4 PF leading. Box C is inductive, and it consumes 2 kW and 3kVA. Determine the power factor of the combination.

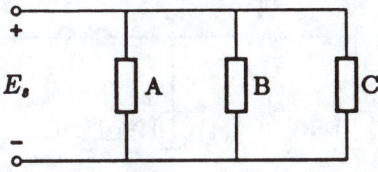

9.19 $E_s = 100\underline{/0°}$ V. Box A consumes 1.2 kVA at 0.7 PF leading. Box B consumes 300 W at 0.5 PF leading. $I = 10\underline{/30°}$ V. The impedance of Box D is $3 + j13$ Ω. Determine the real, reactive, and apparent powers supplied by the voltage source.

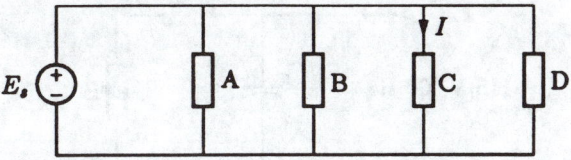

9.20 Box A consumes 2 kW at 0.7 PF leading. Box B consumes 3 kVA at 0.8 PF lagging. Box 3 consumes 1 kVA at unity PF. $E_s = 100\underline{/0°}$ V. Calculate I.

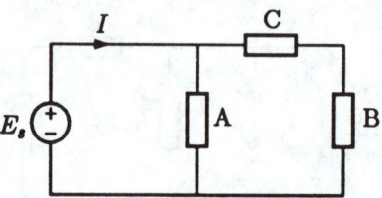

9.21 Box A consumes 6 kW at 0.5 PF lagging. Box B consumes 5 kVA at 0.4 PF leading. Box C consumes 3.5 kVA at 0.8 PF lagging. Box D has a PF of 0.5. The power factor seen by the voltage source is unity. What are the real and reactive power consumed by Box D?

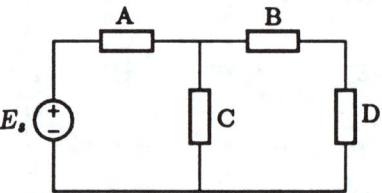

9.22 Use the definition $S = EI^*$ to calculate the complex power *delivered to* each of the five elements in the circuit. Verify that these five complex powers add up to zero.

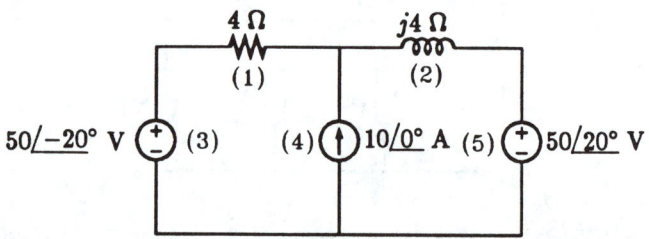

9.23 Box A consumes 1 kW at 0.8 PF lagging. Box B consumes 1 kVAR at 0.6 PF lagging. Calculate the value of C to give the combination a unity power factor.

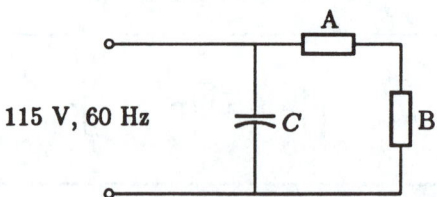

9.24 Box A consumes 400 W at 0.8 PF lagging. Box B consumes 1 kVAR at 0.6 PF leading. The real power consumed by Box C is 500 W. It is also known that $|E_s| = 100$ V and $|I_s| = 20$ A. Determine the impedance of Box C.

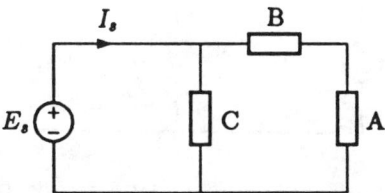

9.25 The ac powers delivered by the current source is 100 W and 100 VAR(L). $|I_s| = 2$ A and $|E_L| = 100$ V. Determine R, X_L, and X_C.

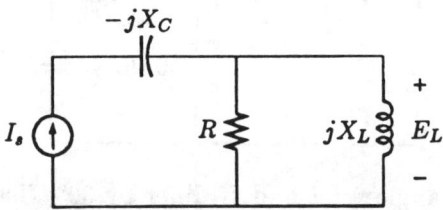

9.26 The apparent power delivered by the voltage source is 350 VA. The reactive power delivered to the capacitance is 100 VAR. Calculate the reactance of the inductance. $|E_s| = 100$ V.

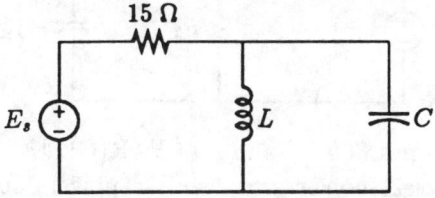

9.27 The power consumed by R is 500 W. Box A consumes 1 kVA at 0.8 PF leading. $|I| = 8$ A. Calculate $|E_s|$.

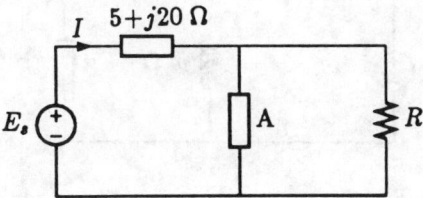

9.28 Box A consumes 300 W and 400 VAR(C). $R = X$. $|E_s| = 200$ V and $|E| = 100$ V. Determine R.

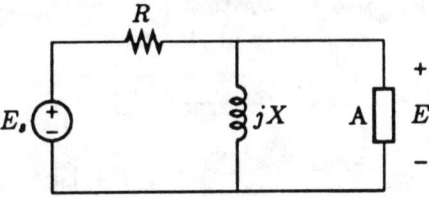

9.29 The voltage source supplies 1 kW at 0.707 PF lagging. Determine R and X_C.

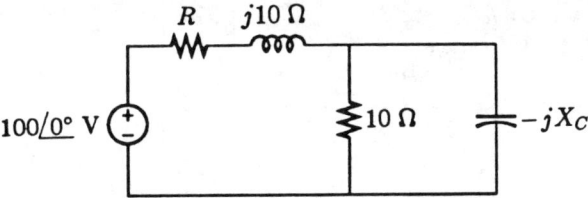

9.30 The phase angles of I and I_1 differ by 30°. The power factor of the combination is 0.95. Determine R_1 and R_2.

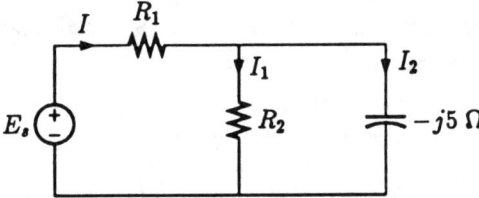

9.31 Box Z consumes 300 W and 400 VAR(C). $|E| = 100$ V. Determine the real power and reactive powers supplied by the source.

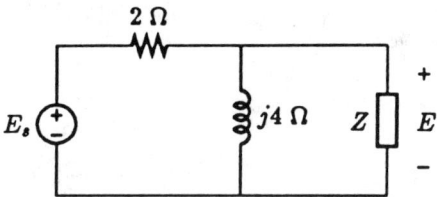

ALTERNATING-CURRENT POWERS

9.32 Based on the measurements given, determine the direction of power flow—whether it is from east to west or west to east. Also determine the amounts supplied, lost in the 1-Ω resistor, and delivered, at this point of the transmission line.

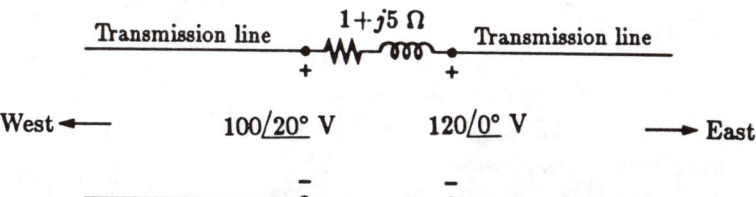

9.33 Box A consumes 2 kVA at a 0.6 PF leading. Box B consumes 400 W at 0.8 PF leading. Determine R, L_1, and L_2 so the current in the middle wire is zero and the power factor of the entire circuit is unity. The frequency is 60 Hz.

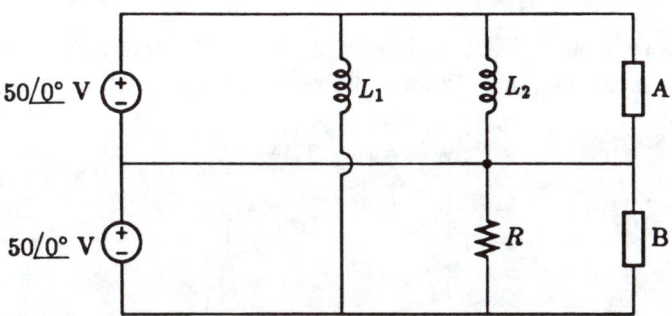

9.34 What is the power available to Z_L?

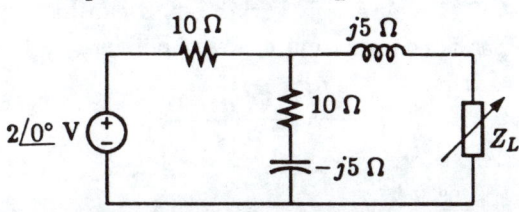

9.35 What is the power available to Z_L?

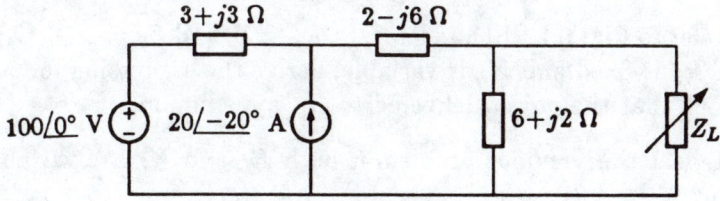

9.36 What is the power available to Z_L?

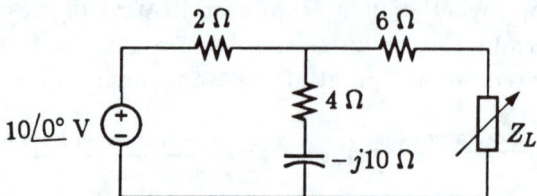

9.37 The value of R may be varied from 0 to ∞. What is the maximum power deliverable to R?

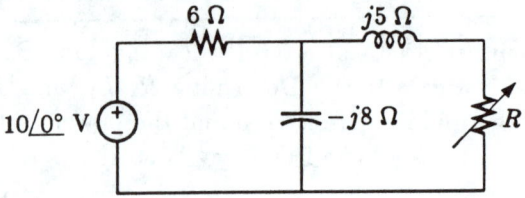

9.38 Both R and X are variable, but $X = 2R$. Determine the value of R such that the power delivered to it is maximum.

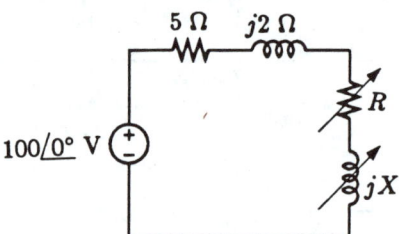

9.39 $Z_L = R + j8 \ \Omega$ where R is variable and can have any positive value. What is the maximum power deliverable to Z_L?

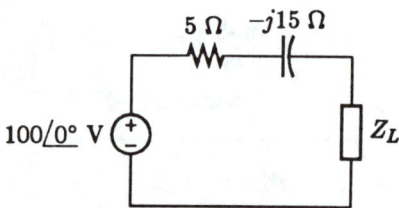

9.40 Refer to Fig. 9.22(b) and let $Z_s = R_s + jX_s$ and $Z_L = R_L + jX_L$. If R_L is fixed and X_L is variable, derive the expression for X_L such that the power delivered to Z_L is maximum.

9.41 Repeat the previous problem if both R_L and X_L are variable but $X_L = kR_L$.

Chapter 10

Two-Port Networks

So far, we have limited our consideration primarily to networks that involve interconnected two-terminal elements. Although controlled sources are four-terminal entities, they are actually special arrangements that include two-terminal sources whose strengths are proportional to some other electric quantities in some other parts of the network. They can, therefore, be treated as two-terminal elements.

Many commonly used electric devices are inherently multiterminal devices—transistors, op amps, transformers, amplifiers, etc. Whenever possible, we would prefer to model these devices using interconnected two-terminal elements and controlled sources. However, in some cases, this is not possible. Further, in some situations, these multiterminal devices are more logically and conveniently handled as multiterminal entities.

There are two types of multiterminal networks that require two very different ways of treatment. In one type, terminals are paired as *terminal pairs*. Fig. 10.1(a) shows a network that has three terminal pairs. Two terminals form a terminal pair only if the same current that flows into one of its terminals also flows out of the other. We shall refer to this condition as the *integrity* of the terminal pair. A terminal pair is also called a *port*.[1] A network formulated in the form of Fig. 10.1(a) is a three-port network and is usually referred to as a *three-port*.

The same network, under a different circumstance, could be

---

[1] This term was borrowed from microwave technology, wherein signals and power are literally delivered or received through a porthole. Circuit theory has adopted this term because it is much shorter then "terminal pair."

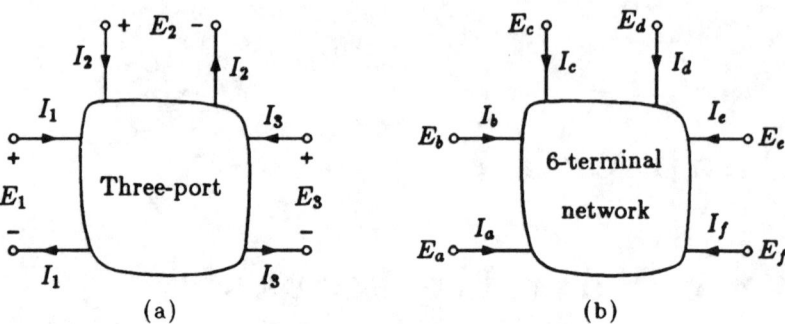

Figure 10.1: A three-port and a six-terminal network.

viewed as a six-terminal network. No current requirement is imposed on any of the terminals. In Fig. 10.1(b), we have six voltages (referred to ground) and six currents. Although KVL and KCL place some restrictions on these two groups of six quantities, they are essentially viewed as independent quantities. Although a three-port can be viewed as a special case of a six- terminal network, the ways these two types of networks should be treated are so different that they should be considered as two different classes of networks.

In this chapter, we shall deal with multiterminal networks on the port basis, especially two-ports. The two-port formalism is very useful in electronics, power transmission, and telecommunication. In Chapter 12, we shall deal with multiterminal networks on the terminal basis.

Matrix notations are used frequently in this and in the following chapters. It is assumed that the student is familiar with their notations and terminology. If this is not the case, the student will find a brief summary of matrix algebra in Appendix A.

10.1 One-port and two-port

A two-terminal network is also known as a one-port. We have dealt with such networks on numerous occasions, although we may not have recognized them as such. Fig. 10.2 represents a general one-port. If N contains no independent sources, then we have either

$$E = ZI \qquad \text{or} \qquad I = YE \tag{10.1}$$

TWO-PORT NETWORKS

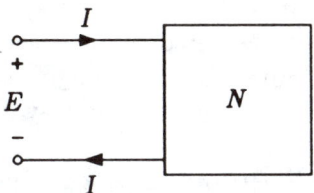

Figure 10.2: A one-port.

where Z and Y are the equivalent impedance or admittance of the one-port respectively. These impedances and admittances are known as the *driving-point* impedances and admittances because they represent the current-voltage relationship at the same terminal pair.

Fig. 10.3(a) represents a general two-port. It is assumed that there are no independent sources in network N. There are four port quantities—E_1, E_2, I_1, and I_2. Although here the notation strongly suggests that these quantities are either dc or phasors, they could also be time functions or functions of the frequency variable.

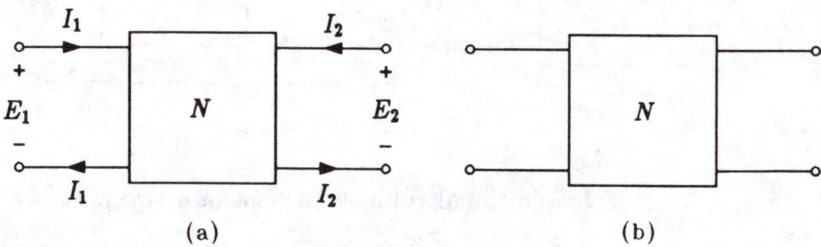

Figure 10.3: (a) Port quantities of a two-port. (b) Simplified representation of a two-port.

We will, however, limit our study to cases in which these port quantities are phasors. (Direct-current problems are actually included in this class of problems since dc quantities are simply phasors that are real and $s = 0$.)

When a network is designated as a two-port, its port quantities are sometimes not indicated on the diagram and they are automatically assumed to have the directions and polarities shown in Fig. 10.3(a). In other words, unless otherwise marked, the left port is port 1 and the right port is port 2; the port voltage rises are from the lower terminal to the upper terminal; and the positive port currents enter the upper terminals. Hence, the two-port in Fig. 10.3(b) implies the notation indicated in Fig. 10.3(a).

Port 1 is frequently considered as the *input* port, and port 2 the *output* port. This is quite natural as engineers are accustomed to working from the left to the right, the way the English language reads.

When we consider a network to be a two-port, we are only concerned about the four quantities shown in Fig. 10.3(a). There are two other voltages as indicated in Fig. 10.4(a)—E_3 and E_4. When N is treated as a two-port, E_3 and E_4 are either undeterminable or irrelevant. This doesn't mean that they don't exist. They are simply of no concern to us. Of course, $E_1 + E_3 - E_2 - E_4 = 0$ is required by KVL. However, the values of E_3 and E_4 are usually determined by factors other than the two-port itself.

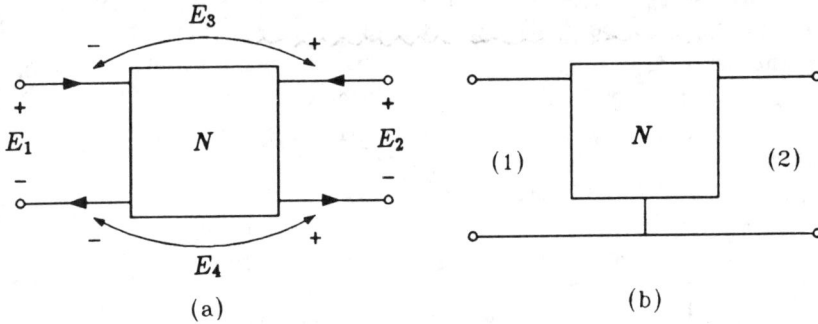

Figure 10.4: Two additional voltages present in a two-port.

There is one situation when a two-port does not have this extra degree of freedom. This occurs when the structure of the two-port itself dictates that $E_4 = 0$. For example, when the bottom terminals of the two-port are short-circuited as shown in Fig. 10.5. Such two-ports occur quite frequently in practice. For example, in an amplifier a large number of branches are all connected to ground. Such a two-port is known as a *grounded* or *three-terminal* two-port.

Figure 10.5: Grounded or three-terminal two-ports.

10.2 Two-port parameters and matrices

In order to formulate a general convention for describing the relationship among the four port quantities of a two-port, we may let any two of the four quantities be the independent variables, and the other two the dependent variables. For instance, if we let E_1 and I_2 be the independent variables, then we have the situation of Fig. 10.6, in which

$$I_{1a} \propto E_1 \quad I_{1b} \propto I_2 \quad E_{2a} \propto E_1 \quad E_{2b} \propto I2 \qquad (10.2)$$

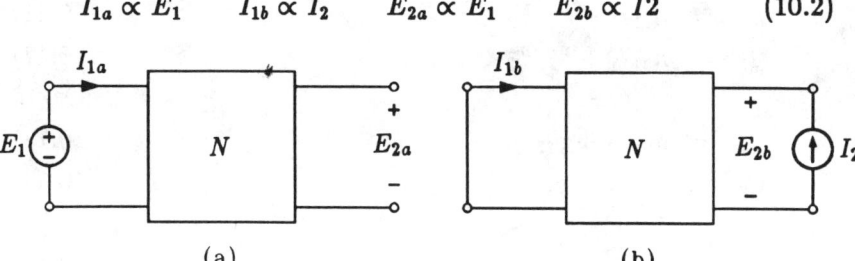

Figure 10.6: Arrangement used to determine the g parameters.

or

$$I_{1a} = g_{11}E_1 \quad I_{1b} = g_{12}I_2 \quad E_{2a} = g_{21}E_1 \quad E_{2b} = g_{22}I_2 \qquad (10.3)$$

where the g's are determined by the content of the network N. By superposition, we have the following relationship.

$$I_1 = g_{11}E_1 + g_{12}I_2 \qquad (10.4)$$

$$E_2 = g_{21}E_1 + g_{22}I_2 \qquad (10.5)$$

Eqs. (10.4) and (10.5) are frequently written in matrix form, that is,

$$\begin{bmatrix} I_1 \\ E_2 \end{bmatrix} = \begin{bmatrix} g_{11} & g_{12} \\ g_{21} & g_{22} \end{bmatrix} \begin{bmatrix} E_1 \\ I_2 \end{bmatrix} \qquad (10.6)$$

By choosing various pairs of port quantities as the independent variables, we can formulate five other pairs of equations similar to Eq. (10.4) and Eq. (10.5). These equations and their parameter notations are summarized in the following:

$$E_1 = z_{11}I_1 + z_{12}I_2 \tag{10.7}$$
$$E_2 = z_{21}I_1 + z_{22}I_2 \tag{10.8}$$

$$I_1 = y_{11}E_1 + y_{12}E_2 \tag{10.9}$$
$$I_2 = y_{21}E_1 + y_{22}E_2 \tag{10.10}$$

$$E_1 = h_{11}I_1 + h_{12}E_2 \tag{10.11}$$
$$I_2 = h_{21}I_1 + h_{22}E_2 \tag{10.12}$$

$$I_1 = g_{11}E_1 + g_{12}I_2 \tag{10.13}$$
$$E_2 = g_{21}E_1 + g_{22}I_2 \tag{10.14}$$

$$E_1 = AE_2 - BI_2 \tag{10.15}$$
$$I_1 = CE_2 - DI_2 \tag{10.16}$$

$$E_2 = \mathcal{A}E_1 - \mathcal{B}I_1 \tag{10.17}$$
$$I_2 = \mathcal{C}E_1 - \mathcal{D}I_1 \tag{10.18}$$

When these relationships are written in matrix form, we have the following:

$$\begin{bmatrix} E_1 \\ E_2 \end{bmatrix} = \begin{bmatrix} z_{11} & z_{12} \\ z_{21} & z_{22} \end{bmatrix} \begin{bmatrix} I_1 \\ I_2 \end{bmatrix} = [z] \begin{bmatrix} I_1 \\ I_2 \end{bmatrix} \tag{10.19}$$

$$\begin{bmatrix} I_1 \\ I_2 \end{bmatrix} = \begin{bmatrix} y_{11} & y_{12} \\ y_{21} & y_{22} \end{bmatrix} \begin{bmatrix} E_1 \\ E_2 \end{bmatrix} = [y] \begin{bmatrix} E_1 \\ E_2 \end{bmatrix} \tag{10.20}$$

$$\begin{bmatrix} E_1 \\ I_2 \end{bmatrix} = \begin{bmatrix} h_{11} & h_{12} \\ h_{21} & h_{22} \end{bmatrix} \begin{bmatrix} I_1 \\ E_2 \end{bmatrix} = [h] \begin{bmatrix} I_1 \\ E_2 \end{bmatrix} \tag{10.21}$$

$$\begin{bmatrix} I_1 \\ E_2 \end{bmatrix} = \begin{bmatrix} g_{11} & g_{12} \\ g_{21} & g_{22} \end{bmatrix} \begin{bmatrix} E_1 \\ I_2 \end{bmatrix} = [g] \begin{bmatrix} E_1 \\ I_2 \end{bmatrix} \tag{10.22}$$

TWO-PORT NETWORKS

$$\begin{bmatrix} E_1 \\ I_1 \end{bmatrix} = \begin{bmatrix} A & B \\ C & D \end{bmatrix} \begin{bmatrix} E_2 \\ -I_2 \end{bmatrix} = [F] \begin{bmatrix} E_2 \\ -I_2 \end{bmatrix} \qquad (10.23)$$

$$\begin{bmatrix} E_2 \\ I_2 \end{bmatrix} = \begin{bmatrix} \mathcal{A} & \mathcal{B} \\ \mathcal{C} & \mathcal{D} \end{bmatrix} \begin{bmatrix} E_1 \\ -I_1 \end{bmatrix} = [\mathcal{F}] \begin{bmatrix} E_1 \\ -I_1 \end{bmatrix} \qquad (10.24)$$

Given a two-port, the meaning of each of the 24 parameters in Eqs. (10.7) through (10.18) can be understood by investigating the equation in which it is defined. For example, parameter z_{11} appears in Eq. (10.7), which reads

$$E_1 = z_{11}I_1 + z_{12}I_2$$

If we set $I_2 = 0$, then z_{11} is the ratio of E_1 to I_1; or

$$z_{11} = \left. \frac{E_1}{I_1} \right|_{I_2=0} \qquad (10.25)$$

Hence, to obtain the z_{11} of a two-port, we open-circuit port 2 and excite port 1 with a source (voltage or current) and compute the ratio of E_1 to I_1. One such arrangement is shown in Fig. 10.7, in which we use a voltage source E_1, and we need to compute I_1. Alternatively, we could excite port 1 with a current source I_1, and then compute E_1. On the other hand, we cannot excite port 2 with either a voltage source or a current source because we would not have the assurance that $I_2 = 0$.

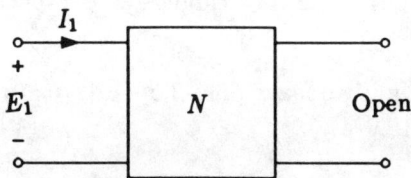

Figure 10.7: An arrangement to find the z_{11} parameter of N.

Similarly we have

$$z_{21} = \left. \frac{E_2}{I_1} \right|_{I_2=0} \qquad z_{12} = \left. \frac{E_1}{I_2} \right|_{I_1=0} \qquad z_{22} = \left. \frac{E_2}{I_2} \right|_{I_1=0} \qquad (10.26)$$

In applying these formulas, we can further simplify the procedure if we choose the excitation to be 1 unit (in this case 1 A). Then we can say that

$$z_{11} = E_1\big|_{I_1=1,\, I_2=0} \qquad z_{21} = E_2\big|_{I_1=1,\, I_2=0} \qquad (10.27)$$

$$z_{12} = E_1\big|_{I_1=0,\, I_2=1} \qquad z_{22} = E_2\big|_{I_1=0,\, I_2=1} \qquad (10.28)$$

Thus z_{11} is the *driving-point impedance* at port 1 with port 2 open circuited. It is also known as the *open-circuit driving-point impedance* at port 1. Parameter z_{22} is known as the *open-circuit driving-point impedance* at port 2. Parameters z_{12} and z_{21} are known as the *open-circuit transfer impedances*. The *z* matrix of Eq. (10.19) is known as the *open-circuit impedance matrix* of the two-port, or simply as the *impedance matrix* of the two-port.

EXAMPLE 1. Obtain the z matrix of the two-port in Fig. 10.8.

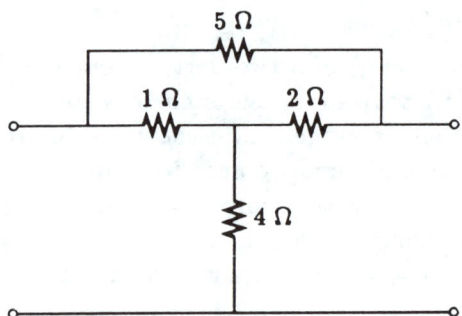

Figure 10.8: A bridged tee two-port.

SOLUTION First we leave port 2 open ($I_2 = 0$) and let $I_1 = 1$ A as shown in Fig. 10.9(a).

$$I_3 = \frac{1}{1+7} \times 1 = \frac{1}{8}$$

$$z_{11} = E_{1a} = 1 \times 4 + \frac{1}{8} \times 7 = \frac{39}{8}$$

$$z_{21} = E_{2a} = 1 \times 4 + \frac{1}{8} \times 2 = \frac{17}{4}$$

TWO-PORT NETWORKS

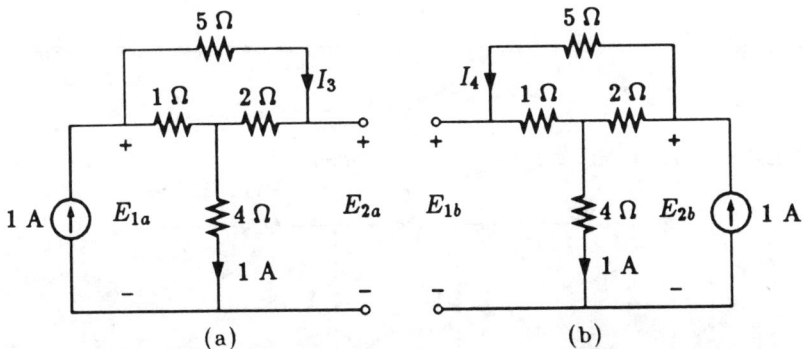

Figure 10.9: Analysis of a two-port with one port open.

Then we leave port 1 open ($I_1 = 0$) and let $I_2 = 1$ A as shown in Fig. 10.9(b).

$$I_4 = \frac{2}{1+7} \times 1 = \frac{1}{4}$$

$$z_{12} = E_{1b} = 1 \times 4 + \frac{1}{4} \times 1 = \frac{17}{4}$$

$$z_{22} = E_{2b} = 1 \times 4 + \frac{1}{4} \times 6 = \frac{11}{2}$$

In this simple circuit, the student should be able to observe that

$$z_{11} = 1 \| (2+5) + 4 = \frac{39}{8}$$

$$z_{22} = 2 \| (1+5) + 4 = \frac{11}{2}$$

Hence,

$$[z] = \begin{bmatrix} \dfrac{39}{8} & \dfrac{17}{4} \\ \dfrac{17}{4} & \dfrac{11}{2} \end{bmatrix} \Omega$$

EXERCISE

10.2.1 Obtain the z matrix of the two-port.

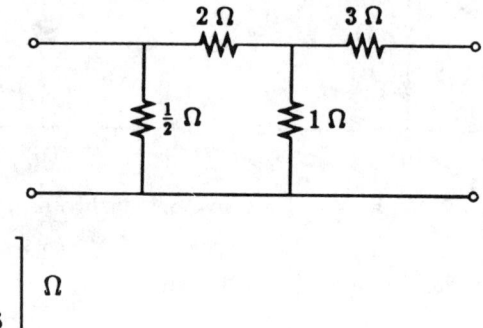

Ans.

$$\frac{1}{7}\begin{bmatrix} 3 & 1 \\ 1 & 26 \end{bmatrix} \Omega$$

The y parameters of Eqs. (10.9) and (10.10) are perhaps the most frequently used two-port parameters. We can easily interpret each of its parameters as

$$y_{11} = \left.\frac{I_1}{E_1}\right|_{E_2=0} \qquad y_{21} = \left.\frac{I_2}{E_1}\right|_{E_2=0} \qquad (10.29)$$

$$y_{12} = \left.\frac{I_1}{E_2}\right|_{E_1=0} \qquad y_{22} = \left.\frac{I_2}{E_2}\right|_{E_1=0} \qquad (10.30)$$

Again, in applying these formulas, we can further simplify the procedure if we choose the excitation to be 1 unit (in this case 1 V). Then we can say that

$$y_{11} = \left.I_1\right|_{E_1=1,\, E_2=0} \qquad y_{21} = \left.I_2\right|_{E_1=1,\, E_2=0} \qquad (10.31)$$

$$y_{12} = \left.I_1\right|_{E_1=0,\, E_2=1} \qquad y_{22} = \left.I_2\right|_{E_1=0,\, E_2=1} \qquad (10.32)$$

Thus, y_{11} is the *driving-point admittance* at port 1 with port 2 short-circuited. It is also known as the *short-circuit driving-point admittance* at port 1. Parameter y_{22} is known as the *short-circuit driving-point admittance* at port 2. Parameters y_{12} and y_{21}

TWO-PORT NETWORKS

are known as the *short-circuit transfer admittances*. The y matrix of Eq. (10.20) is known as the *short-circuit admittance matrix* of the two-port, or simply as the *admittance matrix* of the two-port.

EXAMPLE 2. Determine the y matrix of the two-port in Fig. 10.10.

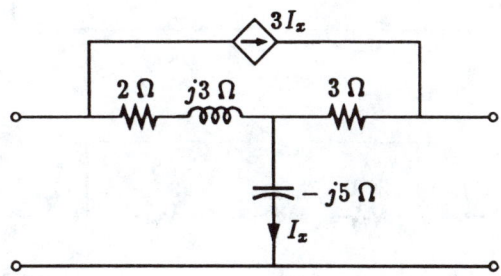

Figure 10.10: The two-port for Example 2.

SOLUTION First, we shall analyze the two-port with $E_2 = 0$ and $E_1 = 1$. The arrangement is shown in Fig. 10.11(a). We shall use node analysis.

$$\frac{E_a - 1}{2 + j3} + \frac{E_a}{-j5} + \frac{E_a}{3} = 0$$

Solving gives

$$E_a = 0.3443 - j0.4519 \quad \Longrightarrow \quad I_x = 0.0904 + j0.0689$$

Thus

$$y_{11} = I_{1a} = \frac{1 - E_a}{2 + j3} + 3I_x = 0.4763 + j0.1248$$

$$y_{21} = I_{2a} = -\frac{E_a}{3} - 3I_x = -0.3859 - j0.0056$$

Second, we analyze the two-port with $E_1 = 0$ and $E_2 = 1$. The arrangement is shown in Fig. 10.11(b). Again, using node analysis, we have

$$\frac{E_b}{2 + j3} + \frac{E_b}{-j5} + \frac{E_b - 1}{3} = 0$$

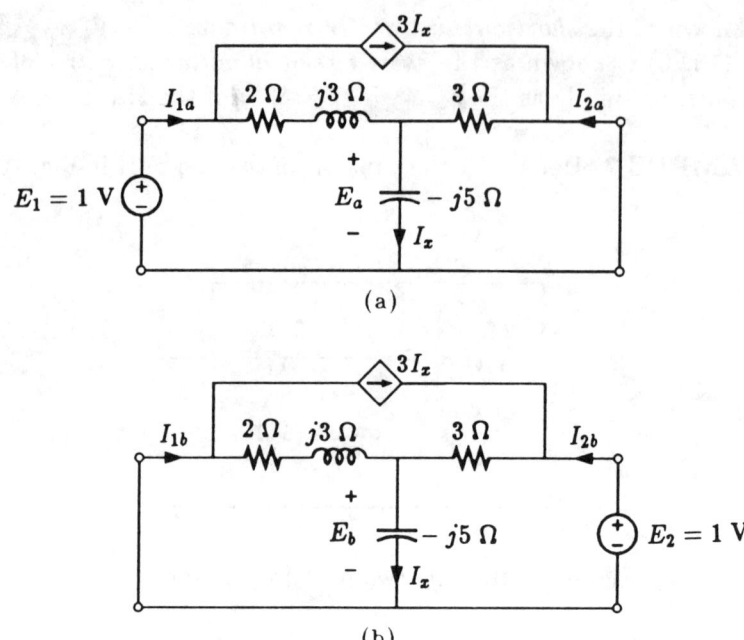

Figure 10.11: Arrangements to determine the y parameters of a two-port.

Solving gives

$$E_b = 0.6815 + j0.0430 \quad \Longrightarrow \quad I_x = -0.0086 + 0.1363$$

Thus,

$$y_{12} = I_{1b} = -\frac{E_b}{2 + j3} + 3I_x = -0.1406 + j0.5595$$

$$y_{22} = I_{2b} = \frac{1 - E_b}{3} - 3I_x = 0.1320 - j0.4232$$

Summarizing,

$$[y] = \begin{bmatrix} 0.476 + j0.125 & -0.141 + j0.560 \\ -0.386 - j0.0056 & 0.132 - j0.423 \end{bmatrix} \mho$$

TWO-PORT NETWORKS

EXERCISES

10.2.2 Obtain the y matrix of the two-port.

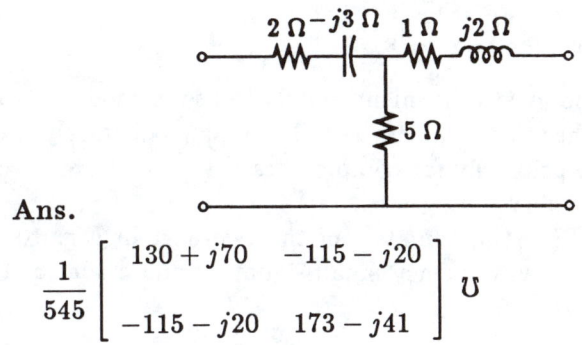

Ans.

$$\frac{1}{545} \begin{bmatrix} 130 + j70 & -115 - j20 \\ -115 - j20 & 173 - j41 \end{bmatrix} \mho$$

10.2.3 Obtain the y matrix of the two-port.

Ans.

$$\begin{bmatrix} 0 & 0 \\ \dfrac{\mu}{r_d} & \dfrac{1}{r_d} \end{bmatrix}$$

The matrices $[h]$ and $[g]$ of Eqs. (10.21) and (10.22) are known as *hybrid matrices* because independent variables are mixtures of a voltage and a current. Their elements may be interpreted in the same manner as the z and y parameters were interpreted. For the h matrix, we may write

$$h_{11} = \left.\frac{E_1}{I_1}\right|_{E_2=0} \qquad h_{21} = \left.\frac{I_2}{I_1}\right|_{E_2=0} \qquad (10.33)$$

$$h_{12} = \left.\frac{E_1}{E_2}\right|_{I_1=0} \qquad h_{22} = \left.\frac{I_2}{E_2}\right|_{I_1=0} \qquad (10.34)$$

and for the g matrix

$$g_{11} = \left.\frac{I_1}{E_1}\right|_{I_2=0} \qquad g_{21} = \left.\frac{E_2}{E_1}\right|_{I_2=0} \qquad (10.35)$$

$$g_{12} = \left.\frac{I_1}{I_2}\right|_{E_1=0} \qquad g_{22} = \left.\frac{E_2}{I_2}\right|_{E_1=0} \qquad (10.36)$$

The h matrix is the most convenient matrix to use to model a transistor and some other electronic devices. The g matrix is rarely used. We include it here primarily for completeness.

EXAMPLE 3. Find the h matrix of the two-port in Fig. 10.12. This two-port is a low-frequency small-signal circuit model of the bipolar junction transistor.

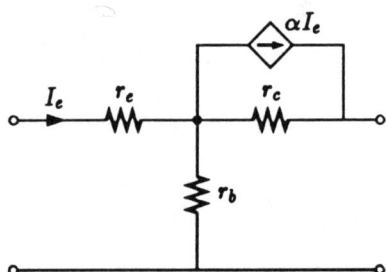

Figure 10.12: A two-port circuit model of a transistor.

SOLUTION First, we let $E_2 = 0$ and $I_1 = 1$ as shown in Fig. 10.13(a).

$$E_x = (1-\alpha)\frac{r_b r_c}{r_b + r_c}$$

$$h_{11} = E_{1a} = r_e + (1-\alpha)\frac{r_b r_c}{r_b + r_c}$$

$$h_{21} = I_{2a} = -(1-\alpha)\frac{r_b}{r_b + r_c} - \alpha = -\frac{\alpha r_c + r_b}{r + b + r_c}$$

Then, we let $I_1 = 0$ and $E_2 = 1$ as shown in Fig. 10.13(a). We have

TWO-PORT NETWORKS

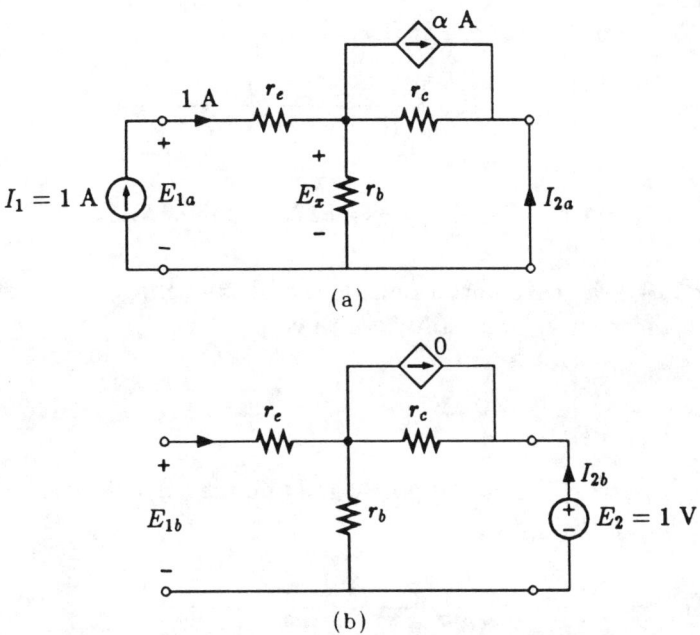

Figure 10.13: Circuits for evaluating the h parameters of a transistor equivalent circuit.

$$h_{12} = E_{1b} = \frac{r_b}{r_b + r_c}$$

$$h_{22} = I_{2b} = \frac{1}{r_b + r_c}$$

EXAMPLE 4. Find the h parameters of the two-port in Fig. 10.14 at 1 MHz.

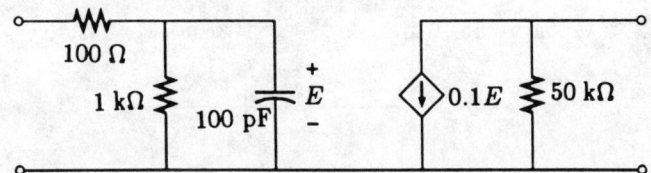

Figure 10.14: The equivalent circuit of an amplifier.

FUNDAMENTALS OF CIRCUIT ANALYSIS

SOLUTION To find h_{11}, we leave port 2 short-circuited and evaluate the input impedance at port 1.

$$h_{11} = \frac{E_1}{I_1}\bigg|_{E_2=0} = 100 + \frac{1}{10^{-3} + j2\pi \times 10^6 \times 10^{-10}}$$

$$= 100 + \frac{10^3}{1 + j0.6283} = 817.0 - j450.5 \; \Omega$$

To find h_{12}, we leave port 1 open and find the voltage ratio between port 1 and port 2. We readily see that

$$h_{12} = \frac{E_1}{E_2}\bigg|_{I_1=0} = 0$$

To find h_{21}, we short-circuit port 2 and find the current ratio between port 2 and port 1, or

$$h_{21} = \frac{I_2}{I_1}\bigg|_{E_2=0} = \frac{0.1}{10^{-3} + j2\pi \times 10^6 \times 10^{-10}}$$

$$= \frac{100}{1 + j0.6283} = 71.7 - j45.05$$

To find h_{22}, we short-circuit port 1 and evaluate the admittance seen at port 2. Since $E_1 = 0$, $I_1 = 0$, and $E = 0$, we have

$$h_{22} = \frac{I_2}{E_2}\bigg|_{I_1=0} = \frac{1}{50 \times 10^3} = 2 \times 10^{-5} \; \mho$$

EXERCISES

10.2.4 Obtain the h matrix of the two-port.

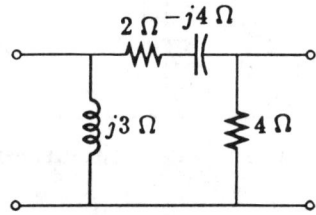

TWO-PORT NETWORKS

Ans.

$$\begin{bmatrix} 3.6 + j4.8 & -0.6 + j1.2 \\ 0.6 - j1.2 & 0.65 + j0.2 \end{bmatrix}$$

10.2.5 Obtain the g matrix of the two-port.

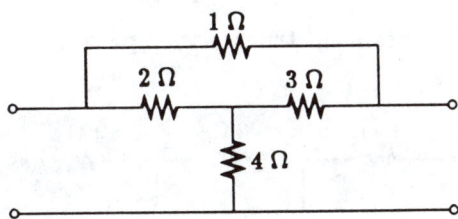

Ans.

$$\frac{1}{16} \begin{bmatrix} 3 & -15 \\ 15 & 13 \end{bmatrix}$$

The $[F]$ matrix of Eq. (10.23) is known as the *transmission matrix* or the *chain matrix*. It is most suitable to be used when two-ports are connected in cascade—the output of one two-port feeds directly into the input of the subsequent two-port. This application also accounts for the minus sign we assign to I_2 in Eq. (10.23), as $-I_2$ would be equal to the next I_1 of the next two-port. A typical example of this arrangement is cable TV. The signal is fed from the transmitter, through sections of cable, with repeaters between adjacent sections, and finally to the consumer. The $[\mathcal{F}]$ matrix has the same significance when the roles of port 1 and port 2 are interchanged. Each of these eight parameters can be individually interpreted as follows:

$$A = \frac{E_1}{E_2}\bigg|_{I_2=0} \qquad C = \frac{I_1}{E_2}\bigg|_{I_2=0} \qquad (10.37)$$

$$B = \frac{E_1}{-I_2}\bigg|_{E_2=0} \qquad D = \frac{I_1}{-I_2}\bigg|_{E_2=0} \qquad (10.38)$$

and

$$A = \left.\frac{E_2}{E_1}\right|_{I_1=0} \qquad C = \left.\frac{I_2}{E_1}\right|_{I_1=0} \qquad (10.39)$$

$$B = \left.\frac{E_2}{-I_1}\right|_{E_1=0} \qquad D = \left.\frac{I_2}{-I_1}\right|_{E_1=0} \qquad (10.40)$$

EXAMPLE 5. Obtain the transmission parameters of the two-port in Fig. 10.15.

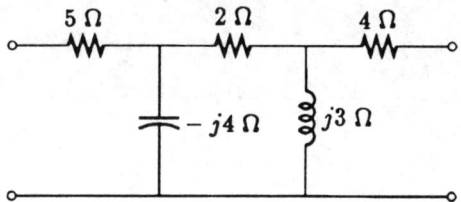

Figure 10.15: Two-port used in Example 5.

SOLUTION To obtain the A and C parameters, we leave port 2 open and use mesh analysis. The arrangement is shown in Fig. 10.16(a). The mesh equations are

$$5I_1 - j4(I_1 - I_2) = 1$$

$$-j4(I_2 - I_1) + (2 + j3)I_2 = 0$$

Solving, we obtain

$$I_1 = \frac{57}{653} + j\frac{4}{653} \qquad \text{and} \qquad I_2 = \frac{52}{653} - j\frac{88}{653}$$

From which

$$E_2 = j3I_2 = \frac{264}{653} + j\frac{156}{653}$$

Thus,

$$A = \frac{E_1}{E_2} = \frac{1}{E_2} = \frac{11}{6} - j\frac{13}{12}$$

TWO-PORT NETWORKS

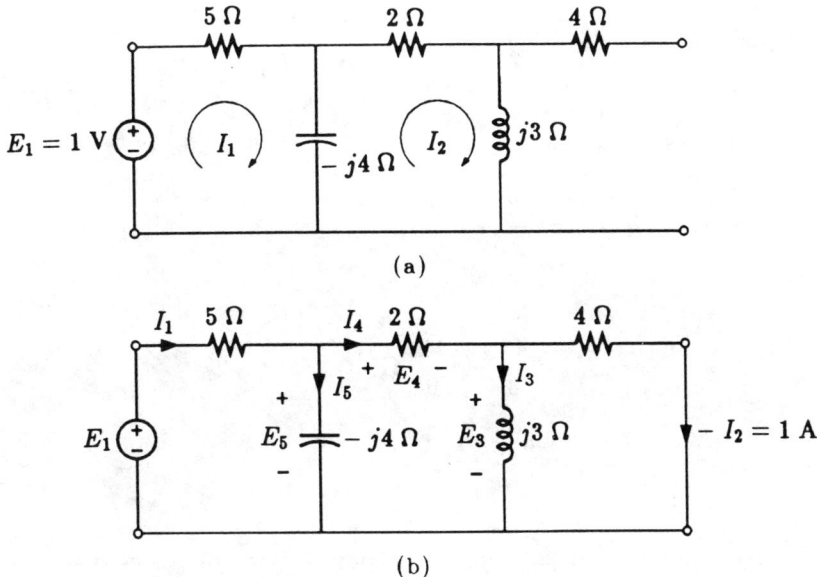

(a)

(b)

Figure 10.16: Arrangements used to obtain the transmission parameters of the two-port in Fig. 11.15.

$$C = \frac{I_1}{E_2} = \frac{1}{6} - j\frac{1}{12} \; \mho$$

To obtain the B and D parameters, we short-circuit port 2 and use the ladder analysis. The arrangement is shown in Fig. 10.16(b). We have successively

$$-I_2 = 1$$

$$E_3 = 4$$

$$I_3 = \frac{E_3}{j3} = -j\frac{4}{3}$$

$$I_4 = -I_2 + I_3 = 1 - j\frac{4}{3}$$

$$E_4 = 2I_4 = 2 - j\frac{8}{3}$$

$$E_5 = E_3 + E_4 = 6 - j\frac{8}{3}$$

$$I_5 = \frac{E_5}{-j4} = \frac{2}{3} + j\frac{3}{2}$$

$$I_1 = I_4 + I_5 = \frac{5}{3} + j\frac{1}{6}$$

$$E_1 = E_5 + 5I_1 = \frac{43}{3} - j\frac{11}{6}$$

Thus,

$$B = \frac{E_1}{-I_2} = E_1 = \frac{43}{3} - j\frac{11}{6}\ \Omega$$

$$D = \frac{I_1}{-I_2} = I_1 = \frac{5}{3} + j\frac{1}{6}$$

EXERCISE

10.2.6 Obtain the F matrix of the following two-port.

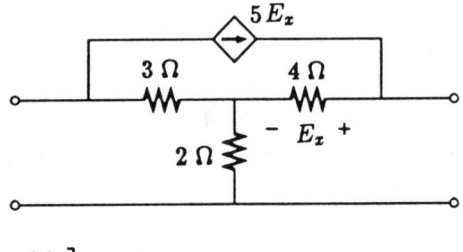

Ans.

$$\frac{1}{38}\begin{bmatrix} 95 & -26 \\ 19 & 34 \end{bmatrix}$$

10.3 Relationships among two-port parameters

When a network has only one port, its electrical property is relatively easy to characterize. There are only two electric quantities associated with the network—the voltage and the current. If we know either the admittance or the impedance of a one-port, then its external property is completely defined. Furthermore, once either of these two quantities is known, the other is readily obtained. Whether we choose to use the admittance or the impedance of a one-port will depend on the particular problem. For example, if we use node analysis, then we would prefer to use the admittances.

The situation with two-ports is analogous. However, since there are four port quantities associated with a two-port, it takes four parameters, or a 2 × 2 matrix, to characterize it. And there are six possible descriptions for every two-port. The question naturally arises as to which of the six matrices should we use for a given two-port.

In theory, it does not matter which of the six matrices is used to describe a two-port, as long as the matrix exists. As we shall soon see, a one-to-one correspondence exists between any two matrices. For practical as well as historical reasons, certain matrices are more commonly used in certain areas.

For example, the transmission matrix $[F]$ is a natural for transmission lines studies. The $[y]$ and the $[z]$ matrices are the most basic two-port matrices. They are the direct extensions of the driving-point admittances and impedances of one-ports and are used in most derivations of basic theorems and theoretical developments, particularly for networks containing resistances, capacitances, and inductances only. The $[h]$ matrix is used widely in electronic circuits primarily because these parameters are, for the most part, easily measured for many electronic devices.

In a broader sense, the six two-port matrices are simply sets of condensed information regarding how the four port quantities of a two-port are related. In the previous section, we showed how each of the 24 two-port parameters can be obtained individually. In practice, we need not always obtain these parameters one at a time. In fact, for various reasons, it may not be feasible to directly obtain some of these parameters.

The relationships of the four port quantities may come in any form, as long as the information enables us to arrive at two independent equations involving all four port quantities. Then we algebraically manipulate these two equations into one of the six standard forms given in Eqs. (10.7) through (10.24). Once that is done, the quantitative information regarding a two-port is presented as one of the six matrices.

EXAMPLE 6. The port quantities of a two-port must satisfy the following equations. Obtain its z matrix.

$$3E_1 + 5E_2 + 2I_1 - 5I_2 = 0$$

$$E_1 + 2E_2 + 4I_1 + 6I_2 = 0$$

SOLUTION Solving for E_2 from both equations and equating, we get

$$E_2 = -\frac{1}{2}(E_1 + 4I_1 + 6I_2) = -\frac{1}{5}(3E_1 + 2I_1 - 5I_2)$$

Solve for E_1 to get

$$E_1 = 16I_1 + 40I_2$$

Substituting gives

$$E_2 = -10I_1 - 23I_2$$

Hence, for the two-port,

$$[z] = \begin{bmatrix} 16 & 40 \\ -10 & -23 \end{bmatrix}$$

EXERCISE

10.3.1 The port quantities of a two-port must satisfy the following equations.

$$5E_1 + 4E_2 = 5I_2$$

$$2E_1 + 4E_2 = 3I_1 - 4I_2$$

Find its z and F matrices.

Ans.

$$[z] = \begin{bmatrix} -1 & 3 \\ 1.25 & -2.5 \end{bmatrix} \Omega \quad [F] = \begin{bmatrix} -0.8 & -1 \\ 0.8 & -2 \end{bmatrix}$$

TWO-PORT NETWORKS

EXAMPLE 7. Obtain the y matrix of the lattice two-port in Fig. 10.17.

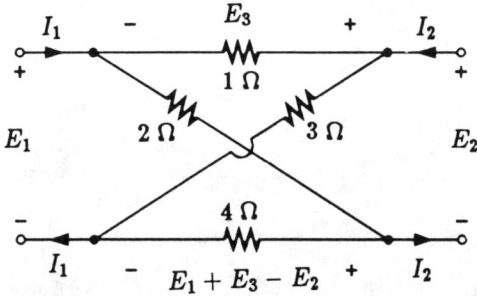

Figure 10.17: A lattice two-port.

SOLUTION Write the KCL equations for three of the four nodes to get

$$I_1 = \frac{E_2 - E_3}{2} - \frac{E_3}{1} \tag{10.41}$$

$$I_2 = \frac{E_3}{1} + \frac{E_1 + E_3}{3} \tag{10.42}$$

$$I_1 = \frac{E_1 + E_3}{3} + \frac{E_1 - E_2 + E_3}{4} \tag{10.43}$$

From Eq. (10.41), we get

$$E_3 = \frac{E_2 - 2I_1}{3} \tag{10.44}$$

Substitute Eq. (10.44) into Eq. (10.42) to get

$$8I_1 + 9I_2 = 3E_1 + 4E_2 \tag{10.45}$$

and into Eq. (10.43) to get

$$50I_1 = 21E_1 - 2E_2 \tag{10.46}$$

Substitute Eq. (10.46) into Eq. (10.45) to get

$$25I_2 = -E_1 - 12E_2 \tag{10.47}$$

From Eqs. (10.46) and (10.47), we get

$$[y] = \begin{bmatrix} \dfrac{21}{50} & -\dfrac{1}{25} \\ -\dfrac{1}{25} & \dfrac{12}{25} \end{bmatrix} \Omega$$

EXERCISE

10.3.2 For the following two-port, write the three node equations, leaving I_1 and I_2 in literal form. Then eliminate E_3 and obtain the y matrix of the two-port.

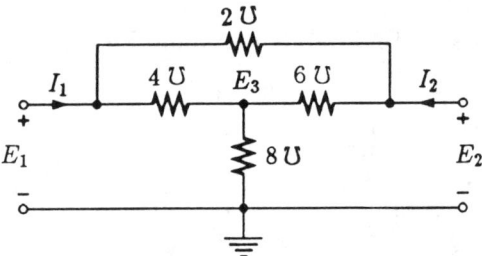

Ans.

$$\begin{bmatrix} \dfrac{46}{9} & -\dfrac{10}{3} \\ -\dfrac{10}{3} & 6 \end{bmatrix} \mho$$

If any one of the six two-port matrices is known, we can simply regard it as containing sufficient information to characterize the two-port. We have two equations implied by the matrix and from them any other matrix (if it exists) can be found simply by manipulating the variables until they appear in the form corresponding to the new matrix.

In doing this, it is helpful to observe that among the 12 equations [Eqs. (10.7) through (10.18)] that define the various matrices, each equation has one of the four port quantities missing. Therefore, we can simply take the known equations and obtain from them two newly arranged equations with the appropriate variables missing.

TWO-PORT NETWORKS

Once we have the equations, each of which contains the right three variables, it's a simple matter to write them in the desired form.

For example, if we have

$$E_1 = z_{11}I_1 + z_{12}I_2 \qquad (10.48)$$

$$E_2 = z_{21}I_1 + z_{22}I_2 \qquad (10.49)$$

and we wish to obtain the h matrix. We see that in the equations using h parameters, the first one contains no I_2 term; the second one contains no E_1 term. To obtain the first of the two h equations, we should eliminate I_2 between Eqs. (10.48) and (10.49). Solving for I_2 from Eqs. (10.48) and (10.49) and equating, we get

$$I_2 = -\frac{z_{11}}{z_{12}}I_1 + \frac{1}{z_{12}}E_1 = -\frac{z_{21}}{z_{22}}I_1 + \frac{1}{z_{22}}E_2 \qquad (10.50)$$

Rearranging Eq. (10.50) so that E_1 alone appears on the left-hand side, we have

$$E_1 = \left[z_{11} - \frac{z_{12}z_{21}}{z_{22}}\right]I_1 + \frac{z_{12}}{z_{22}}E_2 = \frac{|z|}{z_{22}}I_1 + \frac{z_{12}}{z_{22}}E_2 \qquad (10.51)$$

where $|z| = z_{11}z_{22} - z_{12}z_{21}$ is the determinant of the z matrix.

To obtain the second of the two h equations, we seek an equation that contains no E_1 term. Since Eq. (10.49) contains no E_1 term, we merely have to rearrange it so I_2 alone appears on the left-hand side. We get

$$I_2 = -\frac{z_{21}}{z_{22}}I_1 + \frac{1}{z_{22}}E_2 \qquad (10.52)$$

A comparison of Eqs. (10.51) and (10.52) with Eqs. (10.11) and (10.12) enables us to write

$$\begin{bmatrix} h_{11} & h_{12} \\ h_{21} & h_{22} \end{bmatrix} = \begin{bmatrix} \dfrac{|z|}{z_{22}} & \dfrac{z_{12}}{z_{22}} \\ -\dfrac{z_{21}}{z_{22}} & \dfrac{1}{z_{22}} \end{bmatrix} \qquad (10.53)$$

FUNDAMENTALS OF CIRCUIT ANALYSIS

From Eq. (10.53) we see that, if $z_{22} = 0$ for a two-port, then the two-port does not have an h matrix.

By similar manipulations, we can express each of the six two-port matrices in terms of parameters of the other five. These relationships are summarized in Eqs. (10.54) through (10.59) below:

$$[z] = \begin{bmatrix} \dfrac{y_{22}}{|y|} & -\dfrac{y_{12}}{|y|} \\ -\dfrac{y_{21}}{|y|} & \dfrac{y_{11}}{|y|} \end{bmatrix} = \begin{bmatrix} \dfrac{1}{g_{11}} & -\dfrac{g_{12}}{g_{11}} \\ \dfrac{g_{21}}{g_{11}} & \dfrac{|g|}{g_{11}} \end{bmatrix} = \begin{bmatrix} \dfrac{|h|}{h_{22}} & \dfrac{h_{12}}{h_{22}} \\ -\dfrac{h_{21}}{h_{22}} & \dfrac{1}{h_{22}} \end{bmatrix}$$

$$= \begin{bmatrix} \dfrac{A}{C} & \dfrac{|F|}{C} \\ \dfrac{1}{C} & \dfrac{D}{C} \end{bmatrix} = \begin{bmatrix} \dfrac{D}{C} & \dfrac{1}{C} \\ \dfrac{|\mathcal{F}|}{C} & \dfrac{\mathcal{A}}{C} \end{bmatrix} \quad (10.54)$$

$$[y] = \begin{bmatrix} \dfrac{z_{22}}{|z|} & -\dfrac{z_{12}}{|z|} \\ -\dfrac{z_{21}}{|z|} & \dfrac{z_{11}}{|z|} \end{bmatrix} = \begin{bmatrix} \dfrac{|g|}{g_{22}} & \dfrac{g_{12}}{g_{22}} \\ -\dfrac{g_{21}}{g_{22}} & \dfrac{1}{g_{22}} \end{bmatrix} = \begin{bmatrix} \dfrac{1}{h_{11}} & -\dfrac{h_{12}}{h_{11}} \\ \dfrac{h_{21}}{h_{11}} & \dfrac{|h|}{h_{11}} \end{bmatrix}$$

$$= \begin{bmatrix} \dfrac{D}{B} & -\dfrac{|F|}{B} \\ -\dfrac{1}{B} & \dfrac{A}{B} \end{bmatrix} = \begin{bmatrix} \dfrac{\mathcal{A}}{B} & -\dfrac{1}{B} \\ -\dfrac{|\mathcal{F}|}{B} & \dfrac{D}{B} \end{bmatrix} \quad (10.55)$$

$$[h] = \begin{bmatrix} \dfrac{|z|}{z_{22}} & \dfrac{z_{12}}{z_{22}} \\ -\dfrac{z_{21}}{z_{22}} & \dfrac{1}{z_{22}} \end{bmatrix} = \begin{bmatrix} \dfrac{1}{y_{11}} & -\dfrac{y_{12}}{y_{11}} \\ \dfrac{y_{21}}{y_{11}} & \dfrac{|y|}{y_{11}} \end{bmatrix} = \begin{bmatrix} \dfrac{g_{22}}{|g|} & -\dfrac{g_{12}}{|g|} \\ -\dfrac{g_{21}}{|g|} & \dfrac{g_{11}}{|g|} \end{bmatrix}$$

$$= \begin{bmatrix} \dfrac{B}{D} & \dfrac{|F|}{D} \\ -\dfrac{1}{D} & \dfrac{C}{D} \end{bmatrix} = \begin{bmatrix} \dfrac{B}{\mathcal{A}} & \dfrac{1}{\mathcal{A}} \\ -\dfrac{|\mathcal{F}|}{\mathcal{A}} & \dfrac{C}{\mathcal{A}} \end{bmatrix} \quad (10.56)$$

TWO-PORT NETWORKS

$$[g] = \begin{bmatrix} \dfrac{1}{z_{11}} & -\dfrac{z_{12}}{z_{11}} \\ \dfrac{z_{21}}{z_{11}} & \dfrac{|z|}{z_{11}} \end{bmatrix} = \begin{bmatrix} \dfrac{|y|}{y_{22}} & \dfrac{y_{12}}{y_{22}} \\ -\dfrac{y_{21}}{y_{22}} & \dfrac{1}{y_{22}} \end{bmatrix} = \begin{bmatrix} \dfrac{h_{22}}{|h|} & -\dfrac{h_{12}}{|h|} \\ -\dfrac{h_{21}}{|h|} & \dfrac{h_{11}}{|h|} \end{bmatrix}$$

$$= \begin{bmatrix} \dfrac{C}{A} & -\dfrac{|F|}{A} \\ \dfrac{1}{A} & \dfrac{B}{A} \end{bmatrix} = \begin{bmatrix} \dfrac{C}{D} & -\dfrac{1}{D} \\ \dfrac{|\mathcal{F}|}{D} & \dfrac{B}{D} \end{bmatrix} \quad (10.57)$$

$$[F] = \begin{bmatrix} \dfrac{z_{11}}{z_{21}} & \dfrac{|z|}{z_{21}} \\ \dfrac{1}{z_{21}} & \dfrac{z_{22}}{z_{21}} \end{bmatrix} = \begin{bmatrix} -\dfrac{y_{22}}{y_{21}} & -\dfrac{1}{y_{21}} \\ -\dfrac{|y|}{y_{21}} & -\dfrac{y_{11}}{y_{21}} \end{bmatrix} = \begin{bmatrix} \dfrac{1}{g_{21}} & \dfrac{g_{22}}{g_{21}} \\ \dfrac{g_{11}}{g_{21}} & \dfrac{|g|}{g_{21}} \end{bmatrix}$$

$$= \begin{bmatrix} -\dfrac{|h|}{h_{21}} & -\dfrac{h_{11}}{h_{21}} \\ -\dfrac{h_{22}}{h_{21}} & -\dfrac{1}{h_{21}} \end{bmatrix} = \begin{bmatrix} \dfrac{D}{|\mathcal{F}|} & \dfrac{B}{|\mathcal{F}|} \\ \dfrac{C}{|\mathcal{F}|} & \dfrac{A}{|\mathcal{F}|} \end{bmatrix} \quad (10.58)$$

$$[\mathcal{F}] = \begin{bmatrix} \dfrac{z_{22}}{z_{12}} & \dfrac{|z|}{z_{12}} \\ \dfrac{1}{z_{12}} & \dfrac{z_{11}}{z_{12}} \end{bmatrix} = \begin{bmatrix} -\dfrac{y_{11}}{y_{12}} & -\dfrac{1}{y_{12}} \\ -\dfrac{|y|}{y_{12}} & -\dfrac{y_{22}}{y_{12}} \end{bmatrix} = \begin{bmatrix} -\dfrac{|g|}{g_{12}} & -\dfrac{g_{22}}{g_{12}} \\ -\dfrac{g_{11}}{g_{12}} & -\dfrac{1}{g_{12}} \end{bmatrix}$$

$$= \begin{bmatrix} \dfrac{1}{h_{12}} & \dfrac{h_{11}}{h_{12}} \\ \dfrac{h_{22}}{h_{12}} & \dfrac{|h|}{h_{12}} \end{bmatrix} = \begin{bmatrix} \dfrac{D}{|F|} & \dfrac{B}{|F|} \\ \dfrac{C}{|F|} & \dfrac{A}{|F|} \end{bmatrix} \quad (10.59)$$

Within these relationships the following are evident.

$$[z][y] = [U] \qquad (10.60)$$

$$[h][g] = [U] \qquad (10.61)$$

where $[U]$ is the identity or unit matrix. That is, ==$[z]$ and $[y]$ are inverse of each other==. So are $[h]$ and $[g]$. One might anticipate that $[F]$ and $[\mathcal{F}]$ are inverse of each other. This is not the case because of the minus signs we attach to I_1 or I_2. However, we can say that

$$\begin{bmatrix} A & \pm B \\ \pm C & D \end{bmatrix} \begin{bmatrix} \mathcal{A} & \mp \mathcal{B} \\ \mp \mathcal{C} & \mathcal{D} \end{bmatrix} = [U] \qquad (10.62)$$

10.4 Circuit models of two-ports with known parameters

Sometimes it is more expedient to replace a two-port whose parameters matrix is known by a circuit that includes two-terminal elements and controlled sources only. Once this is done, we can analyze the network in which the two-port is imbedded as ordinary circuits. The two-port formalism is no longer needed. Fig. 10.18 shows several such models. The student should verify that each model does have the respective parameter matrix whose parameters are used in the models.

One should also be aware that some of these are three-terminal two-ports while others are four-terminal ones. In replacing a two-port with its model, one should make sure that the integrity of the two-port is not compromised. By "integrity" we mean that the original assumption that each port is a terminal-pair (the current entering a terminal returns through the other terminal) is not violated. In addition, no part of the model should affect the circuit outside the two-port. For instance, when a three-terminal model is taking the place of a two-port, the short circuit of the two bottom terminals should not connect two nodes that are not already short-circuited.[2]

EXAMPLE 8. Find E_1 and E_2 of two-port N in Fig. 10.19. For two-port N

---

[2] The integrity of a two-port can always be preserved by using an ideal transformer as explained in a later section.

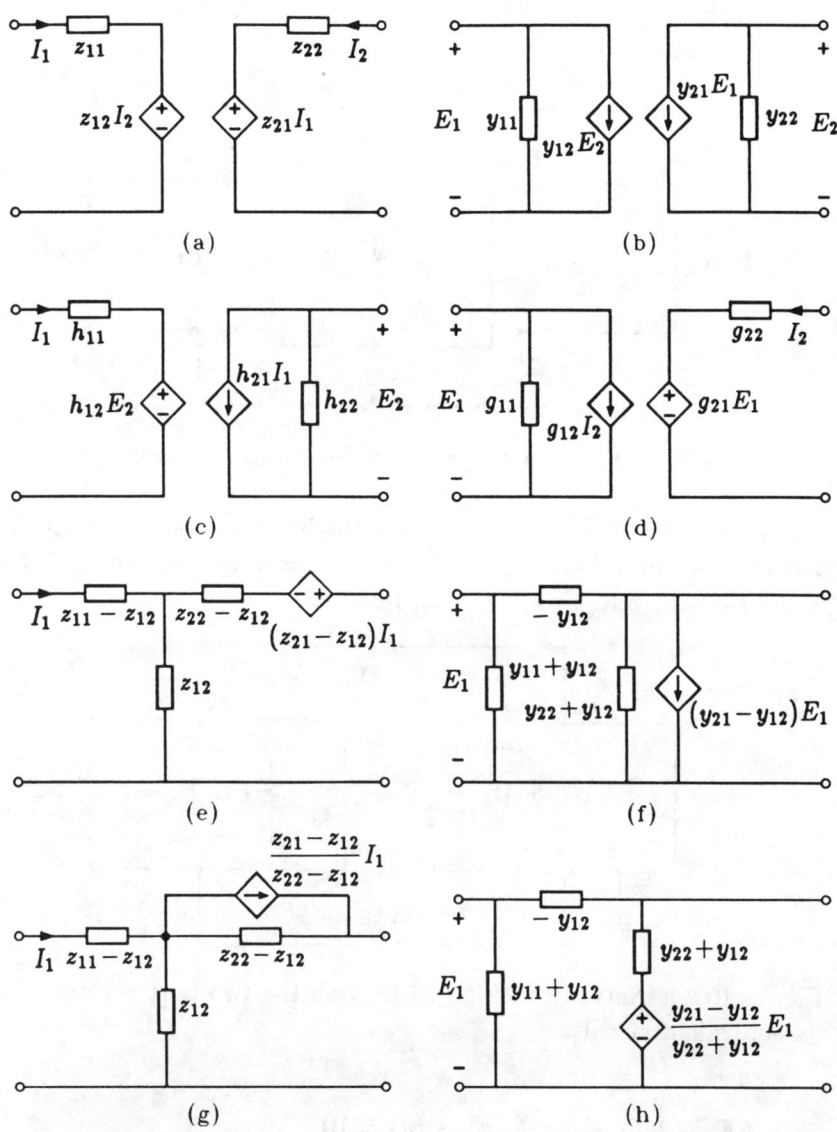

Figure 10.18: Some equivalent circuits for two-ports.

$$[y] = \begin{bmatrix} 5 & -2 \\ -4 & 6 \end{bmatrix} \mho$$

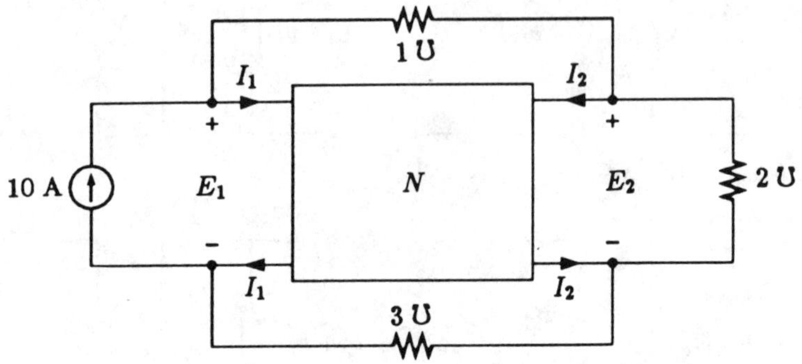

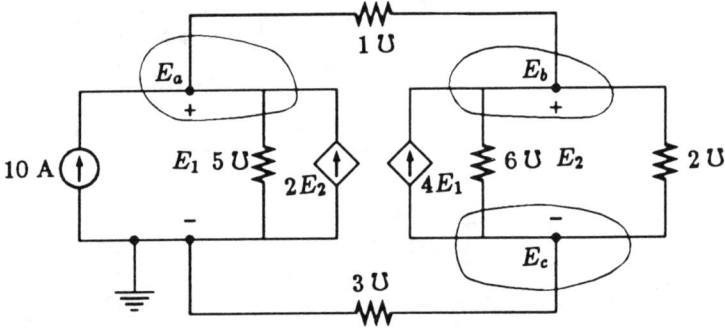

Figure 10.19: A two-port imbedded in a network.

SOLUTION Replacing the two-port with its model in Fig. 10.18(b), we have the circuit in Fig. 10.20. Now we can apply node analysis. With $E_1 = E_a$ and $E_2 = E_b - E_c$, we have

Figure 10.20: Network in Fig. 10.18 with the two-port replaced by its equivalent circuit.

$$5E_a - 2(E_b - E_c) + (E_a - E_b) = 10$$

$$(E_b - E_a) + 6(E_b - E_c) + 2(E_b - E_c) = 4E_a$$

$$2(E_c - E_b) + 6(E_c - E_b) + 4E_a + 3E_c = 0$$

Solving, we obtain

TWO-PORT NETWORKS

$$E_a = \frac{350}{149} \qquad E_b = \frac{230}{149} \qquad E_c = \frac{40}{149}$$

Hence,

$$E_1 = E_a = \frac{350}{149} \text{ V} \qquad \text{and} \qquad E_2 = E_b - E_c = \frac{190}{149} \text{ V}$$

We could also apply node analysis to the original circuit without employing the equivalent circuit. We use the arrangement in Fig. 10.21. The node equations now read

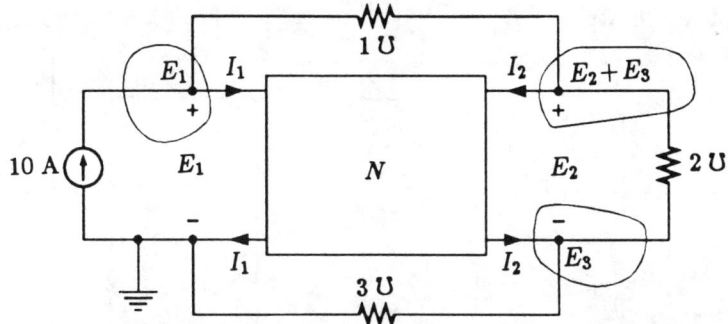

Figure 10.21: Node analysis of the network in Fig. 10.19

$$I_1 + E_1 - (E_2 + E_3) = 10$$

$$E_2 + E_3 - E_1 + I_2 + 2E_2 = 0$$

$$2E_2 + I_2 - 3E_3 = 0$$

For the two-port

$$I_1 = 5E_1 - 2E_2$$

$$I_2 = -4E_1 + 6E_2$$

Now we can solve the five equations simultaneously. The results are

$$E_1 = \frac{350}{149} \text{ V} \qquad E_2 = \frac{190}{149} \text{ V}$$

In this example, it is clear that we cannot use the grounded model in Fig. 10.18(f). If we use that model, the 3-℧ resistor would be short-circuited.

EXAMPLE 9. For two-port N in Fig. 10.22

$$[y] = \begin{bmatrix} 2 & -4 \\ 8 & 10 \end{bmatrix} \mho$$

Find the z matrix of N'.

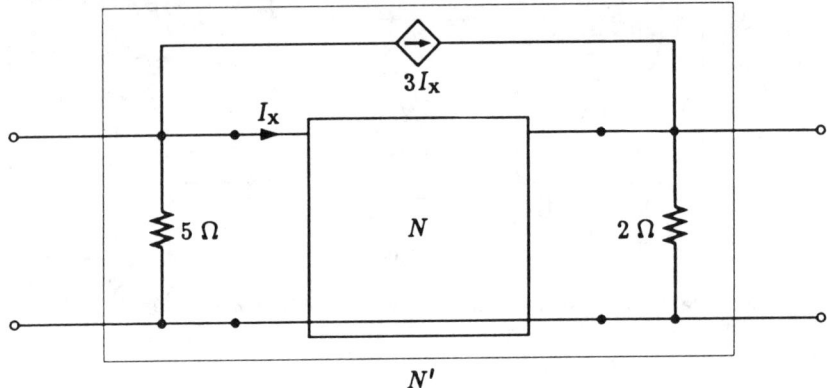

Figure 10.22: A two-port imbedded in a two-port.

SOLUTION With the notations shown in Fig. 10.23, we have, for N,

$$I_x = 2E_1 - 4E_2 = 2E_1' - 4E_2' \qquad (10.63)$$

$$I_2 = 8E_1 + 10E_2 = 8E_1' + 10E_2' \qquad (10.64)$$

KCL requires

$$I_1' = 3I_x + I_x + 0.2E_1' \qquad (10.65)$$

$$I_2' = -3I_x + I_2 + 0.5E_2' \qquad (10.66)$$

Substitute Eqs. (10.63) and (10.64) into Eqs. (10.65) and (10.66) to get

TWO-PORT NETWORKS

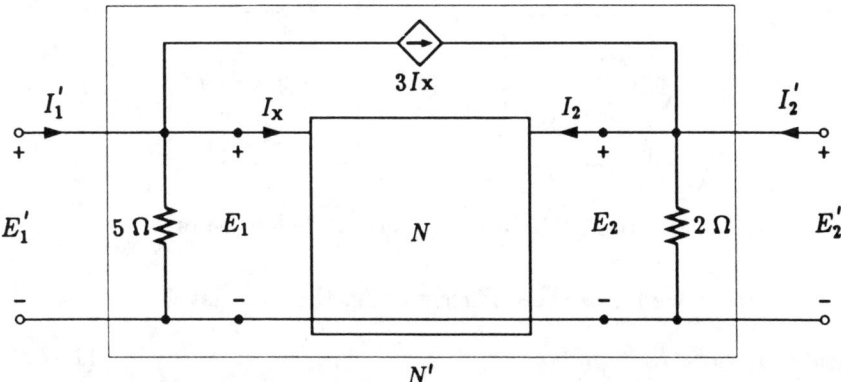

Figure 10.23: Analysis of two-port in Fig. 10.22.

$$I_1' = 4(2E_1' - 4E_2') + 0.2E_1' = 8.2E_1' - 16E_2'$$
$$I_2' = -3(2E_1' - 4E_2') + 8E_1' + 10E_2' + 0.5E_2' = 2E_1' + 22.5E_2'$$

From these two equations, we write

$$[y'] = \begin{bmatrix} 8.2 & -16 \\ 2 & 22.5 \end{bmatrix} \mho$$

Hence,

$$[z'] = [y']^{-1} = \begin{bmatrix} \dfrac{45}{433} & \dfrac{32}{433} \\ -\dfrac{4}{433} & \dfrac{82}{2165} \end{bmatrix} \Omega$$

10.5 Relationships in a terminated two-port

One of the most common applications of two-port parameters is the derivation of certain voltage-current relationships when the network is terminated in a load impedance at one of its ports. To be specific, let us consider the arrangement in Fig. 10.24. One of the possible quantities of interest may be the input impedance

$$Z_{\text{in}} = \frac{E_1}{I_1}$$

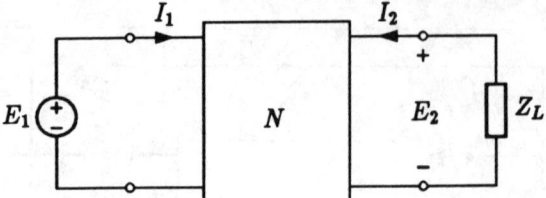

Figure 10.24: A singly terminated two-port.

Suppose we know the g matrix of N, then we have

$$I_1 = g_{11}E_1 + g_{12}I_2 \tag{10.67}$$

$$E_2 = g_{21}E_1 + g_{22}I_2 \tag{10.68}$$

The load impedance Z_L places the constraint

$$E_2 = -I_2 Z_L \tag{10.69}$$

on E_2 and I_2. Substitution of Eq. (10.69) into Eq. (10.68) gives

$$-I_2 Z_L = g_{21}E_1 + g_{22}I_2$$

or

$$I_2 = -\frac{g_{21}}{g_{22} + Z_L}E_1 \tag{10.70}$$

Substitution of Eq. (10.70) into Eq. (10.67) gives

$$I_1 = g_{11}E_1 - \frac{g_{12}g_{21}}{g_{22} + Z_L}E_1 = \frac{g_{11}Z_L + |g|}{g_{22} + Z_L}E_1$$

Hence

$$Z_{\text{in}} = \frac{g_{22} + Z_L}{|g| + g_{11}Z_L} \tag{10.71}$$

When other two-port parameters of N are known, the procedure for finding Z_{in} is similar.

There are several other relationships that may be of interest. They are:

$$\text{Voltage gain} = A_v = \frac{E_2}{E_1}$$

TWO-PORT NETWORKS

Current gain $= A_i = \dfrac{I_2}{I_1}$

Transfer impedance $= Z_{21} = \dfrac{E_2}{I_1}$

Transfer admittance $= Y_{21} = \dfrac{I_2}{E_1}$

The following is a summary of these five relationships expressed in terms of all six sets of two-port parameters. The student should be able to derive any one of these relationships.

$$Z_{\text{in}} = \frac{E_1}{I_1} = \frac{|z| + z_{11}Z_L}{z_{22} + Z_L} = \frac{y_{22} + Y_L}{|y| + y_{11}Y_L} = \frac{|h| + h_{11}Y_L}{h_{22} + Y_L}$$

$$= \frac{g_{22} + Z_L}{|g| + g_{11}Z_L} = \frac{AZ_L + B}{CZ_L + D} = \frac{B + DZ_L}{\mathcal{A} + CZ_L} \quad (10.72)$$

$$A_V = \frac{E_2}{E_1} = \frac{z_{21}Z_L}{|z| + z_{11}Z_L} = \frac{-y_{21}}{y_{22} + Y_L} = \frac{-h_{21}}{|h| + h_{11}Y_L}$$

$$= \frac{g_{21}Z_L}{g_{22} + Z_L} = \frac{Z_L}{B + AZ_L} = \frac{|\mathcal{F}|Z_L}{B + DZ_L} \quad (10.73)$$

$$A_i = \frac{I_2}{I_1} = \frac{-z_{21}}{z_{22} + Z_L} = \frac{y_{21}Y_L}{|y| + y_{11}Y_L} = \frac{h_{21}Y_L}{h_{22} + Y_L}$$

$$= \frac{-g_{21}}{|g| + g_{11}Z_L} = \frac{-1}{D + CZ_L} = \frac{-|\mathcal{F}|}{\mathcal{A} + CZ_L} \quad (10.74)$$

$$Z_{21} = \frac{E_2}{I_1} = \frac{z_{21}Z_L}{z_{22} + Z_L} = \frac{-y_{21}}{|y| + y_{11}Y_L} = \frac{-h_{21}}{h_{22} + Y_L}$$

$$= \frac{g_{21}Z_L}{|g| + g_{11}Z_L} = \frac{Z_L}{D + CZ_L} = \frac{|\mathcal{F}|Z_L}{\mathcal{A} + CZ_L} \quad (10.75)$$

$$Y_{21} = \frac{I_2}{E_1} = \frac{-z_{21}}{|z| + z_{11}Z_L} = \frac{y_{21}Z_L}{y_{22} + Y_L} = \frac{h_{21}Y_L}{|h| + h_{11}Y_L}$$

$$= \frac{-g_{21}}{g_{22} + Z_L} = \frac{-1}{B + AZ_L} = \frac{-|\mathcal{F}|}{B + DZ_L} \quad (10.76)$$

The arrangement in Fig. 10.24 assumes an ideal voltage source at port 1. A more practical situation is that in Fig. 10.25—the doubly terminated two-port. E_s and Z_s is the Thévenin's equivalent of that part of the network looking to the left from port 1. Z_L represents the equivalent impedance seen by the two-port from port 2 to its right. Suppose $[F]$ of N is given and we are interested in the voltage gain E_2/E_s. Then we have

$$E_1 = AE_2 - BI_2 \qquad (10.77)$$

$$I_1 = CE_2 - DI_2 \qquad (10.78)$$

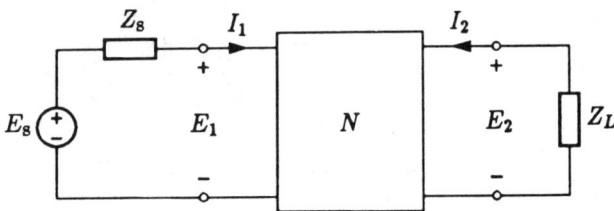

Figure 10.25: A doubly terminated two-port.

We also have

$$E_2 = -I_2 Z_L \quad \text{or} \quad I_2 = -\frac{E_2}{Z_L} \qquad (10.79)$$

$$E_1 = E_s - I_1 Z_s \quad \text{or} \quad I_1 = \frac{E_s - E_1}{Z_s} \qquad (10.80)$$

Substitute Eq. (10.77) into Eq. (10.80) and equate it to Eq. (10.78). We get

$$I_1 = \frac{E_s - (A_2 - BI_2)}{Z_s} = CE_2 - DI_2 \qquad (10.81)$$

Solving for I_2 and using Eq. (10.79), we get

$$I_2 = \frac{AE_2 + CE_2 Z_s - E_s}{B + DZ_s} = -\frac{E_2}{Z_L} \qquad (10.82)$$

Eq. (10.82) involves only E_s and E_2. Hence,

$$\frac{E_2}{E_s} = \frac{Z_L}{AZ_L + B + CZ_LZ_s + DZ_s} \qquad (10.83)$$

is easily obtained.

EXERCISE

10.5.1 Obtain $Y_{21} = I_2/E_s$ for the arrangement in Fig. 10.25 in terms of the two-port open-circuit impedance parameters.

Ans. $-\dfrac{z_{21}}{(z_{22} + Z_L)(z_{11} + Z_s) - z_{12}z_{21}}$

10.6 Special two-ports

There are several special two-ports that either occur frequently in practice or are useful for the study of networks involving two-ports. We shall describe several of them in this section.

10.6.1 Reciprocal two-ports

A two-port is reciprocal if the reciprocity property applies to its two ports. Fig. 10.26 shows one manifestation when a two-port is reciprocal. Reciprocity requires that

$$\frac{I_2}{E_1} = \frac{I_1}{E_2} \qquad (10.84)$$

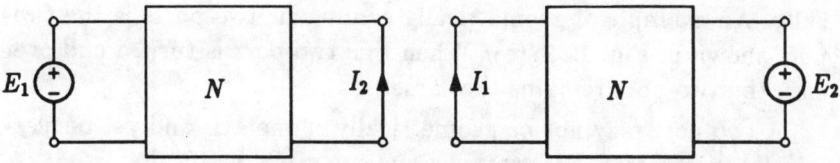

Figure 10.26: N is reciprocal if Eq. (10.85) or (10.86) is satisfied.

In a two-port, Eq.(10.84) implies

$$y_{12} = y_{21} \qquad (10.85)$$

From Eq. (10.55), in terms of other parameters, a two-port is reciprocal if

$$z_{12} = z_{21} \quad \text{or} \quad g_{12} = -g_{21} \quad \text{or} \quad h_{12} = -h_{21}$$

$$\text{or} \quad |F| = 1 \quad \text{or} \quad |\mathcal{F}| = 1 \qquad (10.86)$$

If a two-port contains only resistors, inductors[3], and capacitors it is a reciprocal two-port. A two-port could contain controlled sources and still be reciprocal as long as Eqs. (10.85) and (10.86) are satisfied.

10.6.2 Symmetric two-port

A two-port is symmetric if its port 1 and port 2 can be interchanged without altering its electrical characteristics. In terms of y's, a two-port is symmetric if

$$y_{12} = y_{21} \quad \text{and} \quad y_{11} = y_{22} \qquad (10.87)$$

In terms of other parameters, a two-port is symmetric if, in addition to satisfying the reciprocal conditions [Eqs. (10.85) and (10.86)], its parameters satisfy

$$z_{11} = z_{22} \quad \text{or} \quad |g| = 1 \quad \text{or} \quad |h| = 1$$

$$\text{or} \quad A = D \quad \text{or} \quad \mathcal{A} = \mathcal{D} \qquad (10.88)$$

A two-port is obviously symmetric if it is symmetric geometrically. An example of geometrically symmetric two-ports is the two-port shown in Fig. 10.27(a). When this two-port is turned end over end, the two-port remains the same.

A two-port may not be geometrically symmetric and yet be electrically symmetric. For example, one can easily verify that $z_{11} = z_{22}$ for the two-port in Fig. 10.27(b) which is not geometrically symmetric. [Do this as an exercise.]

---

[3] Including mutual inductances to be describe in Chapter 11.

TWO-PORT NETWORKS

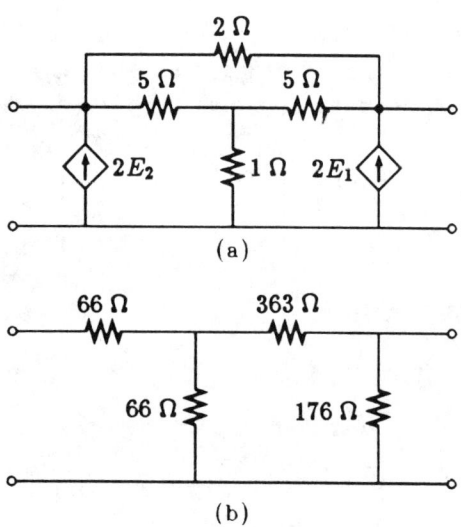

Figure 10.27: Examples of symmetric two-ports.

10.6.3 Tee and pi two-ports

The impedance matrix of the tee two-port in Fig. 10.28(a) can be obtained easily. We get

$$[z] = \begin{bmatrix} z_a + z_b & z_b \\ z_b & z_b + z_c \end{bmatrix} \qquad (10.89)$$

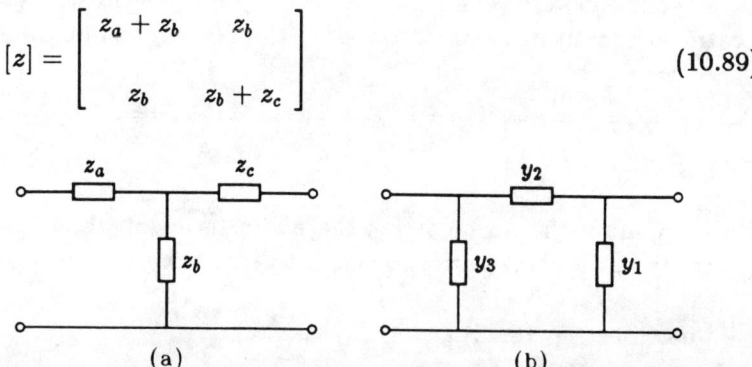

Figure 10.28: The tee and pi two-ports.

The admittance matrix of the pi two-port in Fig. 10.27(b) is also easily obtained. We get

$$[y] = \begin{bmatrix} y_3 + y_2 & -y_2 \\ -y_2 & y_1 + y_2 \end{bmatrix} \qquad (10.90)$$

If we want to make the two two-ports in Fig. 10.28 equivalent, we can simply make their z or y matrices equal to each other. To make their y matrices equal, we require

$$\begin{bmatrix} y_3 + y_2 & -y_2 \\ -y_2 & y_1 + y_2 \end{bmatrix} = \begin{bmatrix} z_a + z_b & z_b \\ z_b & z_b + z_c \end{bmatrix}^{-1}$$

$$= \frac{\begin{bmatrix} z_b + z_c & -z_b \\ -z_b & z_a + z_b \end{bmatrix}}{z_a z_b + z_b z_c + z_c z_a} \tag{10.91}$$

Equating elements and solve, we obtain

$$\begin{aligned} y_1 &= \frac{z_a}{z_a z_b + z_b z_c + z_c z_a} \\ y_2 &= \frac{z_b}{z_a z_b + z_b z_c + z_c z_a} \\ y_3 &= \frac{z_c}{z_a z_b + z_b z_c + z_c z_a} \end{aligned} \tag{10.92}$$

These equations are identical to those in Eq. (8.44). Our present treatment constitutes a proof of the tee-to-pi (or wye-to-delta) equivalence.

EXERCISE

10.6.1 Obtain the formulas for the z's in terms of y's so the two two-ports in Fig. 10.28 are equivalent to each other.

Ans. [See Eq. (8.43).]

10.6.4 Padding

When an impedance is connected in series with a port terminal, its value is added to the driving-point impedance at that port. When an admittance is connected in parallel (or in shunt) at a port, its value is added to the driving-point admittance of that port. Thus, in Fig. 10.29

TWO-PORT NETWORKS

$$[z'] = \begin{bmatrix} z_{11} + z_a & z_{12} \\ z_{21} & z_{22} + z_b \end{bmatrix} \tag{10.93}$$

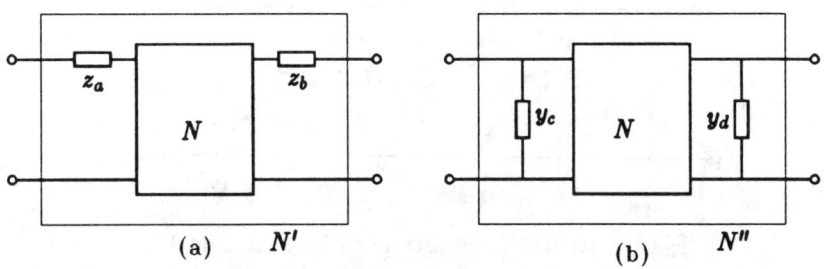

Figure 10.29: Padding of two-ports.

and

$$[y''] = \begin{bmatrix} y_{11} + y_c & y_{12} \\ y_{21} & y_{22} + y_d \end{bmatrix} \tag{10.94}$$

EXERCISE

10.6.2 Obtain the z matrix of N_a and the y matrix of N_b.

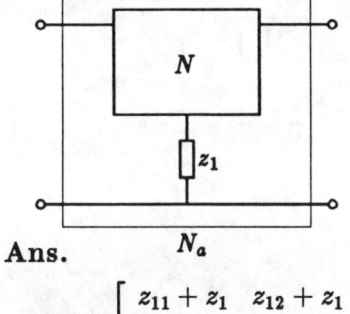
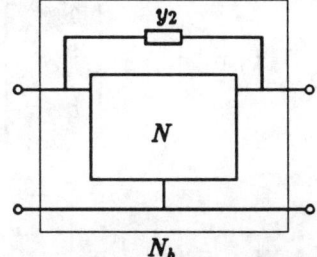

Ans.

$$[z_a] = \begin{bmatrix} z_{11} + z_1 & z_{12} + z_1 \\ z_{21} + z_1 & z_{22} + z_1 \end{bmatrix}$$

$$[y_b] = \begin{bmatrix} y_{11} + y_2 & y_{12} - y_2 \\ y_{21} - y_2 & y_{22} + y_2 \end{bmatrix}$$

EXAMPLE 10. Find the z matrix of the two-port in Fig. 10.30.

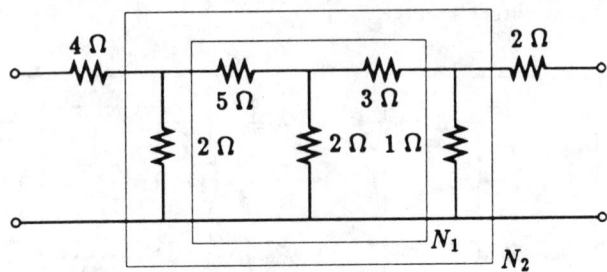

Figure 10.30: Two-port used in Example 10.

SOLUTION We obtain successively,

$$[z_1] = \begin{bmatrix} 7 & 2 \\ 2 & 5 \end{bmatrix}$$

$$[y_1] = [z_1]^{-1} = \frac{1}{31} \begin{bmatrix} 5 & -2 \\ -2 & 7 \end{bmatrix}$$

$$[y_2] = [y_1] + \begin{bmatrix} \frac{1}{2} & 0 \\ 0 & 1 \end{bmatrix} = \begin{bmatrix} \frac{41}{62} & -\frac{2}{31} \\ -\frac{2}{31} & \frac{38}{31} \end{bmatrix}$$

$$[z_2] = [y_2]^{-1} = \begin{bmatrix} \frac{138}{25} & \frac{2}{25} \\ \frac{2}{25} & \frac{41}{50} \end{bmatrix}$$

$$[z] = [z_2] + \begin{bmatrix} 4 & 0 \\ 0 & 2 \end{bmatrix} = \begin{bmatrix} \frac{138}{25} & \frac{2}{25} \\ \frac{2}{25} & \frac{141}{50} \end{bmatrix} \Omega$$

TWO-PORT NETWORKS

10.6.5 Balanced two-port

A two-port is *balanced* if it is symmetric about a horizontal line. Fig. 10.31 is an example of a two-port that is balanced.

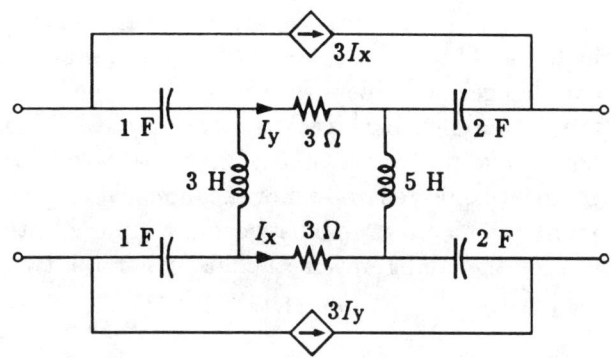

Figure 10.31: A balanced two-port.

There is no special significance when a two-port is balanced, except that in certain situations a balanced network offers some simplicity in the analysis process.

10.6.6 Balanced and unbalanced ladders

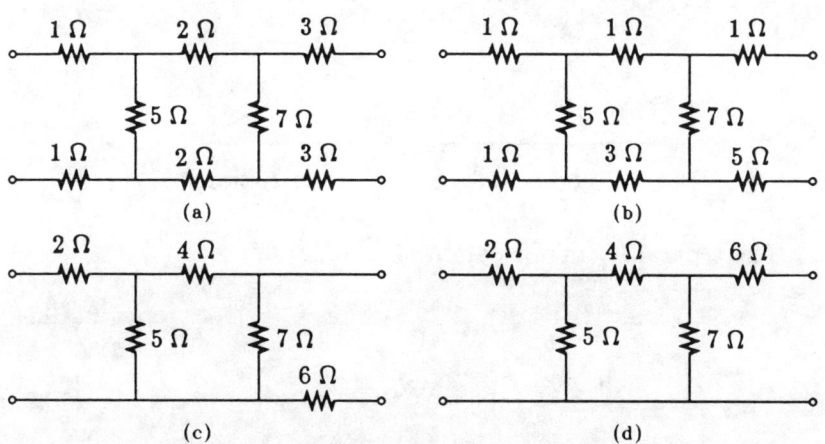

Figure 10.32: Balanced and unbalanced ladders.

The ladder in Fig. 10.32(a) is balanced. The other three ladders in Fig. 10.32 are not balanced. All four ladders are equivalent as two-ports. This is clear because within each mesh of the ladders, the sum of the resistances all add up to the same value. Hence, any mesh current that might have been assigned for a mesh will encounter the same amount of resistance in each mesh. These examples are meant to demonstrate that in a ladder two-port, impedances may be shifted between the top and the bottom horizontal branches at will and this shift will not change the ladders as two-ports.

In certain situations, balanced networks are easier to analyze. The unbalanced two-port, particularly the three-terminal ones [e.g. that in Fig. 10.32(d)], is generally more economical to construct and has a common grounded terminal between the input and the output. They are usually the configuration of choice when the two-ports are actually constructed.

10.6.7 Symmetric lattice

A lattice whose horizontal arms are the same and whose cross arms are also the same, as shown in Fig. 10.33, is both symmetric and balanced. It is usually simply referred to as a *symmetric lattice*.

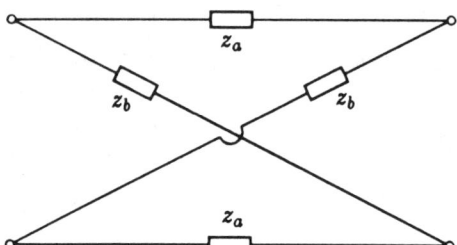

Figure 10.33: A symmetric lattice.

The z parameters of a symmetric lattice are

$$z_{11} = z_{22} = \frac{1}{2}(z_b + z_a) \qquad (10.95)$$

$$z_{12} = z_{21} = \frac{1}{2}(z_b - z_a) \qquad (10.96)$$

and the y parameters are

$$y_{11} = y_{22} = \frac{1}{2}(y_b + y_a) \qquad (10.97)$$

TWO-PORT NETWORKS

$$y_{12} = y_{21} = \frac{1}{2}(y_b - y_a) \tag{10.98}$$

where $y_a = 1/z_a$ and $y_b = 1/z_b$.

Conversely, if a two-port is symmetric and its z or y matrix is known, it is easy to realize it by a symmetric lattice. We simply make

$$z_a = z_{11} - z_{12} \quad \text{and} \quad z_b = z_{11} + z_{12} \tag{10.99}$$

or

$$y_a = y_{11} - y_{12} \quad \text{and} \quad y_b = y_{11} + y_{12} \tag{10.100}$$

Hence, lattices are frequently used as the first step of realizing a desired two-port. However, they are rather impractical to construct and implement. They require large numbers of elements and do not have a common ground. Hence, the initially realized lattices are eventually "unbalanced" whenever possible.

The process of unbalancing a symmetric lattice is somewhat like applying the padding discussed earlier in reverse. If we look at Eqs. (10.95) and (10.96), we see that if z_a and z_b are both increased by the same amount, say z_1, then both z_{11} and z_{22} are increased by z_1, while z_{12} and z_{21} remain unchanged. This is equivalent to padding both ports with z_1 in series. This equivalence is shown in Fig. 10.34. Thus, anything that is in series with all four arms of a symmetric lattice may be removed and replaced by two series branches at both ports.

Similarly, if we look at Eqs. (10.97) and (10.98), we see that as y_a and y_b are both increased by the same amount, say y_1, then both y_{11} and y_{22} are increased by y_1, while y_{12} and y_{21} remain unchange. This is equivalent to padding ports with y_1 in shunt. This equivalence is shown in Fig. 10.35. Anything that is in parallel with all four arms of a symmetric lattice may be removed and replaced by two shunt branches at both ports.

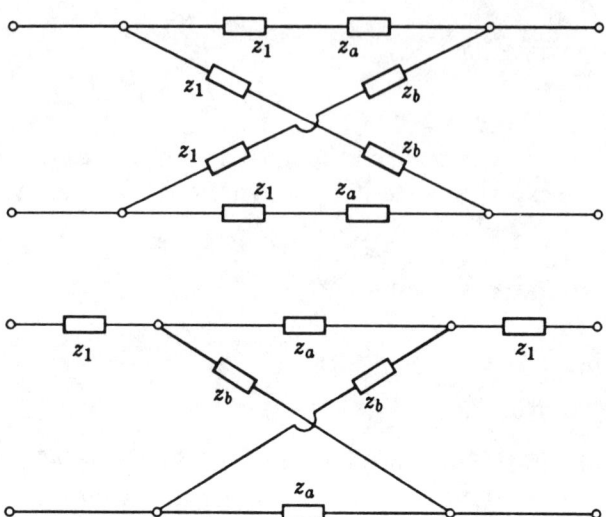

Figure 10.34: Replacement of a series impedance common to all four arms of a symmetric lattice.

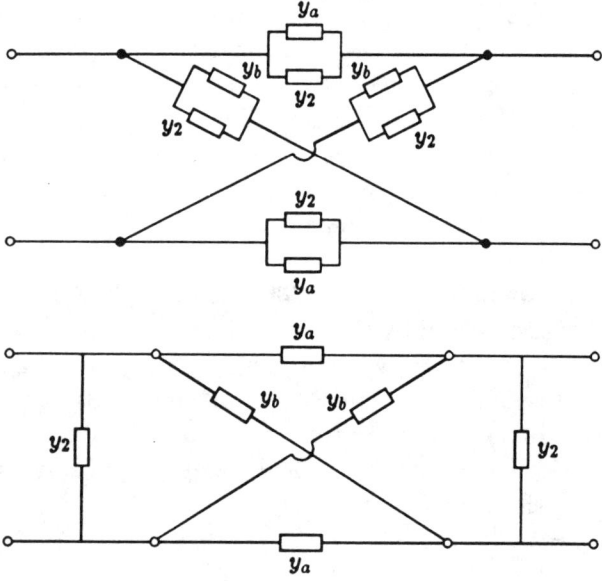

Figure 10.35: Replacement of a parallel admittance common to all four arms of a symmetric lattice.

EXAMPLE 11. Fig. 10.36 shows a sequence of steps in which series and parallel elements in the lattice are replaced by series and shunt elements. In the final step, the ladder is unbalance by combining the impedances in the top and bottom branches.

EXERCISE

10.6.3 Unbalance the symmetric lattice by removing successively a 1-Ω resistance, either in series or in parallel whichever is appropriate, from all four arms of the following lattice. Finally unbalance the ladder.

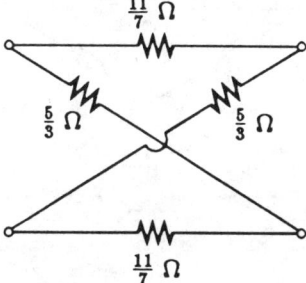

Ans. The final two-port is shown below in which the resistors are 1-Ω each.

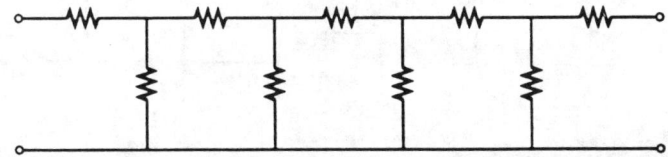

10.6.8 Ideal transformer

An ideal transformer is an idealized two-port that performs several functions in network theory. It isolates two parts of a network or two networks. It fixes the input-output current and voltage ratios. It consumes no real or reactive power. It is the limiting case of mutual inductances to be discussed in Chapter 11. It can be used to model real-world transformers.

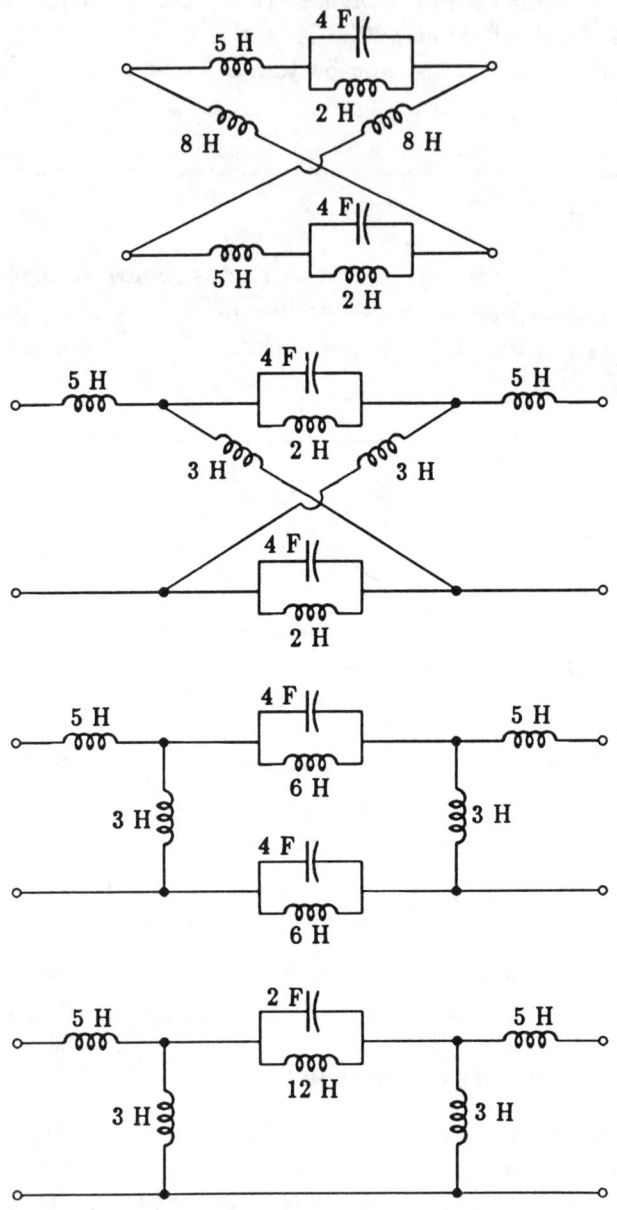

Figure 10.36: The unbalancing of a symmetric lattice.

TWO-PORT NETWORKS

A 1:n ideal transformer, whose symbol is shown in Fig. 10.37, requires

$$e_2 = ne_1 \tag{10.101}$$

$$-i_2 = \frac{1}{n}i_1 \tag{10.102}$$

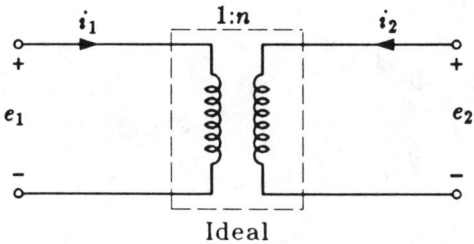

Figure 10.37: An ideal transformer.

where n is known as the *turns ratio*, it's real and may be positive or negative, greater than, equal to, or less than unity.

EXERCISE

10.6.4 Show that the input impedance Z_{in} of a terminated transformer is $Z_{\text{in}} = \dfrac{1}{n^2} Z_L$.

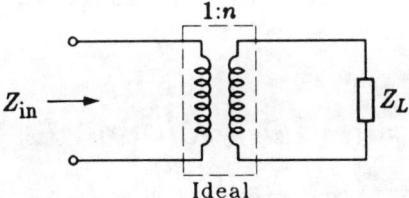

The impedance and admittance matrices of an ideal transformer are undefined. Its other two-port matrices are

$$[h] = \begin{bmatrix} 0 & \dfrac{1}{n} \\ -\dfrac{1}{n} & 0 \end{bmatrix} \tag{10.103}$$

$$[g] = \begin{bmatrix} 0 & -n \\ n & 0 \end{bmatrix} \tag{10.104}$$

$$[F] = \begin{bmatrix} \dfrac{1}{n} & 0 \\ 0 & n \end{bmatrix} \tag{10.105}$$

An ideal transformer connected to one of the ports of a two-port alters its matrices in a very specific way. Referring to Fig. 10.38, we have

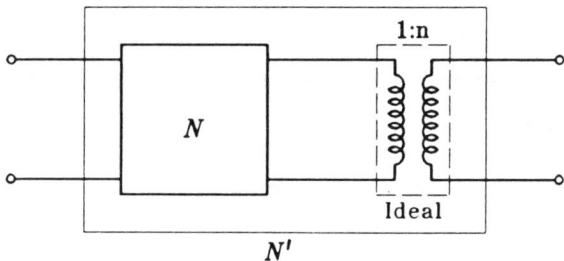

Figure 10.38: An ideal transformer connected to one port of a two-port.

$$E'_1 = E_1 = z_{11}I_1 + z_{12}I_2 = z_{11}I'_1 + z_{12}(nI'_2)$$
$$E'_2 = nE_2 = n(z_{21}I_1 + z_{22}I_2) = n[z_{21}I'_1 + z_{22}(nI'_2)]$$

Hence,

$$[z'] = \begin{bmatrix} z_{11} & nz_{12} \\ nz_{21} & n^2 z_{22} \end{bmatrix} \tag{10.106}$$

In words, $[z']$ is equal to $[z]$ with its second row and second column multiplied by n.

Similarly, we can easily show that

$$[y'] = \begin{bmatrix} y_{11} & \dfrac{1}{n}y_{12} \\ \dfrac{1}{n}y_{21} & \dfrac{1}{n^2}y_{22} \end{bmatrix} \qquad (10.107)$$

In words, $[y']$ is equal to $[y]$ with its second row and second column multiplied by $1/n$.

If $n = 1$, then

$$[z'] = [z] \qquad \text{and} \qquad [y'] = [y] \qquad (10.108)$$

Although a 1:1 ideal transformer does not change the parameter of a two-port, it serves the purpose of ensuring that the integrity of a two-port is not violated. Because any current must flow though either side of an ideal transformer, it forces the current flowing into one terminal of a port to flow out of its other terminal.

EXAMPLE 12. Find the input impedance Z_{in} of the circuit in Fig. 10.39.

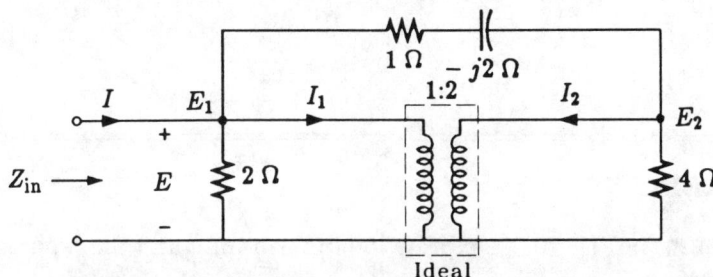

Figure 10.39: Circuit used in Example 12.

SOLUTION Since

$$Z_{\text{in}} = \frac{E}{I} = E\Big|_{I=1}$$

we shall let $I = 1$ and find E. Using node analysis, we have

$$\frac{E_1}{2} + I_1 + \frac{E_1 - E_2}{1 - j2} = 1$$

$$\frac{E_2}{4} + I_2 + \frac{E_2 - E_1}{1 - j2} = 0$$

The ideal transformer requires that

$$E_2 = 2E_1 = 2E \quad \text{and} \quad I_2 = -\frac{I_1}{2}$$

Substitution gives

$$\frac{E}{2} + I_1 - \frac{E}{1 - j2} = 1$$

$$\frac{E}{2} - \frac{I_1}{2} + \frac{E}{1 - j2} = 0$$

Elimination of I_1 gives

$$E = \frac{2 - j4}{5 - j6}$$

Thus,

$$Z_{\text{in}} = 0.573\underline{/-13.2°}\ \Omega$$

EXERCISE

10.6.5 In Fig. 10.38, express $[h']$ in terms of n and the h parameters of N.

Ans.

$$[h'] = \begin{bmatrix} h_{11} & \dfrac{1}{n}h_{12} \\ \dfrac{1}{n}h_{21} & \dfrac{1}{n^2}h_{22} \end{bmatrix}$$

10.7 Interconnection of two-ports

When one-ports are connected in series, the same current flows through all one-ports. Their voltages add and the total impedance is the sum of all impedances. When one-ports are connected in parallel, the same voltage appears across all one-ports. Their currents add and the total admittance is the sum of all admittances.

By connecting the ports of two two-ports either in series or in parallel, there are four possible ways two two-ports can be combined if we simply use the series or parallel interconnections. In this section, the parameters of two-port N_a will all carry the subscript a, those of N_b will carry the subscript b, and those of the combination will have no subscript.

10.7.1 Series-series combination

If we want to connect ports 1 and ports 2 of two two-ports in series, it may appear that we can simply employ the arrangement in Fig. 10.40. There is one possible serious flaw with this arrangement—

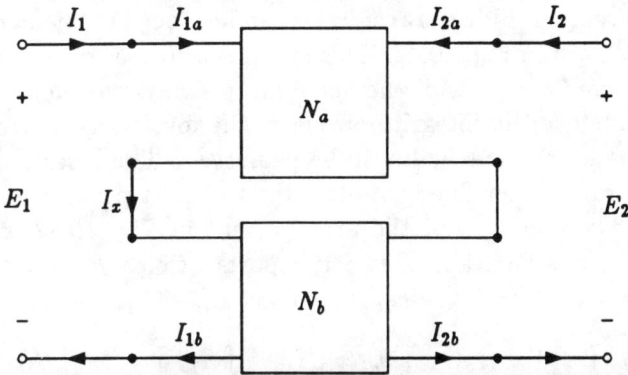

Figure 10.40: An apparent series-series combination of two-ports.

we have no assurance that the integrity of the two-ports is not compromised. In Fig. 10.40, in order for N_a and N_b each to remain a two-port, we must have $I_x = I_{1a} = I_{1b} = I_1$. If this condition is not satisfied, the combined two-port has no direct relationship with the parameters of the individual two-ports.

An example of how such an interconnection may not be a true series-series combination is shown in Fig. 10.41, which might be an

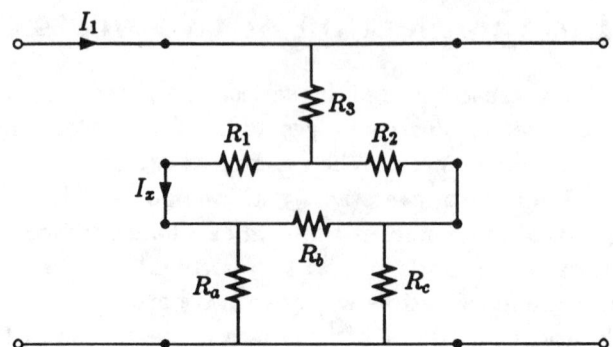

Figure 10.41: Two two-ports that are not truly series-series connected.

attempt to connect the tee two-port and the pi two-port in the series-series manner. Since the value of I_x depends on the values of several resistances, we have no assurance that $I_x = I_1$. Hence, the integrity of both two-ports are no longer preserved and we cannot regard the two two-ports as being series-series connected.

To ensure that the integrity of the constituent two-ports is not compromised, a 1:1 ideal transformer can be placed at one of the four ports, as shown in Fig. 10.42. This ideal transformer may be located at any of the four ports of the network and only one such device is necessary. Once the integrity of one of the four ports is preserved, that of the other three will also be preserved. The insertion of the 1:1 ideal transformer does not alter the two-port parameters of the individual two-ports, and the arrangement in Fig. 10.42 is a true series-series combination of two two-ports. Since $I_1 = I_{1a} = I_{1b}$, $I_2 = I_{2a} = I_{2b}$, $E_1 = E_{1a} + E_{1b}$, and $E_2 = E_{2a} + E_{2b}$, we have

$$\begin{bmatrix} E_1 \\ E_2 \end{bmatrix} = \begin{bmatrix} E_{1a} \\ E_{2a} \end{bmatrix} + \begin{bmatrix} E_{1b} \\ E_{2b} \end{bmatrix} = [z_a] \begin{bmatrix} I_{1a} \\ I_{2a} \end{bmatrix} + [z_b] \begin{bmatrix} I_{1b} \\ I_{2b} \end{bmatrix}$$

$$= \{[z_a] + [z_b]\} \begin{bmatrix} I_1 \\ I_2 \end{bmatrix} = [z] \begin{bmatrix} I_1 \\ I_2 \end{bmatrix} \qquad (10.109)$$

Hence,

$$[z] = [z_a] + [z_b] \qquad (10.110)$$

TWO-PORT NETWORKS

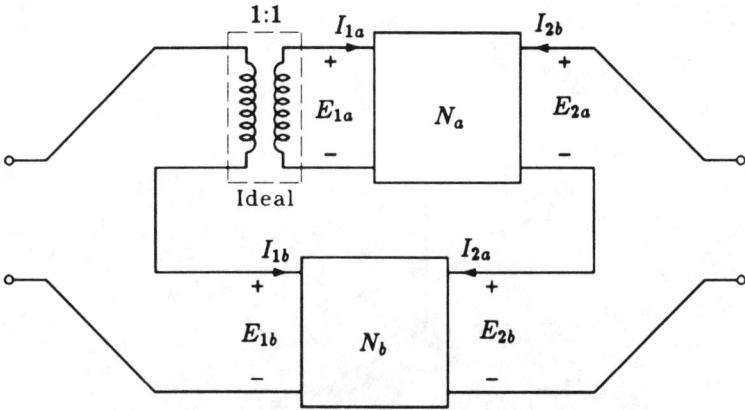

Figure 10.42: Series-series combination of two two-ports.

In words, when two-ports are connected in the series-series manner, their z matrices add.

Often, the series-series combination is simply referred to as the series combination for brevity.

EXAMPLE 13. Find the z matrix of the two-port shown in Fig. 10.43.

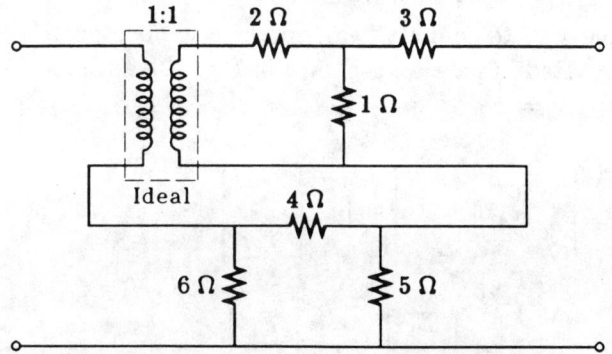

Figure 10.43: Example of series-series combination of two two-ports.

SOLUTION

$$[z_a] = \begin{bmatrix} 3 & 1 \\ 1 & 4 \end{bmatrix}$$

$$[y_b] = \begin{bmatrix} \frac{1}{4}+\frac{1}{6} & -\frac{1}{4} \\ -\frac{1}{4} & \frac{1}{4}+\frac{1}{5} \end{bmatrix}$$

$$[z_b] = [y_b]^{-1} = \begin{bmatrix} \frac{18}{5} & 2 \\ 2 & \frac{10}{3} \end{bmatrix}$$

$$[z] = [z_a] + [z_b] = \begin{bmatrix} \frac{33}{5} & 3 \\ 3 & \frac{22}{3} \end{bmatrix} \Omega$$

10.7.2 Parallel-parallel combination

The proper way to connect two two-ports so both ports 1 and ports 2 are connected in parallel is shown in Fig. 10.44. Since $E_1 = E_{1a} = E_{1b}$, $E_2 = E_{2a} = E_{2b}$, $I_1 = I_{1a} + I_{1b}$, and $I_2 = I_{2a} + I_{2b}$, we have

$$\begin{bmatrix} I_1 \\ I_2 \end{bmatrix} = [y_a] \begin{bmatrix} E_{1a} \\ E_{2a} \end{bmatrix} + [y_b] \begin{bmatrix} E_{1b} \\ E_{2b} \end{bmatrix}$$

$$= \{[y_a] + [y_b]\} \begin{bmatrix} E_1 \\ E_2 \end{bmatrix} = [y] \begin{bmatrix} E_1 \\ E_2 \end{bmatrix} \qquad (10.111)$$

Hence,

$$[y] = [y_a] + [y_b] \qquad (10.112)$$

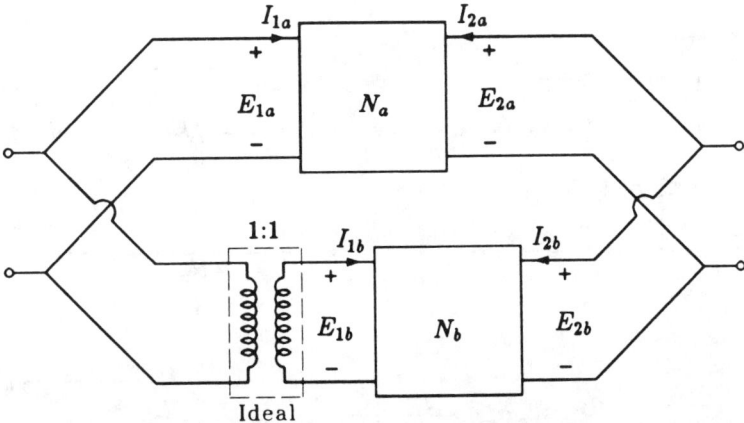

Figure 10.44: Parallel-parallel combination of two two-ports.

In words, when two-ports are connected in the parallel-parallel manner, their y matrices add. Often the parallel-parallel combination is simply referred to as the parallel combination for brevity.

EXAMPLE 14. The two-port in Fig. 10.45(a) is the parallel combination of the two two-ports as shown in Fig. 10.45(b) and (c). For two-port N_a, we have

$$[y_a] = \begin{bmatrix} \dfrac{1}{R_f} & -\dfrac{1}{R_f} \\ -\dfrac{1}{R_f} & \dfrac{1}{R_f} \end{bmatrix}$$

For two-port N_b, comparing it to the circuit in Fig. 10.18(d), we have

$$[g_b] = \begin{bmatrix} \dfrac{1}{R_i} & 0 \\ A_0 & R_o \end{bmatrix}$$

Using Eq. 10.55, we get

$$[y_b] = \begin{bmatrix} \dfrac{1}{R_i} & 0 \\ -\dfrac{A_0}{R_o} & \dfrac{1}{R_o} \end{bmatrix}$$

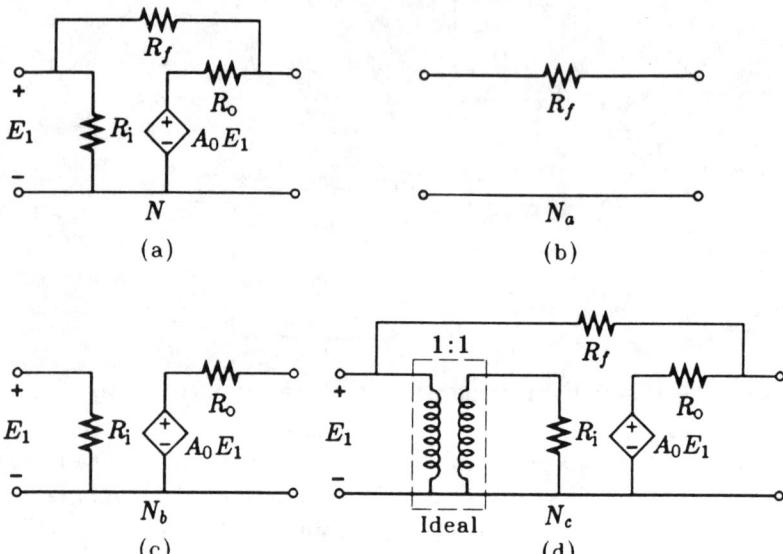

Figure 10.45: A two-port that is the parallel combination of two two-ports.

Hence, for two-port N

$$[y] = \begin{bmatrix} \dfrac{1}{R_f} + \dfrac{1}{R_o} & -\dfrac{1}{R_f} \\ -\dfrac{1}{R_f} - \dfrac{A_0}{R_o} & \dfrac{1}{R_f} + \dfrac{1}{R_o} \end{bmatrix}$$

When we put N_a and N_b in the parallel-parallel combination according to Fig. 10.44, the network in Fig. 10.45(d) results. However, since the bottom terminals of the ideal transformer are short-circuited, the ideal transformer is not necessary and may be eliminated altogether, resulting in the network in Fig. 10.45(a).

10.7.3 Series-parallel combination

Fig. 10.46 shows a hybrid of the series/parallel combination of the ports of two two-ports. Ports 1 are connected in series and ports 2 are connected in parallel. We have $E_1 = E_{1a} + E_{1b}$, $E_2 = E_{2a} = E_{2b}$,

TWO-PORT NETWORKS

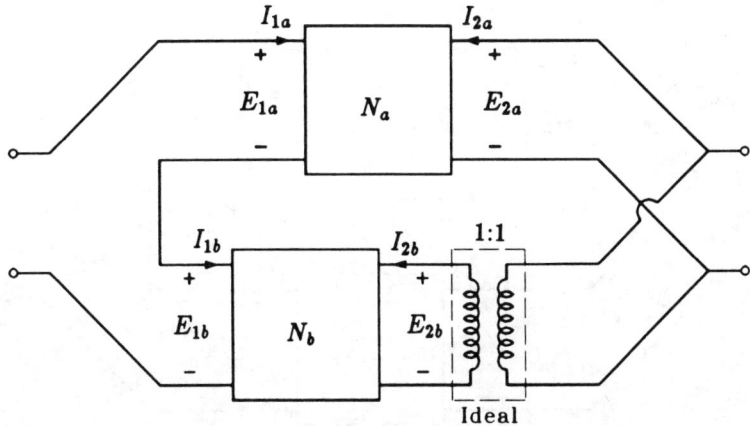

Figure 10.46: The series-parallel combination of two two-ports.

$I_1 = I_{1a} = I_{1b}$, and $I_2 = I_{2a} + I_{2b}$. It follows that when two two-ports are connected this way

$$[h] = [h_a] + [h_b] \qquad (10.113)$$

In words, their h matrices add.

10.7.4 Parallel-series combination

The other hybrid combination of two two-ports is when their ports 1 are connected in parallel while their ports 2 are connected in series as shown in Fig. 10.47. There we have $E_1 = E_{1a} = E_{1b}$, $E_2 = E_{2a} + E_{2b}$, $I_1 = I_{1a} + I_{1b}$, and $I_2 = I_{2a} = I_{2b}$. It follows that when two two-ports are connected this way

$$[g] = [g_a] + [g_b] \qquad (10.114)$$

In other words, when two two-ports are connected in the parallel-series manner, their g matrices add.

We include the two hybrid combinations here mainly for the sake of completeness as well as an academic exercise. In practice, two-ports are seldom combined this way.

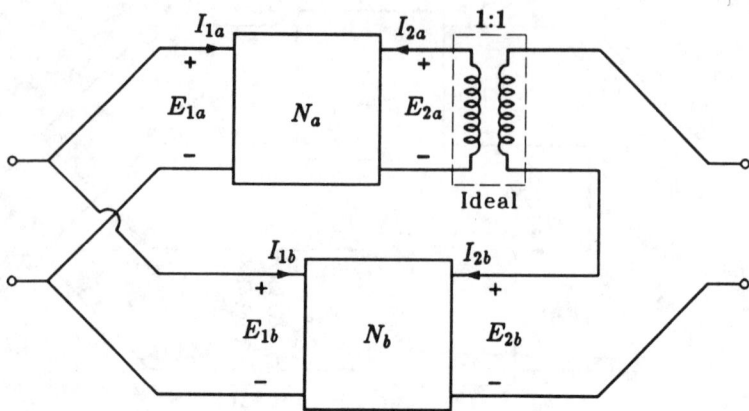

Figure 10.47: The parallel-series combination of two two-ports.

10.7.5 Series and parallel combinations without ideal transformers

As we have shown in Example 14, sometimes two-ports can be combined without using ideal transformers. This happens either when the integrity of the two-ports being combined is guaranteed by the circuits themselves or the interconnection does not have an effect on the integrity of the two-ports. Or, it can be demonstrated that the ideal transformer is superfluous, as in the case of Example 14. We shall describe two tests by which we can ascertain whether or not two-ports can be combined in series or in parallel without transformers.

For the series combination, we refer to Fig. 10.48. We first combine port 1a and port 1b in series and excite port 1 of the series combination, as shown in Fig. 10.48(a). If the combination is such that $V_1 = 0$, then the two terminals connected by the dashed line can be short-circuited without disturbing the current distribution of the network. There is still no current flowing in and out of the terminals of ports 2a and 2b. So, z_{11a}, z_{11b}, z_{21a} and z_{21b} are intact. For the series combination, $z_{11} = z_{11a} + z_{11b}$ and $z_{21} = z_{21a} + z_{21b}$.

If we apply the same test in the reverse direction as shown in Fig. 10.48(b) and if we find $V_2 = 0$, the two terminals connected by the dashed line can be short-circuited without disturbing the current distribution. We then have $z_{12} = z_{12a} + z_{12b}$ and $z_{22} = z_{22a} + z_{22b}$.

TWO-PORT NETWORKS

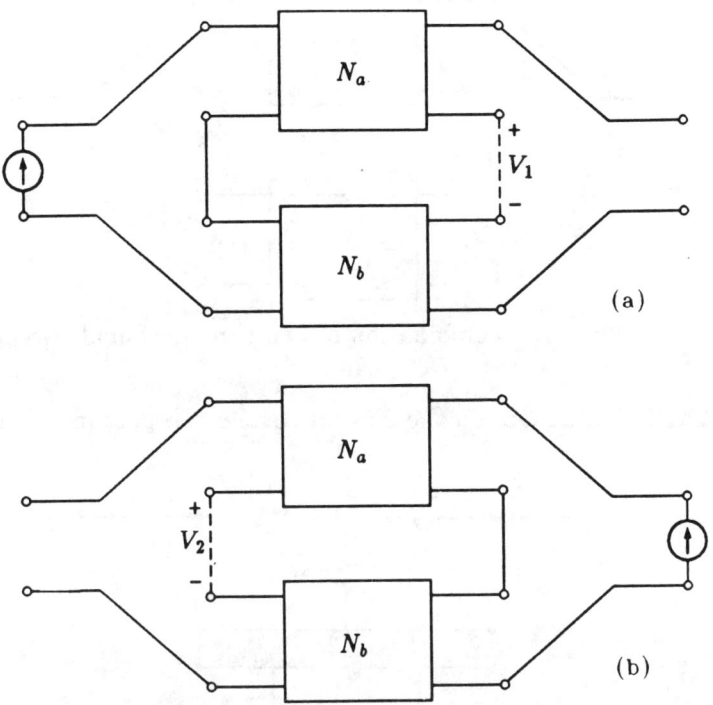

Figure 10.48: Tests for series combination of two-ports without ideal transformers.

If the two two-ports pass both tests, then they may be combined in series without ideal transformers.

Fig. 10.49 is one situation in which these tests are automatically satisfied. This occurs when both two-ports are three-terminal and their ground terminals are connected together. In this situation, the short circuits that are already present in the two-ports guarantee that $V_1 = 0$ and $V_2 = 0$. Thus, two three-terminal two-ports can be combined in series without transformers if they are properly oriented.

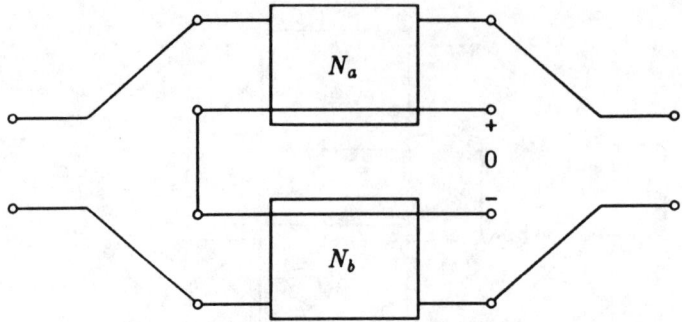

Figure 10.49: Series combination of two three-terminal two-ports.

EXAMPLE 15. Obtain the z matrix of the two-port in Fig. 10.50.

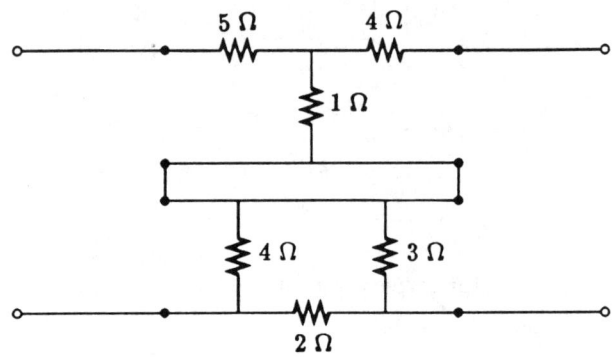

Figure 10.50: Two-port of Example 15.

SOLUTION

$$[z_a] = \begin{bmatrix} 6 & 1 \\ 1 & 5 \end{bmatrix}$$

$$[y_b] = \begin{bmatrix} \dfrac{1}{4} + \dfrac{1}{2} & -\dfrac{1}{2} \\ -\dfrac{1}{2} & \dfrac{1}{2} + \dfrac{1}{3} \end{bmatrix}$$

$$[z_b] = [y_b]^{-1} = \begin{bmatrix} \dfrac{20}{9} & \dfrac{4}{3} \\ \dfrac{4}{3} & 2 \end{bmatrix}$$

$$[z] = [z_a] + [z_b] = \begin{bmatrix} \dfrac{74}{9} & \dfrac{7}{3} \\ \dfrac{7}{3} & 7 \end{bmatrix} \Omega$$

EXERCISE

10.7.1 Find the z matrix of the two-port.

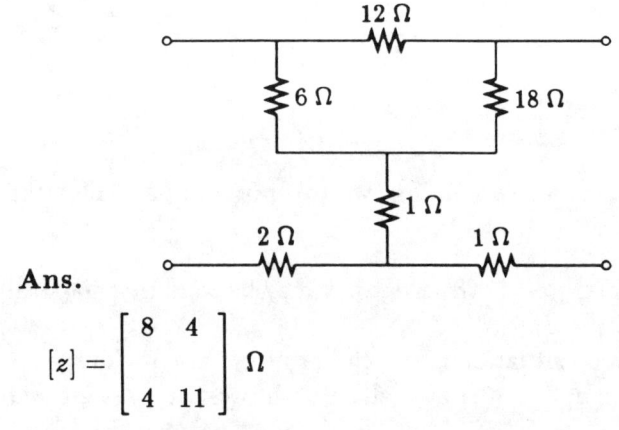

Ans.
$$[z] = \begin{bmatrix} 8 & 4 \\ 4 & 11 \end{bmatrix} \Omega$$

The tests for combining two two-ports in parallel is shown in Fig. 10.51. In Fig. 10.51(a), we first connect port 1a and port 1b in parallel. We then short-circuit port 2a and port 2b. If we can show that $V_3 = 0$, the dashed lines may be made solid without disturbing the current distribution. Hence, y_{11a}, y_{11b}, y_{21a} and y_{21b} remain intact, and $y_{11} = y_{11a} + y_{11b}$ and $y_{21} = y_{21a} + y_{21b}$.

For the other test, we reverse ports 1 and ports 2, as shown in Fig. 10.51(b). If $V_4 = 0$, then port 1a and port 1b may be connected in parallel and, for the combination, $y_{12} = y_{12a} + y_{12b}$ and $y_{22} = y_{22a} + y_{22b}$.

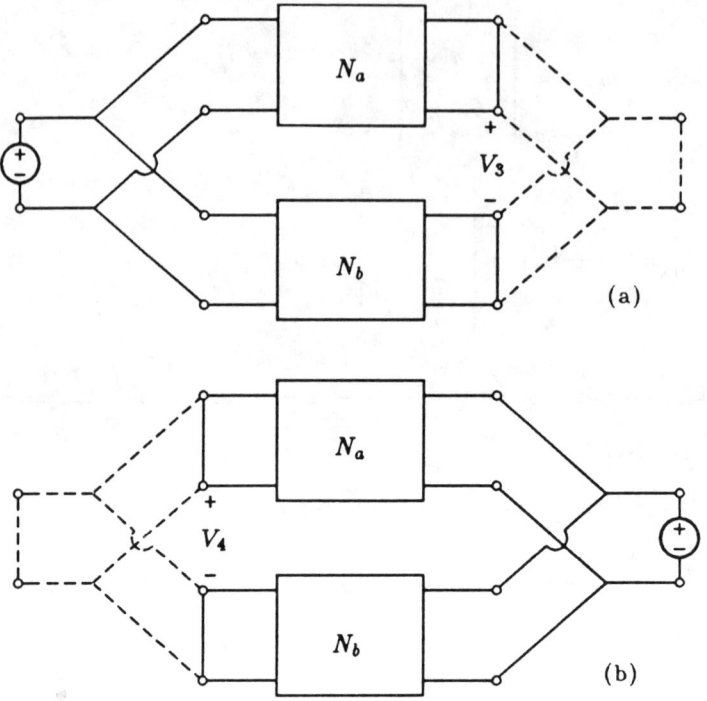

Figure 10.51: Tests for combining two two-ports in parallel without ideal transformers.

If two two-ports pass both tests, they may be combined in parallel without transformers.

A very common situation in which both tests are obviously satisfied is that shown in Fig. 10.52, in which both two-ports are grounded

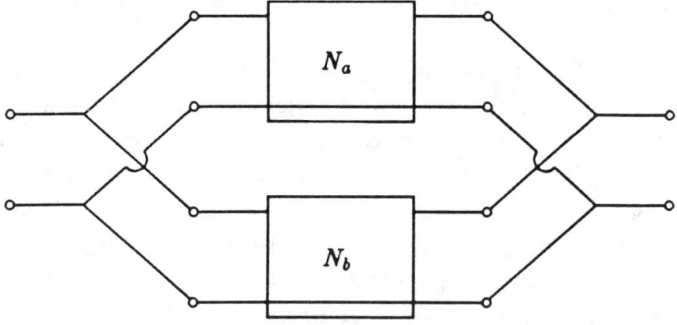

Figure 10.52: Parallel combination of two three-terminal two-ports.

TWO-PORT NETWORKS

and their grounded terminals are connected together. Another example when two two-ports may be combined in parallel without transformers is when both two-ports are both symmetric and balanced.

EXAMPLE 16. Find the y matrix of the two-port of Fig. 10.53.

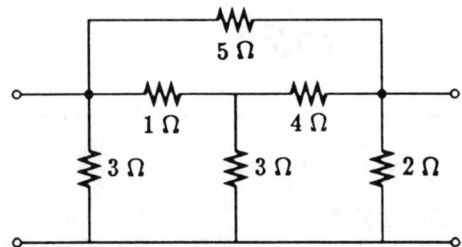

Figure 10.53: Two-port used in Example 16.

SOLUTION This two-port is the parallel combination of a tee two-port and a pi two-port. For the tee

$$[z_a] = \begin{bmatrix} 4 & 3 \\ 3 & 7 \end{bmatrix}$$

$$[y_a] = [z_a]^{-1} = \frac{1}{19}\begin{bmatrix} 7 & -3 \\ -3 & 4 \end{bmatrix}$$

For the pi,

$$[y_b] = \begin{bmatrix} \frac{1}{3}+\frac{1}{5} & -\frac{1}{5} \\ -\frac{1}{5} & \frac{1}{2}+\frac{1}{5} \end{bmatrix} = \begin{bmatrix} \frac{8}{15} & -\frac{1}{5} \\ -\frac{1}{5} & \frac{7}{10} \end{bmatrix}$$

$$[y] = [y_a] + [y_b] = \begin{bmatrix} \frac{257}{285} & -\frac{34}{95} \\ -\frac{34}{95} & \frac{173}{190} \end{bmatrix} \mho$$

EXAMPLE 17. Find the y matrix of the twin-tee two-port in Fig. 10.54.

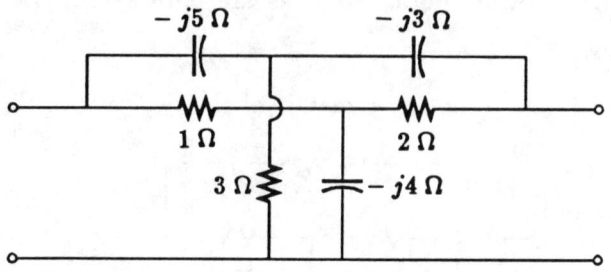

Figure 10.54: A twin-tee two-port.

SOLUTION We have two tee two-ports connected in parallel. We have

$$[z_a] = \begin{bmatrix} 3-j5 & 3 \\ 3 & 3-j3 \end{bmatrix}$$

$$[y_a] = [z_a]^{-1} = \frac{1}{267}\begin{bmatrix} 9+j39 & 15-j24 \\ 15-j24 & 25+j49 \end{bmatrix}$$

$$[z_b] = \begin{bmatrix} 1-j4 & -j4 \\ -j4 & 2-j4 \end{bmatrix}$$

$$[y_b] = [z_b]^{-1} = \frac{1}{74}\begin{bmatrix} 26+j8 & -24+j4 \\ -24+j4 & 25+j2 \end{bmatrix}$$

$$[y] = [y_a] + [y_b] = \begin{bmatrix} 0.3851+j0.2542 & -0.2681-j0.0358 \\ -0.2681-j0.0358 & 0.4315+j0.2105 \end{bmatrix} \mho$$

EXERCISES

10.7.2 Find the y matrix of the bridged-tee two-port.

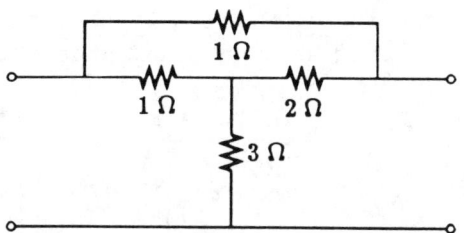

Ans.

$$[y] = \frac{1}{11} \begin{bmatrix} 16 & -14 \\ -14 & 15 \end{bmatrix} \text{ U}$$

10.7.3 The symmetric lattice may be unbalanced into a bridged-tee as shown. First consider the lattice as the parallel combination of two two-ports—one consists of the two 2-F capacitors only and the other the remainder. Each lattice may be unbalanced into three-terminal two-ports. They may then be combined in parallel. Give the values of all elements in the bridged-tee.

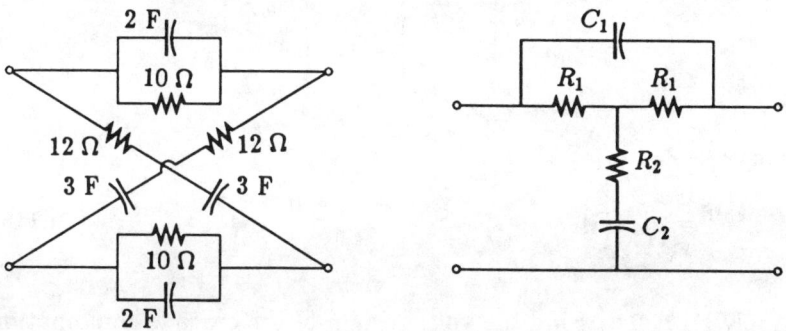

Ans. $C_1 = 1$ F, $C_2 = 6$ F, $R_1 = 10$ Ω, $R_2 = 1$ Ω.

10.7.6 Cascade combination

In addition to the series/parallel methods of interconnecting two-ports, there is another very important interconnection. That is the *cascade* or *tandem* connection. This connection is shown in Fig. 10.55. This connection occurs very frequently. For example, in transmission lines, TV cables, and telephone services, the output of one device is connected to the input of the next device.

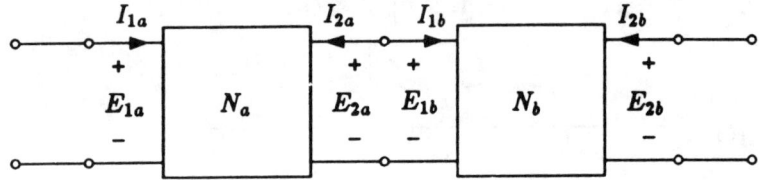

Figure 10.55: The cascade connection of two two-ports.

In Fig. 10.55, we have $E_1 = E_{1a}$, $E_2 = E_{2b}$, $I_1 = I_{1a}$, $I_2 = I_{2b}$, $E_{2a} = E_{1b}$, and $-I_{2a} = I_{1b}$. Thus we have

$$\begin{bmatrix} E_1 \\ I_1 \end{bmatrix} = \begin{bmatrix} E_{1a} \\ I_{1a} \end{bmatrix} = [F_a] \begin{bmatrix} E_{2a} \\ -I_{2a} \end{bmatrix} = [F_a] \begin{bmatrix} E_{1b} \\ I_{1b} \end{bmatrix}$$

$$= [F_a][F_b] \begin{bmatrix} E_{2b} \\ -I_{2b} \end{bmatrix} = [F] \begin{bmatrix} E_2 \\ -I_2 \end{bmatrix}$$

Hence

$$[F] = [F_a][F_b] \tag{10.115}$$

EXAMPLE 18. Find the voltage gain of the cascade combination of two identical amplifiers in Fig. 10.56, (a) when the output terminals are open-circuited and (b) when its output is terminated in a resistance R_L.

SOLUTION Each amplifier is identical to the two-port of Fig. 10.45(c). Hence

TWO-PORT NETWORKS

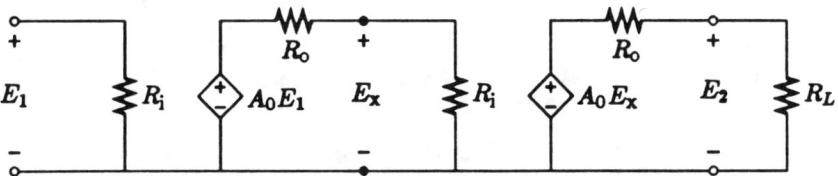

Figure 10.56: Two identical amplifiers in cascade.

$$[y_a] = [y_b] = \begin{bmatrix} \dfrac{1}{R_i} & 0 \\ -\dfrac{A_0}{R_o} & \dfrac{1}{R_o} \end{bmatrix}$$

According to Eq. (10.58), the transmission matrix of each amplifier is

$$[F_a] = [F_b] = \begin{bmatrix} \dfrac{1}{A_0} & \dfrac{R_o}{A_0} \\ \dfrac{1}{A_0 R_i} & \dfrac{R_o}{A_0 R_i} \end{bmatrix}$$

$$[F] = [F_a][F_b] = \dfrac{1}{A_0^2 R_i^2} \begin{bmatrix} R_i(R_i + R_o) & R_i R_o(R_i + R_o) \\ R_i + R_o & R_o(R_i + R_o) \end{bmatrix}$$

(a) From Eq. (10.73), we see that the open-circuit gain of a two-port is equal to the reciprocal of its A parameter. Hence, for the cascaded amplifier,

$$A_v = \dfrac{E_2}{E_1}\bigg|_{I_2=0} = \dfrac{1}{A} = \dfrac{R_i^2 A_0^2}{R_i(R_i + R_o)}$$

(b) From Eq. (10.73), with $Z_L = R_L$, we obtain

$$A_v = \dfrac{E_2}{E_1} = \dfrac{R_L}{A + BR_L} = \dfrac{R_L R_i A_0^2}{(R_i + R_o)(R_o + R_L)}$$

EXAMPLE 19. Find the z matrix of the ladder in Fig. 10.57 by first considering it as the cascade of three two-ports.

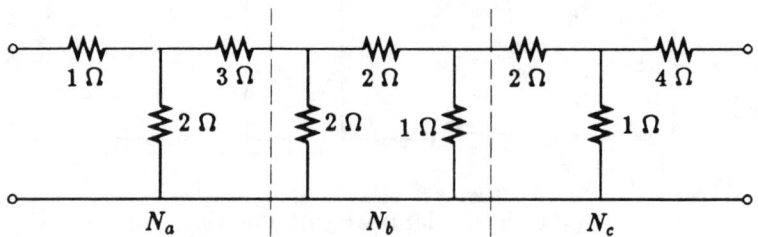

Figure 10.57: A ladder two-port.

SOLUTION

$$[z_a] = \begin{bmatrix} 3 & 2 \\ 2 & 5 \end{bmatrix} \implies [F_a] = \frac{1}{2}\begin{bmatrix} 3 & 11 \\ 1 & 5 \end{bmatrix}$$

$$[y_b] = \begin{bmatrix} 1 & -\frac{1}{2} \\ -\frac{1}{2} & \frac{3}{2} \end{bmatrix} \implies [F_b] = \begin{bmatrix} 3 & 2 \\ \frac{5}{2} & 2 \end{bmatrix}$$

$$[z_c] = \begin{bmatrix} 3 & 1 \\ 1 & 5 \end{bmatrix} \implies [F_c] = \begin{bmatrix} 3 & 14 \\ 1 & 5 \end{bmatrix}$$

$$[F] = [F_a][F_b][F_c] = \begin{bmatrix} \frac{275}{4} & \frac{651}{2} \\ \frac{117}{4} & \frac{277}{2} \end{bmatrix}$$

$$[z] = \frac{1}{117}\begin{bmatrix} 275 & 4 \\ 4 & 554 \end{bmatrix} \Omega$$

10.8 Concluding remarks

In this chapter, we have developed the formalism and techniques of handling two-ports. What we have covered can be extended to

TWO-PORT NETWORKS

networks that have more than two ports. However, networks with more than two ports are used only in more specialized areas, and we will not delve into those topics.

In the next chapter, we will discuss a special two-port—the magnetically coupled coils or mutual inductances.

Problems

10.1 Determine the z matrix of the two-port.

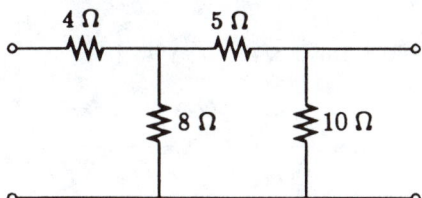

10.2 Determine the z matrix of the two-port.

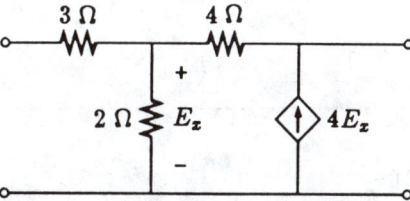

10.3 Determine the y matrix of the two-port.

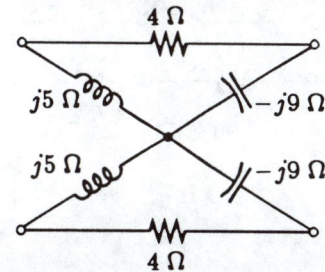

10.4 Determine the z matrix of the two-port.

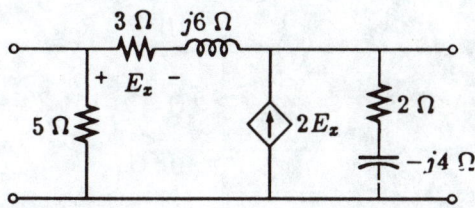

10.5 Determine the z matrix of the two-port.

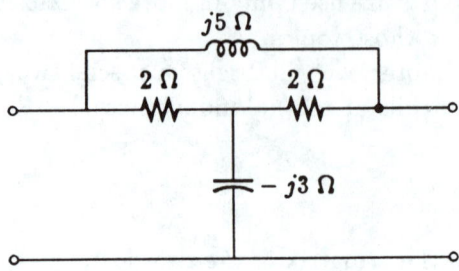

10.6 Determine the y matrix of the two-port.

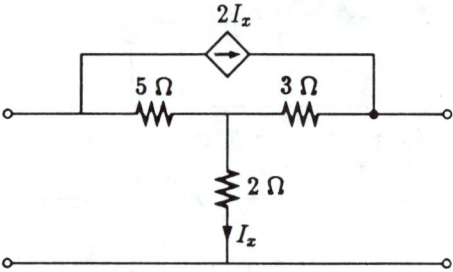

10.7 Determine the z matrix of the two-port.

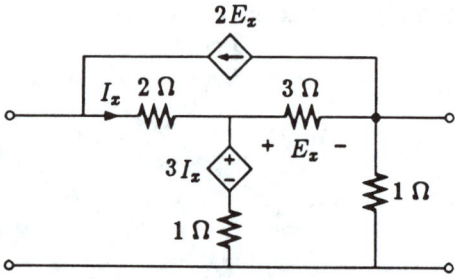

10.8 Determine the h matrix of the two-port.

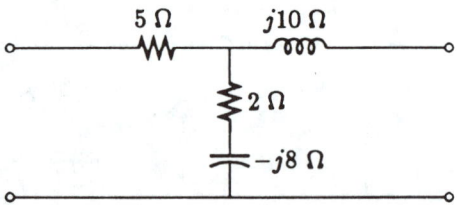

10.9 Determine the g matrix of the two-port.

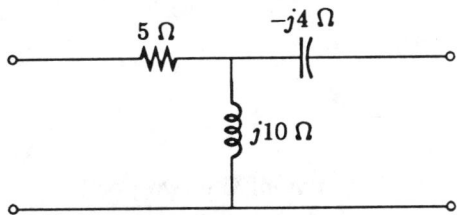

10.10 Determine the g matrix of the two-port.

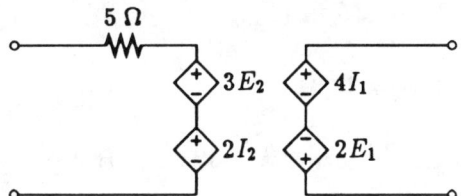

10.11 Determine the g matrix of the two-port.

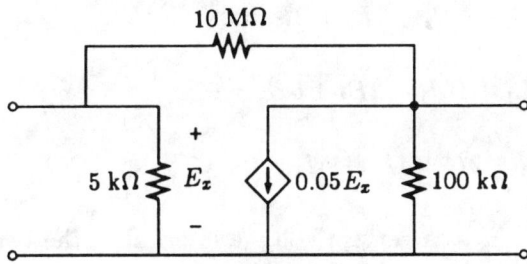

10.12 Determine the F matrix of the two-port.

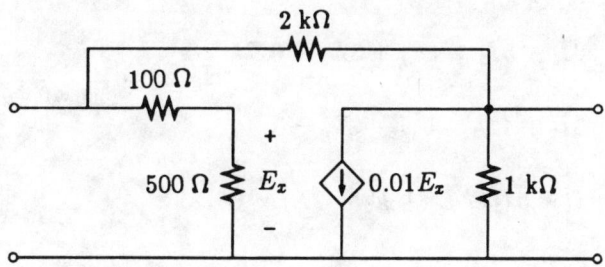

10.13 The port quantities of a two-port must satisfy the equations

$$5E_1 + 6I_2 = I_1$$

$$7E_1 + 8I_2 = E_2$$

Determine the F matrix of the two-port.

10.14 The port quantities of a two-port must satisfy the equations

$$I_1 + 2E_2 = 2I_2 + 4E_1$$

$$E_2 - 2E_1 = 3I_1 - 4I_2$$

Determine the y matrix of the two-port.

10.15 The port quantities of a two-port must satisfy the equations

$$4E_1 + 5E_2 - 5E_3 + 2I_1 = 0$$

$$2E_1 + 10I_1 - 2E_2 + 5E_3 = 0$$

$$E_1 + 2I_1 + 4E_2 + 5I_2 = 0$$

in which E_3 is a certain voltage internal to the two-port. Determine the z matrix of the two-port.

10.16 The port quantities of a two-port must satisfy the equations

$$3I_1 + 5I_2 + 4E_1 + 5E_2 + 6E_x = 0$$

$$5I_1 + 4E_1 - 5E_2 + E_x = 0$$

$$I_1 - 5I_2 - E_1 + E_2 + 2E_x = 0$$

in which E_x is a certain voltage internal to the two-port. Determine the y matrix of the two-port.

TWO-PORT NETWORKS

10.17 The port quantities of a two-port must satisfy the matrix equation

$$\begin{bmatrix} E_1 \\ E_2 \\ 0 \end{bmatrix} = \begin{bmatrix} 1 & 2 & 3 \\ 2 & -5 & 0 \\ -2 & 4 & 1 \end{bmatrix} \begin{bmatrix} I_1 \\ I_2 \\ I_3 \end{bmatrix}$$

in which I_3 is a certain current internal to the two-port. Determine the h matrix of the two-port.

10.18 The y matrix of a two-port is

$$[y] = \begin{bmatrix} 3+j2 & -5 \\ j3 & 4-j2 \end{bmatrix} \mho$$

Obtain its h matrix.

10.19 Determine the y matrix of the two-port.

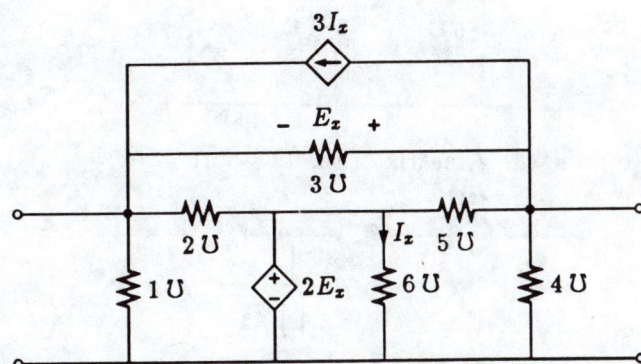

10.20 Determine the z matrix of the two-port.

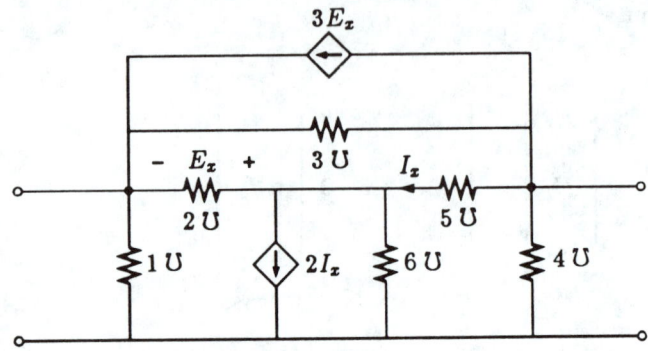

10.21 Determine the y matrix of the two-port.

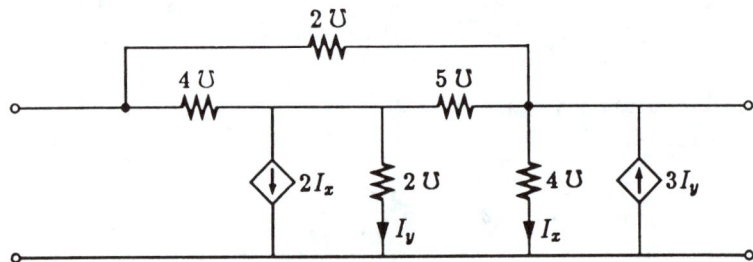

10.22 Determine the y matrix of the two-port.

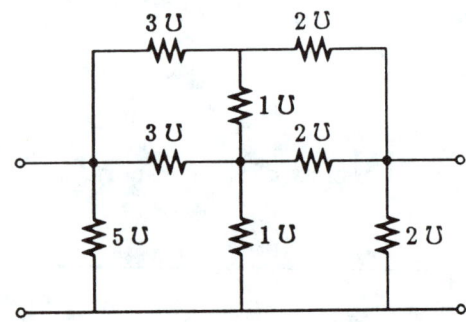

10.23 Determine the z matrix of the two-port.

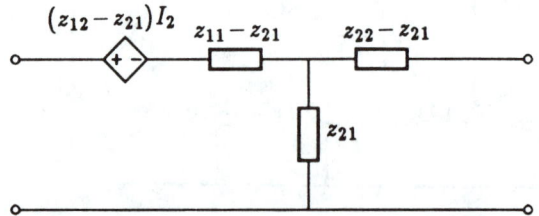

TWO-PORT NETWORKS

10.24 Determine the y matrix of the two-port.

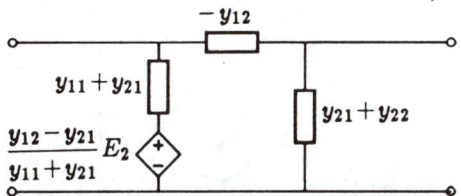

10.25 Make use of the delta-wye and wye-delta transformations to find the equivalent tee and equivalent pi two-ports of the ladder two-port.

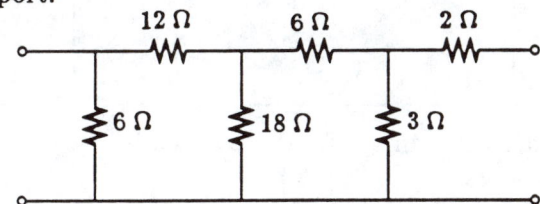

10.26 Obtain the z matrix of the two-port by successively padding the constituent two-ports and inverting their matrices.

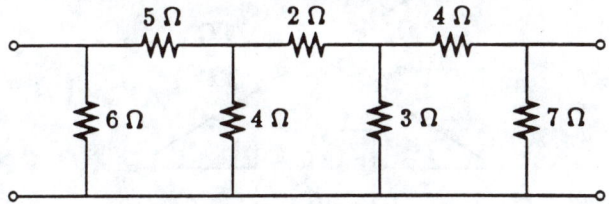

10.27 Determine the y matrix of the two-port.

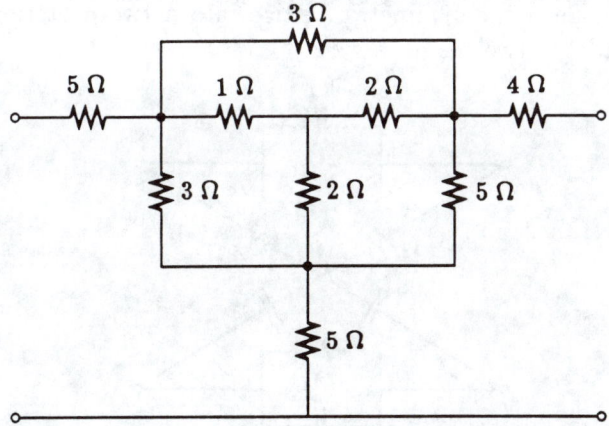

10.28 Determine the y matrix of the two-port.

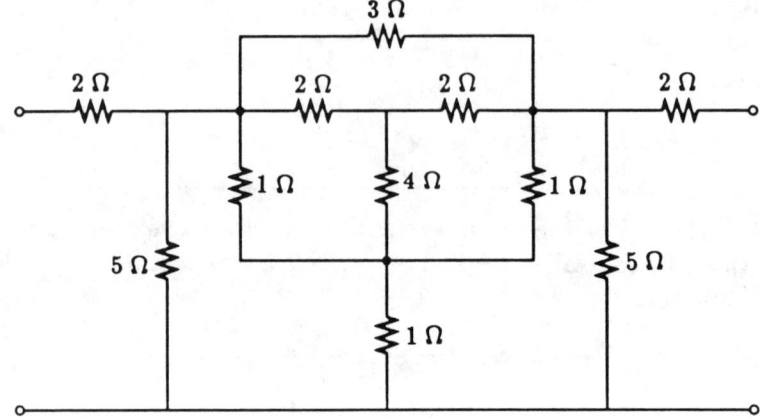

10.29 Unbalance the symmetric lattice into a bridged tee two-port.

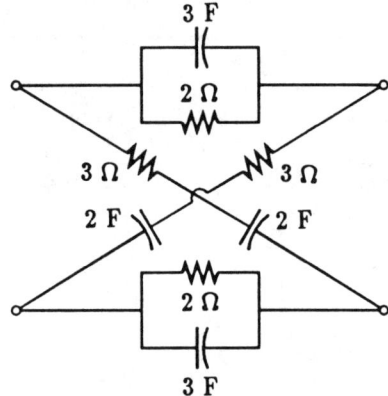

10.30 Unbalance the symmetric lattice into a two-port that is the parallel of two tee networks.

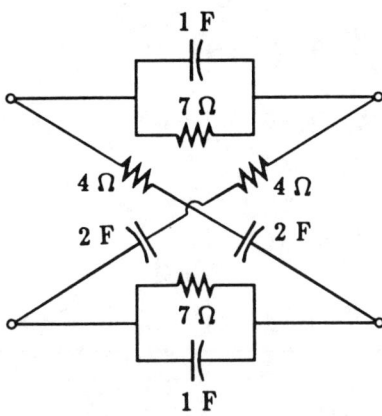

TWO-PORT NETWORKS

10.31 Obtain Z_{in} of the circuit.

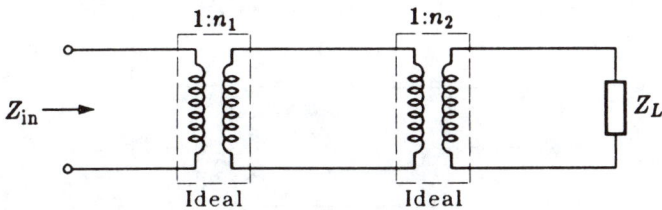

10.32 Determine the y matrix of the two-port.

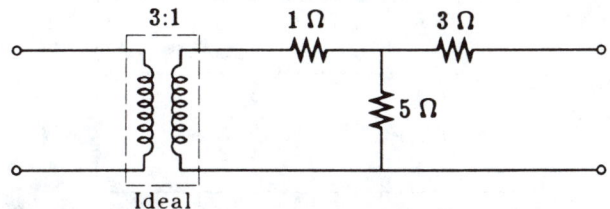

10.33 Determine the z matrix of the two-port.

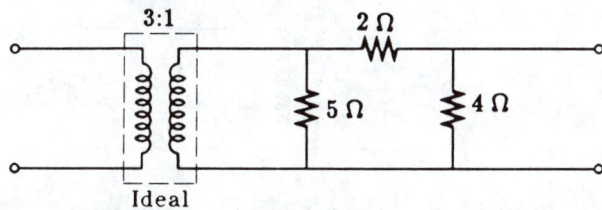

10.34 Determine the z matrix of the two-port.

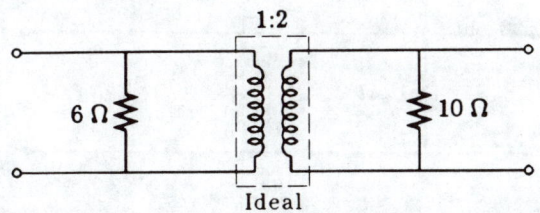

10.35 Obtain Z_{in} of the circuit.

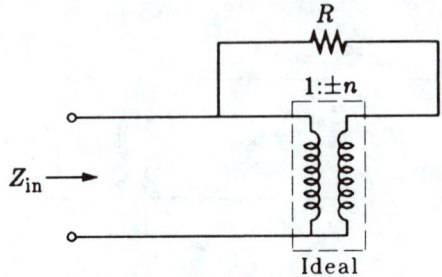

10.36 Obtain Z_{in} of the circuit.

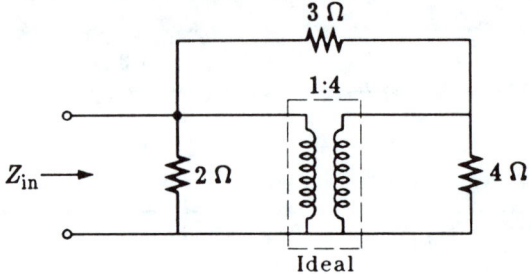

10.37 Obtain Z_{in} of the circuit.

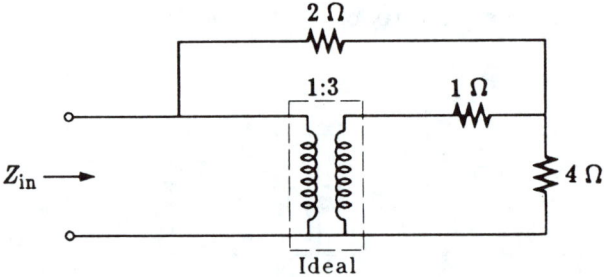

10.38 Determine the z matrix of the two-port.

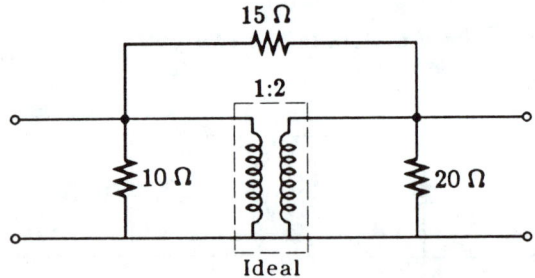

10.39 $E_s = 10\underline{/0°}$ V and $I_s = 5\underline{/30°}$ A. Determine the voltages on the two sides of the ideal transformer, and I_x.

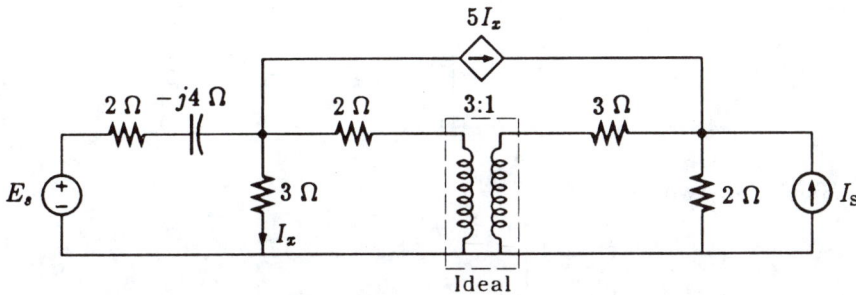

10.40 The turns ratio n of the ideal transformer is variable. Determine its value for which the voltage gain E_2/E_1 is maximum. What is this maximum voltage gain?

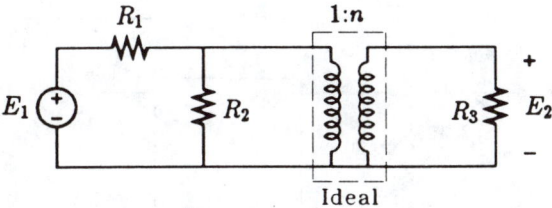

10.41 For two-port N,

$$E_2 = 4E_1 \quad \text{and} \quad I_1 = 4I_2$$

Calculate Z_{in}.

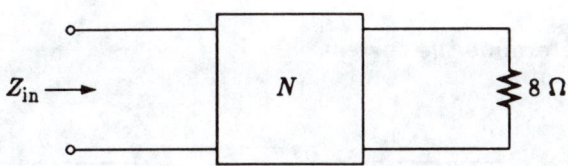

10.42 For two-port N,

$$[h] = \begin{bmatrix} 10 & j2 \\ j10 & 0.1 \end{bmatrix}$$

Determine the voltage gain E_2/E_1.

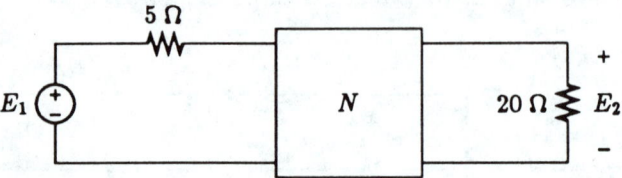

10.43 For two-port N,

$$[z] = \begin{bmatrix} 2 & 3 \\ 5 & 4 \end{bmatrix} \Omega$$

Determine Z_{in}.

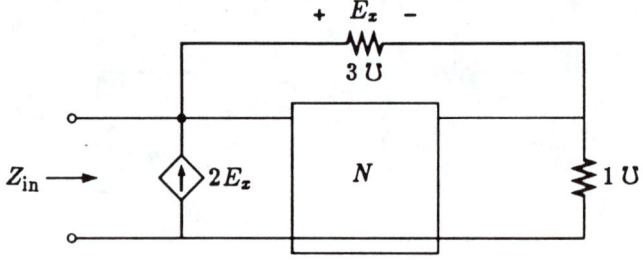

10.44 For two-port N,

$$[h] = \begin{bmatrix} 10 & j2 \\ j10 & 0.1 \end{bmatrix}$$

Determine the current gain I_L/I_s.

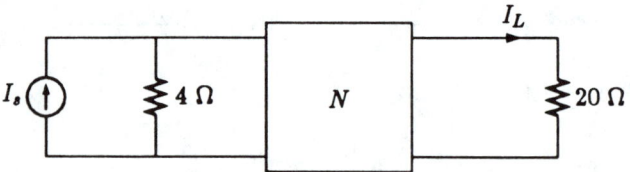

TWO-PORT NETWORKS

10.45 For two-port N,

$$[y] = \begin{bmatrix} 2 & -4 \\ 8 & 10 \end{bmatrix} \mho$$

Determine the impedance matrix of two-port N'.

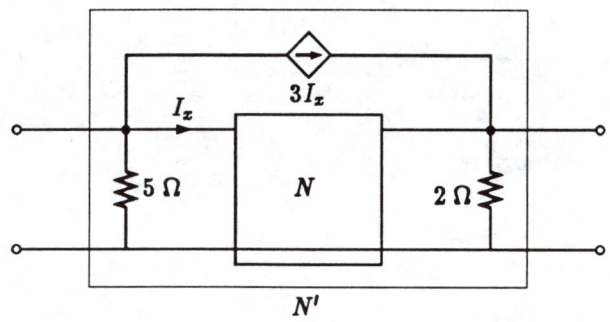

10.46 For two-port N,

$$E_1 = 5I_1 + 10E_2$$

$$I_2 = 4I_1 - 2E_2$$

Determine E_1, E_2, I_1, and I_2.

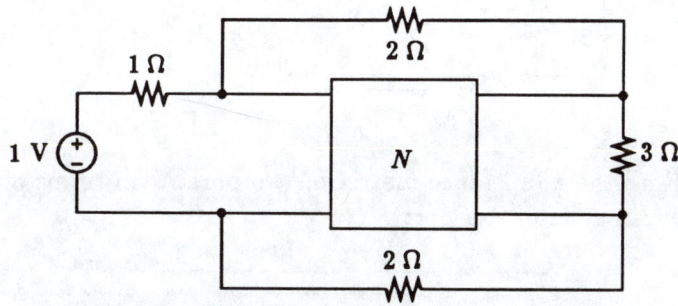

10.47 The g matrix of two-port N is

$$[g] = \begin{bmatrix} 5 & 2 \\ 7 & 9 \end{bmatrix}$$

Determine the admittance matrix of two-port N'.

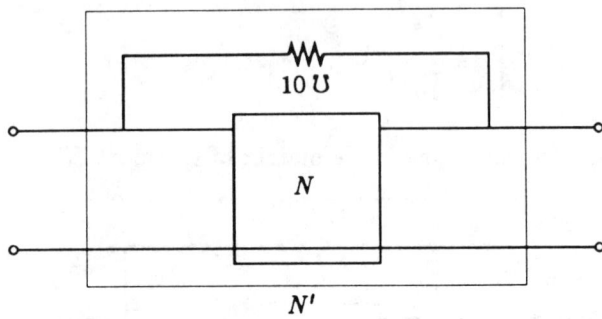

10.48 For two-port N,

$$E_1 = 2E_2 - 3I_2$$

$$I_1 = 4E_2 + I_2$$

Determine the admittance matrix of two-port N'.

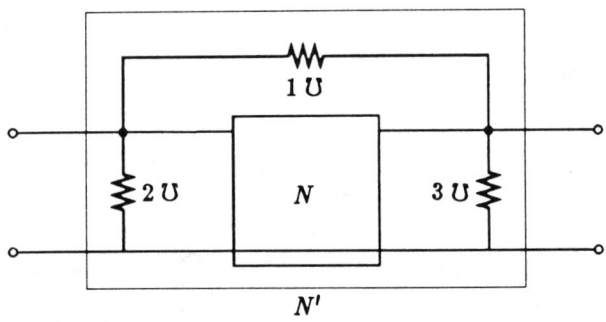

10.49 Find the admittance matrix of two-port N' in terms of the y parameters of two-port N.

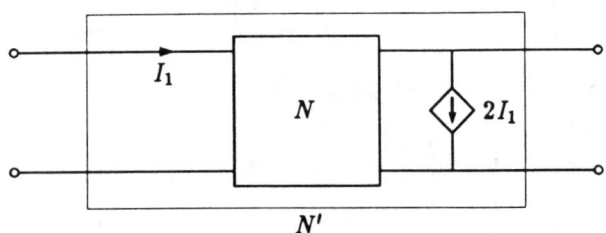

TWO-PORT NETWORKS

10.50 In the following arrangement, assume that all matrices of two-port N are given. Express the corresponding matrices of N' in terms of those of N.

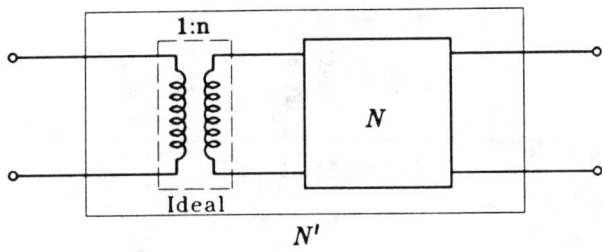

10.51 Determine the y matrix of the two-port.

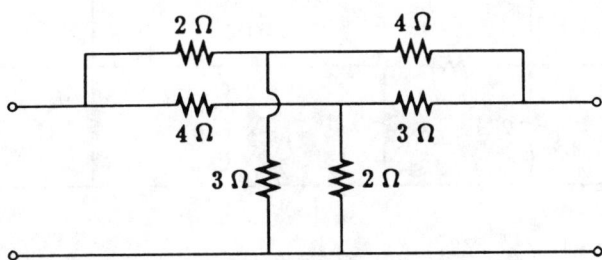

10.52 Determine the z matrix of the two-port. $\omega = 500$ rad/sec.

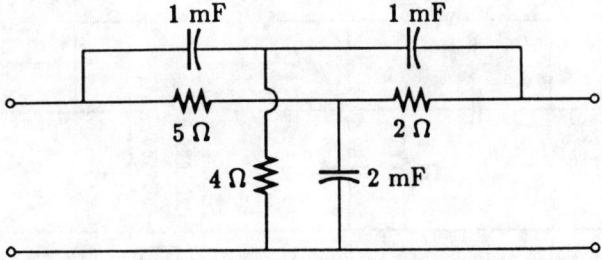

10.53 Determine the h matrix of the two-port.

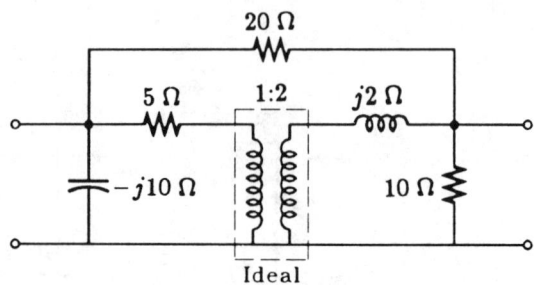

10.54 Determine the y matrix of the two-port.

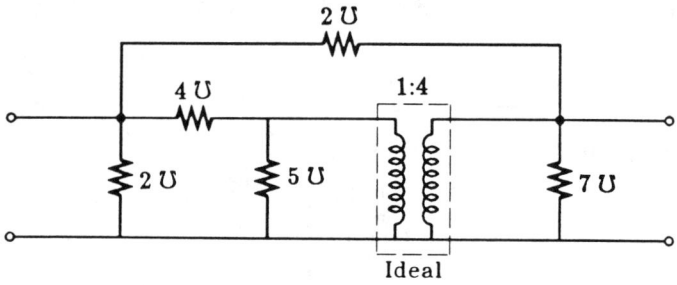

10.55 Determine the y matrix of the two-port.

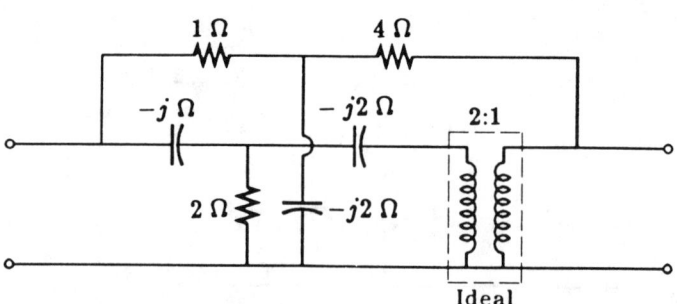

10.56 Determine the y matrix of the two-port.

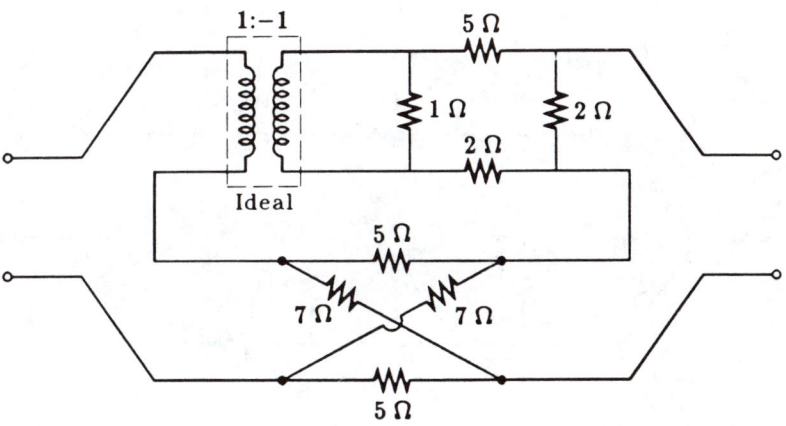

10.57 For the individual two-ports,

$$[g_a] = \begin{bmatrix} 5 & 1 \\ 2 & 4 \end{bmatrix} \qquad [y_b] = \begin{bmatrix} 5 & -2 \\ -1 & 4 \end{bmatrix} \text{℧} \qquad [z_c] = \begin{bmatrix} 4 & 1 \\ 2 & 3 \end{bmatrix} \text{Ω}$$

Determine the z matrix of the overall two-port.

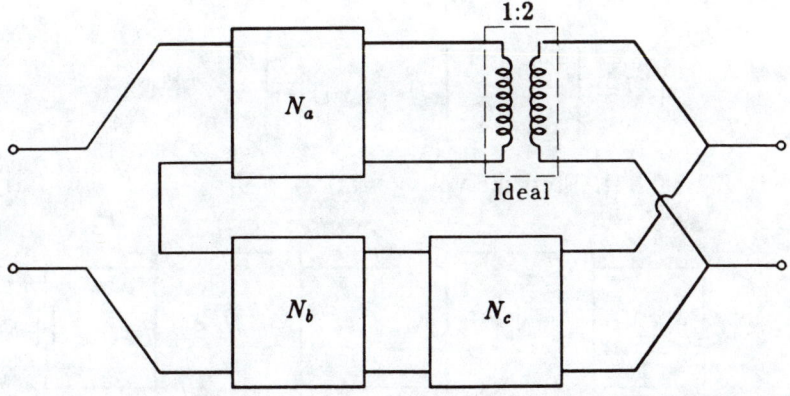

10.58 Determine the h matrix of the two-port.

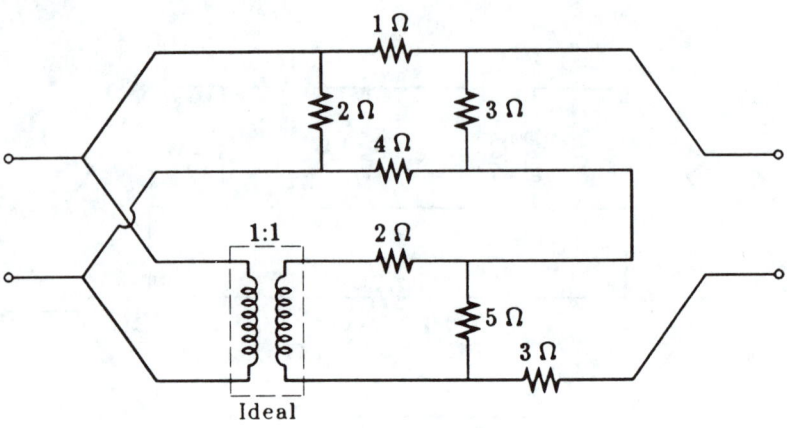

10.59 For the individual two-ports,

$$[y_a] = \begin{bmatrix} 1 & 2 \\ 3 & 4 \end{bmatrix} \mho \qquad [z_b] = \begin{bmatrix} 1 & 0 \\ 5 & 2 \end{bmatrix} \Omega$$

Determine the chain matrix of the overall two-port.

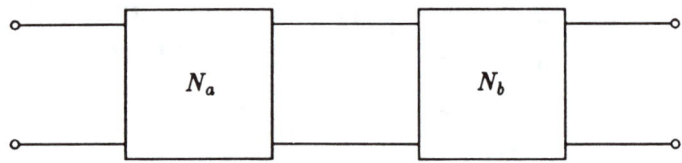

10.60 For the individual two-ports,

$$[y_a] = \begin{bmatrix} 5 & -1 \\ -2 & 4 \end{bmatrix} \mho \qquad [z_b] = \begin{bmatrix} 4 & 2 \\ 2 & 8 \end{bmatrix} \Omega$$

$Z_L = 0.5\ \Omega$. Find the voltage ratio E_2/E_1.

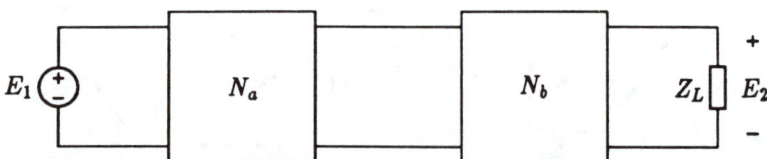

10.61 The resistors are 1-ohm each. Find a symmetric lattice that is equivalent to the two-port.

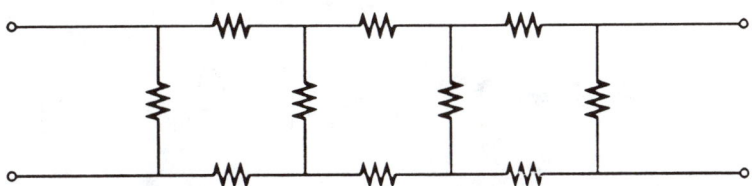

10.62 Divide the two-port into three identical sections. Find the chain matrix of each section. Then obtain the chain matrix of the overall two-port. Finally, obtain the z matrix of the overall two-port.

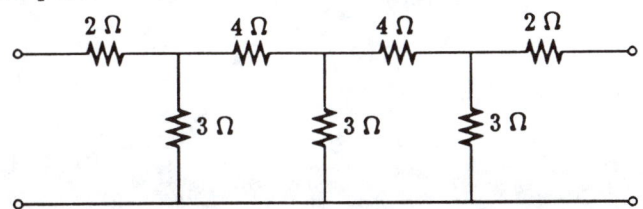

10.63 Divide the two-port into three identical sections. Find the chain matrix of each section. Then obtain the chain matrix of the overall two-port. Finally, obtain the y matrix of the overall two-port.

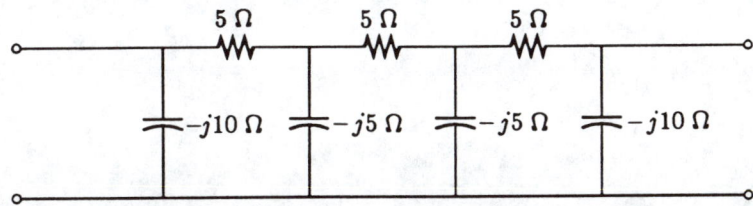

Chapter 11

The Mutual Inductance

In Chapter 5, we introduced the concept of the inductance to relate the voltage and current of an electromagnetic energy storing device—typically a coil, with or without a core. In such a device, a current i produces a flux ϕ. In a linear medium, $\phi \propto i$. The flux linkage, $\Phi = N\phi$, refers to the effects of ϕ being multiplied by an equivalent number of turns of a winding. Faraday's law applied to such a device gives a voltage

$$e = \frac{d\Phi}{dt} = \frac{d(N\phi)}{dt} = L\frac{di}{dt} \qquad (11.1)$$

The inductance

$$L = \frac{N\phi}{i}$$

is the *self inductance* and serves to relate the voltage and current of an inductor. The self inductance is the inductance of a one-port. There is only one current, one voltage, and one inductance.

11.1 The magnetic coupling between two coils

Faraday's law requires that any time the flux in a winding changes, a voltage is induced in the winding. In other words,

$$e = \frac{d\Phi}{dt} = \frac{d(N\phi)}{dt} \qquad (11.2)$$

must be true no matter how ϕ is produced.

When two coils are placed in proximity to each other, the flux produced by the current in one coil may link part of the other coil. The varying of this part of the flux will induce a voltage in the other coil. This effect is the *mutual* electromagnetic coupling of the two coils. Fig. 11.1 shows two coils that are placed in the vicinity of each other. In Fig. 11.1(a), only i_1 is present. Current i_1 produces a flux ϕ_1. Flux ϕ_1 consists of two parts. One part, ϕ_{11}, links only coil 1.[1] The other part, ϕ_{21}, also links coil 2.

$$\phi_1 = \phi_{11} + \phi_{21} \tag{11.3}$$

Under this arrangement, we have

$$e_1 = \frac{d(N_1\phi_1)}{dt} \tag{11.4}$$

$$e_2 = \frac{d(N_2\phi_{21})}{dt} \tag{11.5}$$

In Fig. 11.1(b), only i_2 is present. Current i_2 produces a flux ϕ_2. Flux ϕ_2 may be divided into two parts. One part, ϕ_{22}, links only coil 2. The other part, ϕ_{12}, also links coil 1.

$$\phi_2 = \phi_{22} + \phi_{12} \tag{11.6}$$

Under this arrangement, we have

$$e_2 = \frac{d(N_2\phi_2)}{dt} \tag{11.7}$$

$$e_1 = \frac{d(N_1\phi_{12})}{dt} \tag{11.8}$$

In Fig. 11.1(c), both i_1 and i_2 are present. Combining Eqs. (11.4) and (11.8) as well as Eqs. (11.5) and (11.7), we get

$$e_1 = \frac{d(N_1\phi_1)}{dt} + \frac{d(N_1\phi_{12})}{dt} \tag{11.9}$$

$$e_2 = \frac{d(N_2\phi_{21})}{dt} + \frac{d(N_2\phi_2)}{dt} \tag{11.10}$$

---

[1] This component of ϕ_1 is sometimes called the *leakage* flux because it is viewed as that part of ϕ_1 that failed to also link coil 2.

THE MUTUAL INDUCTANCE

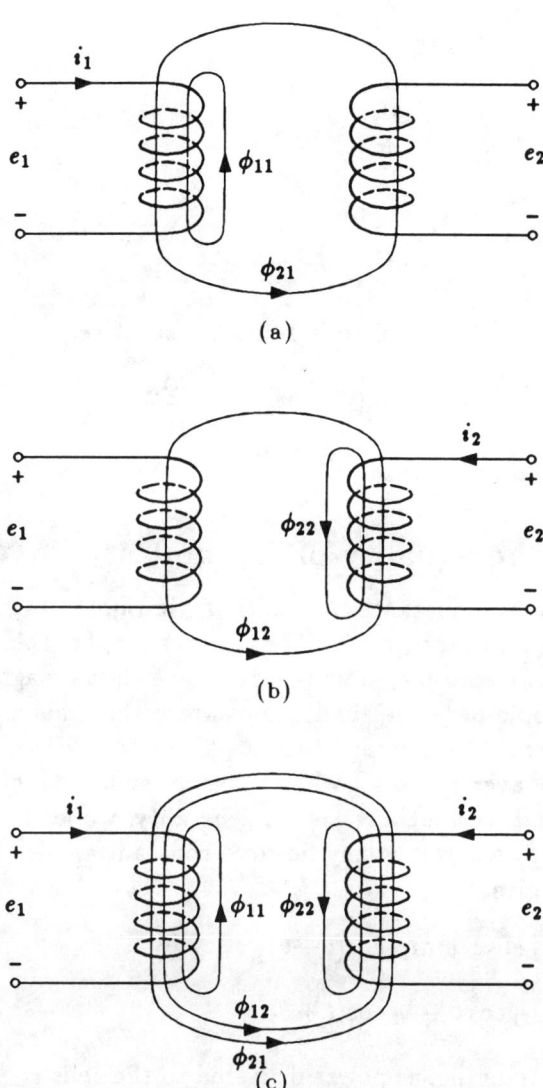

Figure 11.1: A pair of coupled coils.

Since both ϕ_1 and ϕ_{21} are proportional to i_1 and both ϕ_2 and ϕ_{12} are proportional to i_2, Eqs. (11.9) and (11.10) may be written as

$$e_1 = L_1 \frac{di_1}{dt} + M_{12} \frac{di_2}{dt} \tag{11.11}$$

$$e_2 = M_{21} \frac{di_1}{dt} + L_2 \frac{di_2}{dt} \tag{11.12}$$

in which

$$L_1 = \frac{N_1 \phi_1}{i_1} \quad \text{and} \quad L_2 = \frac{N_2 \phi_2}{i_2} \tag{11.13}$$

are the *self inductances* of coils 1 and 2 respectively and

$$M_{12} = \frac{N_1 \phi_{12}}{i_2} \quad \text{and} \quad M_{21} = \frac{N_2 \phi_{21}}{i_1} \tag{11.14}$$

are the *mutual inductances*.

11.1.1 The equality of M_{12} and M_{21}

The two mutual inductances M_{12} and M_{21} in Eq. (11.14) are equal to each other in a linear medium. The rigorous proof of this assertion is an element of electromagnetic theory. Since electromagnetic theory is not our topic here, we shall demonstrate that this assertion can also be inferred if we accept that a pair of coupled coils does not consume any average power when it carries sinusoidal currents. For simplicity and without any loss of generality, we let $i_1 = \sin t$ and $i_2 = \sin(t + \theta)$ where θ is any nonzero constant angle. Eqs. (11.11) and (11.12) give

$$e_1 = L_1 \cos t + M_{12} \cos(t + \theta) \tag{11.15}$$

$$e_2 = M_{21} \cos t + L_2 \cos(t + \theta) \tag{11.16}$$

The total instantaneous power delivered to the coils is

$$p(t) = L_1 \cos t \sin t + M_{12} \cos(t + \theta) \sin t$$

$$+ M_{21} \cos t \sin(t + \theta) + L_2 \cos(t + \theta) \sin(t + \theta)$$

$$= L_1 \cos t \sin t + M_{12} \cos \theta \cos t \sin t - M_{12} \sin \theta \sin^2 t$$

THE MUTUAL INDUCTANCE

$$+ M_{21} \cos\theta \sin t \cos t + M_{21} \sin\theta \cos^2 t + L_2 \cos(t+\theta)\sin(t+\theta) \quad (11.17)$$

When we integrate $p(t)$ over a complete period of 2π, the only terms that contribute to the integral are the sine and cosine squared terms. The average power is given by

$$P_{\text{avg}} = \frac{\sin\theta}{2\pi} \int_0^{2\pi} \left[M_{21} \cos^2 t - M_{12} \sin^2 t \right] dt$$

$$= \frac{\sin\theta}{2}(M_{21} - M_{12}) \quad (11.18)$$

Hence if we accept the assertion that a pair of coupled coils is lossless, then $P_{\text{avg}} = 0$ and

$$M_{12} = M_{21} = M \quad (11.19)$$

So, we may drop the subscript for the mutual inductances. Also, networks including mutual inductances are *reciprocal*.

11.1.2 The signs of the mutual terms

In Fig. 11.1, if the pitch or direction of winding of one of the coils is reversed, as shown in Fig. 11.2, we see now ϕ_{12} and ϕ_{21} are opposing

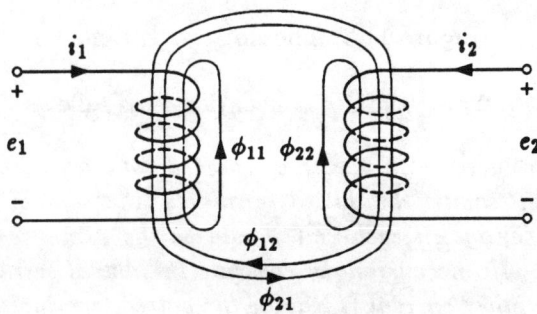

Figure 11.2: A pair of coupled coils with one winding reversed.

each other instead of aiding each other as in Fig. 11.1. Thus, we have

$$e_1 = \frac{d(N_1 \phi_1)}{dt} - \frac{d(N_1 \phi_{12})}{dt} \quad (11.20)$$

$$e_2 = -\frac{d(N_2\phi_{21})}{dt} + \frac{d(N_2\phi_2)}{dt} \qquad (11.21)$$

and

$$e_1 = L_1\frac{di_1}{dt} - M_{12}\frac{di_2}{dt} \qquad (11.22)$$

$$e_2 = -M_{21}\frac{di_1}{dt} + L_2\frac{di_2}{dt} \qquad (11.23)$$

Our discussion regarding the equality of M_{12} and M_{21} in the previous subsection is not affected by this sign reversal.

In practice, it is not feasible to show the relative directions of the mutual flux in coupled coils in a circuit diagram. The information regarding the signs of the mutual terms is conveyed by the *dot convention*. The two coupled coils of Fig. 11.1 are represented by those in Fig. 11.3(a), and those in Fig. 11.2 by that in Fig. 11.3(b).

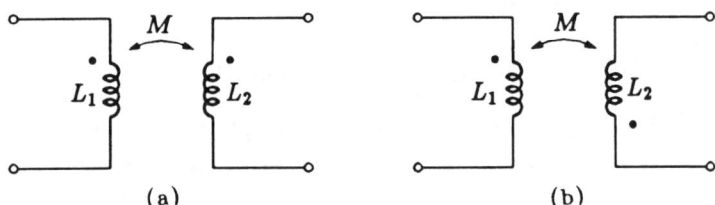

Figure 11.3: The dot convention.

The rule governing the dot convention is as follows:

If the two currents in a pair of coupled coils are either both entering or both leaving the dotted terminals, the sign of the mutual term of the voltage in each coil should be the same as the sign of the self term. If one current is entering the dotted terminal in one coil while the other current is leaving the dotted terminal of the other coil, the sign of the mutual term of the voltage in each coil should be the opposite of the self term.

For example, in Fig. 11.4(a), if we follow the dot convention, we will have

$$e_a = L_a\frac{di_a}{dt} - M\frac{di_b}{dt} \qquad (11.24)$$

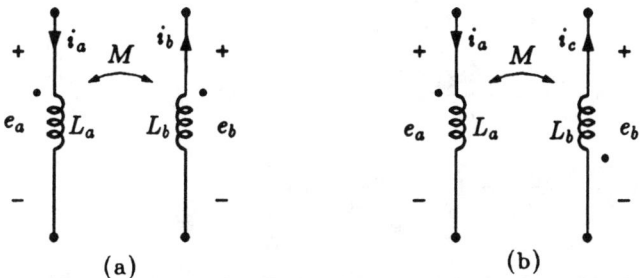

Figure 11.4: Two pairs of couples coils with dots.

$$e_b = M\frac{di_a}{dt} - L_b\frac{di_b}{dt} \tag{11.25}$$

For the same pair of coils with one of the currents reversed, as in Fig. 11.4(b), we will have

$$e_a = L_a\frac{di_a}{dt} + M\frac{di_c}{dt} \tag{11.26}$$

$$e_b = -M\frac{di_a}{dt} - L_b\frac{di_c}{dt} \tag{11.27}$$

EXAMPLE 1. Write the mesh equations for the circuit in Fig. 11.5.

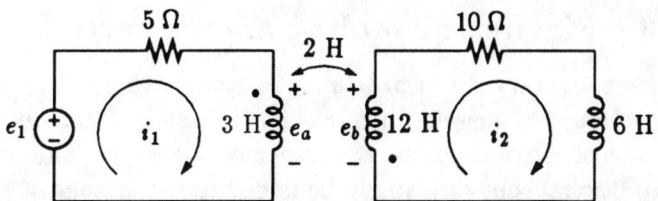

Figure 11.5: Circuit used in Example 1.

SOLUTION For the coupled coils, we have

$$e_a = 3\frac{di_1}{dt} + 2\frac{di_2}{dt}$$

$$e_b = -12\frac{di_2}{dt} - 2\frac{di_1}{dt}$$

Hence, the mesh equations are

$$5i_1 + 3\frac{di_1}{dt} + 2\frac{di_2}{dt} = e_1$$

$$12\frac{di_2}{dt} + 2\frac{di_1}{dt} + 10i_2 + 6\frac{di_2}{dt} = 0$$

EXERCISE

11.1.1 We wish to represent the coupled coils on the left by its circuit symbols shown on the right in the following diagram. Where should the dots be placed?

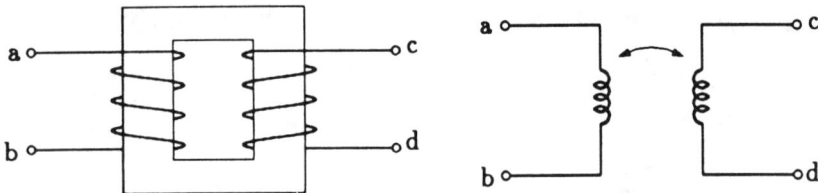

Ans. Dots should be placed either both at the top terminals or both at the bottom terminals of the coils.

11.1.3 Magnetic coupling among several coils

If we place several coils in proximity to one another, there could be magnetic coupling among all coils. Thus, each coil may experience the effects of currents in several other or all coils. The effects of these additional coils can simply be added to the voltage of each coil until all mutual effects have been included.

To illustrate the procedure for including the magnetic effects on a coil due to several other coils, we look at the situation in Fig. 11.6. There are four coils and magnetic coupling is present between any two coils. We can write

$$e_1 = L_1\frac{di_1}{dt} - M_{12}\frac{di_2}{dt} + M_{13}\frac{di_3}{dt} - M_{14}\frac{di_4}{dt}$$

$$e_2 = L_2\frac{di_2}{dt} - M_{12}\frac{di_1}{dt} - M_{23}\frac{di_3}{dt} + M_{24}\frac{di_4}{dt}$$

THE MUTUAL INDUCTANCE

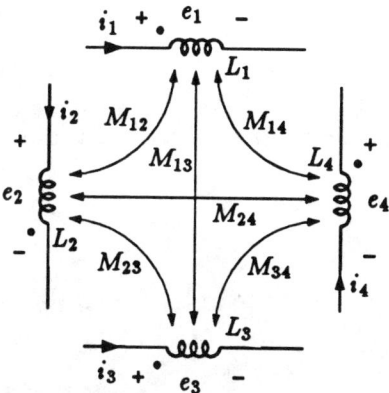

Figure 11.6: Mutual coupling among four coils.

$$e_3 = L_3 \frac{di_3}{dt} + M_{13}\frac{di_1}{dt} - M_{23}\frac{di_2}{dt} - M_{34}\frac{di_4}{dt}$$

$$e_4 = -L_4 \frac{di_4}{dt} + M_{14}\frac{di_1}{dt} - M_{24}\frac{di_2}{dt} + M_{34}\frac{di_3}{dt}$$

EXAMPLE 2. Find the equivalent inductance L_{eq} between terminals A and B for the circuit in Fig. 11.7.

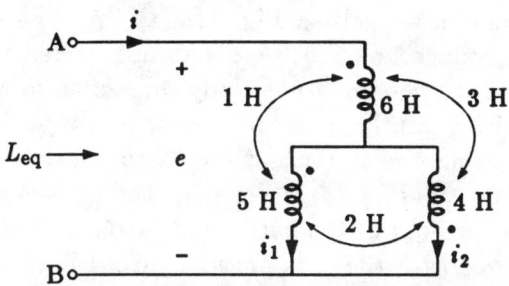

Figure 11.7: Circuit used in Example 2.

SOLUTION We write the KVL equations for the circuit. We get

$$e = 6\frac{di}{dt} + 1\frac{di_1}{dt} - 3\frac{di_2}{dt} + 5\frac{di_1}{dt} + 1\frac{di}{dt} - 2\frac{di_2}{dt}$$

$$e = 6\frac{di}{dt} + 1\frac{di_1}{dt} - 3\frac{di_2}{dt} + 4\frac{di_2}{dt} - 3\frac{di}{dt} - 2\frac{di_1}{dt}$$

KCL requires

$$i = i_1 + i_2$$

Eliminating i_1 and i_2 from these three equations, we get

$$e = \frac{49}{13}\frac{di}{dt}$$

Hence,

$$L_{eq} = \frac{49}{13} \text{ H}$$

It is not always possible to place one single set of dots on more than two coils such that the dot convention can be followed to give all correct signs for all mutual terms. There are two methods of accommodating this type of situation. One is to make the mutual term itself negative when the dots are placed contrary to where that ought to be. The other is to use one pair of symbols for each mutual inductance. We shall adopt the second method.

Fig. 11.8 is an example of this method of indicating the relative signs of the mutual and self terms when there are more than two coils. There are three coils in Fig. 11.8(a). A close examination of the fluxes produced by the three assumed currents will readily reveal that it is not possible to use only three dots to represent the correct polarities of all three mutual fluxes. For M_{12}, the dots must be placed one at the top and one at the bottom terminals of L_1 and L_2. We use solid dots for M_{12}. For M_{23}, the dots must be placed both at the top or both at the bottom of L_2 and L_3. We use hollow dots for M_{23}. For M_{13}, the dots must be placed both at the top or both at the bottom of L_1 and L_3. We use crosses for M_{13}. Thus, we have the markings in Fig. 11.8(b).

THE MUTUAL INDUCTANCE

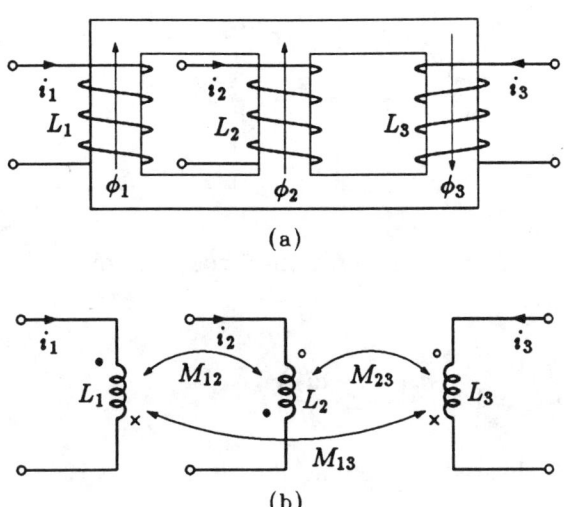

Figure 11.8: Multiple sets of dots for three coupled coils.

EXERCISE

11.1.2 Find the equivalent inductance seen between terminals A and B in the following circuit.

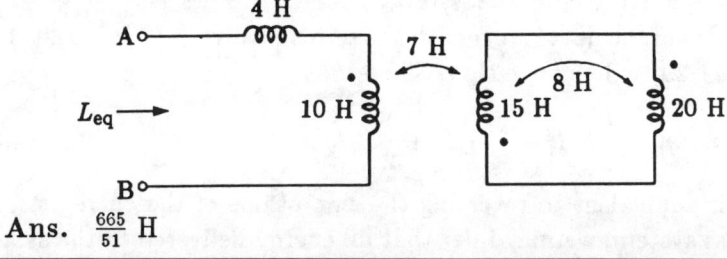

Ans. $\frac{665}{51}$ H

11.1.4 Energy delivered to a pair of coupled coils

We refer to Fig. 11.9. With the assumption that $M_{12} = M_{21} = M$, the instantaneous power delivered to each terminal pair is

$$p_1 = e_1 i_1 = \left(L_1 \frac{di_1}{dt} + M \frac{di_2}{dt} \right) i_1 \qquad (11.28)$$

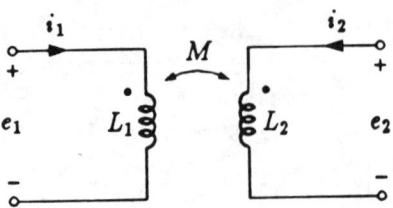

Figure 11.9: A pair of coupled coils.

and

$$p_2 = e_2 i_2 = \left(M\frac{di_1}{dt} + L_2\frac{di_2}{dt}\right) i_2 \qquad (11.29)$$

The energy delivered to the pair of coils at $t = t_0$ is

$$w(t_0) = \int_{-\infty}^{t_0} (p_1 + p_2)\, dt$$

$$= \int_{t=-\infty}^{t=t_0} \left[L_1 i_1\, di_1 + M\, d(i_1 i_2) + L_2 i_2\, di_2\right] \qquad (11.30)$$

Since $w(-\infty) = 0$, we can say that

$$w(t_0) = \frac{1}{2}L_1 I_1^2 + M I_1 I_2 + \frac{1}{2}L_2 I_2^2 \qquad (11.31)$$

where $I_1 = i_1(t_0)$ and $I_2 = i_2(t_0)$.

If one of the dots is reversed, the terms involving M in Eqs. (11.28) and (11.29) will change sign. Then

$$w(t_0) = \frac{1}{2}L_1 I_1^2 - M I_1 I_2 + \frac{1}{2}L_2 I_2^2 \qquad (11.32)$$

This is equivalent to reversing the sign of one of the currents. In a lossless system, we may infer that all energy delivered to the system is stored. Similar to the self inductances, in the case of mutual inductances, the energy is stored in the electromagnetic field.

The energy delivered to a lossless system must be nonnegative for any values of I_1 and I_2, positive or negative. Since the middle term of either Eq. (11.31) or Eq. (11.32) can be negative, the quantity

$$w = \frac{1}{2}L_1 I_1^2 - M|I_1 I_2| + \frac{1}{2}L_2 I_2^2 \qquad (11.33)$$

THE MUTUAL INDUCTANCE

must be nonnegative. Rewrite Eq. (11.33) as

$$w = \left(\sqrt{\frac{L_1}{2}}I_1 - \sqrt{\frac{L_2}{2}}I_2\right)^2 + (\sqrt{L_1 L_2} - M)|I_1 I_2| \qquad (11.34)$$

The first quantity is always nonnegative, and it can be made zero by choosing appropriate values for I_1 and I_2. Hence, for w to be nonnegative, it is necessary that

$$\sqrt{L_1 L_2} - M \geq 0 \quad \Longrightarrow \quad M \leq \sqrt{L_1 L_2} \qquad (11.35)$$

Therefore, there is an upper bound for the value of the mutual inductance associated with two self inductances. This upper bound is the geometric mean of the two self inductances.

11.1.5 Coefficient of coupling

The degree of electromagnetic coupling between two coils is usually measured by the *coefficient of coupling*, k, where

$$k = \frac{M}{\sqrt{L_1 L_2}} \qquad (11.36)$$

From Eqs. (11.13) and (11.14), we see

$$k = \sqrt{\frac{\phi_{21}\phi_{12}}{\phi_1 \phi_2}} \qquad (11.37)$$

Hence the coefficient of coupling is a measure of how high fractions of the fluxes produced by the currents that also link the other coils. In other words, it is an indication of how high ϕ_{21} is a part of ϕ_1, and ϕ_{12} of ϕ_2. Physically, this coefficient is a measure of how closely the two magnetic paths assumed by the fields produced by the two currents coincide with each other.

Clearly, $k \leq 1$. If two coils have a high k (close to unity), we shall describe them to be *closely* or *tightly* coupled. If k is low (close to zero), they are *loosely* coupled.

11.1.6 Mutual inductance in ac circuits

In ac steady-state analysis, in which electrical quantities are represented by phasors, the effects of electromagnetic coupling is represented by the mutual reactance or mutual impedance exactly like the self-inductance is represented by its reactance or impedance. Thus, the pair of coupled coils in Fig. 11.10(a) is represented by the circuit in Fig. 11.10(b). The dot convention applies to the signs of the self

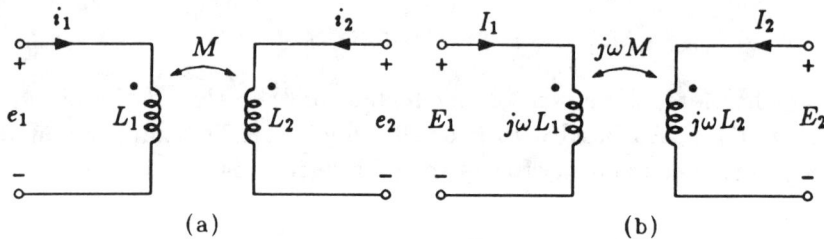

Figure 11.10: (a) A pair of coupled coils. (b) Circuit in (a) represented in the phasor domain.

and mutual terms in a voltage equation exactly as it was described in the time domain.

In the phasor domain, the circuit in Fig. 11.10(b) gives the following equations:

$$E_1 = j\omega L_1 I_1 + j\omega M I_2 = jX_1 I_1 + jX_m I_2 \tag{11.38}$$

$$E_2 = j\omega M I_1 + j\omega L_2 I_2 = jX_m I_1 + jX_2 I_2 \tag{11.39}$$

The circuit may also be treated as a two-port whose z matrix is

$$[z] = \begin{bmatrix} j\omega L_1 & j\omega M \\ j\omega M & j\omega L_2 \end{bmatrix} = j \begin{bmatrix} X_1 & X_m \\ X_m & X_2 \end{bmatrix} \tag{11.40}$$

11.2 Illustrative examples

In this section, we shall present several examples of how circuit problems with mutual inductances are typically handled. We shall dwell on the analysis of circuits in the phasor domain.

EXAMPLE 3. Find the equivalent impedances Z_a and Z_b of the series connected inductances with mutual coupling as shown in Fig. 11.11.

THE MUTUAL INDUCTANCE

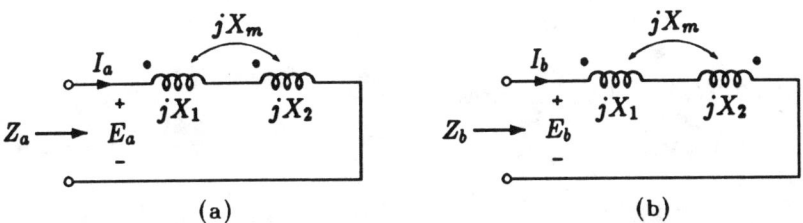

Figure 11.11: Series inductances with mutual coupling.

SOLUTION For (a),

$$E_a = jX_1 I_a + jX_m I_a + jX_2 I_a + jX_m I_a$$

Hence

$$Z_a = \frac{E_a}{I_a} = jX_1 + jX_2 + j2X_m$$

For (b),

$$E_b = jX_1 I_b - jX_m I_b + jX_2 I_b - jX_m I_b$$

Hence

$$Z_b = \frac{E_b}{I_b} = jX_1 + jX_2 - j2X_m$$

EXAMPLE 4. Find the equivalent impedance of the parallel combination of two inductances with mutual coupling as shown in Fig. 11.12.

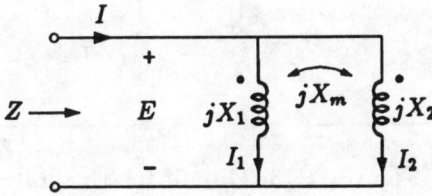

Figure 11.12: Parallel inductances with mutual coupling.

SOLUTION KVL requires

$$E = jX_1 I_1 + jX_m I_2$$

$$E = jX_2 I_2 + jX_m I_1$$

KCL requires

$$I = I_1 + I_2$$

Eliminating I_1 and I_2, we get

$$Z = \frac{E}{I} = j\frac{X_1 X_2 - X_m^2}{X_1 + X_2 - 2X_m}$$

EXERCISE

11.2.1 What would be the equivalent impedance if one of the dots of the coupled coils in Fig. 11.12 is placed at the opposite end of the coil?

Ans. $j\dfrac{X_1 X_2 - X_m^2}{X_1 + X_2 + 2X_m}$

EXAMPLE 5. Derive the expression for the input impedance Z_{in} of the circuit in Fig. 11.13.

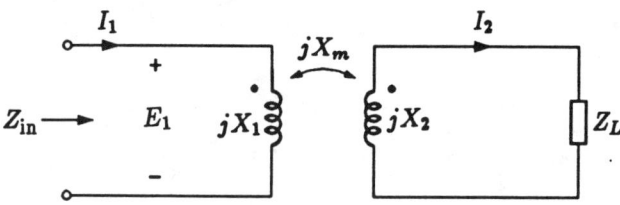

Figure 11.13: Circuit for Example 5.

THE MUTUAL INDUCTANCE

SOLUTION The mesh equations are

$$E_1 = jX_1 I_1 - jX_m I_2$$

$$jX_2 I_2 - jX_m I_1 + Z_L I_2 = 0$$

Eliminating I_2, we get

$$Z_{in} = \frac{E_1}{I_1} = jX_1 + \frac{X_m^2}{jX_2 + Z_L}$$

EXERCISE

11.2.2 What would be the input impedance Z_{in} in Fig. 11.13 if one of the dots is changed to the other terminal?

Ans. Unchanged

EXAMPLE 6. Determine E_2 for the circuit in Fig. 11.14.

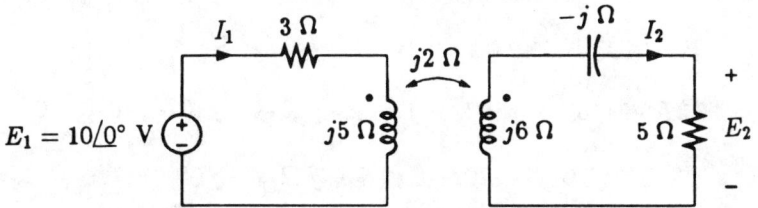

Figure 11.14: Circuit for Example 6.

SOLUTION The mesh equations are

$$3I_1 + j5I_1 - j2I_2 = 10$$

$$j6I_2 - j2I_1 - jI_2 + 5I_2 = 0$$

Solving, we obtain

$$I_1 = 1.039 - j1.406 \quad \text{and} \quad I_2 = 0.489 - j0.0733$$

Hence,

$$E_2 = 5I_2 = 2.445 - j0.367 = 2.742\underline{/-8.53°} \text{ V}$$

EXAMPLE 7. Use mesh analysis to determine the three mesh currents in the circuit of Fig. 11.15.

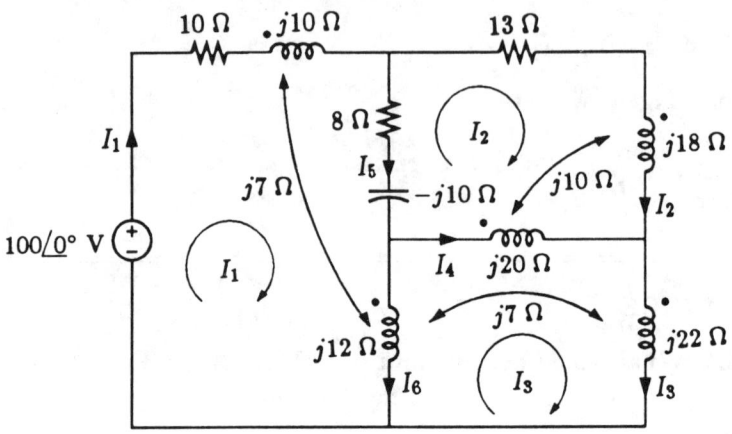

Figure 11.15: Circuit for Example 7.

SOLUTION We shall first use the branch currents to write the mesh equations. They are

$$10I_1 + j10I_1 + j7I_6 + 8I_5 - j10I_5 + j12I_6 + j7I_1 + j7I_3 = 100$$

$$13I_2 + j18I_2 + j10I_4 - j20I_4 - j10I_2 + j10I_5 - 8I_5 = 0$$

$$j20I_4 + j10I_2 + j22I_3 + j7I_6 - j12I_6 - j7I_1 - j7I_3 = 0$$

Now we substitute

$$I_5 = I_1 - I_2 \qquad I_4 = I_3 - I_2 \qquad I_5 = I_1 - I_2$$

into the mesh equations and gather like terms. We obtain

$$(18 + j26)I_1 + (-8 + j10)I_2 - j12I_3 = 100$$

$$(-8 + j10)I_1 + (21 + j8)I_2 - j10I_3 = 0$$

$$-j12I_1 - j10I_2 + j40I_3 = 0$$

Solving yields

$$I_1 = 1.683 - j2.490 = 3.005\underline{/-55.94°} \text{ A}$$

$$I_2 = -0.546 + j1.366 = 1.472\underline{/-111.80°} \text{ A}$$

$$I_3 = 0.368 - j1.088 = 1.149\underline{/-71.30°} \text{ A}$$

EXAMPLE 8. Use node analysis to find the three node voltages for the circuit in Fig. 11.16.

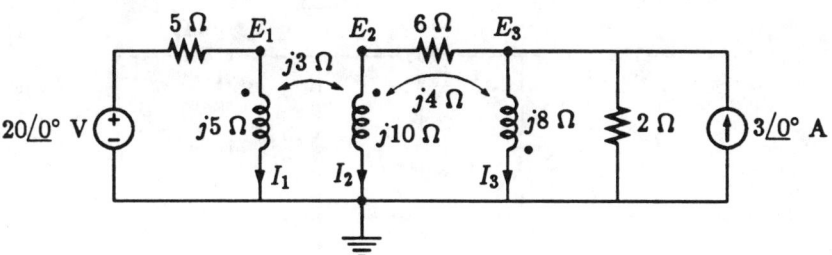

Figure 11.16: Circuit for Example 8.

SOLUTION We first write the three node equations.

$$\frac{E_1 - 20}{5} + I_1 = 0$$

$$\frac{E_2 - E_3}{6} + I_2 = 0$$

$$\frac{E_3 - E_2}{6} + I_3 + \frac{E_3}{2} = 3$$

For the three coils, we have

$$E_1 = j5I_1 + j3I_2$$

$$E_2 = j10I_2 + j3I_1 - j4I_3$$

$$E_3 = j8I_3 - j4I_2$$

Solving the six equations simultaneously, we get

$$E_1 = 10.466 + j9.468 = 14.113\underline{/42.13°} \text{ V}$$

$E_2 = 5.605 + j4.362 = 7.103\underline{/37.89°}$ V

$E_3 = 5.474 + j2.366 = 5.964\underline{/23.38°}$ V

EXAMPLE 9. Find the z matrix of the two-port in Fig. 11.17.

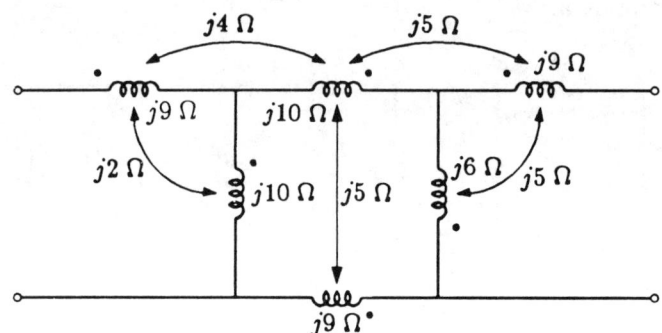

Figure 11.17: Two-port used in Example 9.

SOLUTION We set up the circuit for analyzing the two-port as follows.

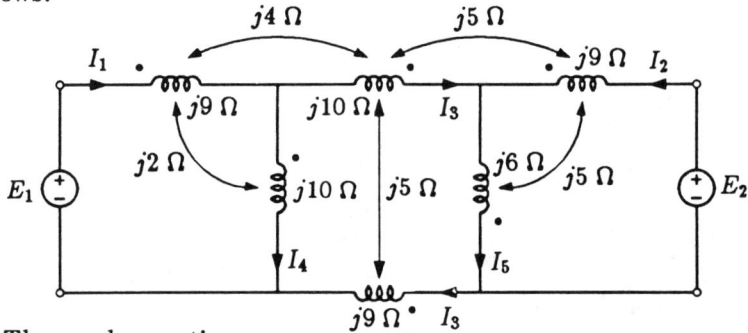

The mesh equations are

$E_1 = j8I_1 - j4I_3 + j2I_4 + j10I_4 + j2I_1$

$E_2 = j9I_2 + j5I_3 + j5I_5 + j6I_5 + j5I_2$

$0 = j9I_3 - j5I_3 - j10I_4 - j2I_1$

$\qquad + j10I_3 - j4I_1 + j5I_2 - j5I_3 + j6I_5 + j5I_2$

THE MUTUAL INDUCTANCE

where $I_4 = I_1 - I_3$ and $I_5 = I_2 + I_3$. Substitute these into the mesh equations and gather like terms; they read

$$E_1 = j22I_1 - j16I_3$$

$$E_2 = j25I_2 + j16I_3$$

$$0 = -j16I_1 + j16I_2 + j25I_3$$

Solve for I_3 from the third equation and substitute it into the first two equations, and rearrange to get

$$E_1 = j\frac{294}{25}I_1 + j\frac{256}{25}I_2$$

$$E_2 = j\frac{256}{25}I_1 + j\frac{369}{25}I_2$$

Hence,

$$[z] = \frac{j}{25}\begin{bmatrix} 294 & 256 \\ 256 & 369 \end{bmatrix} \ \Omega$$

11.3 Equivalent circuits of coupled coils

In ac analysis, it is sometimes more convenient to replace the mutual effects by controlled sources. The two-port in Fig. 11.18(a) has the z matrix

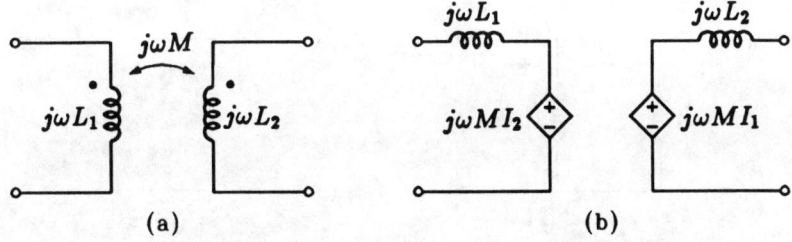

Figure 11.18: Equivalent circuit of coupled coils.

$$[z] = j\omega \begin{bmatrix} L_1 & M \\ M & L_2 \end{bmatrix} \quad (11.41)$$

The two-port in Fig. 11.18(b) also has the same z matrix. Hence the two two-ports are equivalent.

Likewise, the two two-ports in Fig. 11.19 are equivalent. They both have the following z matrix.

$$[z] = j\omega \begin{bmatrix} L_1 & -M \\ -M & L_2 \end{bmatrix} \quad (11.42)$$

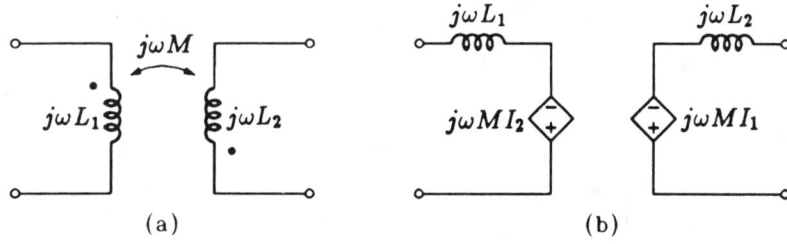

Figure 11.19: Equivalent circuit of coupled coils

If one terminal of each of a pair of coupled coils are connected together, they form a three-terminal two-port as shown in Fig. 11.20(a). The three-terminal network is equivalent to a simple tee, as shown

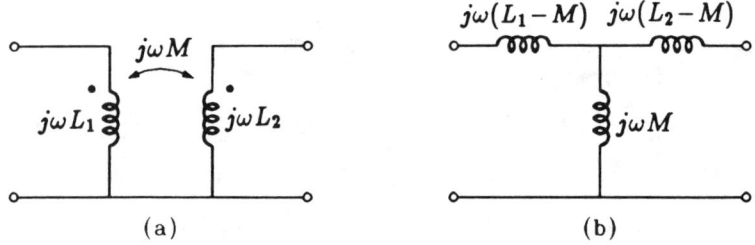

Figure 11.20: Three-terminal coupled coil pair and its equivalent.

in Fig. 11.20(b). These two two-ports both have the z matrix of Eq. (11.41).

THE MUTUAL INDUCTANCE

If the dots are reversed, the vertical branch of the equivalent tee becomes negative. The two two-ports in Fig. 11.21 are equivalent as they both have the z matrix given in Eq. (11.42).

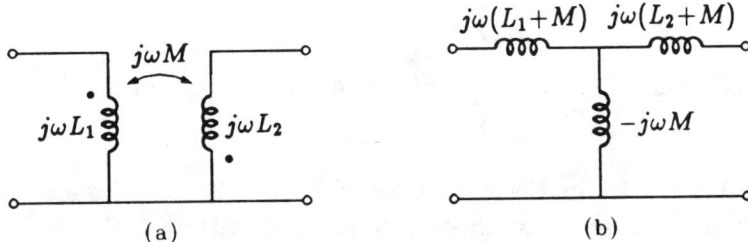

Figure 11.21: Another three-terminal coupled coil pair and its equivalent.

In some networks, this type of equivalence can be used to reduce a number of magnetically coupled coils to all plain inductances. The following are two such examples.

EXAMPLE 10. Find the equivalent impedance of the two parallel coupled coils shown in Fig. 11.22(a). (This circuit is identical to that in Example 4.)

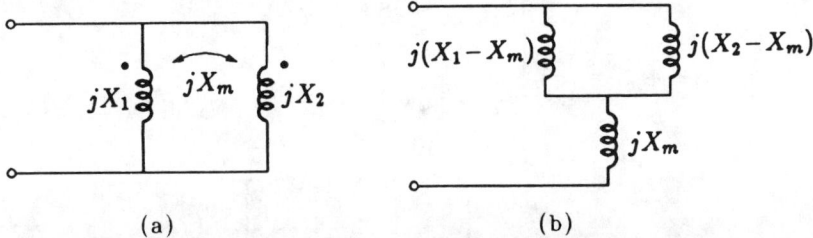

Figure 11.22: Equivalent circuit of a pair of parallel coupled coils.

SOLUTION From Fig. 11.20, it is seen that the two circuits in Fig. 11.22 are equivalent. From Fig. 11.22(b), we get

$$Z_{eq} = jX_m + \frac{(jX_1 - jX_m)(jX_2 - jX_m)}{jX_1 + jX_2 - j2X_m}$$

$$= j\left[X_m + \frac{X_1 - X_m)(X_2 - X_m)}{X_1 + X_2 - 2X_m}\right] = j\frac{X_1X_2 - X_m^2}{X_1 + X_2 - 2X_m}$$

which is identical to that obtained in Example 4.

EXAMPLE 11. Find the z matrix of the two-port in Fig. 11.23(a).

SOLUTION In the remainder of Fig. 11.23, we eliminate one mutual inductance in each step. In (b), the $j5$-Ω mutual inductance is replaced by the $-j5$-Ω inductance and the rest of the tee. In (c), the $j4$-Ω mutual inductance is replaced by another tee with $j4$-Ω inductance in the center branch. At the same time, the -5-Ω and the $j8$-Ω are combined as they are in series. In (d), the $j4$-Ω and the $j10$-Ω are combined. Finally, in (e), the $j7$-Ω mutual inductance is replaced by another tee with $-j7$-Ω in its center branch.

For the equivalent circuit in Fig. 11.23(e), we have

$$[y_a] = -j\begin{bmatrix} \frac{1}{10} + \frac{1}{9} & -\frac{1}{9} \\ -\frac{1}{9} & \frac{1}{21} + \frac{1}{9} \end{bmatrix}$$

$$[z_a] = [y_a]^{-1} = j\begin{bmatrix} \frac{15}{2} & \frac{21}{4} \\ \frac{21}{4} & \frac{399}{40} \end{bmatrix}$$

$$[z_b] = [z_a] - j\begin{bmatrix} 7 & 7 \\ 7 & 7 \end{bmatrix} = j\begin{bmatrix} \frac{1}{2} & -\frac{7}{4} \\ -\frac{7}{4} & \frac{119}{40} \end{bmatrix}$$

$$[z] = [z_b] + j\begin{bmatrix} 15 & 0 \\ 0 & 2 \end{bmatrix} = j\begin{bmatrix} \frac{31}{2} & -\frac{7}{4} \\ -\frac{7}{4} & \frac{199}{40} \end{bmatrix} \Omega$$

THE MUTUAL INDUCTANCE

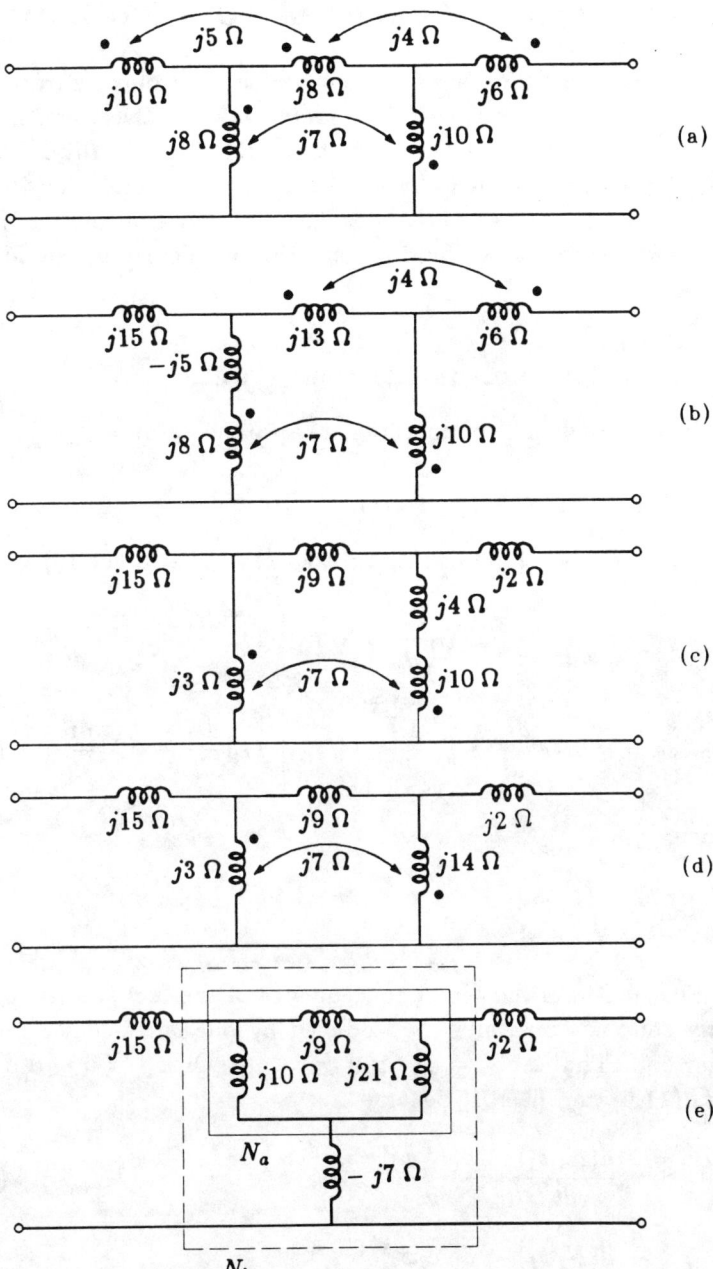

Figure 11.23: Elimination of mutual inductances.

11.4 Limiting cases of coupled coils

Now we shall examine three cases in which some of the parameters of a pair of coils approach certain extreme values. These limiting cases are idealized situations and cannot be attained in reality. However, they can be approached closely. In many situations, the difference between a real device and the idealized device is so negligible that, for practical purposes, we could replace the real device by the idealized device.

11.4.1 The voltage transformer

When the coefficient of coupling, k, approaches unity, we have

$$M = \sqrt{L_1 L_2} \tag{11.43}$$

The time-domain relationship in Eqs. (11.11) and (11.12) becomes

$$e_1 = L_1 \frac{di_1}{dt} + \sqrt{L_1 L_2} \frac{di_2}{dt} = \sqrt{L_1} \left[\sqrt{L_1} \frac{di_1}{dt} + \sqrt{L_2} \frac{di_2}{dt} \right] \tag{11.44}$$

$$e_2 = \sqrt{L_1 L_2} \frac{di_1}{dt} + L_2 \frac{di_2}{dt} = \sqrt{L_2} \left[\sqrt{L_1} \frac{di_1}{dt} + \sqrt{L_2} \frac{di_2}{dt} \right] \tag{11.45}$$

Hence,

$$\frac{e_2}{e_1} = \sqrt{\frac{L_2}{L_1}} \tag{11.46}$$

Eq. (11.46) is the direct consequence of the fact that in a pair of unity-coupled coils, all flux produced by one current also links the other coil. That is, $\phi_1 = \phi_{21}$ and $\phi_2 = \phi_{12}$. Or $\phi_{11} = 0$ and $\phi_{22} = 0$. Eqs. (11.9) and (11.10) become

$$e_1 = \frac{d(N_1 \phi_1)}{dt} + \frac{d(N_1 \phi_2)}{dt} \tag{11.47}$$

$$e_2 = \frac{d(N_2 \phi_1)}{dt} + \frac{d(N_2 \phi_2)}{dt} \tag{11.48}$$

Hence,

THE MUTUAL INDUCTANCE

$$\frac{e_2}{e_1} = \frac{N_2}{N_1} = n \tag{11.49}$$

Eq. (11.49) is the reason that n is called the *turns ratio*. A pair of unity-coupled coils constrains their voltages such that they are always proportional to each other according to Eq. (11.49). This device is called a *voltage transformer*. A voltage transformer places no constraint on the two currents. They may have any values and are determined by something external to the coils.

11.4.2 The current transformer

If we take a pair of unity-coupled coils and short circuit one of their terminal pairs, say terminal pair 2, as shown in Fig. 11.24, we have

$$e_2 = 0 \quad \text{and} \quad e_1 = 0 \tag{11.50}$$

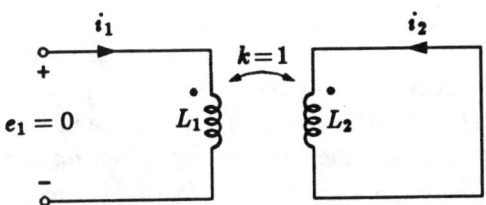

Figure 11.24: A current transformer.

Either Eq. (11.44) or (11.45) leads to

$$\frac{di_1}{dt} = -\sqrt{\frac{L_2}{L_1}} \frac{di_2}{dt} \tag{11.51}$$

It can be shown using electromagnetic theory that, in a pair of unity-coupled coils, $\sqrt{L_2/L_1} = N_2/N_1 = n$. Hence,

$$\frac{di_1}{dt} = -n\frac{di_2}{dt} \quad \text{and} \quad i_1 = -ni_2 \tag{11.52}$$

This device constrains the current ratio in the two coils (while holding both voltages zero) and is called a *current transformer*. The practical version of this device is used in ac instrumentation where kiloamperes are measured by scaling it down to milliamperes so it can be read by practical milliameters.

11.4.3 The ideal transformer

If, in addition to $k = 1$, we let $L_1 \to \infty$ and $L_2 \to \infty$ but hold

$$\frac{L_2}{L_1} = n^2$$

then, in addition to Eq. (11.49) in which

$$e_2 = ne_1 \tag{11.53}$$

we also have

$$i_2 = -\frac{1}{n}i_1 \tag{11.54}$$

which can be arrived at by making e_1 or e_2 negligible compared to the quantities on the right-hand side of Eqs. (11.44) and (11.45). This device is the *ideal transformer* defined in Subsection 10.6.8.

Since we have defined the ideal transformer in the context of two-ports, the voltages and currents at its terminal pairs follow what are implied in the two-port notation. Therefore, it is not necessary to place the dots at the terminals of an ideal transformer. Since in the development of this section, we have used Eqs. (11.11) and (11.12), which are associated with the mutual inductance in Fig. 11.1 or Fig. 11.3(a), it is implicit that the coils be dotted either both at the top or both at the bottom. If an ideal transformer is the limiting case of a pair of coils whose associated dots are opposite to what we have used here, we can simply reverse the sign of the turns ratio. Hence, if we found it necessary to represent a situation in which $E_2 = -nE_1$ and $I_2 = \frac{1}{n}I_1$, we simply call this a $1:-n$ ideal transformer.

11.5 Summary

When two coils are placed in proximity to each other, the voltage in each coil is made up of two parts—one due to the flux produced by its own current and the other due to the flux produced by the other current. We have developed a convention to quantitatively deal with this mutual effect. The dot convention enables us to represent the relative directions of the mutual fluxes by circuit symbols.

The tightness of the coupling of two coils is represented by the coefficient of coupling. The phenomenon of magnetic coupling is the basis of many forms or transformers used many electrical applications.

Problems

11.1 Determine $i(t)$ and $e(t)$.

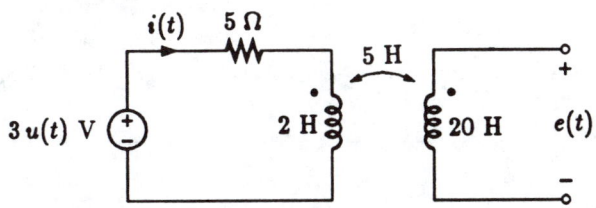

11.2 Determine $e(t)$.

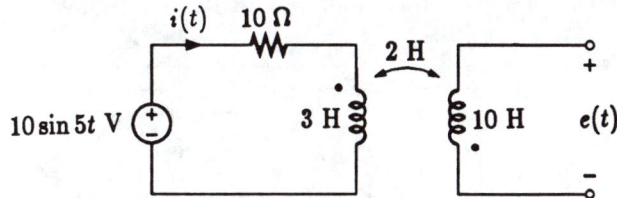

11.3 Determine $i(t)$.

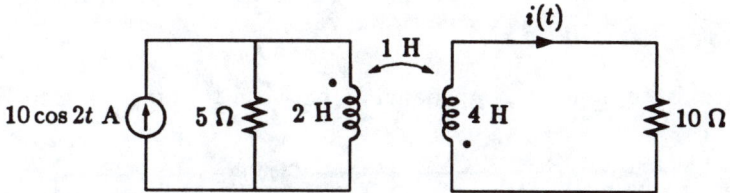

11.4 Determine the equivalent L_{eq}.

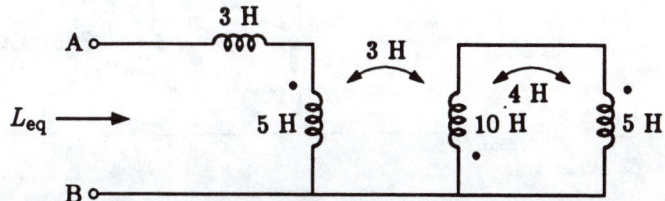

11.5 Determine the impedance Z between terminals A and B.

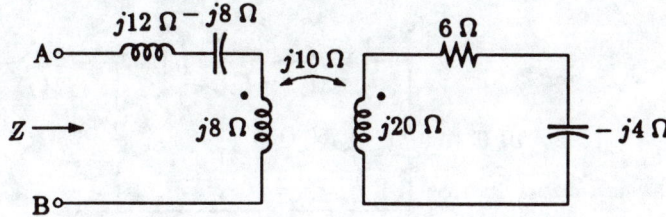

11.6 Calculate E_2.

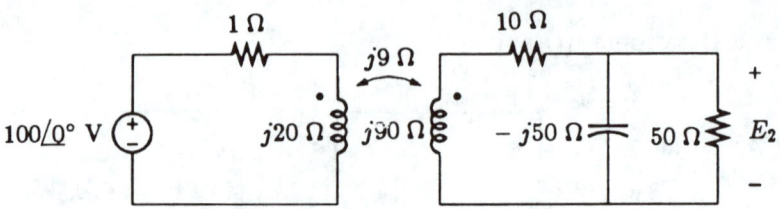

11.7 Obtain the Thévenin's equivalent of the two-terminal network N with respect to terminals A and B.

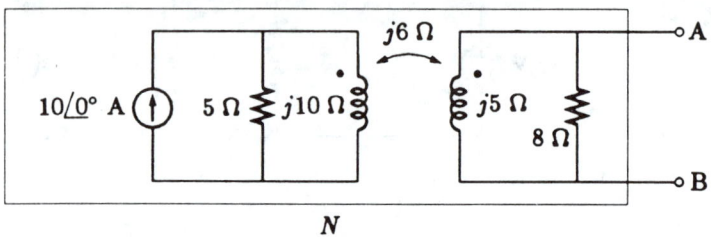

11.8 Determine the impedance Z between terminals A and B.

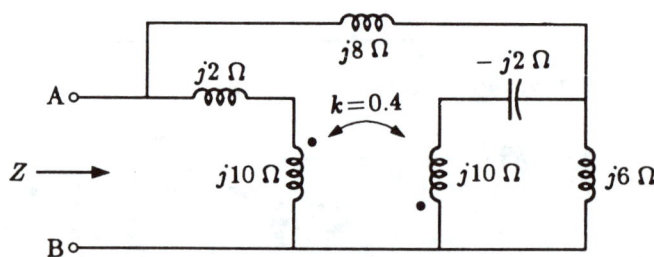

11.9 Calculate I_2.

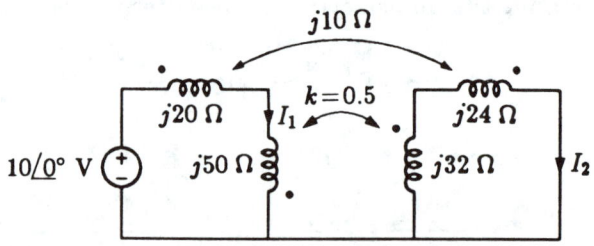

11.10 Calculate I_1, I_2, I_3, and I_4.

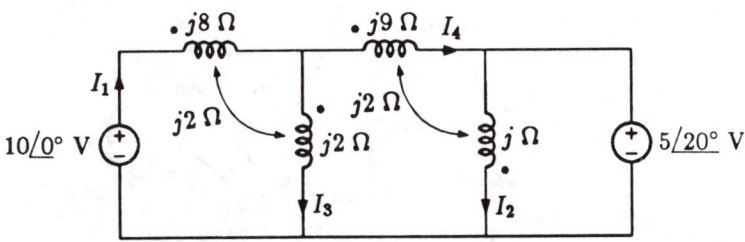

11.11 The coefficient of coupling of each mutual inductance is 0.5. Calculate I_1 and I_2.

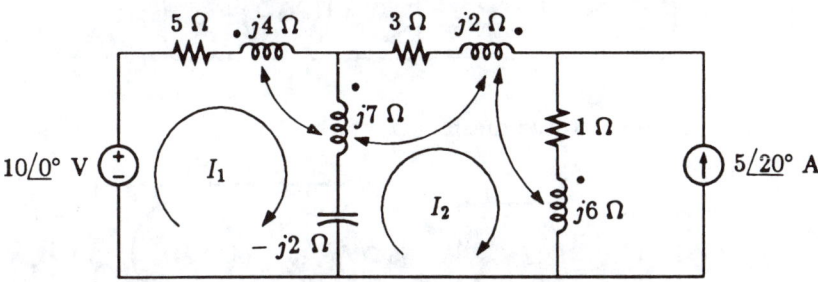

11.12 Calculate E_a, E_b, and E_c.

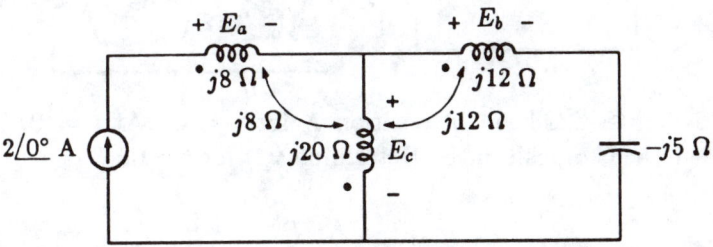

11.13 Calculate E_a, E_b, and E_c.

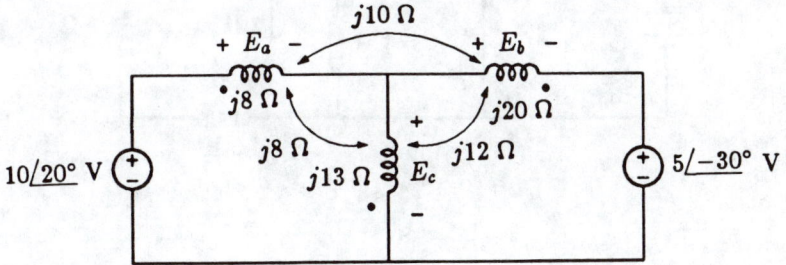

△ **11.14** Calculate the three mesh currents.

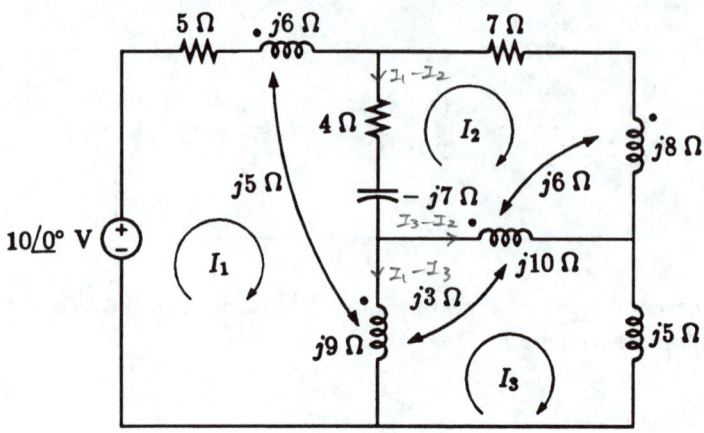

11.15 Calculate the three mesh currents.

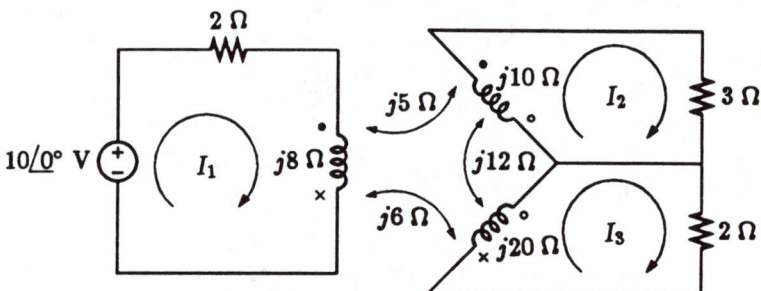

11.16 Switch S has been in position A for $t < 0$. At $t = 0$, it is switched to position B. Determine $i(t)$ for $t > 0$.

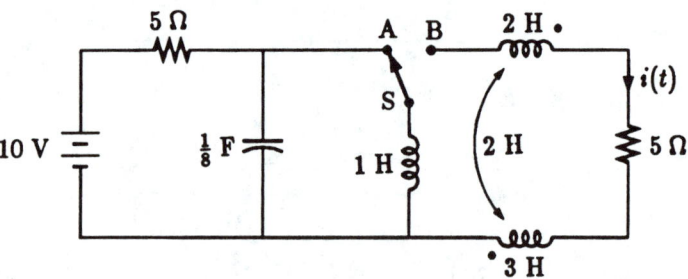

THE MUTUAL INDUCTANCE

11.17 Calculate the energy stored in the electromagnetic field for (a) $I_1 = 10$ A, $I_2 = 5$ A; and (b) $I_1 = 10$ A, $I_2 = -5$ A.

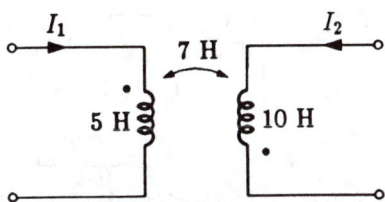

11.18 In a pair of coupled coils, $\phi_{21} = 0.9\phi_1$ and $\phi_{12} = 0.75\phi_2$. Calculate the coefficient of coupling k.

11.19 Calculate I_2.

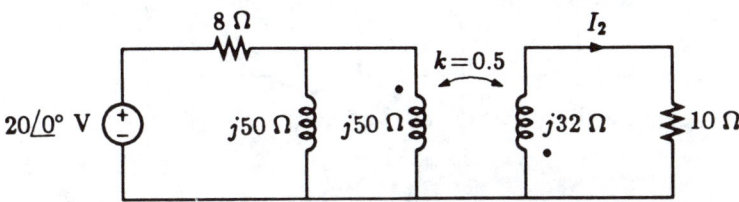

11.20 In the loosely coupled coils, ω is variable. Determine the value of ω at which the voltage gain $|E_2/E_1|$ is maximum. Also, calculate this maximum gain.

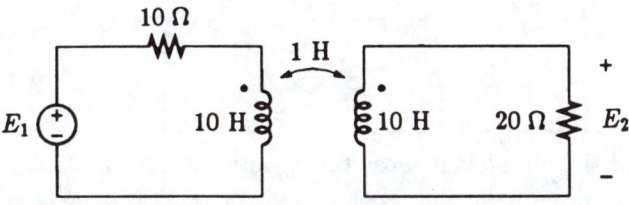

11.21 Determine the z matrix of the two-port.

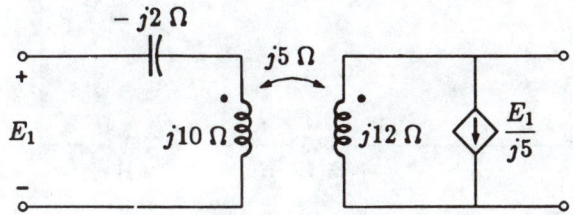

11.22 Determine the z matrix of the two-port.

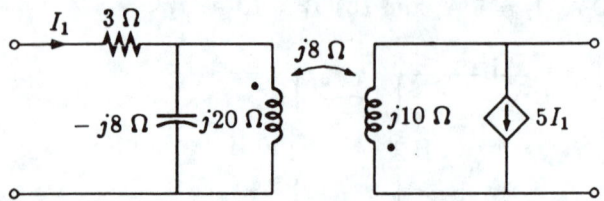

11.23 Determine the z matrix of the two-port.

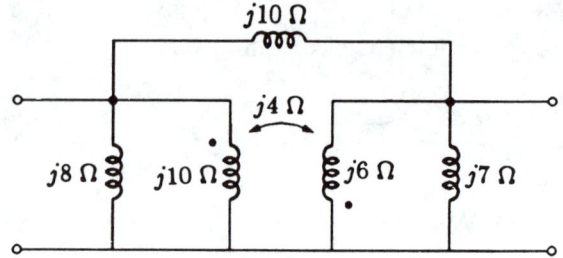

11.24 Determine the y matrix of the two-port.

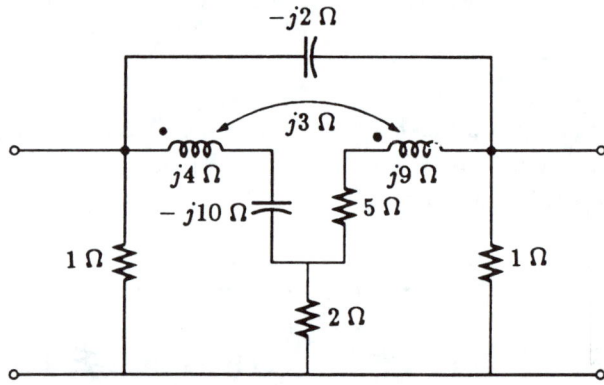

11.25 Find the equivalent inductance between terminals A and B.

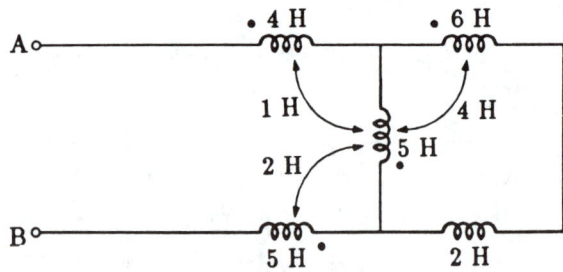

11.26 Find the equivalent inductance between terminals A and B.

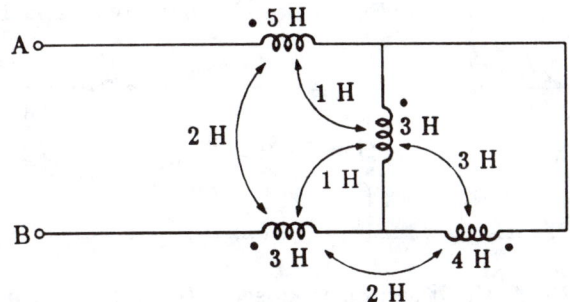

11.27 Determine the y matrix of the two-port.

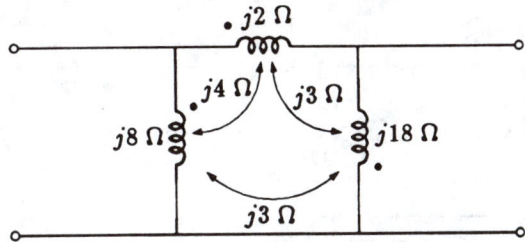

11.28 Calculate accurately the voltage ratio E_2/E_1 for (a) $k = 0.1$, (b) $k = 0.5$, (c) $k = 0.95$, (d) $k = 0.99$, and (e) $k = 1$.

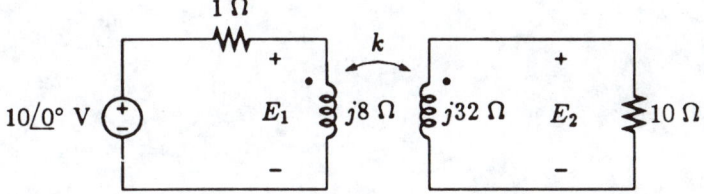

11.29 Calculate $i_2(t)$ and $e_2(t)$ for (a) $R = \infty$, (b) $R = 5\,\Omega$, and (c) $R = 0$.

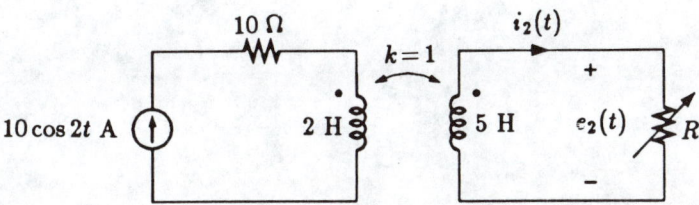

11.30 $E_1 = 25\underline{/0°}$ V. When switch S is open, we have $|I_1| = 1$ A and $|E_2| = 40$ V. With switch S closed, calculate I_1 and E_2.

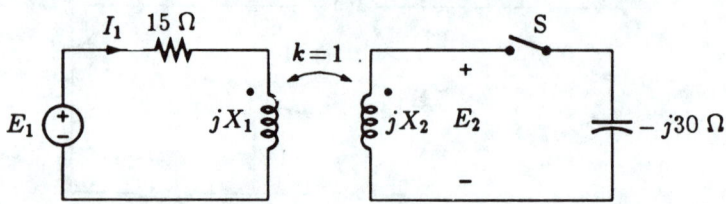

11.31 $E_1 = 20\underline{/0°}$ V. If switch is closed, $|I_1| = 4$ A. If it is open, $|I_1| = 2$ A. Calculate E_2 if switch S is replaced by a 10-Ω resistor.

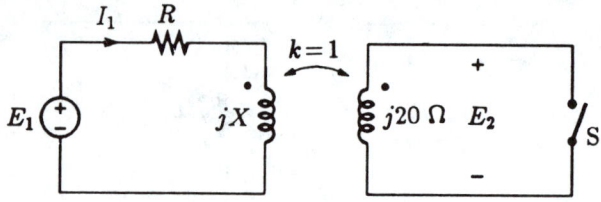

Chapter 12

Multiterminal Networks

In Chapter 10, we mentioned that a multiterminal network may be treated either on the port (terminal-pair) basis or on the terminal basis. We dwelt on the former in Chapter 10—particularly two-ports. In this chapter, we shall see how a device or subnetwork may be treated on the terminal basis.

There are situations in which it is more convenient to treat a network or a device on the terminal basis rather than on the port basis. A prime example of such a device is the transistor, which is a three-terminal device. A variety of orientations can be made out of such a single device. The advantage of treating a network or device on the terminal basis is that no commitment is made in advance as to how the terminals are paired or grouped. For instance, we are free to designate any of the terminals of a multiterminal device to be the ground terminal. Our ability to handle subnetwork and devices on the terminal basis also has a great many of applications in computer-aided circuit analysis.

12.1 The indefinite admittance matrix

A multiterminal network may be handled on the impedance or admittance basis, as well as on some other hybrid or composite basis. However, the admittance basis is by far the most convenient to use for all practical considerations. We shall limit our discussion to this type of formalism.

Let us look at an n-terminal network that is linear and contains no independent source. Let n external voltages (E_1, E_2,..., E_n, with respect to ground) be applied to the network as shown in Fig. 12.1. We can obtain a set of equations for the terminal current in the form

$$I_1 = y_{11}E_1 + y_{12}E_2 + \cdots + y_{1n}E_n$$
$$I_2 = y_{21}E_1 + y_{22}E_2 + \cdots + y_{2n}E_n \qquad (12.1)$$
$$\cdots\cdots\cdots\cdots\cdots\cdots\cdots\cdots\cdots\cdots\cdots$$
$$I_n = y_{n1}E_1 + y_{n2}E_2 + \cdots + y_{nn}E_n$$

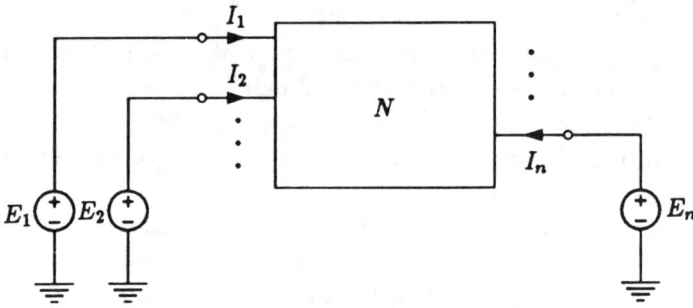

Figure 12.1: An n-terminal network.

If we write Eq. (12.1) in matrix form, it becomes

$$\begin{bmatrix} I_1 \\ I_2 \\ \vdots \\ I_n \end{bmatrix} = \begin{bmatrix} y_{11} & y_{12} & \cdots & y_{1n} \\ y_{21} & y_{22} & \cdots & y_{2n} \\ \vdots & \vdots & & \vdots \\ y_{n1} & y_{n2} & \cdots & y_{nn} \end{bmatrix} \begin{bmatrix} E_1 \\ E_2 \\ \vdots \\ E_n \end{bmatrix} \qquad (12.2)$$

We shall call the matrix containing the admittances the *indefinite admittance matrix* (IAM)[1] of the n-terminal network N, or

$$[y_i] = \begin{bmatrix} y_{11} & y_{12} & \cdots & y_{1n} \\ y_{21} & y_{22} & \cdots & y_{2n} \\ \vdots & \vdots & & \vdots \\ y_{n1} & y_{n2} & \cdots & y_{nn} \end{bmatrix} \qquad (12.3)$$

---

[1] Also known as the terminal admittance matrix.

MULTITERMINAL NETWORK

We may interpret each of the elements of $[y_i]$ by referring to Eq. (12.1) in much the same way as we interpret the two-port parameters from their defining equations. That is

$$y_{ij} = \left. \frac{I_i}{E_j} \right|_{E_k=0,\ k \neq j} \tag{12.4}$$

In other words, y_{ij} is the ratio of the current flowing into terminal i to the voltage at terminal j with all terminals except terminal j grounded.

When a network is designated as an n-terminal network, its terminal voltage and currents are frequently omitted from the diagram. Each terminal is simply given a number to avoid cluttering up the diagram. The voltage with respect to ground of terminal k is understood to be E_k and the current *entering* the network at terminal k is I_k. Hence, the voltages and currents for N in Fig. 12.1 are automatically implied in the network shown in Fig. 12.2.

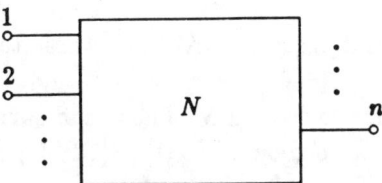

Figure 12.2: Abbreviated representation of the n-terminal network in Fig. 12.1.

Also, throughout this chapter, we shall adopt the convention that electric quantities in or parameters of a network shall have the same subscript or prime(s) as those of the network when such identification will not cause any confusion or ambiguity. Thus the IAM of N_a shall be denoted by $[y_i]_a$ with elements y_{ija} and that of N' shall be denoted by $[y_i]'$ with elements y'_{ij}.

EXAMPLE 1. Obtain the IAM of the three-terminal network of Fig. 12.3(a).

SOLUTION To obtain the elements of the IAM, we excite the network at one terminal at a time, with all other terminals grounded. Thus, we have the situations in Fig. 12.3(b) to (d), in which each

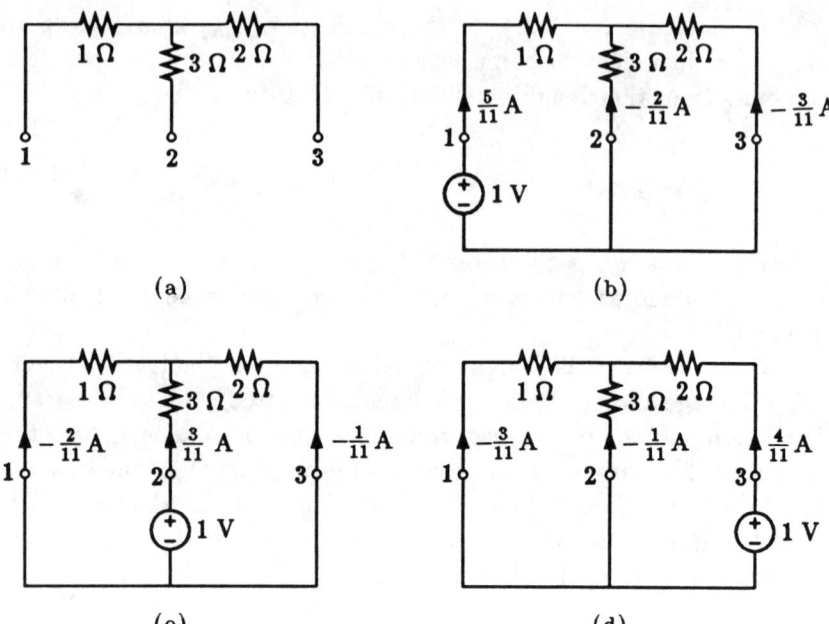

Figure 12.3: Evaluation of the IAM of a three-terminal network.

exciting voltage is made to be 1 V. Thus each current value is equal to an admittance value defined in Eq. (12.4). From the current values indicated in Fig. 12.3, we may write

$$[y_i] = \frac{1}{11} \begin{bmatrix} 5 & -2 & -3 \\ -2 & 3 & -1 \\ -3 & -1 & 4 \end{bmatrix} \mho$$

EXAMPLE 2. Obtain the IAM of the four-terminal network in Fig. 12.4.

SOLUTION In order to obtain the elements of the IAM column by column, we set up the four circuits of Fig. 12.5. In each of the circuits, one of the terminals is excited by a 1-volt source with all other terminals short-circuited to the minus terminal of the voltage source.

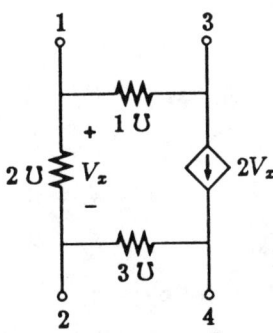

Figure 12.4: A four-terminal network.

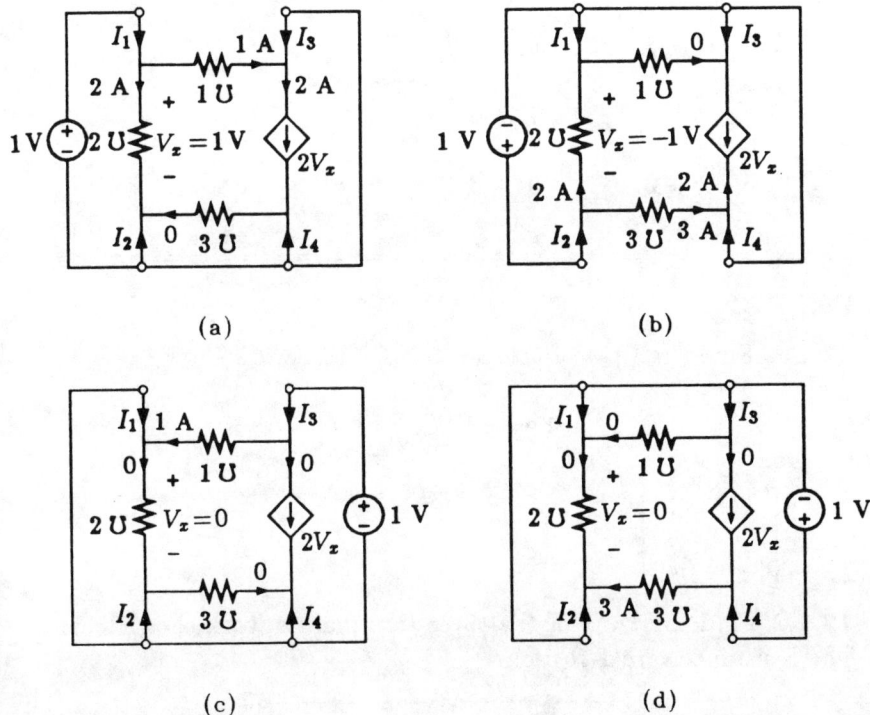

Figure 12.5: Analysis to obtain the IAM of the four-terminal network in Fig. 12.4.

In Fig. 12.5(a), $E_1 = 1$, $E_2 = E_3 = E_4 = 0$. We get
$$I_1 = 3 = y_{11} \quad I_2 = -2 = y_{21} \quad I_3 = 1 = y_{31} \quad I_4 = -2 = y_{41}$$
In Fig. 12.5(b), $E_2 = 1$, $E_1 = E_3 = E_4 = 0$. We get
$$I_1 = -2 = y_{12} \quad I_2 = 5 = y_{22} \quad I_3 = -2 = y_{32} \quad I_4 = -1 = y_{42}$$
In Fig. 12.5(c), $E_3 = 1$, $E_1 = E_2 = E_4 = 0$. We get
$$I_1 = -1 = y_{13} \quad I_2 = 0 = y_{23} \quad I_3 = 1 = y_{33} \quad I_4 = 0 = y_{43}$$
In Fig. 12.5(d), $E_4 = 1$, $E_1 = E_2 = E_3 = 0$. We get
$$I_1 = 0 = y_{14} \quad I_2 = -3 = y_{24} \quad I_3 = 0 = y_{34} \quad I_4 = 3 = y_{44}$$
Hence,
$$[y_i] = \begin{bmatrix} 3 & -2 & -1 & 0 \\ -2 & 5 & 0 & -3 \\ 1 & -2 & 1 & 0 \\ -2 & -1 & 0 & 3 \end{bmatrix} \mho$$

EXERCISE

12.1.1 Find the IAM of the two-terminal network shown.

1 o——[y]——o 2

Ans.
$$\begin{bmatrix} y & -y \\ -y & y \end{bmatrix}$$

12.1.2 Find the IAM of the three-terminal network shown below. The resistances are 1-Ω each.

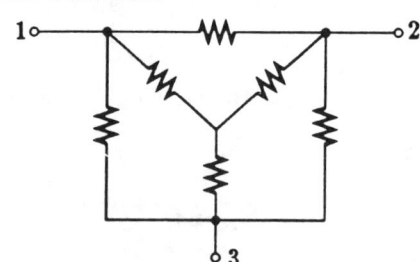

Ans.

$$\frac{1}{3}\begin{bmatrix} 8 & -4 & -4 \\ -4 & 8 & -4 \\ -4 & -4 & 8 \end{bmatrix} \mho$$

12.2 Properties of the IAM

The IAM has several properties and features that make it quite different from the admittance matrices obtained on the port basis. These properties and features are not only interesting in themselves but also useful in manipulating an IAM or processing a number of them.

12.2.1 The zero-sum property

The elements of the IAM of a multiterminal network are not independent. In Fig. 12.1, KCL requires that

$$I_1 + I_2 + \cdots + I_n = 0 \tag{12.5}$$

for any values of the E's. If we set all voltages except one, say E_j, to zero, we can conclude from Eq. (12.1) that

$$I_1 + I_2 + \cdots + I_n = (y_{1j} + y_{2j} + \cdots + y_{nj}) E_j = 0 \tag{12.6}$$

Hence

$$y_{1j} + y_{2j} + \cdots + y_{nj} = 0, \quad j = 1, 2, \ldots, n \tag{12.7}$$

Furthermore, if we let all voltages be equal, say to E_0, then all terminals are at the same potential and every terminal current will be zero. Each equation of Eq. (12.1) yields

$$I_i = (y_{i1} + y_{i2} + \cdots + y_{in}) E_0 = 0 \tag{12.8}$$

Hence,

$$y_{i1} + y_{i2} + \cdots + y_{in} = 0, \quad i = 1, 2, \cdots n \tag{12.9}$$

Eqs. (12.7) and (12.9) state that every row and every column of an IAM must add up to zero algebraically. This is evident in the IAM's of Examples 1 and 2. The IAM of a network is also known as the *zero-sum matrix*.

Since the IAM is a zero-sum matrix, its determinant is always zero. The inverse of an IAM does not exist. Electrically, this means that if all terminal currents of a multiterminal network are known, its terminal voltages are not uniquely defined.

We can take advantage of this zero-sum property in two ways. First, if the dimension of an IAM is $n \times n$, once we know $n-1$ elements of any row or column, the remaining one can simply be made the negative of the sum of the $n-1$ known ones. Hence, once we know any $(n-1) \times (n-1)$ block of an IAM, we can simply fill in the remainder of the IAM by making the matrix zero-sum.

Second, every time we obtain an IAM, if the elements fail to satisfy the zero-sum property, the matrix is obviously flawed.

If we take a three-terminal network and let $E_3 = 0$ as shown in Fig. 12.6, we can regard the network as a three-terminal two-port. The y matrix of the two-port is simply the upper left 2×2

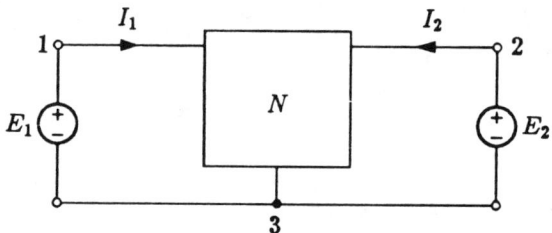

Figure 12.6: A three-terminal network as a two-port.

submatrix of the IAM of the three-terminal network. On the other hand, if the y matrix of a three-terminal two-port is known, then the IAM of the network can be obtained simply by bordering the y matrix with an additional row and an additional column in such a way that Eqs. (12.7) and (12.9) are satisfied.

EXAMPLE 3. Obtain the IAM of the three-terminal network in Fig. 12.7. The circuit is the low-frequency equivalent circuit of a transistor.

MULTITERMINAL NETWORK

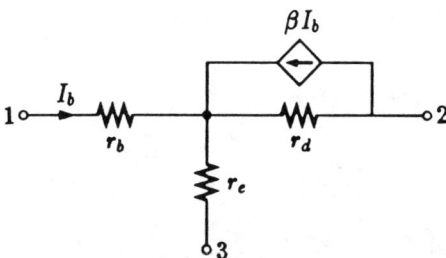

Figure 12.7: A equivalent circuit of a transistor.

SOLUTION We analyze the two-port in Fig. 12.8 first. KCL and Ohm's law require

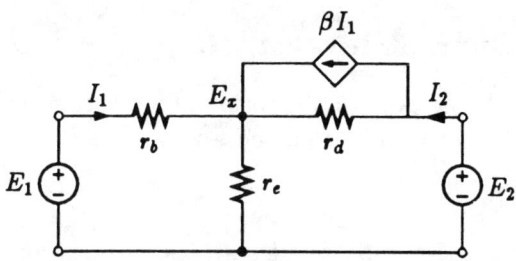

Figure 12.8: Network in Fig. 12.7 rearranged as a two-port.

$$I_1 = \frac{E_1 - E_z}{r_b}$$

$$I_2 = \frac{E_2 - E_z}{r_d} + \beta I_1$$

$$\frac{E_z - E_1}{r_b} + \frac{E_z - E_2}{r_d} + \frac{E_z}{r_e} = \beta I_1$$

Eliminating E_z and rearranging, we obtain

$$I_1 = \frac{r_e + r_d}{\Delta} E_1 - \frac{r_e}{\Delta} E_2$$

$$I_2 = \frac{\beta r_d - r_e}{\Delta} E_1 + \frac{r_e + r_b}{\Delta} E_2$$

where $\Delta = (1 - \beta) r_e r_d + r_e r_b + r_b r_d$. Hence, for the two-port

$$[y] = \frac{1}{\Delta} \begin{bmatrix} r_e + r_d & -r_e \\ \beta r_d - r_e & r_e + r_b \end{bmatrix}$$

and for the three-terminal network in Fig. 12.7,

$$[y_i] = \frac{1}{\Delta} \begin{bmatrix} r_e + r_d & -r_e & -r_d \\ \beta r_d - r_e & r_e + r_b & -\beta r_d + r_b \\ -(1+\beta)r_d & -r_b & (1+\beta)r_d + r_b \end{bmatrix}$$

12.2.2 Current invariance

If all terminal voltages in an n-terminal network are changed by the same amount, all terminal currents will remain unchanged. From Eq. (12.2)

$$\begin{bmatrix} y_{11} & y_{12} & \cdots & y_{1n} \\ y_{21} & y_{22} & \cdots & y_{2n} \\ \vdots & \vdots & & \vdots \\ y_{n1} & y_{n2} & \cdots & y_{nn} \end{bmatrix} \begin{bmatrix} E_1 + E_0 \\ E_2 + E_0 \\ \vdots \\ E_n + E_0 \end{bmatrix}$$

$$= \begin{bmatrix} y_{11} & y_{12} & \cdots & y_{1n} \\ y_{21} & y_{22} & \cdots & y_{2n} \\ \vdots & \vdots & & \vdots \\ y_{n1} & y_{n2} & \cdots & y_{nn} \end{bmatrix} \begin{bmatrix} E_1 \\ E_2 \\ \vdots \\ E_n \end{bmatrix}$$

$$+ \begin{bmatrix} y_{11} & y_{12} & \cdots & y_{1n} \\ y_{21} & y_{22} & \cdots & y_{2n} \\ \vdots & \vdots & & \vdots \\ y_{n1} & y_{n2} & \cdots & y_{nn} \end{bmatrix} \begin{bmatrix} E_0 \\ E_0 \\ \vdots \\ E_0 \end{bmatrix} = \begin{bmatrix} I_1 \\ I_2 \\ \vdots \\ I_n \end{bmatrix} \quad (12.10)$$

Since

MULTITERMINAL NETWORK

$$\begin{bmatrix} y_{11} & y_{12} & \cdots & y_{1n} \\ y_{21} & y_{22} & \cdots & y_{2n} \\ \vdots & \vdots & & \vdots \\ y_{n1} & y_{n2} & \cdots & y_{nn} \end{bmatrix} \begin{bmatrix} E_0 \\ E_0 \\ \vdots \\ E_0 \end{bmatrix} = 0$$

because of the zero-sum property.

The current invariance with respect to a uniform change in terminal voltages is also quite evident from the circuit consideration. Increasing every voltage by E_0 is equivalent to the arrangement in Fig. 12.9. The introduction of E_0 clearly does not affect the volt-

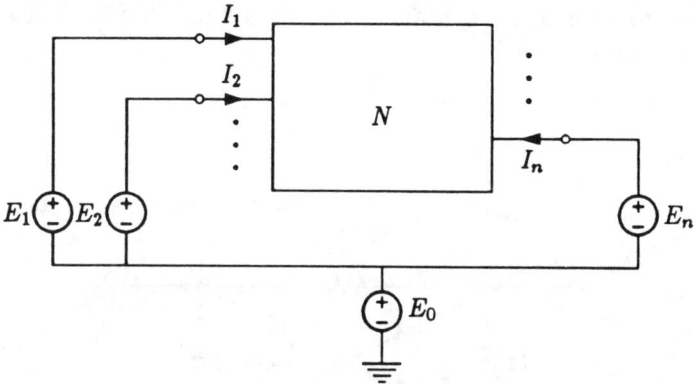

Figure 12.9: All voltages of an n-terminal network are increased by E_0.

age differences among the n terminals. Therefore, all currents are unaffected.

12.2.3 The IAM of a network with no internal node

The IAM of an n-terminal network that has no internal node—all its branches are connected between two of the n terminals—is particularly simple to obtain. First, let's look at the contribution of a passive branch or a branch that can be represented by an admittance. Let the branch admittance be y_k, connected between terminals i and j. This admittance will appear in the IAM as follows.

$$\begin{array}{c} \begin{array}{cc} i & j \\ \downarrow & \downarrow \end{array} \\ \begin{array}{c} i \to \\ \\ j \to \\ \\ \end{array} \left[\begin{array}{ccccc} \vdots & & \vdots & \\ \cdots & y_k & \cdots & -y_k & \cdots \\ & \vdots & & \vdots & \\ \cdots & -y_k & \cdots & y_k & \cdots \\ & \vdots & & \vdots & \end{array}\right] \end{array} \quad (12.11)$$

Hence, if a network is made up of only branches connected between the n terminals, we can simply set an $n \times n$ matrix and enter the contribution of each branch, one at a time, until all branches have been accounted for.

EXERCISE

12.2.1 Obtain the IAM of the following four-terminal network.

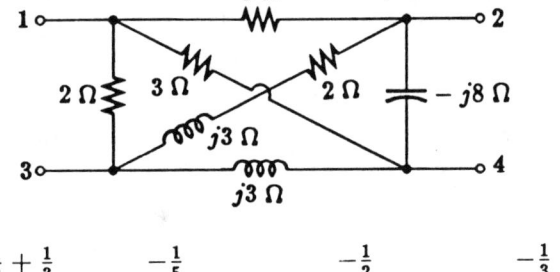

Ans.

$$\left[\begin{array}{cccc} \frac{1}{5}+\frac{1}{2}+\frac{1}{3} & -\frac{1}{5} & -\frac{1}{2} & -\frac{1}{3} \\ -\frac{1}{5} & \frac{1}{5}+\frac{1}{-j8}+\frac{1}{2+j3} & -\frac{1}{2+j3} & -\frac{1}{-j8} \\ -\frac{1}{2} & -\frac{1}{2+j3} & \frac{1}{j3}+\frac{1}{2}+\frac{1}{2+j3} & -\frac{1}{j3} \\ -\frac{1}{3} & -\frac{1}{-j8} & -\frac{1}{j3} & \frac{1}{-j8}+\frac{1}{j3}+\frac{1}{3} \end{array}\right] \mho$$

Note that each diagonal element is equal to the sum of the admittances of all branches connected to the corresponding terminal. Each off-diagonal element is the negative of the admittance connected between two terminals.

MULTITERMINAL NETWORK

For a voltage-controlled current source, we refer to Fig. 12.10. For the contribution of such a controlled source, we write the equations

$$I_c = -I_d = y_m(E_a - E_b) \qquad (12.12)$$

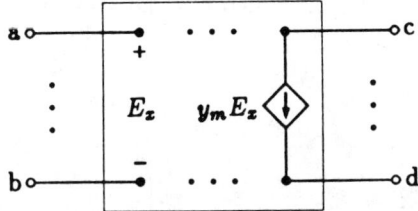

Figure 12.10: A voltage-controlled current source in an multiterminal network.

Hence, the contribution of the controlled source is

$$
\begin{array}{c}
\begin{array}{cc} a & b \\ \downarrow & \downarrow \end{array} \\
\begin{array}{c} c\rightarrow \\ \\ d\rightarrow \end{array}
\left[
\begin{array}{ccccc}
\vdots & & \vdots & \\
\cdots & y_m & \cdots & -y_m & \cdots \\
\vdots & & \vdots & \\
\cdots & -y_m & \cdots & y_m & \cdots \\
\vdots & & \vdots & \\
\end{array}
\right]
\end{array}
\qquad (12.13)
$$

If we have a current-controlled current source as shown in Fig. 12.11, we have

$$I_a = -I_b = I_x = y_k(E_a - E_b) \qquad (12.14)$$

$$I_c = -I_d = \alpha I x = \alpha y_k(E_a - E_b) \qquad (12.15)$$

Hence the contribution of the controlled source is

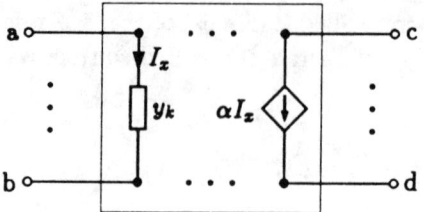

Figure 12.11: A current-controlled current source in a multiterminal network.

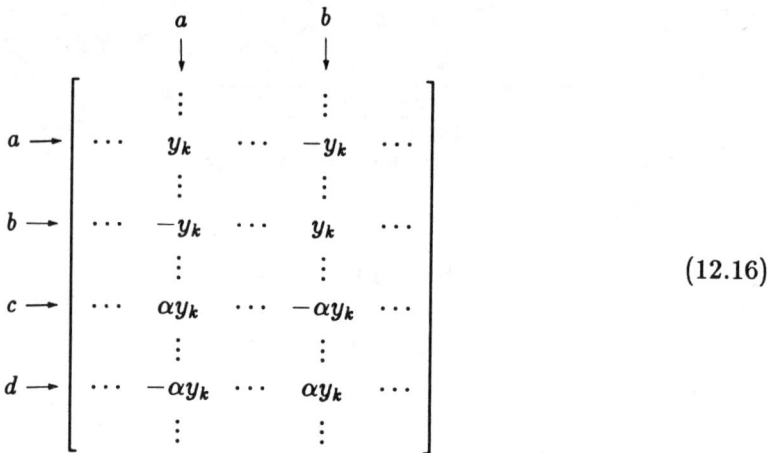

(12.16)

The entries in Eq. (12.16) can be viewed as being made up of two separate groups. One is the contribution of the admittance y_k connected between terminals a and b, as in Eq. (12.11). The other is the contribution of a voltage-controlled current source with $y_m = \alpha y_k$ as in Eq. (12.13).

EXERCISE

12.2.2 Obtain the IAM of the following four-terminal network.

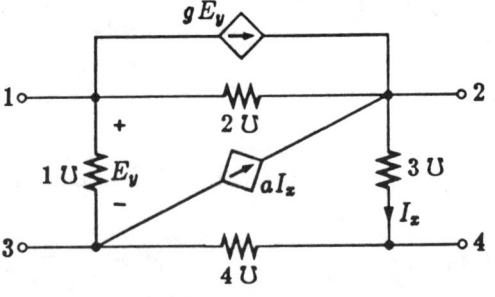

MULTITERMINAL NETWORK

Ans.

$$\begin{bmatrix} 2+1+g & -2 & -1-g & 0 \\ -2-g & 2+3-3a & g & -3+3a \\ -1 & 3a & 1+4 & -4-3a \\ 0 & -3 & -4 & 3+4 \end{bmatrix} U$$

12.3 Modification and applications of the IAM

There are situations in which IAM's require some modifications. This is sometimes necessary so that a multiterminal network can be interconnected with other multiterminal networks. Or, it will facilitate the analysis of a network with multiterminal networks imbedded in it. We shall describe some of these techniques.

12.3.1 Grounding of a terminal

If one of the terminals, say the n-th, is grounded, then $E_n = 0$. The arrangement, shown in Fig. 12.12, can be considered as a grounded $(n-1)$-port network. There are $n-1$ port voltages and $n-1$ port currents. The current flowing to ground is

$$-I_n = I_1 + I_2 + \cdots + I_{n-1} \qquad (12.17)$$

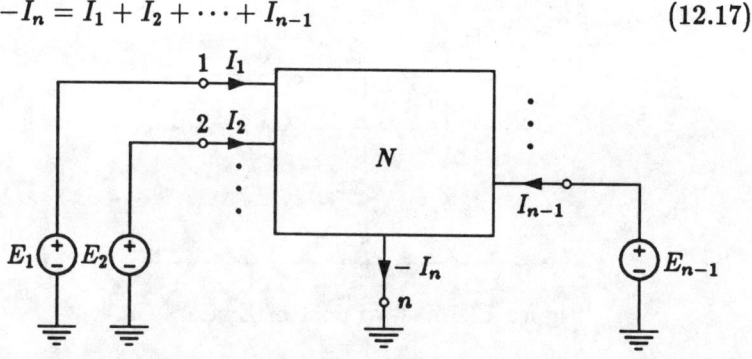

Figure 12.12: An n-terminal network with one terminal grounded.

and is usually of no particular interest in such a situation. The integrity of the ports, which we emphasized greatly in Chapter 10, becomes moot as $-I_n$ can be arbitrarily divided and allotted to various ports.

The y matrix of the $(n-1)$-port is simply the IAM of the n-terminal network with the n-th row and the n-th column deleted, or

$$\begin{bmatrix} I_1 \\ I_2 \\ \vdots \\ I_{n-1} \end{bmatrix} = \begin{bmatrix} y_{11} & y_{12} & \cdots & y_{1(n-1)} \\ y_{21} & y_{22} & \cdots & y_{2(n-1)} \\ \vdots & \vdots & & \vdots \\ y_{(n-1)1} & y_{(n-1)2} & \cdots & y_{(n-1)(n-1)} \end{bmatrix} \begin{bmatrix} E_1 \\ E_2 \\ \vdots \\ E_{n-1} \end{bmatrix} \quad (12.18)$$

Note that the admittance matrix in Eq. (12.18) is not zero-sum. It is the short-circuit admittance matrix of an $(n-1)$-port network—an extension of the two-port y matrix. It has all the attributes of the two-port y matrix. For instance, it may usually be inverted to form the open-circuit impedance matrix of the $(n-1)$-port network.

The notion that we may take an n-terminal network and commit one of its terminals to ground is particularly useful when $n = 3$. We may commit any one of its three terminals to ground and form a grounded two-port. The two-port y matrix can be obtained by deleting the row and column corresponding to the terminal that has been grounded.

EXAMPLE 4. Obtain the y matrix of the two-port of Fig. 12.13.

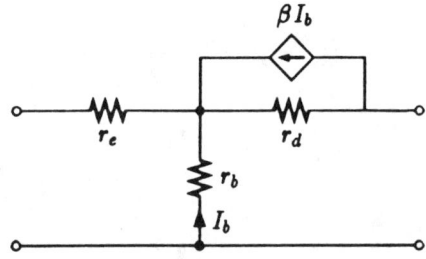

Figure 12.13: Two-port of Example 4.

SOLUTION This two-port is simply the three-terminal network in Fig. 12.7 with its terminal 1 grounded. The y matrix is simply the

MULTITERMINAL NETWORK

IAM of Example 3 with first row and first column deleted. However, after this deletion, the remaining 2 × 2 matrix would have the old terminal 2 as port 1 and the old terminal 3 as port 2. Since this ordering is opposite to that of the two-port in Fig. 12.13, we need to interchange the two rows and the two columns. Hence,

$$[y] = \frac{1}{\Delta} \begin{bmatrix} (1+\beta)r_d & -r_b \\ -\beta r_d + r_b & r_e + r_b \end{bmatrix}$$

EXAMPLE 5. Two-port N of Fig. 12.14(a) has the admittance matrix

$$[y] = \begin{bmatrix} 3 & -1 \\ -2 & 4 \end{bmatrix} \mho$$

If a new two-port, N', is formed by rearranging its terminals as shown in Fig. 12.14(b), what is the y matrix of N'?

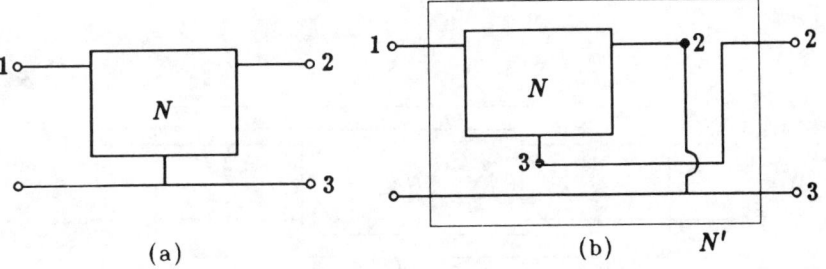

Figure 12.14: Rearrangement of a three-terminal two-port.

SOLUTION The IAM of the three-terminal network N, with the common terminal designated as terminal 3, is

$$[y_i] = \begin{bmatrix} 3 & -1 & -2 \\ -2 & 4 & -2 \\ -1 & -3 & 4 \end{bmatrix}$$

To obtain $[y']$, we simply delete the second row and the second column, or

$$[y'] = \begin{bmatrix} 3 & -2 \\ -1 & 4 \end{bmatrix}$$

12.3.2 Reordering of terminals

It is sometimes necessary or desirable to reorder the terminals of a multiterminal network. For example, in Fig. 12.15, terminal 3 of N becomes terminal 1 of N', and vice versa. Matrix $[y_i]'$ can be obtained by interchanging rows 1 and 3 and columns 1 and 3 of $[y_i]$. For instance, if

$$[y_i] = \begin{bmatrix} 2 & 4 & -1 & -5 \\ 0 & 1 & 2 & -3 \\ -3 & -1 & 0 & 4 \\ 1 & -4 & -1 & 4 \end{bmatrix} \quad (12.19)$$

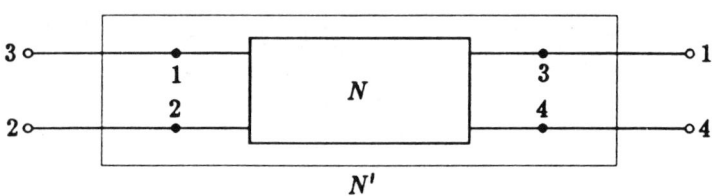

Figure 12.15: Renumbering of two terminals.

then

$$[y_i]' = \begin{bmatrix} 0 & -1 & -3 & 4 \\ 2 & 1 & 0 & -3 \\ -1 & 4 & 2 & -5 \\ -1 & -4 & 1 & 4 \end{bmatrix} \quad (12.20)$$

MULTITERMINAL NETWORK

To avoid confusion, it is advisable to perform this reordering in two steps—first interchange the rows and then interchange the columns.

If the reordering is more than just the interchange of two terminal numbers, this type of interchange of rows and columns may take too many steps. For example, in Fig. 12.16, the terminal numbers of N''

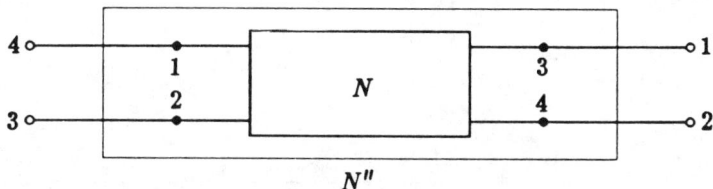

Figure 12.16: Renumbering of all terminals.

and those of N are pretty much randomly related. The most reliable method to obtain $[y_i]''$ from $[y_i]$ is to keep track of the subscripts. In Fig. 12.16, we have $1'' = 3$, $2'' = 4$, $3'' = 2$, and $4'' = 1$. Hence $y''_{11} = y_{33}$, $y''_{12} = y_{34}$, $y''_{13} = y_{32}$, $y''_{14} = y_{31}$, etc. Following this pattern, we have

$$[y_i]'' = \begin{bmatrix} y_{33} & y_{34} & y_{32} & y_{31} \\ y_{43} & y_{44} & y_{42} & y_{41} \\ y_{23} & y_{24} & y_{22} & y_{21} \\ y_{13} & y_{14} & y_{12} & y_{11} \end{bmatrix} \quad (12.21)$$

12.3.3 Terminal combination

If a new $(n-1)$-terminal network is formed from an n-terminal network by combining terminals j and k, then

$$E_{j,k} = E_j = E_k \quad \text{and} \quad I_{j,k} = I_j + I_k \quad (12.22)$$

From Eqs. (12.1) and (12.2), it is easy to see that we can obtain the IAM of the new $(n-1)$-terminal network from the IAM of the original n-terminal network by combining the jth and kth rows and the jth and the kth columns.

As an example, if the IAM of the four-terminal network N is given by Eq. (12.19) and a new three-terminal network N_a is formed by combining terminals 1 and 4 as shown in Fig. 12.17, then rows 1 and 4 and columns 1 and 4 combine, or

$$[y_i]_a = \begin{bmatrix} 2 & 0 & -2 \\ -3 & 1 & 2 \\ 1 & -1 & 0 \end{bmatrix} \qquad (12.23)$$

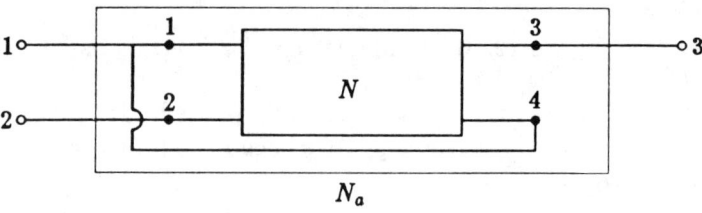

Figure 12.17: Terminal combination.

Again, it is usually advisable to do this in two steps—first combine rows and then combine columns.

EXERCISE

12.3.1 The IAM of the five-terminal network N is

$$[y_i] = \begin{bmatrix} 2 & -4 & -1 & 5 & -2 \\ -1 & 4 & -2 & 0 & -1 \\ 4 & -3 & -2 & 1 & 0 \\ -4 & -6 & 10 & -2 & 2 \\ -1 & 9 & -5 & -4 & 1 \end{bmatrix}$$

Obtain the IAM of the three-terminal network N'.

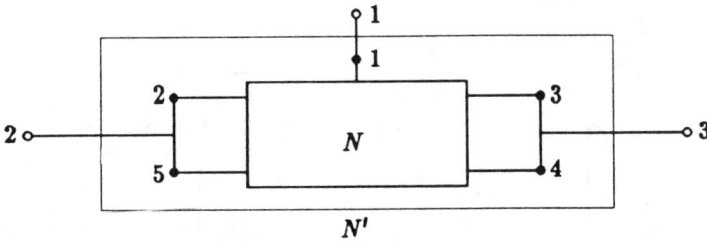

Ans.

$$\begin{bmatrix} 2 & -6 & 4 \\ -2 & 13 & -11 \\ 0 & -7 & 7 \end{bmatrix}$$

12.3.4 Parallel combination

Unlike networks treated on the port basis, networks treated on the terminal basis can always be connected in parallel without using ideal transformers. In combining two n-terminal networks in parallel, it would be straightforward if we prepared both networks so that all terminals connected in parallel had like terminal numbers. Fig. 12.18 shows two four-terminal networks, N_a and N_b, interconnected in such a way that like numbered terminals are connected in parallel. When this is done, then

$$E_j = E_{ja} = E_{jb} \qquad I_j = I_{ja} + I_{jb} \quad j = 1, 2, \ldots, n \qquad (12.24)$$

It is easy to see that

$$[y_i] = [y_i]_a + [y_i]_b \qquad (12.25)$$

EXAMPLE 6. Find the IAM of the three-terminal network in Fig. 12.19 by considering it as the parallel combination of two three-terminal networks.

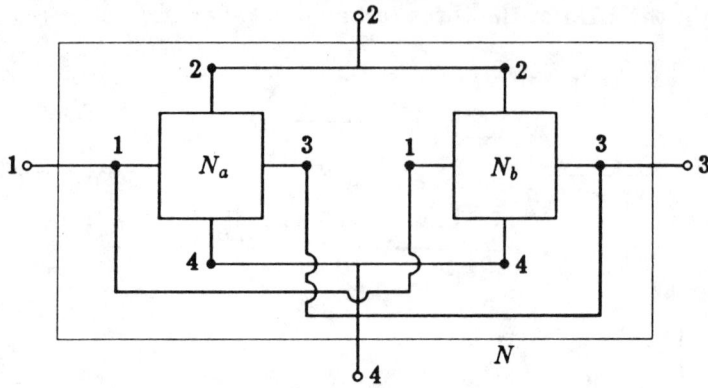

Figure 12.18: Parallel combination of two four-terminal networks.

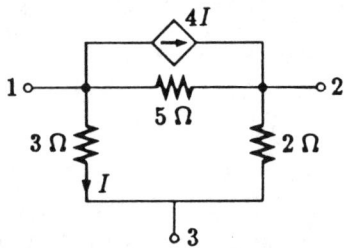

Figure 12.19: A three-terminal network.

SOLUTION We decompose the network in Fig. 12.19 into two networks as shown in Fig. 12.20. For the network in (a), we have

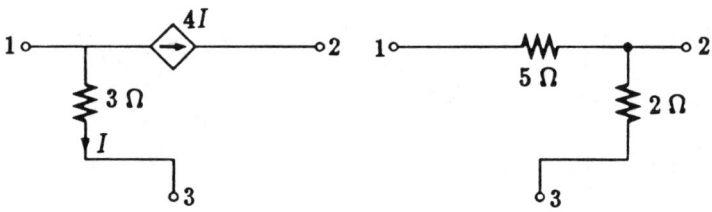

Figure 12.20: Network of Fig. 12.19 decomposed into two networks.

$$I = \frac{1}{3}(E_1 - E_3)$$

$$I_1 = 5I = \frac{5}{3}(E_1 - E_3)$$

MULTITERMINAL NETWORK

$$I_2 = -4I = -\frac{4}{3}(E_1 - E_3)$$

$$I_3 = -I = -\frac{1}{3}(E_1 - E_3)$$

Hence

$$[y_i]_a = \frac{1}{3}\begin{bmatrix} 5 & 0 & -5 \\ -4 & 0 & 4 \\ -1 & 0 & 1 \end{bmatrix}$$

For the network in (b), we have

$$I_1 = \frac{1}{5}(E_1 - E_2)$$

$$I_3 = \frac{1}{2}(E_3 - E_2)$$

$$I_2 = -I_1 - I_3 = -\frac{1}{5}E_1 + \frac{7}{10}E_2 - \frac{1}{2}E_3$$

Thus,

$$[y_i]_b = \frac{1}{10}\begin{bmatrix} 2 & -2 & 0 \\ -2 & 7 & -5 \\ 0 & -5 & 5 \end{bmatrix}$$

For the network in Fig. 12.19

$$[y] = [y_i]_a + [y_i]_b = \begin{bmatrix} \dfrac{28}{15} & -\dfrac{1}{5} & -\dfrac{5}{3} \\ -\dfrac{23}{15} & \dfrac{7}{10} & \dfrac{5}{6} \\ -\dfrac{1}{3} & -\dfrac{1}{2} & \dfrac{5}{6} \end{bmatrix} \mho$$

12.3.5 Interconnecting multiterminal networks

With proper preparation, two interconnected multiterminal networks can sometimes be considered as the parallel combination of two multiterminal networks. We shall illustrate this idea with an example. In Fig. 12.21, two three-terminal networks, N_a and N_b, are inter-

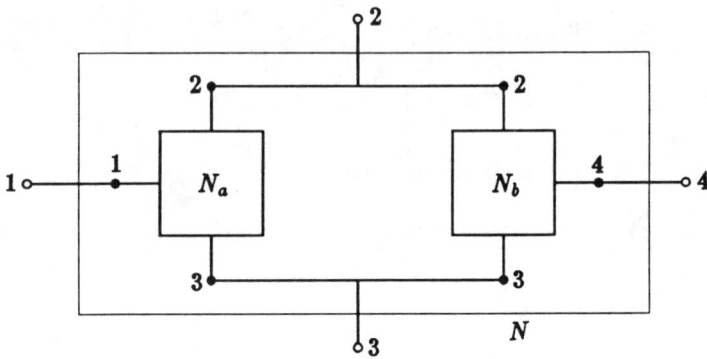

Figure 12.21: Interconnected three-terminal networks.

connected to form a four-terminal network N. If the three-terminal networks were not utilized this way, their IAM's would be 3×3 in dimension. But we can consider N_a as a four-terminal network with one terminal missing—terminal 4. Hence, $I_{4a} = 0$, and E_{4a} has no effect on I_{1a}, I_{2a}, and I_{3a}; and we have

$$[y_i]_a = \begin{bmatrix} y_{11a} & y_{12a} & y_{13a} & 0 \\ y_{21a} & y_{22a} & y_{23a} & 0 \\ y_{31a} & y_{32a} & y_{33a} & 0 \\ 0 & 0 & 0 & 0 \end{bmatrix} \qquad (12.26)$$

in which the non-zero block contains the elements of the original 3×3 matrix. What we have done is to take the original 3×3 matrix (reorder its rows and columns if necessary) and augment it into a 4×4 matrix by bordering it with an additional row and an additional column of zeros.

Similarly, we can consider N_b as a four-terminal network with one terminal missing—terminal 1. Hence,

$$[y_i]_b = \begin{bmatrix} 0 & 0 & 0 & 0 \\ 0 & y_{22b} & y_{23b} & y_{24b} \\ 0 & y_{32b} & y_{33b} & y_{34b} \\ 0 & y_{42b} & y_{43b} & y_{44b} \end{bmatrix} \qquad (12.27)$$

where the non-zero block contains the elements of the original 3 × 3 IAM of N_b.

For the four-terminal network N, we have

$$[y_i] = [y_i]_a + [y_i]_b$$

$$= \begin{bmatrix} y_{11a} & y_{12a} & y_{13a} & 0 \\ y_{21a} & y_{22a} + y_{22b} & y_{23a} + y_{23b} & y_{24b} \\ y_{31a} & y_{32a} + y_{32b} & y_{33a} + y_{33b} & y_{34b} \\ 0 & y_{42b} & y_{43b} & y_{44b} \end{bmatrix} \qquad (12.28)$$

The key to this scheme is that the IAM's of both N_a and N_b have been prepared such that their interconnected terminals have the same terminal numbers. Hence, both IAM's should have been reordered such that their like-numbered terminals are connected together to form the like-numbered terminals of the combination.

EXERCISE

12.3.2 The IAM's of the two three-terminal networks, N_a and N_b are

$$[y_i]_a = \begin{bmatrix} -2 & 3 & -1 \\ 3 & -1 & -2 \\ -1 & -2 & 3 \end{bmatrix} \qquad [y_i]_b = \begin{bmatrix} 6 & -3 & -3 \\ -2 & 4 & -2 \\ -4 & -1 & 5 \end{bmatrix}$$

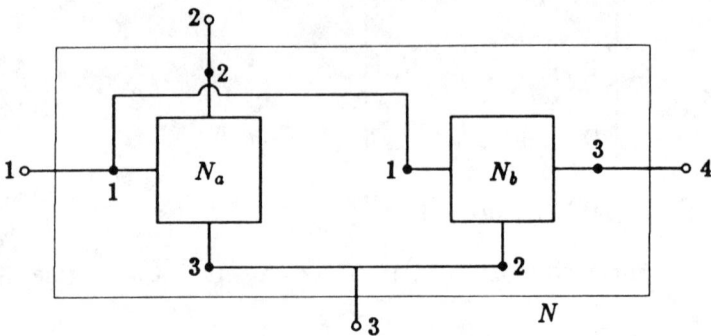

Ans.

$$\begin{bmatrix} 4 & 3 & -4 & -3 \\ 3 & -1 & -2 & 0 \\ -3 & -2 & 7 & -2 \\ -4 & 0 & -1 & 5 \end{bmatrix}$$

12.3.6 Internalizing a number of terminals

One of the most powerful techniques for manipulating the IAM is to take an n-terminal network and reduce it to an m-terminal one by leaving the last $n - m$ terminals unconnected. The process is schematically shown in Fig. 12.22 and may be described as *internalizing* or *suppressing* terminals $m + 1$ through n. In internalizing these terminals, we require that

$$I_{m+1} = I_{m+2} = \cdots = I_n = 0 \tag{12.29}$$

and $E_{m+1}, E_{m+2}, \ldots, E_n$ become irrelevant to N'.

We may partition the IAM after the m-th row and the m-th column, as shown.

MULTITERMINAL NETWORK

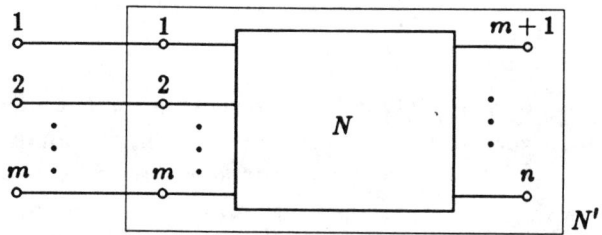

Figure 12.22: Internalizing a number of terminals.

$$\begin{bmatrix} y_{11} & \cdots & y_{1m} & y_{1(m+1)} & \cdots & y_{1n} \\ \vdots & & \vdots & \vdots & & \vdots \\ y_{m1} & \cdots & y_{mm} & y_{m(m+1)} & \cdots & y_{mn} \\ \hdashline y_{(m+1)1} & \cdots & y_{(m+1)m} & y_{(m+1)(m+1)} & \cdots & y_{(m+1)n} \\ \vdots & & \vdots & \vdots & & \vdots \\ y_{n1} & \cdots & y_{nm} & y_{n(m+1)} & \cdots & y_{nn} \end{bmatrix}$$

$$= \begin{bmatrix} [Y_{11}] & [Y_{12}] \\ [Y_{21}] & [Y_{22}] \end{bmatrix} \quad (12.30)$$

Then we have

$$\begin{bmatrix} I_1 \\ \vdots \\ I_m \end{bmatrix} = [Y_{11}] \begin{bmatrix} E_1 \\ \vdots \\ E_m \end{bmatrix} + [Y_{12}] \begin{bmatrix} E_{m+1} \\ \vdots \\ E_n \end{bmatrix} \quad (12.31)$$

$$[0] = [Y_{21}] \begin{bmatrix} E_1 \\ \vdots \\ E_m \end{bmatrix} + [Y_{22}] \begin{bmatrix} E_{m+1} \\ \vdots \\ E_n \end{bmatrix} \quad (12.32)$$

Solving for $[E_{m+1} \ldots E_n]_t$ from Eq. (12.32) and substituting it into Eq. (12.31), we obtain

$$\begin{bmatrix} I_1 \\ \vdots \\ I_m \end{bmatrix} = \left\{ [Y_{11}] - [Y_{12}][Y_{22}]^{-1}[Y_{21}] \right\} \begin{bmatrix} E_1 \\ \vdots \\ E_m \end{bmatrix} \quad (12.33)$$

FUNDAMENTALS OF CIRCUIT ANALYSIS

Hence for N'

$$[y_i]' = [Y_{11}] - [Y_{12}][Y_{22}]^{-1}[Y_{21}] \qquad (12.34)$$

The process of Eq. (12.34) is known as *pivotal condensation* after the m-th row and the m-th column. This technique becomes particularly attractive when n is considerably larger than m. Instead of dealing with a complex network with fewer (m) terminals, we may first consider all internal nodes as terminals and deal with a simpler network with larger number (n) of terminals. The $n \times n$ IAM is relatively simple to obtain. We can use the process described in Subsection 12.2.3. Then we internalize all nodes with no external connections. The $m \times m$ IAM of the m-terminal network is readily obtained by the process of pivotal condensation as given in Eq. (12.34).

EXAMPLE 7. Obtain the IAM of the three-terminal network in Fig. 12.23.

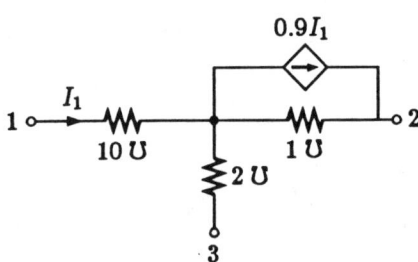

Figure 12.23: Network of Example 7.

SOLUTION The given network may be temporarily considered as a four-terminal network, and it may further be considered as the parallel combination of the two four-terminal networks shown in Fig. 12.24. The IAM's of the individual networks are readily obtained. We have

$$[y_i]_a = \begin{bmatrix} 0 & 0 & 0 & 0 \\ 0 & 1 & 0 & -1 \\ 0 & 0 & 2 & -2 \\ 0 & -1 & -2 & 3 \end{bmatrix} \qquad [y_i]_b = \begin{bmatrix} 10 & 0 & 0 & -10 \\ -9 & 0 & 0 & 9 \\ 0 & 0 & 0 & 0 \\ -1 & 0 & 0 & 1 \end{bmatrix}$$

MULTITERMINAL NETWORK

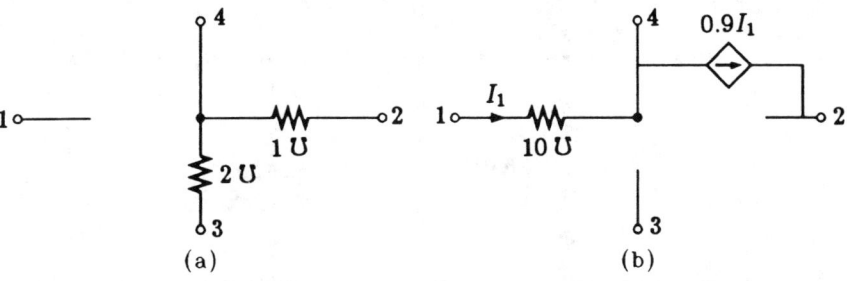

Figure 12.24: Two four-terminal networks from the network of Fig. 12.23.

$$[y_i]_a + [y_i]_b = \begin{bmatrix} 10 & 0 & 0 & -10 \\ -9 & 1 & 0 & 8 \\ 0 & 0 & 2 & -2 \\ -1 & -1 & -2 & 4 \end{bmatrix}$$

$$[y_i] = \begin{bmatrix} 10 & 0 & 0 \\ -9 & 1 & 0 \\ 0 & 0 & 2 \end{bmatrix} - \begin{bmatrix} -10 \\ 8 \\ -2 \end{bmatrix} [4]^{-1} \begin{bmatrix} -1 & -1 & -2 \end{bmatrix}$$

$$= \begin{bmatrix} 7.5 & -2.5 & -5 \\ -7 & 3 & 4 \\ -0.5 & -0.5 & 1 \end{bmatrix} \mho$$

EXAMPLE 8. Obtain the y matrix of the two-port shown in Fig. 12.25. The resistances are 1-Ω each.

SOLUTION We first consider the network to be a nine-terminal one. Its IAM is readily found to be

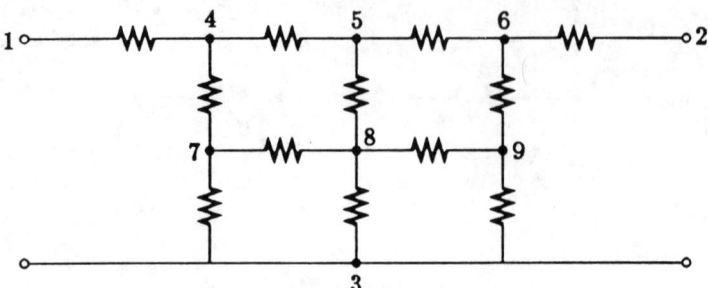

Figure 12.25: Network used in Example 8.

$$\begin{bmatrix} 1 & 0 & 0 & -1 & 0 & 0 & 0 & 0 & 0 \\ 0 & 1 & 0 & 0 & 0 & -1 & 0 & 0 & 0 \\ 0 & 0 & 3 & 0 & 0 & 0 & -1 & -1 & -1 \\ -1 & 0 & 0 & 3 & -1 & 0 & -1 & 0 & 0 \\ 0 & 0 & 0 & -1 & 3 & -1 & 0 & -1 & 0 \\ 0 & -1 & 0 & 0 & -1 & 3 & 0 & 0 & -1 \\ 0 & 0 & -1 & -1 & 0 & 0 & 3 & -1 & 0 \\ 0 & 0 & -1 & 0 & -1 & 0 & -1 & 4 & -1 \\ 0 & 0 & -1 & 0 & 0 & -1 & 0 & -1 & 3 \end{bmatrix}$$

The IAM of the three-terminal network is

$$[y_i] = \begin{bmatrix} 1 & 0 & 0 \\ 0 & 1 & 0 \\ 0 & 0 & 3 \end{bmatrix} - \begin{bmatrix} -1 & 0 & 0 & 0 & 0 & 0 \\ 0 & 0 & -1 & 0 & 0 & 0 \\ 0 & 0 & 0 & -1 & -1 & -1 \end{bmatrix}$$

$$\cdot \begin{bmatrix} 3 & -1 & 0 & -1 & 0 & 0 \\ -1 & 3 & -1 & 0 & -1 & 0 \\ 0 & -1 & 3 & 0 & 0 & -1 \\ -1 & 0 & 0 & 3 & -1 & 0 \\ 0 & -1 & 0 & -1 & 4 & -1 \\ 0 & 0 & -1 & 0 & -1 & 3 \end{bmatrix}^{-1} \begin{bmatrix} -1 & 0 & 0 \\ 0 & 0 & 0 \\ 0 & -1 & 0 \\ 0 & 0 & -1 \\ 0 & 0 & -1 \\ 0 & 0 & -1 \end{bmatrix}$$

$$= \begin{bmatrix} \dfrac{191}{368} & -\dfrac{39}{368} & -\dfrac{19}{46} \\ -\dfrac{39}{368} & \dfrac{191}{368} & -\dfrac{19}{46} \\ -\dfrac{19}{46} & -\dfrac{19}{46} & \dfrac{19}{23} \end{bmatrix}$$

For the two-port,

$$[y] = \begin{bmatrix} \dfrac{191}{368} & -\dfrac{39}{368} \\ -\dfrac{39}{368} & \dfrac{191}{368} \end{bmatrix} \mho$$

EXERCISE

12.3.3 Obtain the IAM of the three-terminal network shown below by internalizing terminals 4 and 5.

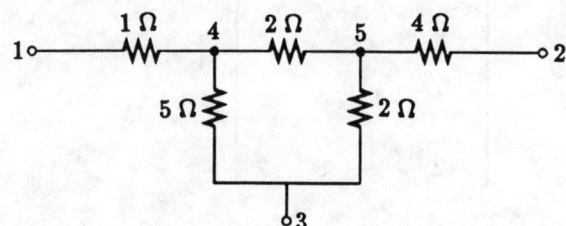

Ans.

$$\begin{bmatrix} \dfrac{1}{3} & -\dfrac{1}{15} & -\dfrac{4}{15} \\ -\dfrac{1}{15} & \dfrac{29}{150} & -\dfrac{19}{150} \\ -\dfrac{4}{15} & -\dfrac{19}{150} & \dfrac{59}{150} \end{bmatrix} \mho$$

12.4 Networks with multiterminal subnetworks

The analysis of a network with an n-terminal subnetwork imbedded in it is analogous to the analysis of a network with a two-port imbedded in it. The n-terminal subnetwork basically imposes certain constraints on a number of node voltages and terminal currents. Incorporating these constraints, which may be quantitatively described by the IAM of the subnetwork on the network, together with any other analysis techniques one may choose to use, is really the heart of the problem. We shall illustrate this idea with several examples.

Frequently, in working with a network with a multiterminal subnetwork imbedded in it, it is expedient to arbitrarily ground one of the terminals of the subnetwork. This commits one of the terminal voltages to zero; the current flowing into that terminal becomes irrelevant. Thus, the number of unknowns is reduced by two.

EXAMPLE 9. The four-terminal network N (It is not a two-port!) in Fig. 12.26 has the following IAM

$$[y_i] = \begin{bmatrix} 1 & 2 & -4 & 1 \\ 2 & 5 & -6 & -1 \\ -1 & -2 & 5 & -2 \\ -2 & -5 & 5 & 2 \end{bmatrix} \mho$$

Determine I_1 and I_3.

MULTITERMINAL NETWORK

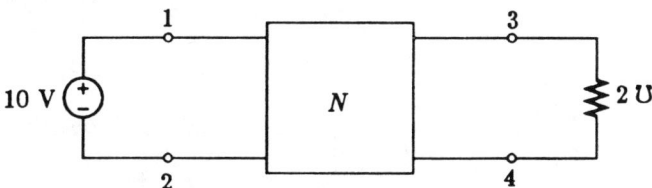

Figure 12.26: Circuit used in Example 9.

SOLUTION We arbitrarily let $E_2 = 0$. Then $E_1 = 10$. Network N requires

$$I_1 = 10 - 4E_3 + E_4$$

$$I_3 = -10 + 5E_3 - 2E_4$$

$$I_4 = -20 + 5E_3 + 2E_4$$

We also have

$$I_3 = -I_4 = 2(E_4 - E_3)$$

Solving these equations simultaneously, we obtain

$$I_1 = 0.75 \text{ A} \quad \text{and} \quad I_3 = -0.5 \text{ A}$$

EXAMPLE 10. The four-terminal network N in Fig. 12.27 has the following IAM

$$[y_i] = \begin{bmatrix} 5 & -2 & 1 & -4 \\ -1 & 1 & -2 & 2 \\ -2 & 3 & 4 & -5 \\ -2 & -2 & -3 & 7 \end{bmatrix} \mho$$

Determine all four terminal currents.

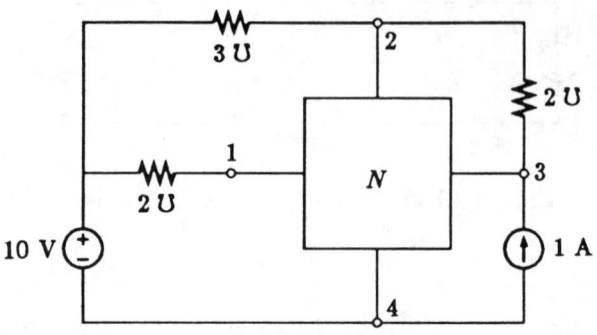

Figure 12.27: Circuit used in Example 10.

SOLUTION We arbitrarily ground terminal 4 and, thus, $E_4 = 0$. Network N requires

$$I_1 = 5E_1 - 2E_2 + E_3$$

$$I_2 = -E_1 + E_2 - 2E_3$$

$$I_3 = -2E_1 + 3E_2 + 4E_3$$

KCL requires

$$I_1 = 2(10 - E_1)$$

$$I_2 = 3(10 - E_2) + 2(E_3 - E_2)$$

$$I_3 = 1 + 2(E_2 - E_3)$$

Solution gives

$$I_1 = \frac{236}{23} \text{ A} \qquad I_2 = -\frac{205}{23} \text{ A} \qquad I_3 = \frac{177}{23} \text{ A}$$

Finally,

$$I_4 = -I_1 - I_2 - I_3 = -\frac{208}{23} \text{ A}$$

MULTITERMINAL NETWORK

EXAMPLE 11. The four-terminal network N in Fig. 12.28 has the following IAM.

$$[y_i] = \begin{bmatrix} 10 & -5 & -4 & -1 \\ -4 & 6 & -2 & 0 \\ -1 & -2 & 7 & -4 \\ -5 & 1 & -1 & 5 \end{bmatrix} \text{ U}$$

Find the Thévenin's equivalent circuit of the two-terminal network N' with respect to terminals A and B.

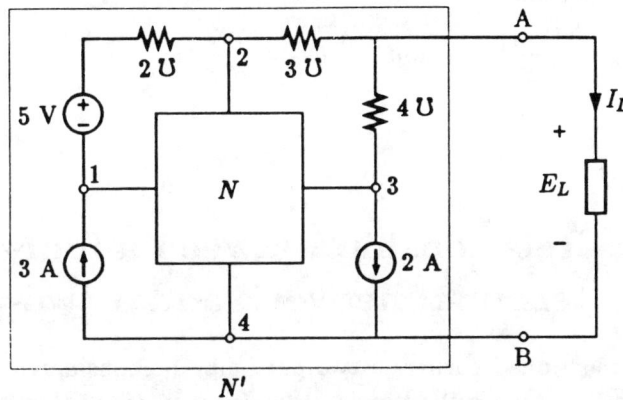

Figure 12.28: Circuit used in Example 11.

SOLUTION We first set $E_4 = 0$. We shall attempt to find the relationship that must be satisfied by I_L and E_L as shown in Fig. 12.28. We have the following equations. Network N requires

$$I_1 = 10E_1 - 5E_2 - 4E_3$$

$$I_2 = -4E_1 + 6E_2 - 2E_3$$

$$I_3 = -E_1 - 2E_2 + 7E_3$$

KCL requires

$$I_1 = 3 + 2(E_2 - 5 - E_1)$$

$$I_2 = 2(5 + E_1 - E_2) + 3(E_L - E_2)$$

$$I_3 = 4(E_L - E_3) - 2$$

$$I_L = 3(E_2 - E_L) + 4(E_3 - E_L)$$

Solving these seven equations to eliminate I_1, I_2, I_3, E_1, E_2, and E_3, we arrive at the relationship

$$E_L = \frac{500}{578} - \frac{209}{578} I_L$$

Comparing this equation with that given by the Thévenin's equivalent

$$E_L = E_{oc} - R_{eq} I_L$$

we readily obtain

$$E_{oc} = \frac{500}{578} \text{ V} \quad \text{and} \quad R_{eq} = \frac{209}{578} \text{ }\Omega$$

12.5 Relationship between a four-terminal network and a two-port

As was mentioned earlier, a two-port may be considered a special case of a four-terminal network. We also mentioned that two-ports are vastly different from four-terminal networks because of the necessity to maintain the integrity of each port as a terminal pair.

In this section, we shall show that when a four-terminal network is made into a two-port, its two-port y matrix can indeed be expressed in terms of the IAM elements. The result also shows that this relationship is not at all simple. Hence, from practical points of view, two-ports and four-terminal networks should be treated as if they were different entities.

We shall use the arrangement in Fig. 12.29 in which N is a four-terminal network and N' is a two-port. We choose terminal 1 to be the upper terminal of port 1 and terminal 2 the upper terminal of port 2. Thus I_1 and I_2 are common to both N and N'. We shall use E's to denote the terminal voltages, and V's the port voltages.

For simplicity, we arbitrarily choose $E_4 = 0$. Then we have

MULTITERMINAL NETWORK

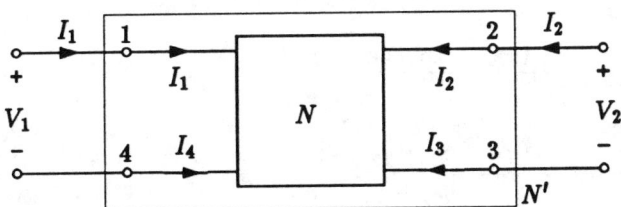

Figure 12.29: A four-terminal network made into a two-port.

$V_1 = E_1$

$E_2 = V_2 + E_3$

From the IAM

$$I_1 = y_{11}V_1 + y_{12}(V_2 + E_3) + y_{13}E_3 \qquad (12.34)$$

$$I_2 = y_{21}V_1 + y_{22}(V_2 + E_3) + y_{23}E_3 \qquad (12.35)$$

$$-I_2 = y_{31}V_1 + y_{32}(V_2 + E_3) + y_{33}E_3 \qquad (12.36)$$

Solving for E_3 from Eq. (12.36), we get

$$E_3 = -\frac{y_{31}V_1 + I_2 + y_{32}V_2}{y_{32} + y_{33}} \qquad (12.37)$$

Substitute Eq. (12.37) into Eq. (12.35) and rearrange to get

$$I_2 = \frac{y_{21}(y_{32} + y_{33}) - y_{31}(y_{22} + y_{23})}{y_{22} + y_{23} + y_{32} + y_{33}}V_1$$

$$+ \frac{y_{22}y_{33} - y_{23}y_{32}}{y_{22} + y_{23} + y_{32} + y_{33}}V_2 \qquad (12.38)$$

Substitute Eqs. (12.37) and (12.38) into Eq. (12.34) and rearrange to get

$$I_1 = \left[y_{11} - \frac{(y_{21} + y_{31})(y_{12} + y_{13})}{y_{22} + y_{23} + y_{32} + y_{33}}\right]V_1$$

$$+ \frac{y_{12}(y_{23} + y_{33}) - y_{13}(y_{22} + y_{32})}{y_{22} + y_{23} + y_{32} + y_{33}}V_2 \qquad (12.39)$$

Hence,

$$[y] = \begin{bmatrix} y_{11} - \dfrac{(y_{21}+y_{31})(y_{13}+y_{12})}{y_{22}+y_{23}+y_{32}+y_{33}} & y_{12} - \dfrac{(y_{22}+y_{32})(y_{13}+y_{12})}{y_{22}+y_{23}+y_{32}+y_{33}} \\ y_{21} - \dfrac{(y_{21}+y_{31})(y_{22}+y_{23})}{y_{22}+y_{23}+y_{32}+y_{33}} & y_{22} - \dfrac{(y_{22}+y_{32})(y_{22}+y_{23})}{y_{22}+y_{23}+y_{32}+y_{33}} \end{bmatrix}$$

Because of the zero-sum property of the four-terminal IAM, $[y]$ may also be written as

$$[y] = \begin{bmatrix} y_{11} + \dfrac{(y_{11}+y_{41})(y_{11}+y_{14})}{y_{21}+y_{24}+y_{31}+y_{34}} & y_{12} + \dfrac{(y_{21}+y_{42})(y_{11}+y_{14})}{y_{21}+y_{24}+y_{31}+y_{34}} \\ y_{21} + \dfrac{(y_{11}+y_{14})(y_{21}+y_{24})}{y_{21}+y_{24}+y_{31}+y_{34}} & y_{22} + \dfrac{(y_{12}+y_{42})(y_{21}+y_{24})}{y_{21}+y_{24}+y_{31}+y_{34}} \end{bmatrix}$$

as well as many other combinations.

These relationships clearly bear out the fact that it is impractical to handle two-ports in terms of their corresponding four-terminal y parameters. However, these developments are of scholastic interest and are a good exercise to satisfy one's intellectual curiosity.

12.6 Summary

When multiterminal devices are not utilized on the port basis, the IAM provides the easiest formalism to characterize the device. When such a device is used in conjunction with other devices or when it is imbedded in a network, using the IAM will enable us to follow systematic procedures to analyze these arrangements.

These methods are useful in incorporating electronic devices in a circuit and in computer-assisted analysis of networks containing multiterminal devices or subnetworks.

Problems

12.1 Find the IAM of the three-terminal network.

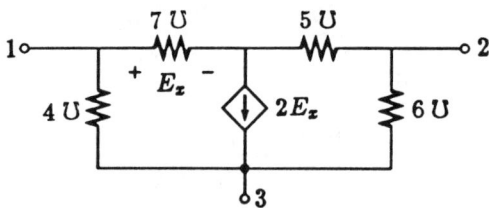

△ **12.2** Find the IAM of the three-terminal network.

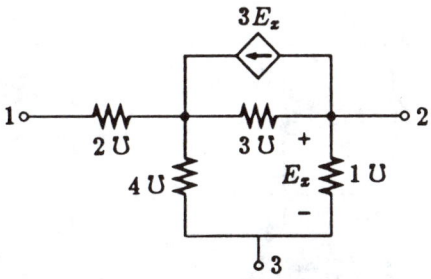

△ **12.3** Find the IAM of the three-terminal network.

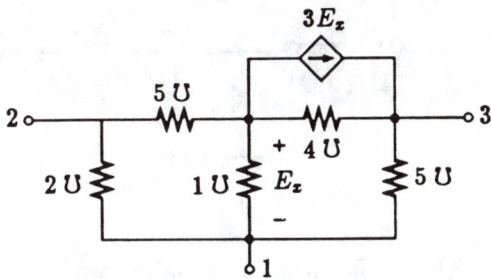

12.4 Find the IAM of the three-terminal network.

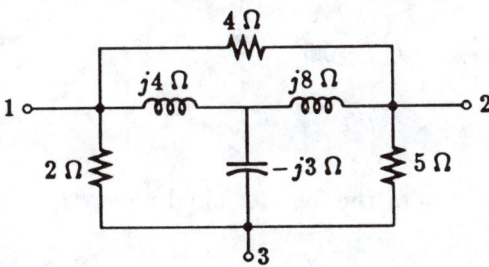

12.5 Find the IAM of the four-terminal network. The resistances are 1-Ω each.

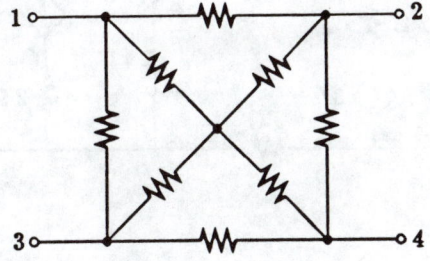

12.6 Find the IAM of the three-terminal network.

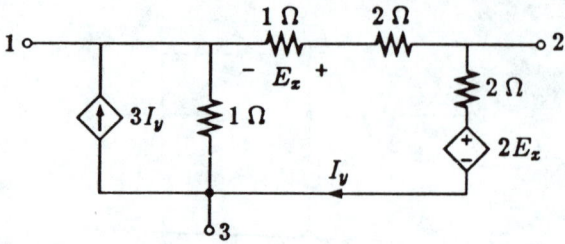

12.7 Find the IAM of the three-terminal network.

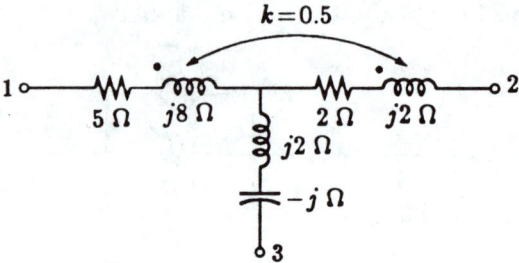

12.8 Find the IAM of the three-terminal network.

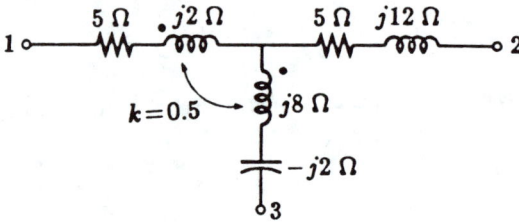

12.9 Find the IAM of the four-terminal network.

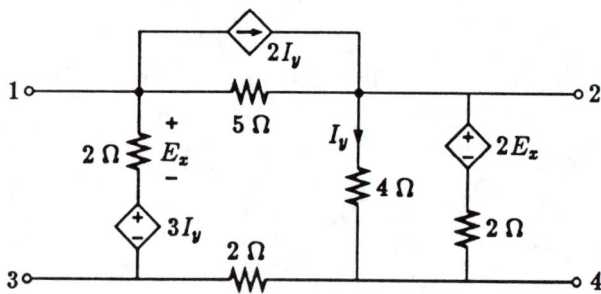

△12.10 First, find the IAM of the three-terminal network in (a). Then find the z matrix of the two-port in (b).

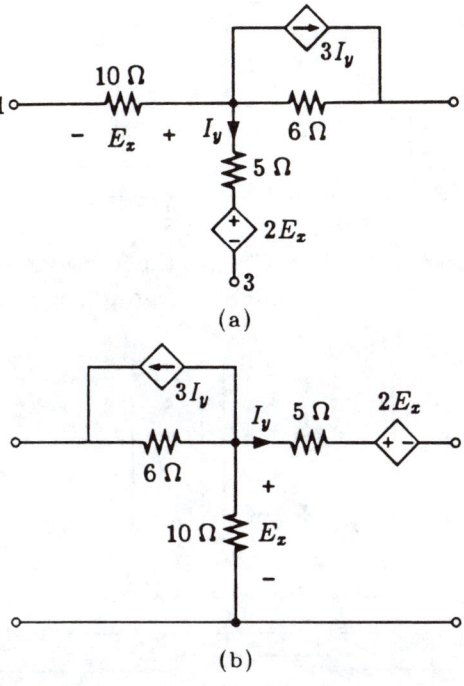

12.11 First find the IAM of the three-terminal network on the left. Then find the z matrix of the two-port on the right.

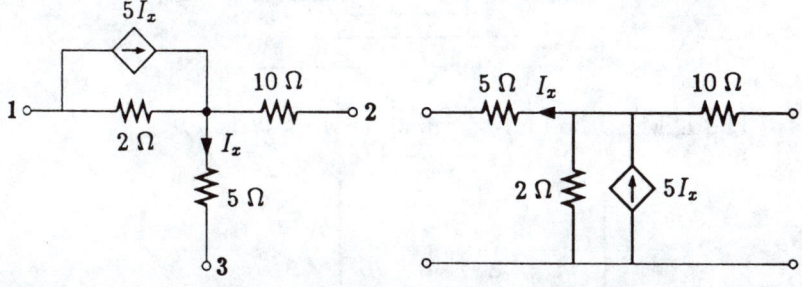

12.12 The IAM of the three-terminal network N is

$$[y_i] = \begin{bmatrix} 50 & -15 & -35 \\ -28 & 45 & -17 \\ -22 & -30 & 52 \end{bmatrix} \mho$$

Find the y matrix of two-port N'.

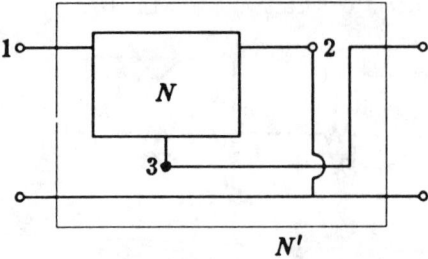

N'

12.13 The four-terminal network N has the following voltage-current relationship when its terminal 4 is grounded.

$$\begin{bmatrix} I_1 \\ I_2 \\ I_3 \end{bmatrix} = \begin{bmatrix} 2 & 4 & -1 \\ 6 & -3 & 8 \\ 1 & 2 & -4 \end{bmatrix} \begin{bmatrix} E_1 \\ E_2 \\ E_3 \end{bmatrix}$$

Determine the y matrix of two-port N'.

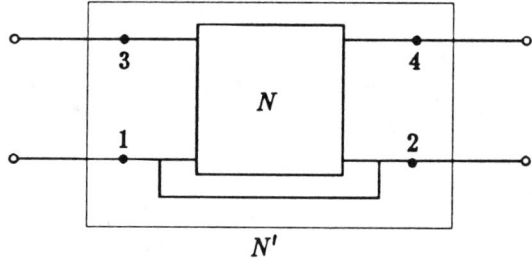

N'

12.14 The IAM of the four-terminal network N is

$$[y_i] = \begin{bmatrix} 3 & -2 & -2 & 1 \\ 4 & 8 & -6 & -6 \\ 5 & 0 & 2 & -7 \\ -12 & -6 & 6 & 12 \end{bmatrix} \text{U}$$

Find the Thévenin's equivalent of network N' with respect to terminals A and B.

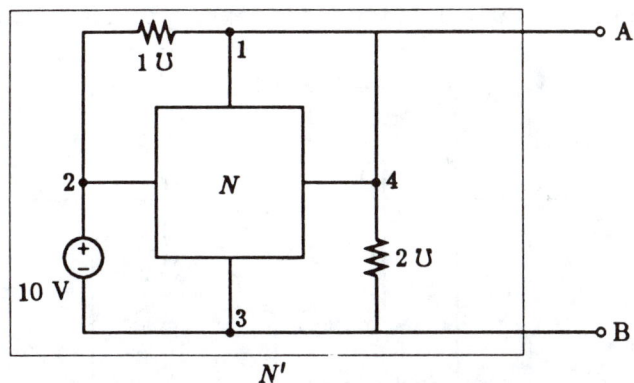

△ **12.15** The IAM of the six-terminal network N is

$$[y_i] = \begin{bmatrix} 1 & -2 & 5 & -1 & -3 & 0 \\ -2 & 3 & -2 & 4 & -1 & -2 \\ 3 & -4 & -1 & 5 & 0 & -3 \\ 2 & -3 & -2 & 2 & 2 & -1 \\ 5 & 1 & -2 & -4 & -3 & 3 \\ -9 & 5 & 2 & -6 & 5 & 3 \end{bmatrix} \text{U}$$

Find the y matrix of two-port N'.

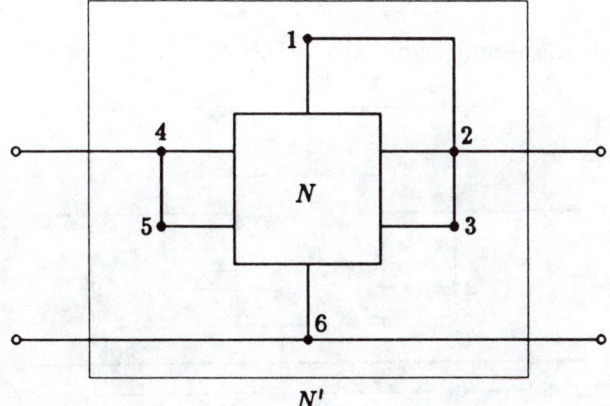

▲ **12.16** The IAM of the six-terminal network N is given in the previous problem. Find the admittance between terminals A and B.

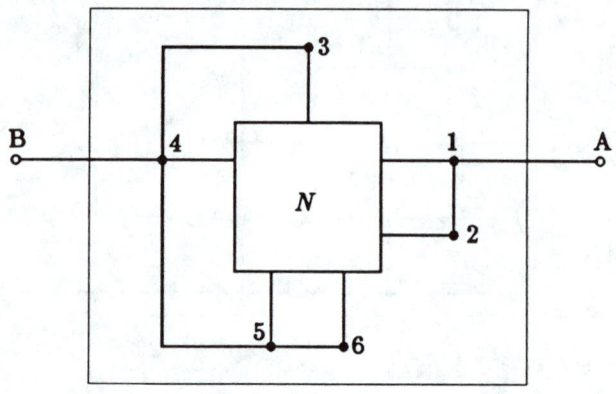

N'

▲ **12.17** The IAM of the seven-terminal network N is

$$[y_i] = \begin{bmatrix} -2 & 0 & 4 & 3 & -1 & -4 & 0 \\ 1 & 2 & -1 & 4 & -2 & 3 & -7 \\ -3 & 1 & -1 & -2 & 4 & 1 & 0 \\ -5 & -3 & -2 & 5 & 4 & -1 & 2 \\ 1 & 1 & -2 & -2 & 3 & -2 & 1 \\ 2 & -2 & 3 & -1 & 4 & -3 & -3 \\ 6 & 1 & -1 & -7 & -12 & 6 & 7 \end{bmatrix} \mho$$

Find the y matrix of two-port N'.

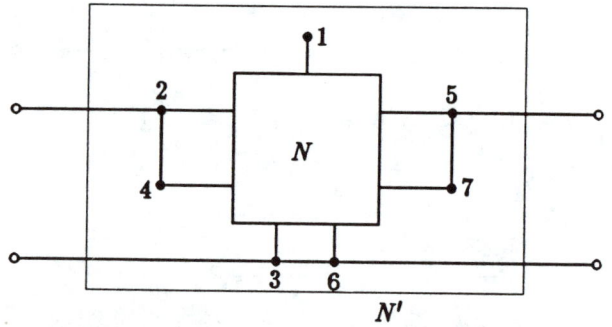

N'

12.18 The IAM of the five-terminal network N is

$$[y_i] = \begin{bmatrix} 5 & -2 & -1 & 0 & -2 \\ -1 & 4 & -2 & -1 & 0 \\ -2 & 0 & 4 & -1 & -1 \\ -2 & -1 & 0 & 2 & 1 \\ 0 & -1 & -1 & 0 & 2 \end{bmatrix} \text{U}$$

Determine the admittance between terminals A and B.

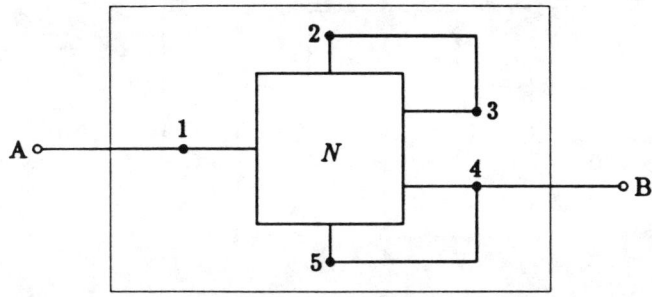

12.19 The IAM of the four-terminal network N is given in the previous problem. Determine the y matrix of two-port N'.

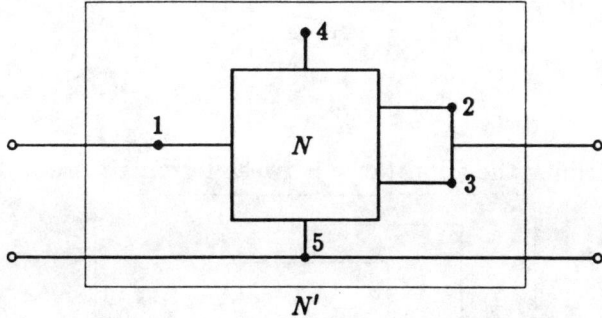

12.20 The IAM of the four-terminal network N is

$$[y_i] = \begin{bmatrix} 2 & -1 & -4 & 3 \\ 1 & 2 & -1 & -2 \\ -3 & 1 & 2 & 0 \\ 0 & -2 & 3 & -1 \end{bmatrix} \mho$$

Determine I and the four terminal currents of N.

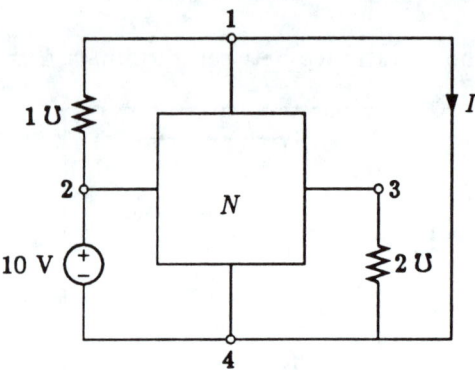

△ **12.21** The IAM of the four-terminal network N is

$$[y_i] = \begin{bmatrix} 2 & -4 & 6 & -4 \\ -5 & 3 & -1 & 3 \\ 9 & -5 & -1 & -3 \\ -6 & 6 & -4 & 4 \end{bmatrix} \mho$$

Determine the admittance between terminals A and B.

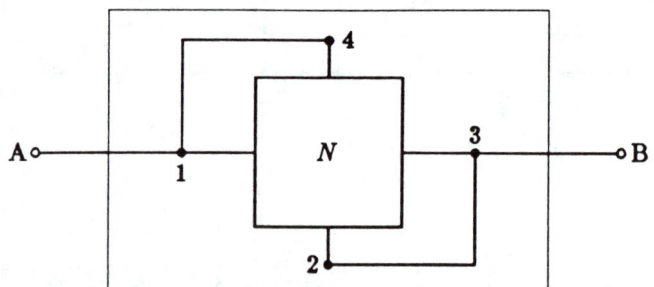

12.22 The current-voltage relationship of the five-terminal network N with terminal 5 grounded is

$$\begin{bmatrix} I_1 \\ I_2 \\ I_3 \\ I_4 \end{bmatrix} = \begin{bmatrix} 1 & 2 & -1 & 4 \\ 2 & -5 & 7 & -2 \\ 3 & 4 & 3 & -2 \\ -5 & 0 & -2 & 4 \end{bmatrix} \begin{bmatrix} E_1 \\ E_2 \\ E_3 \\ E_4 \end{bmatrix}$$

Determine the y matrix of two-port N'.

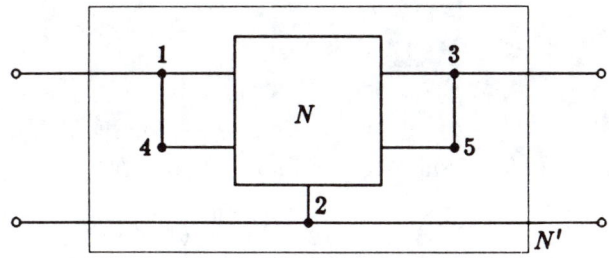

12.23 The IAM of the two four-terminal networks, N_1 and N_2, are respectively

$$[y_i]_1 = \begin{bmatrix} 3 & 2 & -1 & -4 \\ -1 & 1 & -3 & 3 \\ 1 & 0 & -2 & 1 \\ -3 & -3 & 6 & 0 \end{bmatrix} \mho \quad [y_i]_2 = \begin{bmatrix} 8 & -2 & -1 & -5 \\ -2 & 0 & 3 & -1 \\ -4 & 3 & -2 & 3 \\ -2 & -1 & 0 & 3 \end{bmatrix} \mho$$

Determine the IAM of the three-terminal network N.

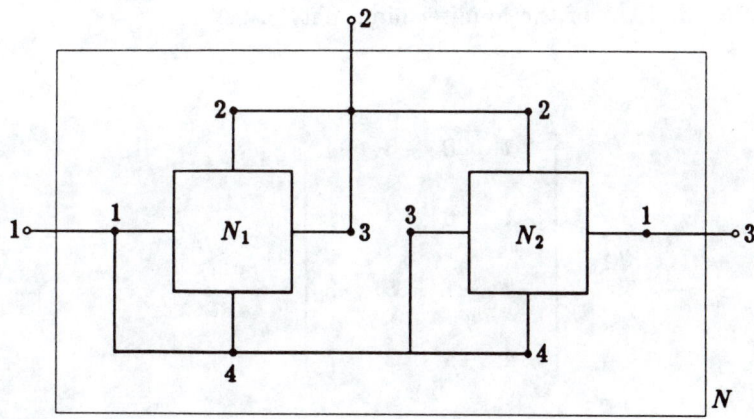

▲ 12.24 The IAM of the three-terminal network N_1 and that of the four-terminal network N_2 are respectively

$$[y_i]_1 = \begin{bmatrix} 3 & -2 & -1 \\ -1 & 4 & -3 \\ -2 & -2 & 4 \end{bmatrix} \mho \quad [y_i]_2 = \begin{bmatrix} 5 & -2 & -2 & -1 \\ -1 & 6 & -3 & -2 \\ -3 & -1 & 4 & 0 \\ -1 & -3 & 1 & 3 \end{bmatrix} \mho$$

Find the IAM of the five-terminal network N.

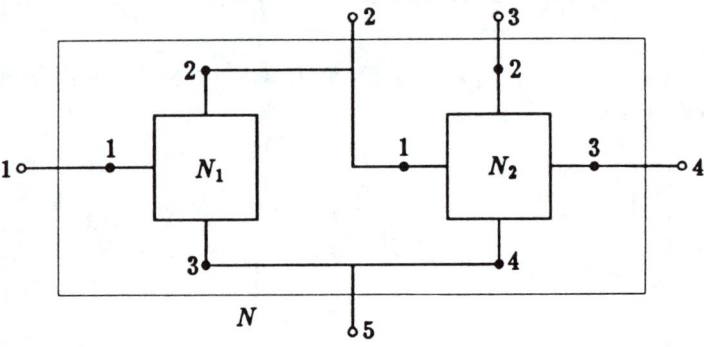

12.25 the IAM of the four-terminal network N is

$$[y_i] = \begin{bmatrix} 7 & 0 & -3 & -4 \\ -1 & 6 & -4 & -1 \\ -2 & -1 & 5 & -2 \\ -4 & -5 & 2 & 7 \end{bmatrix} \mho$$

Determine I.

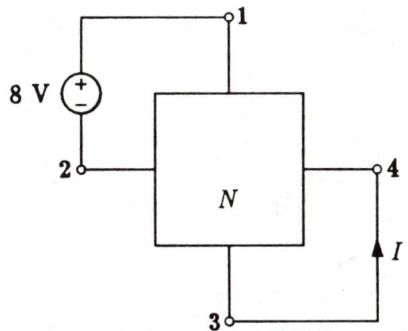

△**12.26** The IAM of the six-terminal network N is

$$[y_i] = \begin{bmatrix} 4 & 3 & -2 & -1 & -3 & -1 \\ 0 & 2 & -3 & 1 & 4 & -4 \\ -2 & -1 & 4 & 2 & 0 & -3 \\ 5 & -3 & 2 & -3 & 1 & -2 \\ 1 & 2 & -3 & 4 & -2 & -2 \\ -8 & -3 & 2 & -3 & 0 & 12 \end{bmatrix} \mho$$

Determine the admittance between terminals A and B.

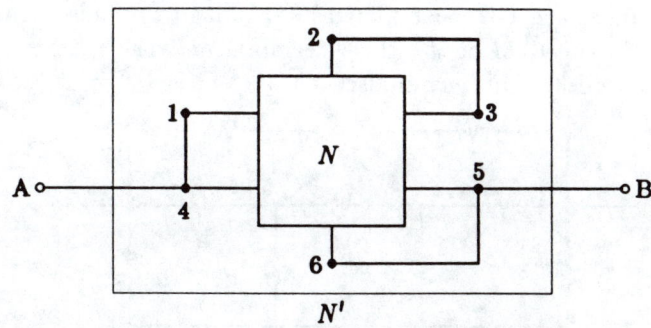

12.27 The six-terminal network is the same one given in the previous problem. Obtain the y matrix of two-port N'.

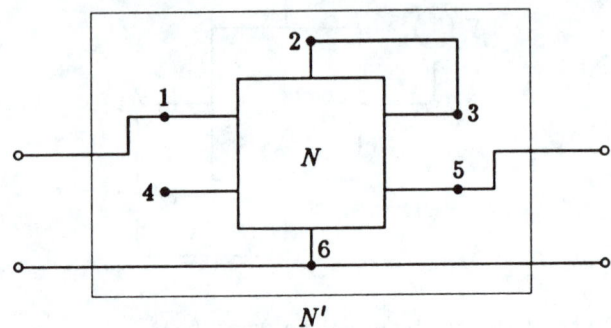

12.28 First obtain the IAM of the four-terminal network. Then internalize terminal 4 to obtain the IAM of the three-terminal network with terminals 1, 2, and 3 as its terminals.

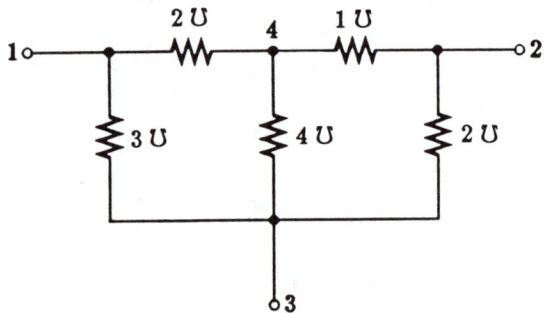

12.29 First obtain the IAM of the five-terminal network. The resistances are 1-Ω each. Then internalize terminals 4 and 5 to obtain the IAM of the three-terminal network with terminals 1, 2, and 3 as its terminals.

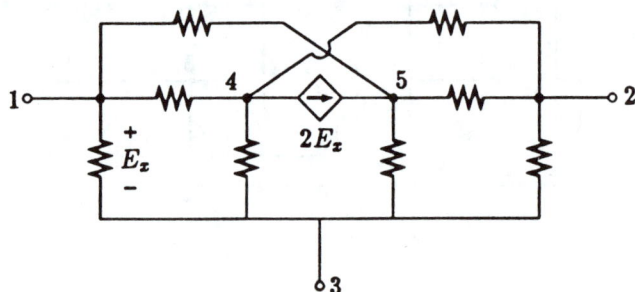

MULTITERMINAL NETWORK

△12.30 First obtain the IAM of the five-terminal network. Then internalize terminals 4 and 5 to obtain the IAM of the three-terminal network with terminals 1, 2, and 3 as its terminals.

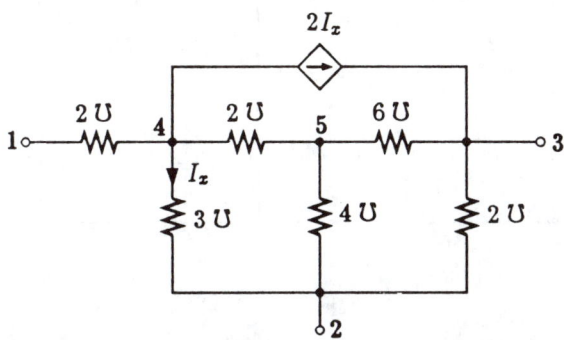

△12.31 First, remove the two short circuits to obtain a seven-terminal network. Obtain the IAM of the seven-terminal network. Then replace the short circuits and internalize terminals 4 through 7 to obtain the IAM of the three-terminal network with terminals 1, 2, and 3 as its terminals.

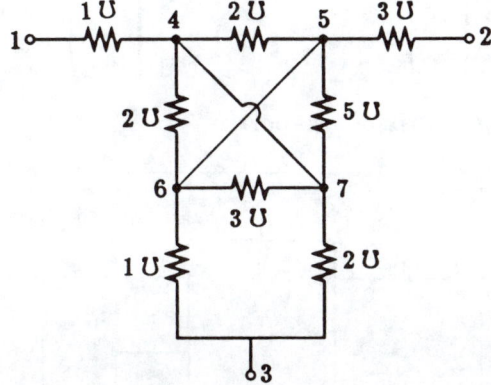

△12.32 First obtain the IAM of the six-terminal network. Then internalize terminals 4 through 6 to obtain the IAM of the three-terminal network with terminals 1, 2, and 3 as its terminals.

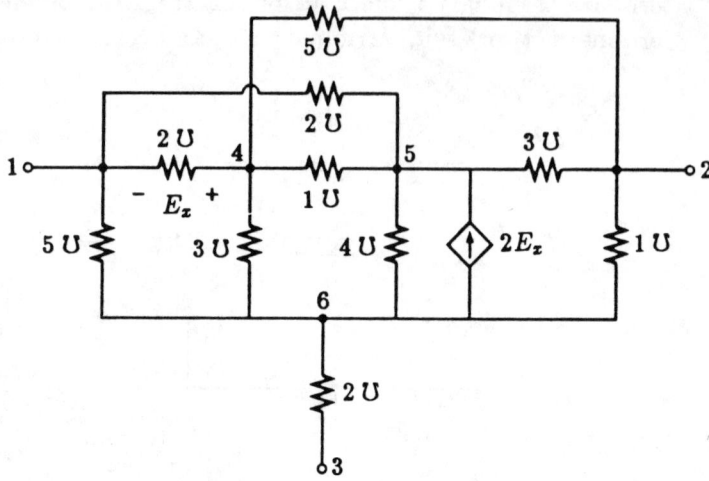

12.33 The current-voltage relationship of the five-terminal network N with terminal 1 grounded is given by

$$\begin{bmatrix} I_2 \\ I_3 \\ I_4 \\ I_5 \end{bmatrix} = \begin{bmatrix} 1 & -2 & 1 & 3 \\ 0 & 4 & -2 & 1 \\ 3 & 2 & 1 & 0 \\ -2 & 1 & 4 & -3 \end{bmatrix} \begin{bmatrix} E_2 \\ E_3 \\ E_4 \\ E_5 \end{bmatrix}$$

Find the y matrix of two-port N'.

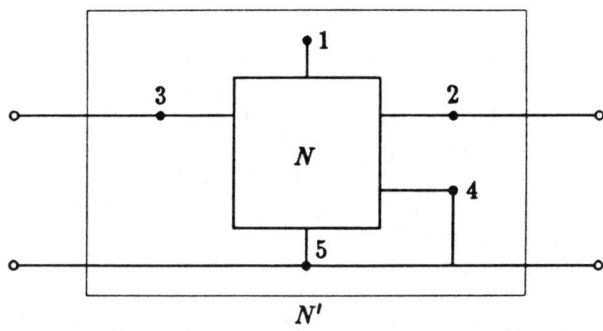

12.34 The IAM of the four-terminal network N is

MULTITERMINAL NETWORK

$$[y_i] = \begin{bmatrix} 4 & -1 & -3 & 0 \\ -2 & 6 & -1 & -3 \\ 0 & -2 & 5 & -3 \\ -2 & -3 & -1 & 6 \end{bmatrix} \mho$$

Find the z matrix of two-port N'.

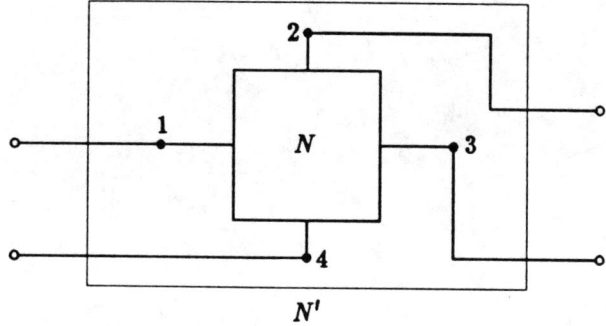

Appendix A

Matrix Algebra

A.1 Definitions

A *matrix* is a rectangular array of elements. Each element is denoted by two subscripts. The first indicates the row and the second indicates the column in which the element appears. Thus matrix $[A]$ represents

$$[A] = \begin{bmatrix} a_{11} & a_{12} & a_{13} & \cdots & a_{1n} \\ a_{21} & a_{22} & a_{23} & \cdots & a_{2n} \\ \vdots & \vdots & \vdots & & \vdots \\ a_{m1} & a_{m2} & a_{m3} & \cdots & a_{mn} \end{bmatrix} = [a_{ij}] \qquad (A.1)$$

The matrix in Eq. (A.1) is said to have a dimension of $m \times n$.

A *row matrix* (also called a row vector) has a dimension of $1 \times n$. Thus

$$[\alpha] = \begin{bmatrix} \alpha_{11} & \alpha_{12} & \alpha_{13} & \cdots & \alpha_{1n} \end{bmatrix} \qquad (A.2)$$

is a row matrix.

A *column matrix* (also called a column vector) has a dimension of $m \times 1$. Thus

$$[\beta] = \begin{bmatrix} \beta_{11} \\ \beta_{21} \\ \beta_{31} \\ \vdots \\ \beta_{m1} \end{bmatrix} \qquad (A.3)$$

is a column matrix.

Two matrices are said to be *equal* if all their corresponding elements are equal. Thus, if

$$a_{ik} = b_{ik}, \quad \text{for all } i \text{ and } k$$

then

$$[A] = [B]$$

If a matrix has only a single element, then it becomes a *scalar*.

The *determinant* of a square matrix is the determinant formed with the elements of the matrix. In other words, the determinant of $[A]$ is

$$|A| = \begin{vmatrix} a_{11} & a_{12} & a_{13} & \cdots & a_{1n} \\ a_{21} & a_{22} & a_{23} & \cdots & a_{2n} \\ \vdots & \vdots & \vdots & & \vdots \\ a_{n1} & a_{n2} & a_{n3} & \cdots & a_{nn} \end{vmatrix} \tag{A.4}$$

The *minor* of the ij element is the determinant formed from the original determinant by deleting the ith row and the jth column. Thus, for the determinant

$$|A| = \begin{vmatrix} a_{11} & a_{12} & a_{13} & a_{14} \\ a_{21} & a_{22} & a_{23} & a_{24} \\ a_{31} & a_{32} & a_{33} & a_{34} \\ a_{41} & a_{42} & a_{43} & a_{44} \end{vmatrix} \tag{A.5}$$

its 11, 23, and 41 minors are

$$M_{11} = \begin{vmatrix} a_{22} & a_{23} & a_{24} \\ a_{32} & a_{33} & a_{34} \\ a_{42} & a_{43} & a_{44} \end{vmatrix} \tag{A.6}$$

$$M_{23} = \begin{vmatrix} a_{11} & a_{12} & a_{14} \\ a_{31} & a_{32} & a_{34} \\ a_{41} & a_{42} & a_{44} \end{vmatrix} \tag{A.7}$$

$$M_{41} = \begin{vmatrix} a_{12} & a_{13} & a_{14} \\ a_{22} & a_{23} & a_{24} \\ a_{32} & a_{33} & a_{34} \end{vmatrix} \tag{A.8}$$

respectively.

The *cofactor* of the ij element of a determinant is denoted by A_{ij} and

$$A_{ij} = (-1)^{i+j} M_{ij} \qquad (A.9)$$

We may obtain the value of a determinant by summing up the products of its element and their cofactors along any row or any column. Hence we get

$$|A| = \sum_{i=1}^{n} a_{ij} A_{ij}, \text{ for any } j; \text{ or } \sum_{j=1}^{n} a_{ij} A_{ij}, \text{ for any } i \qquad (A.10)$$

A.2 Algebraic rules of matrices

The addition of matrices follows the usual rule of addition *element by element*. Hence the sum of two matrices is a matrix whose elements are the sums of the corresponding elements of the addends. Thus, if

$$[A] + [B] = [C]$$

we have

$$c_{ik} = a_{ik} + b_{ik}, \quad \text{for all } i \text{ and } k \qquad (A.11)$$

The subtraction of two matrices is performed similarly. If

$$[A] - [B] = [C]$$

then

$$c_{ik} = a_{ik} - b_{ik}, \quad \text{for all } i \text{ and } k \qquad (A.12)$$

The multiplication of a matrix by a scalar is obtained by multiplying each element of the matrix by the scalar. Thus if

$$[B] = \alpha[A]$$

then

$$b_{ij} = \alpha a_{ij}, \quad \text{for all } i \text{ and } j \qquad (A.13)$$

If $[A]$ is $k \times m$ and $[B]$ is $m \times n$, then $[A]$ and $[B]$ are conformable in that order and the multiplication of $[A]$ and $[B]$ is defined. We obtain the ij element of the product $[A][B]$ by summing up the products of the elements from the ith row of $[A]$ and element from the jth column of $[B]$. That is to say, if

$$[C] = [A][B]$$

then

$$c_{ij} = a_{i1}b_{1j} + a_{i2}b_{2j} + a_{i3}b_{3j} + \cdots + a_{im}b_{mj} = \sum_{k=1}^{m} a_{ik}b_{kj} \qquad (A.14)$$

For example,

$$\begin{bmatrix} a_{11} & a_{12} & a_{13} \\ a_{21} & a_{22} & a_{23} \\ a_{31} & a_{32} & a_{33} \end{bmatrix} \begin{bmatrix} x_{11} & x_{12} \\ x_{21} & x_{22} \\ x_{31} & x_{32} \end{bmatrix}$$

$$= \begin{bmatrix} a_{11}x_{11} + a_{12}x_{21} + a_{13}x_{31} & a_{11}x_{12} + a_{12}x_{22} + a_{13}x_{32} \\ a_{21}x_{11} + a_{22}x_{21} + a_{23}x_{31} & a_{21}x_{12} + a_{22}x_{22} + a_{23}x_{32} \\ a_{31}x_{11} + a_{32}x_{21} + a_{33}x_{31} & a_{31}x_{12} + a_{32}x_{22} + a_{33}x_{32} \end{bmatrix} \qquad (A.15)$$

Multiplication is generally not commutative, so that generally we have

$$[A][B] \neq [B][A] \qquad (A.16)$$

Indeed, even if $[A]$ and $[B]$ are conformable, they may not be conformable in the reversed order.

A.3 Special matrices

1 *Unit matrix or identity matrix,* $[U]$

$$[U] = \begin{bmatrix} 1 & 0 & 0 & \cdots & 0 \\ 0 & 1 & 0 & \cdots & 0 \\ \vdots & \vdots & \vdots & & \vdots \\ 0 & 0 & 0 & \cdots & 1 \end{bmatrix} \qquad (A.17)$$

2 *Scalar matrix,* $[K]$

MATRIX ALGEBRA

$$[K] = k[U] = \begin{bmatrix} k & 0 & 0 & \cdots & 0 \\ 0 & k & 0 & \cdots & 0 \\ \vdots & \vdots & \vdots & & \vdots \\ 0 & 0 & 0 & \cdots & k \end{bmatrix} \quad (A.18)$$

3 *Diagonal matrix*, $[D]$

$$[D] = \begin{bmatrix} d_1 & 0 & 0 & \cdots & 0 \\ 0 & d_2 & 0 & \cdots & 0 \\ \vdots & \vdots & \vdots & & \vdots \\ 0 & 0 & 0 & \cdots & d_n \end{bmatrix} \quad (A.19)$$

A diagonal matrix is frequently written as

$$[D] = \mathrm{diag}[d_1 \; d_2 \; d_3 \; \cdots \; d_n]$$

to save space.

4 A matrix is *symmetric* if

$$a_{ij} = a_{ji}, \quad \text{for all } i \text{ and } j$$

5 A matrix is *skew-symmetric* if

$$a_{ij} = -a_{ji}, \quad \text{for all } i \text{ and } j$$

This definition requires that all diagonal elements be zero.

6 The *transpose* of a matrix is the matrix with its rows and columns interchanged. If we denote the transpose of $[A]$ by $[A]_t$ and if

$$[B] = [A]_t$$

Then

$$b_{ij} = a_{ji} \quad (A.20)$$

For example,

$$\begin{bmatrix} 1 & 3 & -2 \\ x & 5 & 0 \end{bmatrix}_t = \begin{bmatrix} 1 & x \\ 3 & 5 \\ -2 & 0 \end{bmatrix}$$

A column matrix is frequently written as the transpose of a row matrix to save space. For example,

$$\begin{bmatrix} \beta_1 \\ \beta_2 \\ \vdots \\ \beta_m \end{bmatrix} = \begin{bmatrix} \beta_1 & \beta_2 & \cdots & \beta_m \end{bmatrix}_t$$

7 The *adjoint* of a square matrix is a new matrix formed by placing the cofactor of the ij element of $[A]$ in the ji position. Denoting the adjoint of $[A]$ by $\mathrm{adj}[A]$ and letting

$$[B] = \mathrm{adj}[A]$$

we get

$$b_{ij} = A_{ji} \tag{A.21}$$

For example, if we have

$$[A] = \begin{bmatrix} a_{11} & a_{12} & a_{13} \\ a_{21} & a_{22} & a_{23} \\ a_{31} & a_{32} & a_{33} \end{bmatrix}$$

then we get

$$\mathrm{adj}[A] = \begin{bmatrix} \begin{vmatrix} a_{22} & a_{23} \\ a_{32} & a_{33} \end{vmatrix} & -\begin{vmatrix} a_{12} & a_{13} \\ a_{32} & a_{33} \end{vmatrix} & \begin{vmatrix} a_{12} & a_{13} \\ a_{22} & a_{23} \end{vmatrix} \\ -\begin{vmatrix} a_{21} & a_{23} \\ a_{31} & a_{33} \end{vmatrix} & \begin{vmatrix} a_{11} & a_{13} \\ a_{31} & a_{33} \end{vmatrix} & -\begin{vmatrix} a_{11} & a_{13} \\ a_{31} & a_{23} \end{vmatrix} \\ \begin{vmatrix} a_{22} & a_{23} \\ a_{32} & a_{33} \end{vmatrix} & -\begin{vmatrix} a_{11} & a_{12} \\ a_{31} & a_{32} \end{vmatrix} & \begin{vmatrix} a_{11} & a_{12} \\ a_{21} & a_{22} \end{vmatrix} \end{bmatrix} \tag{A.22}$$

8 A square matrix is *singular* if its determinant is zero. Otherwise it is *nonsingular*.

9 The *inverse* of a nonsingular matrix $[A]$ is another matrix whose product with $[A]$ (in either order) is an identity matrix. The inverse of $[A]$ is denoted by $[A]^{-1}$.

The inverse of a matrix is unique, and is equal to its adjoint multiplied by a scalar equal to the reciprocal of its determinant.

$$[A]^{-1} = \frac{\mathrm{adj}[A]}{|A|} \tag{A.23}$$

MATRIX ALGEBRA

This can be shown to be true as we look at the product

$$[A]\{\mathrm{adj}[A]\} = [B] \tag{A.24}$$

The element b_{ii} is the sum of the products of the elements of the ith row of $[A]$ and the elements of the ith column of $\mathrm{adj}[A]$. But the latter are the cofactors of the elements in the ith row of $[A]$. Hence $b_{ii} = |A|$. On the other hand, the element b_{ij} ($j \neq j$) is the sum of the products of the elements of the ith row of $[A]$ and the elements of the jth column of $\mathrm{adj}[A]$. But the latter are the cofactors of the elements of the jth row of $[A]$. Hence this sum is none other than the value of the determinant whose ith row and jth row are identical. The value of such a determinant is, of course, zero. Hence,

$$[A]\{\mathrm{adj}[A]\} = \mathrm{diag}[\,|A|\ \ |A|\ \cdots\ |A|\,]$$

Therefore

$$[A]\left[\frac{\mathrm{adj}[A]}{|A|}\right] = [U]$$

and

$$[A]^{-1} = \frac{\mathrm{adj}[A]}{|A|}$$

As an example, we have

$$\begin{bmatrix} 5 & 3 \\ 2 & -1 \end{bmatrix}^{-1} = \frac{1}{-11}\begin{bmatrix} -1 & -3 \\ -2 & 5 \end{bmatrix} = \begin{bmatrix} \frac{1}{11} & \frac{3}{11} \\ \frac{2}{11} & -\frac{5}{11} \end{bmatrix}$$

As another example, if

$$[A] = \begin{bmatrix} 4 & 6 & 0 \\ 8 & -1 & -5 \\ -2 & 3 & 2 \end{bmatrix}$$

then

$$|A| = 16$$

$$\text{adj}[A] = \begin{bmatrix} \begin{vmatrix} -1 & -5 \\ 3 & 2 \end{vmatrix} & -\begin{vmatrix} 6 & 0 \\ 2 & 2 \end{vmatrix} & \begin{vmatrix} 6 & 0 \\ -1 & -5 \end{vmatrix} \\ -\begin{vmatrix} 8 & -5 \\ -2 & 2 \end{vmatrix} & \begin{vmatrix} 4 & 0 \\ -2 & 2 \end{vmatrix} & -\begin{vmatrix} 4 & 0 \\ 8 & -5 \end{vmatrix} \\ \begin{vmatrix} 8 & -1 \\ -2 & 3 \end{vmatrix} & -\begin{vmatrix} 4 & 6 \\ -2 & 3 \end{vmatrix} & \begin{vmatrix} 4 & 6 \\ 8 & -1 \end{vmatrix} \end{bmatrix}$$

$$= \begin{bmatrix} 13 & -12 & -30 \\ -6 & 8 & 20 \\ 22 & -24 & -52 \end{bmatrix}$$

Hence

$$[A]^{-1} = \begin{bmatrix} \frac{13}{16} & -\frac{3}{4} & -\frac{15}{8} \\ -\frac{3}{8} & \frac{1}{2} & \frac{5}{4} \\ \frac{11}{8} & -\frac{3}{2} & -\frac{13}{4} \end{bmatrix}$$

A.4 Some useful theorems

1 Matrix multiplications are associative, or

$$([A][B])[C] = [A]([B][C]) = [A][B][C] \tag{A.25}$$

2 Matrix multiplications are distributive with respect to addition or substraction, or

$$[A]([B] + [C]) = [A][B] + [A][C] \tag{A.26}$$

3 $\quad ([A][B])_t = [B]_t[A]_t \tag{A.27}$

4 $\quad ([A][B])^{-1} = [B]^{-1}[A]^{-1} \tag{A.28}$

MATRIX ALGEBRA

5 The determinant of the product of two square matrices is equal to the product of their determinants, or

$$|[A][B]| = |A||B| \tag{A.29}$$

6 The determinant of an $n \times n$ matrix multiplied by a scalar α is $\alpha^n|A|$, or

$$|\alpha[A]| = \alpha^n|A| \tag{A.30}$$

7 The effect of multiplying a matrix by a diagonal matrix from the left is that every element in the ith row is multiplied by the ith element of the diagonal matrix. For example,

$$\begin{bmatrix} d_1 & 0 & 0 \\ 0 & d_2 & 0 \\ 0 & 0 & d_3 \end{bmatrix} \begin{bmatrix} x_{11} & x_{12} \\ x_{21} & x_{22} \\ x_{31} & x_{32} \end{bmatrix} = \begin{bmatrix} d_1 x_{11} & d_1 x_{12} \\ d_2 x_{21} & d_2 x_{22} \\ d_3 x_{31} & d_3 x_{32} \end{bmatrix} \tag{A.31}$$

8 The effect of multiplying a matrix by a diagonal matrix from the right is that every element in the ith column is multiplied by the ith element of the diagonal matrix. For example,

$$\begin{bmatrix} x_{11} & x_{12} \\ x_{21} & x_{22} \\ x_{31} & x_{32} \end{bmatrix} \begin{bmatrix} d_1 & 0 \\ 0 & d_2 \end{bmatrix} = \begin{bmatrix} d_1 x_{11} & d_2 x_{12} \\ d_1 x_{21} & d_2 x_{23} \\ d_1 x_{31} & d_2 x_{32} \end{bmatrix} \tag{A.32}$$

9 If $[A]$ is square and $[D]$ is diagonal, as given by Eq. (A.19), then

$$|[D][A]| = d_1 d_2 \cdots d_n |A| \tag{A.33}$$

A.5 Matrix notation in a set of linear simultaneous equations

The set of equations

$$\begin{aligned} a_{11}x_1 + a_{12}x_2 + \cdots + a_{1n} &= y_1 \\ a_{21}x_1 + a_{22}x_2 + \cdots + a_{2n} &= y_2 \\ \cdots\cdots\cdots\cdots\cdots\cdots\cdots\cdots \\ a_{m1}x_1 + a_{m2}x_2 + \cdots + a_{mn} &= y_m \end{aligned} \tag{A.34}$$

may be written in matrix form as

$$\begin{bmatrix} a_{11} & a_{12} & \cdots & a_{1n} \\ a_{21} & a_{22} & \cdots & a_{2n} \\ \vdots & \vdots & & \vdots \\ a_{m1} & a_{m2} & \cdots & a_{mn} \end{bmatrix} \begin{bmatrix} x_1 \\ x_2 \\ \vdots \\ x_n \end{bmatrix} = \begin{bmatrix} y_1 \\ y_2 \\ \vdots \\ y_m \end{bmatrix} \quad (A.35)$$

or as

$$[A][X] = [Y] \quad (A.36)$$

If $m = n$ and $|A| \neq 0$, then a solution exists for x's in terms of y's. This solution is accomplished by the evaluation of the inverse of $[A]$, because

$$[X] = [A]^{-1}[Y] \quad (A.37)$$

A.6 Partitioning of matrices

We may sometimes perform certain operations involving matrices by subdividing or partitioning the matrices into smaller components called *submatrices*. Each submatrix then behaves like a single element for these operations. Specifically, let us take

$$[A] = \begin{bmatrix} a_{11} & a_{12} & \cdots & a_{1n} \\ a_{21} & a_{22} & \cdots & a_{2n} \\ \vdots & \vdots & & \vdots \\ a_{n1} & a_{n2} & \cdots & a_{nn} \end{bmatrix} \quad (A.38)$$

We indicate the partitioning of this matrix by dashed lines, as follows

$$[A] = \left[\begin{array}{ccc|ccc} a_{11} & \cdots & a_{1r} & a_{1(r+1)} & \cdots & a_{1n} \\ \vdots & & \vdots & \vdots & & \vdots \\ a_{s1} & \cdots & a_{sr} & a_{s(r+1)} & \cdots & a_{sn} \\ \hline a_{(s+1)1} & \cdots & a_{(s+1)r} & a_{(s+1)(r+1)} & \cdots & a_{(s+1)n} \\ \vdots & & \vdots & \vdots & & \vdots \\ a_{n1} & \cdots & a_{nr} & a_{n(r+1)} & \cdots & a_{nn} \end{array}\right]$$

$$= \begin{bmatrix} \alpha_{11} & \alpha_{12} \\ \alpha_{21} & \alpha_{22} \end{bmatrix} \quad (A.39)$$

where

$$\alpha_{11} = \begin{bmatrix} a_{11} & \cdots & a_{1r} \\ \vdots & & \vdots \\ a_{s1} & \cdots & a_{sr} \end{bmatrix} \quad \alpha_{12} = \begin{bmatrix} a_{1(r+1)} & \cdots & a_{1n} \\ \vdots & & \vdots \\ a_{s(r+1)} & \cdots & a_{sn} \end{bmatrix}$$

$$\alpha_{21} = \begin{bmatrix} a_{(s+1)1} & \cdots & a_{(s+1)r} \\ \vdots & & \vdots \\ a_{n1} & \cdots & a_{nr} \end{bmatrix} \quad \alpha_{22} = \begin{bmatrix} a_{(s+1)(r+1)} & \cdots & a_{(s+1)n} \\ \vdots & & \vdots \\ a_{n(r+1)} & \cdots & a_{nn} \end{bmatrix}$$

If a second matrix $[B]$ is partitioned along its rows in the same way that $[A]$ is partitioned along its columns, then $[A]$ and $[B]$ are partitioned conformally with respect to the product $[A][B]$. We may evaluate the product by regarding the submatrices as matrix elements. For example, let

$$[A][B] = \begin{bmatrix} a_{11} & a_{12} & \vdots & a_{13} \\ a_{21} & a_{22} & \vdots & a_{23} \\ \cdots & \cdots & \cdots & \cdots \\ a_{31} & a_{32} & \vdots & a_{33} \\ a_{41} & a_{42} & \vdots & a_{43} \end{bmatrix} \begin{bmatrix} b_{11} & \vdots & b_{12} & b_{13} \\ b_{21} & \vdots & b_{22} & b_{23} \\ \cdots & \cdots & \cdots & \cdots \\ b_{31} & \vdots & b_{32} & b_{33} \end{bmatrix} \quad \text{(A.40)}$$

This yields

$$\alpha_{11} = \begin{bmatrix} a_{11} & a_{12} \\ a_{21} & a_{22} \end{bmatrix} \quad \alpha_{12} = \begin{bmatrix} a_{13} \\ a_{23} \end{bmatrix}$$

$$\alpha_{21} = \begin{bmatrix} a_{31} & a_{32} \\ a_{41} & a_{42} \end{bmatrix} \quad \alpha_{22} = \begin{bmatrix} a_{33} \\ a_{43} \end{bmatrix}$$

and

$$\beta_{11} = \begin{bmatrix} b_{11} \\ b_{21} \end{bmatrix} \quad \beta_{12} = \begin{bmatrix} b_{12} & b_{13} \\ b_{22} & b_{23} \end{bmatrix}$$

$$\beta_{21} = [b_{31}] \quad \beta_{22} = \begin{bmatrix} b_{32} & b_{33} \end{bmatrix}$$

Hence

$$[A][B] = \begin{bmatrix} \alpha_{11} & \alpha_{12} \\ \alpha_{21} & \alpha_{22} \end{bmatrix} \begin{bmatrix} \beta_{11} & \beta_{12} \\ \beta_{21} & \beta_{22} \end{bmatrix}$$

or

$$[A][B] = \begin{bmatrix} \alpha_{11}\beta_{11} + \alpha_{12}\beta_{21} & \alpha_{11}\beta_{12} + \alpha_{12}\beta_{22} \\ \alpha_{21}\beta_{11} + \alpha_{22}\beta_{21} & \alpha_{21}\beta_{12} + \alpha_{22}\beta_{22} \end{bmatrix} \quad (A.41)$$

Each of the α and β matrix products is found to be conformable. Notice that the order among the submatrices is important. Observe also that conformability requires that rows of $[B]$ be grouped in the same way as the columns of $[A]$, but the subdivisions of the columns of $[B]$ need not be related to the subdivisions of the rows of $[A]$.

Appendix B

Answers to Problems

Chapter 1

1.1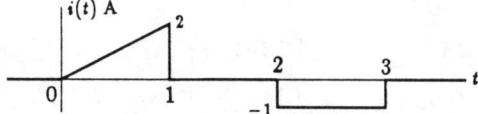

1.2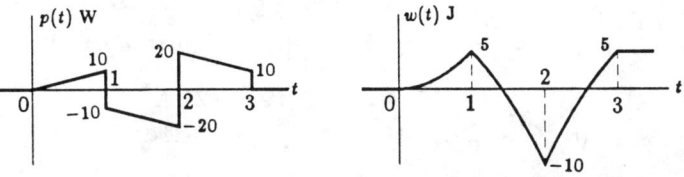

1.3 −0.1247 C **1.4** 25 J **1.5** 110 W, 55 W, −121 W, −44 W
1.6 −150 W, 50 W, −30 W, 130 W
1.7 1.6694×10^5 C, 2.003×10^6 J **1.8** 1.8 W, 0, 2.7 W
1.9 −50 W, −5 W, −3 W, 30 W, −48 W, −12 W, −24 W, 128 W, −32 W, 16 W
1.10 (a) No inconsistency, (b) and (c) contradictory

Chapter 2

2.1 −200 Ω, 500 Ω, −0.6 MΩ, −0.5 V, −10 A, −1 Ω, −0.4 sin t V, ϵ^{-2t} A
2.2 1 A, 0, 13 A, 19 A
2.3 8 A, 11 A, −19 A, 1 A, −11 A, −2 A
2.4 85 V, 15 A **2.5** 28 V **2.6** −1 V, 22 A
2.7 −7 A **2.8** 1.5 A **2.9** −13 A, 18 A, −1 A
2.10 $E_1 = -80$ V, $E_2 = 120$ V, $E_3 = 20$ V, $E_4 = 40$ V, $E_5 = -120$ V, $E_6 = 40$ V

2.11 0 **2.12** -0.25 A **2.13** 20 V **2.14** 2.9 A
2.15 18 A **2.16** $-\frac{45}{31}$ A **2.17** 77 V **2.18** 15 V, -9 A
2.19 $e_1 = -e_9 + e_7 + e_2$, $e_3 = -e_4 + e_2$, $e_5 = -e_8 + e_4$
$e_6 = -e_9 + e_7$, $e_{10} = -e_8 - e_7$, $i_2 = i_3 - i_1$, $i_4 = -i_3 - i_5$,
$i_7 = -i_1 - i_6 - i_{10}$, $i_8 = i_{10} - i_5$, $i_9 = -i_1 - i_6$
2.20 No. of twigs = 10, no. of links = 10 **2.21** $\frac{7802}{953}$ Ω
2.22 $\frac{32}{15}$ A **2.23** -0.06 A **2.24** $-\frac{270}{187}$ V
2.25 $\frac{45}{11}$ A **2.26** 143.57 V **2.27** $R_1R_4 = R_2R_3$
2.28 1.25 A **2.29** $\frac{60}{11}$ Ω **2.30** $\frac{45}{254}$ A
2.31 $\frac{1015}{254}$ A, $\frac{75}{127}$ A **2.32** $-\frac{1080}{107}$ V, 21 V
2.33 $i_1 = \frac{645}{314}\epsilon^{-t}$ A, $i_2 = \frac{370}{314}\epsilon^{-t}$ A, $i_3 = \frac{275}{314}\epsilon^{-t}$ A,
$i_4 = \frac{150}{314}\epsilon^{-t}$ A, $i_5 = \frac{125}{314}\epsilon^{-t}$ A
2.34 411.68 W **2.35** $\frac{25}{37}$ A **2.36** $\frac{40}{9}$ A **2.37** $\frac{20}{103}$ A
2.38 $\frac{103}{7}$ V **2.39** 0.5 A **2.40** $\frac{100}{11}$ V, $-\frac{44}{7}$ Ω
2.41 $\frac{28}{201}$ A, **2.42** $-\frac{6000}{137}$ V **2.43** $\frac{2000}{3131}$ V
2.44 $\frac{35}{58}$ A **2.45** $\frac{20}{33}$ V **2.46** 1.888 V, 2.256 V
2.47 $\frac{547}{32}$ V **2.48** $\frac{5}{18}$ A **2.49** -21 A
2.50 -33 V **2.51** $\frac{186}{503}$ A **2.52** $\frac{1278}{503}$ V
2.53 $\frac{484}{93}$ Ω **2.54** $\frac{25}{36}$ V **2.55** $-\frac{816}{389}$ V
2.56 $-\frac{400}{159}$ V **2.57** $\frac{46}{43}$ A **2.58** $\frac{2036}{1539}$ Ω
2.59 $-R\left(\frac{e_1}{R_1} + \frac{e_2}{R_2}\right)$ **2.60** $-\frac{5}{6}$ mA
2.61 $\frac{R_3+R_4}{R_3(R_1+R_2)}(e_1R_2 + e_2R_1)$ **2.62** $0.24 \sin t$ mA

Chapter 3

3.1 $P_{4\Omega} = 4$ W, $P_{12\Omega} = 108$ W, $P_{8\Omega} = 128$ W
3.2 $-\frac{41}{10}$ V, $\frac{7}{5}$ V **3.3** $\frac{104}{19}$ V, $\frac{197}{38}$ V
3.4 $\frac{80}{13}$ V, $-\frac{105}{26}$ V, $\frac{205}{39}$ V **3.5** $\frac{912}{121}$ A, $-\frac{862}{121}$ A, $-\frac{1955}{121}$ A
3.6 $-\frac{1960}{1369}$ A, $\frac{2420}{1369}$ A, $-\frac{385}{1369}$ A **3.7** $\frac{7}{4}$ A, $\frac{21}{4}$ A
3.8 $-\frac{2}{3}$ A, $-\frac{16}{15}$ A **3.9** $-\frac{47}{553}$ V, $\frac{16}{79}$ V, $\frac{523}{1106}$ V
3.10 $\frac{156}{181}$ V, $\frac{241}{181}$ V, $-\frac{91}{181}$ V, $\frac{306}{181}$ V, $\frac{150}{181}$ V
3.11 $\frac{182}{23}$ V, $\frac{248}{23}$ V **3.12** $\frac{719}{716}$ V, $\frac{469}{716}$ V, $\frac{727}{716}$ V

ANSWERS TO PROBLEMS 691

3.13 $\frac{252}{179}$ A 3.14 $\frac{5}{3}$ V, 1 V 3.15 $-\frac{620}{21}$ V *(260)*

3.16 $\frac{25}{21}$ V 3.17 $\frac{10}{47}$ V, $\frac{8}{47}$ V 3.18 $\frac{25}{14}$ V, $-\frac{5}{14}$ V

3.19 $\frac{41}{7}$ V 3.20 $-\frac{217}{5}$ A 3.21 $-\frac{25}{121}$ V, $\frac{310}{121}$ V

3.22 -1.8 V, -1.2 V 3.23 $\frac{44}{57}$ V, $\frac{164}{57}$ V

3.24 0.36 V, 1.48 V 3.25 0.8583 mV, 0.8579 V

3.26 $-\frac{55}{21}$ V, $-\frac{6}{7}$ V 3.27 $\frac{403}{398}$ A, $-\frac{37}{398}$ A, $\frac{92}{199}$ A

3.28 $\frac{183}{199}$ A, $\frac{275}{199}$ A, $\frac{147}{398}$ A 3.29 $-\frac{40}{47}$ A 3.30 $\frac{65}{9}$ A

3.31 $-\frac{13}{30}$ A 3.32 $\frac{176}{75}$ A 3.33 $-\frac{46}{35}$ A 3.34 $\frac{283}{129}$ A

3.35 $-\frac{6}{17}$ A 3.36 -2.9 A 3.37 -2 A, 4 A

3.38 $-\frac{164}{129}$ A 3.39 -1 A 3.40 $\frac{1145}{369}$ A 3.41 $\frac{140}{27}$ V

3.42 0 3.43 $\frac{122}{17}$ A 3.44 $\frac{17}{59}$ A, $-\frac{3}{59}$ A

Chapter 4

4.1 2.8 A 4.2 -82 V 4.3 $-\frac{5}{22}$ A 4.4 $\frac{10}{9}$ A

4.5 $-6t + \frac{2}{15} \sin 2t$ A 4.6 $-0.6t^2 + \frac{2}{3} \cos 2t$ A

4.7 $\frac{166}{65}$ A 4.8 70 V 4.9 $E = \frac{21}{71}E_1 + \frac{105}{71}I_1 + \frac{45}{71}I_2$

4.10 $\frac{24}{11}$ A 4.11 $15E_L = 332 - 44I_L$

4.12 $15E_L = 94 - 18I_L$ 4.13 $3E_L = 19 - 10I_L$

4.14 $E_L = 40 - 10I_L$ 4.15 $14E_L = 115 - I_L$

4.16 $E_{oc} = -1$ V, $R_{eq} = 2.5$ Ω 4.17 $I_{sc} = \frac{23}{6}$ A, $R_{eq} = 2$ Ω

4.18 $I_{sc} = \frac{71}{33}$ A, $R_{eq} = \frac{33}{14}$ Ω 4.19 $E_{oc} = \frac{95}{11}$ V, $R_{eq} = \frac{6}{11}$ Ω

4.20 $I_{sc} = \frac{115}{3}$ A, $E_{oc} = \frac{230}{11}$ V 4.21 $I_{sc} = -8$ A, $R_{eq} = 5$ Ω

4.22 $E_{oc} = 52.5$ V, $I_{sc} = -131.25$ A

4.23 $E_{oc} = 40$ V, $R_{eq} = 10$ Ω 4.24 A short circuit

4.25 $E_{oc} = 5$ V, $R_{eq} = 1.6$ Ω 4.26 A $\frac{30}{7}$-Ω resistor

4.27 $I_{sc} = 15$ A, $R_{eq} = 0.25$ Ω 4.28 -57 Ω 4.29 $\frac{179}{20}$ Ω

4.30 $\frac{2}{3}$ Ω 4.31 $\frac{34}{21}$ Ω 4.32 $E_{oc} = -\frac{10}{3}$ V, $R_{eq} = -\frac{8}{3}$ Ω

4.33 40 V 4.34 $\frac{30}{7}$ A 4.35 $\frac{88}{3}$ A 4.36 $\frac{15}{44}$ A

4.37 $-\frac{21}{13}$ A 4.38 5 A 4.39 $E_{oc} = \frac{17}{3}$ V, $R_{eq} = \frac{38}{9}$ Ω

4.40 $\frac{639}{526}$ A 4.41 $\frac{100}{11}$ V 4.42 $\frac{2968}{215}$ V 4.43 $-\frac{6}{17}$ A

4.44 $-\frac{21}{16}$ A   **4.45** -2.75 A   **4.46** $\frac{323}{307}$ A, $-\frac{239}{307}$ A
4.47 $-\frac{261}{47}$ A, $-\frac{139}{47}$ A   **4.48** $\frac{4}{3}$ A   **4.49** $\frac{22}{45}$ A
4.50 9.5 A   **4.52** 7 A   **4.53** $\frac{5}{44}$ A   **4.54** -7.1 V
4.55 8 Ω   **4.56** -7.6 V   **4.57** 0.48 A   **4.58** 14 V
4.59 -0.4 V   **4.60** $\frac{70}{19}$ Ω, 16.92 W   **4.61** 48.66 W
4.62 0.5553 W

Chapter 5

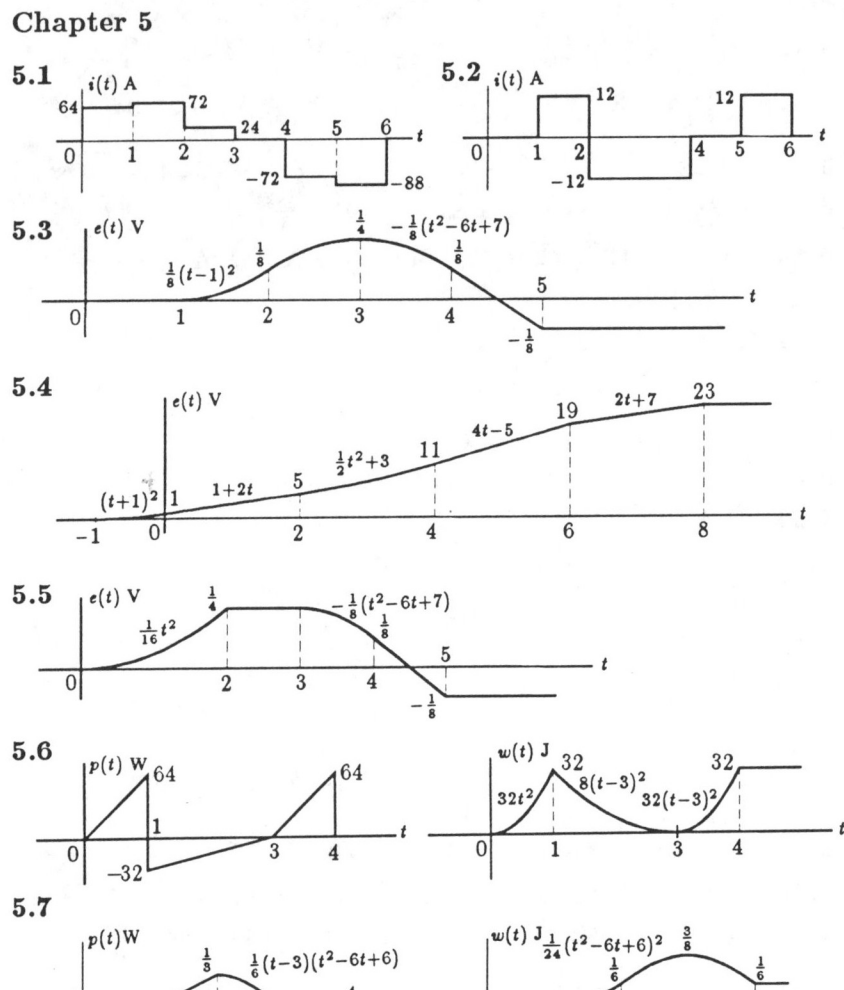

ANSWERS TO PROBLEMS

5.8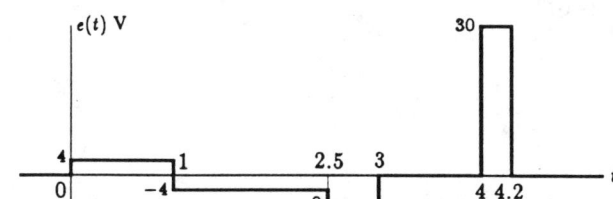

5.9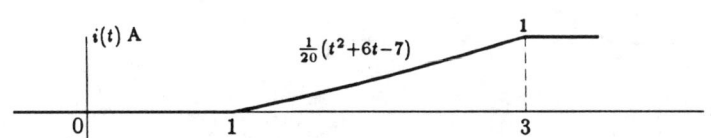

5.10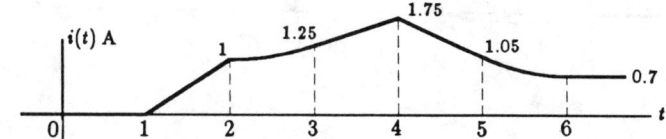

5.11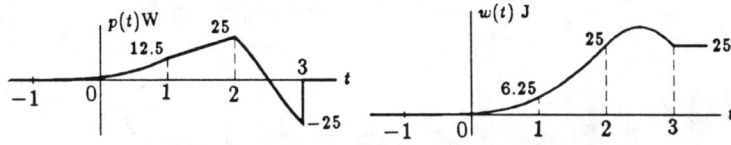

5.12 $\frac{12}{7}$ H   **5.13** 5 V   **5.14** $\frac{16}{17}\cos t$ V   **5.15** $\frac{122}{83}$ F

5.16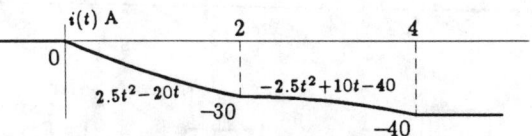

5.17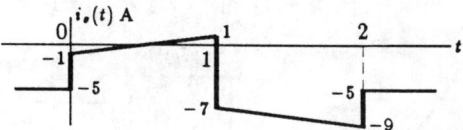

5.18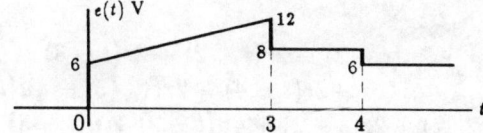

5.19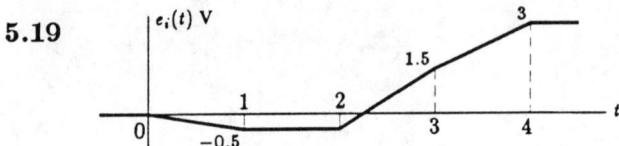

5.20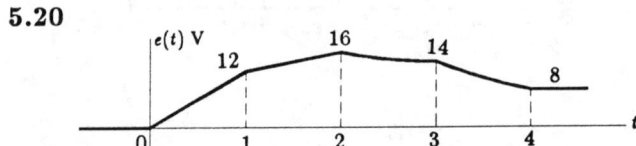

5.21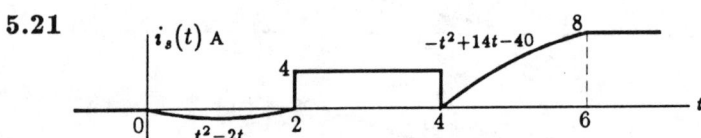

5.22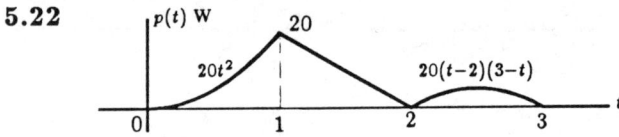

5.23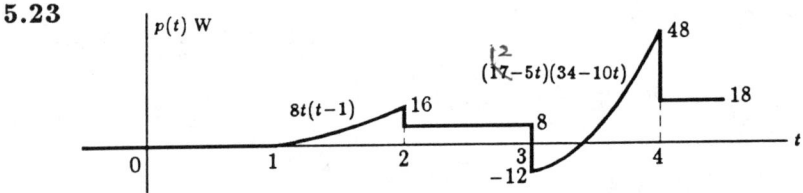

5.24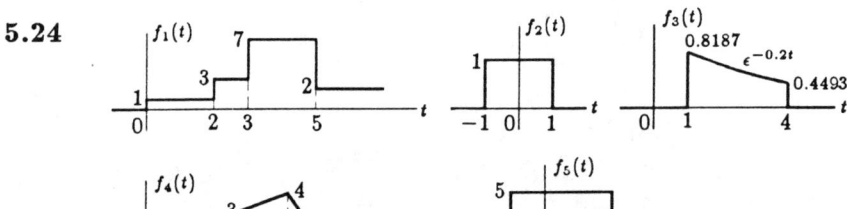

5.25 $g_1(t) = 2u(t-1) + 2u(t-2) - 2r(t-2) + 2r(t-3)$
$\qquad\qquad + 2r(t-4) - 2r(t-5) - 4u(t-5),$
$\quad g_2(t) = 2u(t) - 2r(t) + 2r(t-2) + 4u(t-2) + u(t-3)$
$\qquad\qquad - 1.5r(t-3) + 1.5r(t-5),$

$$g_3(t) = 2u(t) - 4u(t-1) + 7u(t-2) - 4r(t-2)$$
$$+ 4r(t-3) - r(t-4) + r(t-5),$$
$$g_4(t) = 2u(t) + 2r(t-1) - r(t-2) - r(t-3) - 5u(t-3)$$

5.26 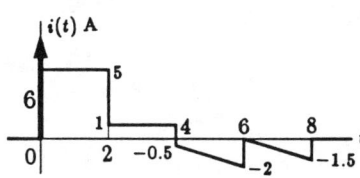 **5.27**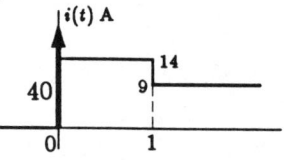

5.28 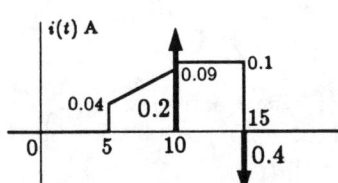 **5.29**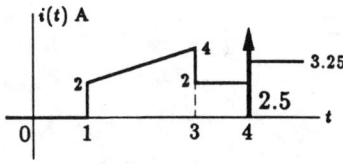

5.30 (a) 3, (b) 2, (c) 8, (d) 2.5, (e) $r(t) - 2$, (f) 36, (g) $u(t-5)$,
(h) 49, (i) 1, (j) 0.5, (k) $\frac{1}{3}(t^3 - 8)u(t-2)$, ($\ell$) 0,
(m) $10u(t)$, (n) 0.03369

5.31 (a) $-3\delta(t-5) - u(t-5)$, (b) $u(t-5)$,
(c) $\sqrt{t}\cos(t-1)u_1(t-1) + \frac{1}{2}\delta(t-1)$, (d) $3\cos 3t\, u(t)$,
(e) $\delta(t-1)$, (f) $0.9093\,\delta(t-1) + 2\cos 2t\, u(t-1)$

Chapter 6

6.1 $(1 - \epsilon^{-20t})u(t) - (1 - \epsilon^{-20(t-1)})u(t-1)$ A

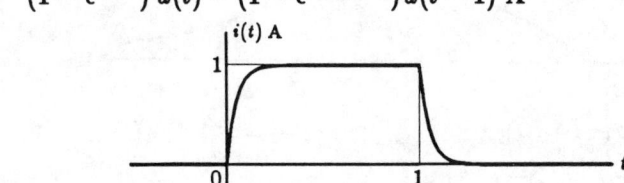

6.2 $0.4(1 - \epsilon^{-10t})$ A, $0 < t < 0.1$; $0.6873\epsilon^{-10t}$ A, $0.1 < t$

6.3 $6\epsilon^{-\frac{1}{4.5}t}$ V

6.4 $e_C = 6(1 - \epsilon^{-0.5t})$ V, $i = 0$, $0 < t < 4$;
$e_C = 5.188\epsilon^{-1000(t-4)}$ V, $i = 5.188\epsilon^{-1000(t-4)}$ A, $4 < t$

6.5 $[20t - 40(1 - \epsilon^{-0.5t})]u(t) - [40(t-2) - 80(1 - \epsilon^{-0.5(t-2)})]u(t-2)$
$+ [20(t-4) - 40(1 - \epsilon^{-0.5(t-4)})]u(t-4)$ V

6.6 $C(t + RC\epsilon^{-\frac{1}{RC}t} - RC)u(t)$ **6.7** $\frac{10}{3} + \frac{20}{3}\epsilon^{-5t}$ V

6.8 $\frac{1}{9}(10\epsilon^{-20t} - \epsilon^{-2t})u(t)$ A   **6.9** $50(\epsilon^{-t} - \epsilon^{-2t})u(t)$ V

6.10 $[-\frac{1}{17}\epsilon^{-1.25t} + \frac{1}{17}\cos 5t + \frac{4}{17}\sin 5t]u(t)$ A

6.11 $[6t^2 - 72t + 432 - 432\epsilon^{-t/6}]u(t)$ V

6.12 $2(1 - \epsilon^{-0.5t})u(t) - 2(1 - \epsilon^{-0.5(t-2)})u(t-2)$
$+ 2(1 - \epsilon^{-0.5(t-3)})u(t-3) - 2(1 - \epsilon^{-0.5(t-4)})u(t-4)$ A

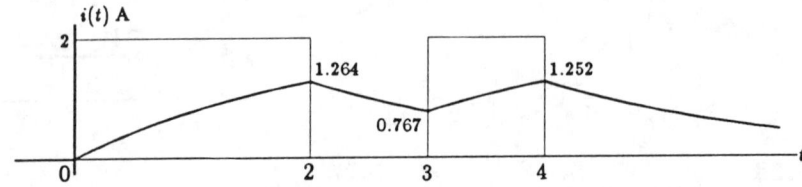

6.13 $10(1 - 10\epsilon^{-5(t-1)})[u(t-1) - u(t-2)]$ W

6.14 $-\frac{27}{5}\epsilon^{-0.9t}$ A   **6.15** $[\frac{1}{15}t^2 - \frac{4}{45}t + \frac{53}{135} - \frac{53}{135}\epsilon^{-\frac{3}{2}t}]u(t)$ A

6.16 $[0.2\sin(10t + 20°) - 0.4\cos(10t + 20°) + 0.3042\epsilon^{-5t}]u(t)$ A

6.17 $[\frac{2}{125}\cos 3t + \frac{39}{125}\sin 3t - \frac{2}{125}\epsilon^{-4t}]u(t)$ A

6.18 $0.5(1 - \epsilon^{-0.4t})u(t) - (0.5 - 0.3\epsilon^{-0.4(t-1)})u(t-1)$ A

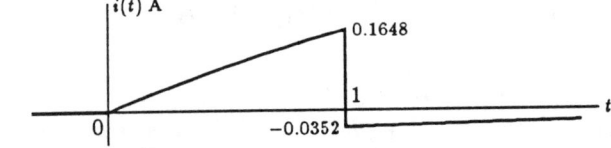

6.19 1.188 C

6.20 $5(1 - \epsilon^{-0.5(t-1)})u(t-1) - 2(1 - \epsilon^{-0.5(t-4)})u(t-4)$
$- 3(1 - \epsilon^{-0.5(t-6)})u(t-6)$ A

6.21 $\frac{11}{6}(1 - \epsilon^{-\frac{3}{8}t})u(t)$ A   **6.22** $(\epsilon^{-2t} - \epsilon^{-1.5t})u(t)$ A

6.23 $[-5t + 10.5(1 - \epsilon^{-2t})]u(t)$ A

6.24 $6.4\epsilon^{-0.4(t-1)}u(t-1) - 16(1 - \epsilon^{-0.4(t-1)})u(t-1)$
$+ 16(1 - \epsilon^{-0.4(t-2)})u(t-2)$ V

6.25 $(t - \frac{3}{5} + \frac{3}{5}\epsilon^{-\frac{5}{3}t})u(t) - (t - \frac{13}{5} + \frac{3}{5}\epsilon^{-\frac{5}{3}(t-2)})u(t-2)$ A

6.26 $-\frac{46}{19}(1 - \epsilon^{-\frac{475}{42}t})u(t)$ A

6.27 $(\epsilon^{-5t} - \epsilon^{-2.5t})u(t) - (\epsilon^{-5(t-2)} - \epsilon^{-2.5(t-2)})u(t-2)$ V

6.28 $(3 - 1.5\epsilon^{-\frac{1}{3}t})u(t)$ A   **6.29** $2\delta(t) - 2\epsilon^{-t}u(t) + 25(1 - \epsilon^{-t})u(t)$ V

6.30 $1.6(1 - \epsilon^{-2.5t})u(t)$ A   **6.31** $0.6452(1 - \epsilon^{-1.177t})u(t)$ A

ANSWERS TO PROBLEMS

6.32 $-\frac{5}{3}(1 - \epsilon^{-\frac{60}{13}t})u(t)$ V **6.33** $(1 - \epsilon^{-\frac{1}{0.9}t})u(t)$ V

6.34 $7u(t) + 3\delta(t) - 2\epsilon^{-5t}u(t)$ V
6.35 -1.25 A, 1.25 A, 0, -8.75 V, 2.5 V
6.36 $0, 0, \frac{5}{16}$ A, 0, 7.5 V, 2.5 V, $\frac{5}{3}$, 0.5, 2.5
6.37 $-2\epsilon^{-3t}$ A   **6.38** $28.8\delta(t) + 6 - 1.2\epsilon^{-4t}$ A
6.39 $0.8\epsilon^{-0.2t}$ A   **6.40** $-2.5\epsilon^{-0.5t}$ A   **6.41** $4\epsilon^{-0.2t}$ V, 2 A
6.42 $i_1(t) = 5 - 1.839\epsilon^{-2(t-1)}$ A, $1 < t < 2$;
$i_1(t) = 4.751\epsilon^{-2(t-2)}$ A, $2 < t$; $i_2(t) = 3.161$ A, $1 < t$
6.43 2 A, $1.2\epsilon^{-t}$ A   **6.44** $100\epsilon^{-15t}$ A
6.45 $(\frac{25}{3}\epsilon^{-2t} - \frac{10}{3}\epsilon^{-\frac{1}{2}t})u(t)$ V   **6.46** $(6 - 7.5\epsilon^{-1.5t})u(t)$ V
6.47 $0.3\delta(t)$ A   **6.48** $2.058\epsilon^{-0.4t}\sin 0.5831t\, u(t)$ A
6.49 $(\cos t - \epsilon^{-2.5t})u(t)$ V   **6.50** $3.676\epsilon^{-0.75t}\sinh 0.6801t\, u(t)$ A
6.51 $[25 + 5\epsilon^{-1.25t}\cosh 0.75t - \frac{275}{3}\epsilon^{-1.25t}\sinh 0.75t]u(t)$ V
6.52 $\epsilon^{-0.25(t-2)}[\cos 0.2041(t-2) - 1.2247\sin 0.2041(t-2)]u(t-2)$ A
6.53 $-2\epsilon^{-1.25(t-1)}[\cosh\frac{\sqrt{17}}{4}(t-1) - \frac{5}{\sqrt{17}}\sinh\frac{\sqrt{17}}{4}(t-1)]u(t-1)$ A
6.54 $[(0.0824\cos 2.437t - 0.0645\sin 2.437t)\epsilon^{-1.25t}$
$\qquad +0.0434\sin 6t - 0.0824\cos 6t]u(t)$ A
6.55 $-10.67\epsilon^{-2.5t}\sinh 2.062t$ V   **6.56** $-0.560\epsilon^{-0.7t}\sin 0.714t$ A
6.57 $2\delta(t)$ A, $2\delta(t)$ A, $4\delta(t)$ V, 0.5 A, 20 V
6.58 $C_1C_2R_1R_2\frac{d^2i_1}{dt^2} + [C_1R_1 + C_2(R_1 + R_2)]\frac{di_1}{dt} + i_1$
$\qquad\qquad = C_1C_2R_1\frac{d^2e_s}{dt^2} + (C_1 + C_2)\frac{de_s}{dt}$
$C_1C_2R_1R_2\frac{d^2i_2}{dt^2} + [C_1R_1 + C_2(R_1 + R_2)]\frac{di_2}{dt} + i_2 = C_2\frac{de_s}{dt}$

Chapter 7

7.1 (a) $14.92 - j26.10$, (b) $0.0107 + j0.2763$, (c) $4.169 - j2.063$,
(d) 2 or $-1 \pm j1.732$, (e) $73.12 + j61.204$, (f) $1.353 + j0$,
(g) $1.572 \pm j1.317$, (h) $1.196 - j2.383$, (i) $-0.3282 + j1.685$,
(j) $\pm(2.279 + j0.439)$, (k) ~~$15.41 + j23.05$~~ $30.03 + j44.93$
7.2 $\frac{10}{7}\epsilon^{-2t}$ V   **7.3** $\frac{5}{3}\epsilon^{-t}$ A, 0, $\frac{5}{3}\epsilon^{-t}$ A
7.4 $0.6812\epsilon^{j(-t-139.76°)}$ A, $9.003\epsilon^{j(-t+42.78°)}$ V
7.5 $(-0.5582 - j1.1877)\epsilon^{-j2t}$ A, $(16.295 + j7.542)\epsilon^{-j2t}$ V
7.6 $\frac{6+j4}{13}\epsilon^{j2t}$ A   **7.7** $\frac{12+j14}{17}\epsilon^{j3t} + \frac{-370+j1192}{1033}\epsilon^{j(2t+30°)}$ V
7.8 $j\frac{2}{3} - \frac{1}{3}(7+j11)\epsilon^{j(10t+20°)}$ A, $j\frac{10}{3} + (-\frac{40}{3} + j60)\epsilon^{j(10t+20°)}$ V
7.9 $2.960\epsilon^{j(2t-109.77°)}$ A, $2.164\epsilon^{j(2t+92.83°)}$ A

7.10 $(6.778 - j7.378)\epsilon^{(-2+j3)t}$ V, $(-6.292 - j19.733)\epsilon^{(-2+j3)t}$ V

7.11 $8.192\epsilon^{-2t}\cos(10t + 65.32°)$ A, $16.246\epsilon^{-2t}\cos(10t + 62.55°)$ V

7.12 $4.190\epsilon^{-t}\cos(3t - 27.07°)$ A

7.13 $\frac{20}{11}\epsilon^{-0.5t}$ A, $0.158\epsilon^{(-j2t+j164.73°)}$ A, $0.481\epsilon^{(-1+j)t-j110.22°}$ A

7.14 $\frac{305}{39} + j\frac{100}{39}$ Ω   **7.15** $0.2766 + j0.0066$ ℧

7.16 $8.646\cos(3t - 26.57°) + 1.265\cos(2t - 18.43°)$ V

7.17 $e_1 = 41.62\cos(5t + 68.03°) + j69.36\cos(5t+68.03°)$ V,
$e_2 = 69.36\cos(5t + 68.03°) + j59.28\cos(5t+43.15°)$ V

7.18 $\frac{s^2+14s-2}{(3s-1)(s+8)}$ ℧   **7.19** $\frac{4}{8s^3+6s^2+17s+6}$

7.20 $\frac{s^3}{5s^3+6s^2+5s+2}$ ℧   **7.21** $0, -\frac{5}{2}$

Chapter 8

8.1 (a) $14.35\cos(\omega t - 37.61°)$, (b) $3.910\cos(\omega t - 161.60°)$,
(c) $24.98\cos(\omega t - 35.82°)$, (d) $38.59\cos(\omega t - 134.73°)$,
(e) $3.453\cos(\omega t + 162.47°)$

8.2 (a) $32.96\epsilon^{-3t}\cos(\omega t - 54.79°)$, (b) $14.37\epsilon^{-6t}\cos(\omega t + 104.10°)$,
(c) $33.88\epsilon^{-\alpha t}\cos(\omega t - 41.85°)$

8.3 (a) 3.944, (b) 1.549, (c) 0.8165, (d) 3.266

8.4 $2.236\cos(100t - 83.43°)$ A   **8.5** $0.2\cos(100t + 70°)$ A

8.6 $0.5\cos(3t - 11.63°)$ A   **8.7** $0.867\sin(5t - 11.79°)$ A

8.8 $1.961\sin(10t + 11.31°)$ A

8.9 $1162.8\cos(100t - 93.95°)$ V, $1264.9\cos(100t - 71.57°)$ V

8.10 $7.071\cos(10t - 45°) + 20\cos 5t$ V

8.11 $1.857\cos(2t - 1.80°) + 4.152\cos(t - 41.63°)$ A

8.12 $-0.0778 + 0.4121\sin(10t + 35.94°)$ A

8.13 $5.607\cos(3t - 127.41°) + 9.940\cos(2t + 61.72°)$ V

8.14 $9.546 - j1.331$ Ω   **8.15** 5.629 Ω, 99.93 μF

8.16 9.444 Ω, 0.1353 H

8.17 1.4615 Ω   j4.6154 Ω   0.06236 ℧ $-j0.1969$ ℧

8.18 $5.999 + j4.560$ Ω   **8.19** 2.25 Ω, 3 H, 6 H

8.20 $2.714 - j3.836$ Ω   **8.21** $7.144 + j4.884$ V

8.22 $I_1 = 3.399\underline{/55.56°}$ A,
$I_2 = 2.108\underline{/-41.57°}$ A,
$I_3 = 4.300\underline{/37.12°}$ A

8.23 $I_1 = 18.26\underline{/11.74°}$ A,
$I_2 = 32.92\underline{/158.05°}$ A,
$I_3 = 11.95\underline{/148.87°}$ A,
$I_4 = 10.14\underline{/103.87°}$ A,
$I = 20\underline{/128.31°}$ A

8.24 $I_1 = 2.110\underline{/-3.29°}$ A,
$I_2 = 2.896\underline{/2.39°}$ A,
$I_3 = 3.180\underline{/72.02°}$ A,
$I_4 = 3.477\underline{/-56.64°}$ A

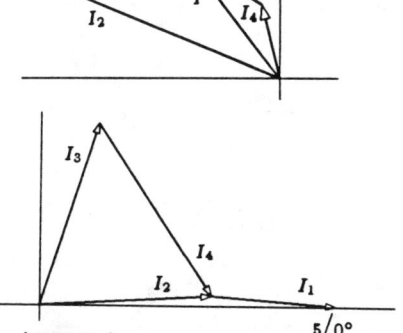

8.25 $5.600 + j3.008$ A, $2.435 - j0.896$ A

8.26 (a) $\frac{64}{41} - j\frac{285}{41}$, $10.25 - j5$, $4 + j8.2$ Ω
(b) $-6 + j36$, $9.6 - j13.2$, $22.5 + j3.75$ Ω
(c) $\frac{36}{65} - j\frac{2}{65}$, $\frac{12}{65} - j\frac{34}{65}$, $\frac{8}{13} + j\frac{4}{13}$ Ω
(d) $\frac{180}{401} + j\frac{3600}{401}$, $-\frac{160}{401} - j\frac{3200}{401}$, $\frac{1440}{401} - j\frac{72}{401}$ Ω

8.27 $1.719 - j2.539$ V   **8.28** $0.8 - j0.4$ A
8.29 $3.269 + j6.346$ V   **8.30** $20.34 + j5.142$ V
8.31 $14.50 + j25.22$ A   **8.32** $3.028 + j0.019$ A
8.33 $137.81 - j47.68$ V, $159.36 - j61.16$ V   **8.34** $-2 - j6$ V
8.35 $4.912 - j17.883$ V   **8.36** $12.75 - j3.75$ V
8.37 $3.931 - j6.827$ Ω   **8.38** $2.386 - j8.197$ Ω
8.39 $-\frac{48}{241} + j\frac{180}{241}$ V   **8.40** $-0.4345 + j0.5882$ V

8.41 $E_1 = 255.92\underline{/14.46°}$ V,
$E_2 = 159.87\underline{/-36.87°}$ V,
$E_3 = 319.75\underline{/53.13°}$ V,
$E_4 = 119.9\underline{/-126.87°}$ V,
$E_5 = 161.25\underline{/-156.6°}$ V

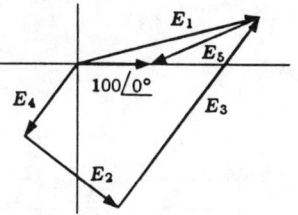

8.42 $I_1 = 7.594 - j2.027$ A, $I_2 = 7.389 - j2.776$ A,
$I_3 = 0.204 + j0.709$ A, $I_4 = 12.389 + j5.885$ A
8.43 $E_{oc} = 6.83 + j11.83$ V, $Z_{eq} = 1 - j$ Ω
8.44 $E_{oc} = 9.904 + j0.194$ V, $Z_{eq} = 0.194 - j3.002$ Ω
8.45 $I_{sc} = 18.510 - j9.471$ A, $Z_{eq} = 4.443 - j0.874$ Ω
8.46 $I_{sc} = 0.875 - j0.793$ A, $Z_{eq} = 2.826 + j0.411$ Ω
8.47 16 V   **8.48** 6 Ω   **8.49** 2 or 14 Ω
8.50 4.513 Ω   **8.51** 4.947 or 20.891 Ω   **8.52** 7.60 Ω
8.53 2.021 Ω   **8.54** 5.620 Ω   **8.55** 6.149 Ω

8.56 19.596 Ω   **8.57** 4.166 Ω   **8.58** 2.646 Ω, 61.41°
8.59 21.33 Ω   **8.60** 3.977 Ω
8.61
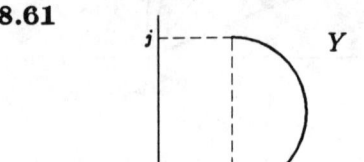
8.62
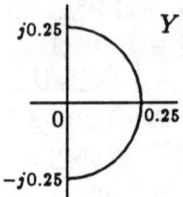

8.63
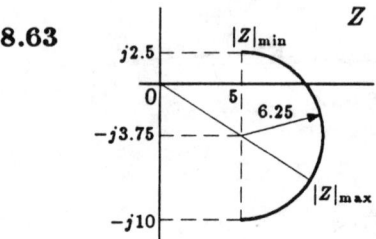
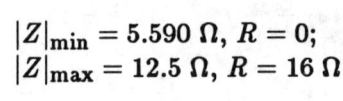
$|Z|_{min} = 5.590$ Ω, $R = 0$;
$|Z|_{max} = 12.5$ Ω, $R = 16$ Ω

8.64
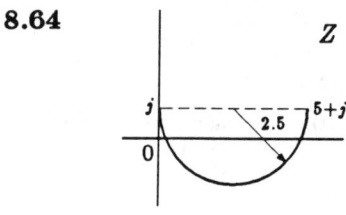
$|Z|_{min} = 0.1926$ Ω

8.65
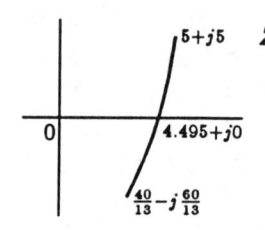
$|Z|_{min} = 4.384$ Ω

8.66
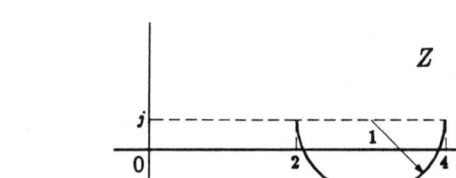

8.67
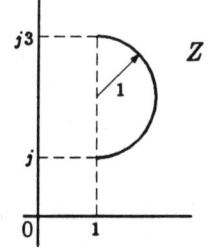

8.68 (a) 35.59 kHz (b) 0.373 (c) 11.8 and 107.3 kHz
(d) 7.43 and 172.73 kHz
8.69 (a) 0.1 F, 10 V (b) 99 mF, 10.05 V
8.70 (a) 2.236 rad/sec, 2236.1 A (b) 2.2428 and 2.2294 rad/sec
8.71 0.9247 rad/sec, 14.45 V   **8.72** $\sqrt{\dfrac{CR_2^2 - L}{CL(CR_1^2 - L)}}$
8.73 2×10^5 rad/sec

Chapter 9

9.1 335.57 W **9.2** 41.67 W
9.3 $P_{5\Omega} = 31.36$ W, $P_{4\Omega} = 17.37$ W, $Q_L = 11.58$ VAR(L),
$P_s = 48.47$ W, $Q_s = 11.58$ VAR(L)
9.4 $P_{5\Omega} = 2.657$ W; $P_{10\Omega} = 16.235$ W; $Q_L = 4.252$ VAR(L);
voltage source: -6.568 W, 3.164 VAR(L);
current source: 25.457 W, 1.089 VAR(L)
9.5 303.56 W, 414.92 VAR(C) **9.6** -49.82 W, 194.03 VAR(L)
9.7 1339.7 W, 157.9 VAR(L) **9.8** $-8 + j6$ A
9.9 $-1 - j4$ A **9.10** 42.09 W, 84.19 VAR(L)
9.11 $P_{10\Omega} = 250$ W, $P_{8\Omega} = 1512.5$ W, $P_L = 125$ VAR(L),
$P_C = 1512.5$ VAR(C), current source: 671.32 W,
228.53 VAR(C), voltage source: 1091.17 W, 1158.96 VAR(C)
9.12 44.276 W, 44.276 VAR(L)
9.13 $P_{3\Omega} = 34.03$ W, $P_{1.5\Omega} = 16.07$ W, $P_{1\Omega} = 0.9$ W, $P_{4\Omega} = 4$ W,
$Q_{-j\Omega} = 102.10$ VAR(C), $Q_{-j0.5\Omega} = 48.20$ VAR(C),
$Q_{j3\Omega} = 0.30$ VAR(L), voltage source: 55 W, 150 VAR(C)
9.14 75 V, 1.5 A **9.15** 47.17 V **9.16** $45 - j80.33$ A
9.17 $11.98\underline{/-10.23°}$ A **9.18** 0.866
9.19 2174.6 W, 1146.3 VAR(C), 2458.2 VA **9.20** $54 + j2.4$ A
9.21 4.566 kW, 7.909 kVAR(C)
9.22 $S_1 = 688.24 + j0$ VA, $S_2 = 0 + j346.22$ VA,
$S_3 = 277.18 - j594.41$ VA, $S_4 = -840.86 - j200$ VA,
$S_5 = -124.56 + j448.19$ VA
9.23 351 μF **9.24** $1.389 + j5.084$ or $11.49 - j9.89$ Ω
9.25 100 Ω, 57.74 Ω, 18.30 Ω **9.26** 18.21 Ω
9.27 219.62 V **9.28** 32.57 Ω **9.29** 0, 10 Ω
9.30 1.638 Ω, 2.887 Ω **9.31** 1200 W, 2100 VAR(L)
9.32 West to east, supplied from west: 740.2 W, 403.3 VAR(C);
supplied to east: 669.1 W, 758.6 VAR(C);
lost: 71.1 W, 355.3 VAR(L)
9.33 3.125 Ω, 44.21 mH, 5.101 mH **9.34** 0.0556 W
9.35 806.4 W **9.36** 2.756 W **9.37** 3.890 W
9.38 2.408 Ω **9.39** 367.6 W **9.40** $X_L = -X_s$
9.41 $\sqrt{\dfrac{R_s^2 + X_s^2}{1 + k^2}}$

Chapter 10

10.1 $\frac{1}{23}\begin{bmatrix} 212 & 80 \\ 80 & 130 \end{bmatrix} \Omega$ **10.2** $\frac{1}{7}\begin{bmatrix} 19 & -2 \\ -34 & -6 \end{bmatrix} \Omega$

10.3 $\begin{bmatrix} \frac{1}{8} - j\frac{1}{10} & -\frac{1}{8} \\ -\frac{1}{8} & \frac{1}{8} + j\frac{1}{10} \end{bmatrix} \mho$

10.4 $\begin{bmatrix} 4.432 - j0.568 & 0.682 - j0.228 \\ 4.773 - j1.591 & 1.364 + j0.455 \end{bmatrix} \Omega$

10.5 $\frac{1}{41}\begin{bmatrix} 66 - j103 & 16 - j143 \\ 16 - j143 & 66 - j103 \end{bmatrix} \Omega$ **10.6** $\frac{1}{31}\begin{bmatrix} 11 & 8 \\ -8 & -3 \end{bmatrix} \mho$

10.7 $\frac{1}{5}\begin{bmatrix} 2 & 7 \\ 4 & 4 \end{bmatrix} \Omega$ **10.8** $\begin{bmatrix} 30 - j15 & -1.5 - j2.5 \\ 1.5 + j2.5 & 0.25 - j0.25 \end{bmatrix}$

10.9 $\begin{bmatrix} 0.04 - j0.08 & -0.8 - j0.4 \\ 0.8 + j0.4 & 4 - j2 \end{bmatrix}$ **10.10** $\frac{1}{17}\begin{bmatrix} 7 & -2 \\ -6 & -8 \end{bmatrix}$

10.11 $\begin{bmatrix} 6.951 \times 10^{-4} & -9.901 \times 10^{-3} \\ -4.951 \times 10^{3} & 9.901 \times 10^{4} \end{bmatrix}$

10.12 $-\frac{1}{47}\begin{bmatrix} 9 & 6000 \\ 43 \times 10^{-3} & 13 \end{bmatrix}$ **10.13** $\frac{1}{7}\begin{bmatrix} 1 & 8 \\ 5 & -2 \end{bmatrix}$

10.14 $\begin{bmatrix} -10 & 5 \\ -7 & \frac{7}{2} \end{bmatrix}$ **10.15** $\frac{1}{7}\begin{bmatrix} -14 & 5 \\ 0 & -10 \end{bmatrix}$

10.16 $\begin{bmatrix} -\frac{29}{36} & \frac{23}{18} \\ -\frac{7}{20} & -\frac{1}{10} \end{bmatrix}$ **10.17** $\begin{bmatrix} 3 & 2 \\ 0.4 & -0.2 \end{bmatrix}$

10.18 $\frac{1}{13}\begin{bmatrix} 3 - j2 & 15 - j10 \\ 6 + j9 & 82 + j19 \end{bmatrix}$ **10.19** $\begin{bmatrix} 46 & -43 \\ -29 & 38 \end{bmatrix} \mho$

10.20 $\frac{1}{189}\begin{bmatrix} 23 & -8 \\ 11 & 8.5 \end{bmatrix} \Omega$ **10.21** $\begin{bmatrix} \frac{50}{11} & -\frac{10}{11} \\ -6 & 14 \end{bmatrix} \mho$

10.22 $\frac{1}{41}\begin{bmatrix} 316 & -90 \\ -90 & 186 \end{bmatrix} \mho$

10.23 $\begin{bmatrix} z_{11} & z_{12} \\ z_{21} & z_{22} \end{bmatrix}$ **10.24** $\begin{bmatrix} y_{11} & y_{12} \\ y_{21} & y_{22} \end{bmatrix}$

10.25

10.26 $\frac{1}{1531}\begin{bmatrix} 4974 & 504 \\ 504 & 4886 \end{bmatrix} \Omega$ **10.27** $\begin{bmatrix} 0.1237 & -0.0677 \\ -0.0677 & 0.1312 \end{bmatrix} \mho$

ANSWERS TO PROBLEMS

10.28 $\begin{bmatrix} 0.3366 & -0.0762 \\ -0.0762 & 0.3360 \end{bmatrix}$ ℧

10.29 (circuit: $\frac{3}{2}$ F capacitor in parallel with two 2 Ω resistors in series, with $\frac{1}{2}$ Ω shunt in middle, and 4 F shunt at output)

10.30 (circuit: 1 F and 1 F capacitors, 7 Ω, 7 Ω resistors, $\frac{1}{2}$ Ω shunt, 2 F shunt, 4 Ω shunt)

10.31 $\frac{Z_L}{n_1^2 n_2^2}$ **10.32** $\begin{bmatrix} \frac{8}{207} & -\frac{5}{69} \\ -\frac{5}{69} & \frac{6}{23} \end{bmatrix}$ ℧   **10.33** $\frac{1}{11}\begin{bmatrix} 270 & 60 \\ 60 & 28 \end{bmatrix}$ Ω

10.34 $\frac{1}{17}\begin{bmatrix} 30 & 60 \\ 60 & 120 \end{bmatrix}$ Ω   **10.35** $\frac{R}{(1\mp n)^2}$ **10.36** $\frac{2}{15}$ Ω

10.37 $\frac{2}{5}$ Ω   **10.38** $\frac{1}{11}\begin{bmatrix} 30 & 60 \\ 60 & 120 \end{bmatrix}$ Ω

10.39 $2.393 + j2.184$ V, $0.798 + j0.728$ V, $0.310 + j0.373$ A

10.40 $\sqrt{\frac{(R_1+R_2)R_3}{R_1 R_2}}, \frac{1}{2}\sqrt{\frac{R_2 R_3}{R_1(R_1+R_2)}}$

10.41 -0.5Ω **10.42** $-j0.4494$ **10.43** 13 Ω

10.44 $-j\frac{20}{221}$ **10.45** $\begin{bmatrix} \frac{45}{433} & \frac{32}{433} \\ -\frac{4}{433} & \frac{82}{2165} \end{bmatrix}$ Ω

10.46 $E_1 = \frac{113}{149}$ V, $E_2 = \frac{33}{745}$ V, $I_1 = \frac{47}{745}$ A, $I_2 = \frac{122}{745}$ A

10.47 $\frac{1}{9}\begin{bmatrix} 121 & -88 \\ -97 & 91 \end{bmatrix}$ ℧   **10.48** $\frac{1}{3}\begin{bmatrix} 8 & 11 \\ -4 & 14 \end{bmatrix}$ ℧

10.49 $\begin{bmatrix} y_{11} & y_{12} \\ y_{21} + 2y_{11} & y_{22} + 2y_{21} \end{bmatrix}$

10.50 $[z'] = \begin{bmatrix} \frac{1}{n^2} z_{11} & \frac{1}{n} z_{12} \\ \frac{1}{n} z_{21} & z_{22} \end{bmatrix}$ $[y'] = \begin{bmatrix} n^2 y_{11} & n y_{12} \\ n y_{21} & y_{22} \end{bmatrix}$

$[h'] = \begin{bmatrix} \frac{1}{n^2} h_{11} & \frac{1}{n} h_{12} \\ \frac{1}{n} h_{21} & h_{22} \end{bmatrix}$ $[g'] = \begin{bmatrix} n^2 g_{11} & n g_{12} \\ n g_{21} & g_{22} \end{bmatrix}$

$[F'] = \begin{bmatrix} \frac{1}{n}A & \frac{1}{n}B \\ nC & nD \end{bmatrix}$ $[\mathcal{F}'] = \begin{bmatrix} n\mathcal{A} & \frac{1}{n}\mathcal{B} \\ n\mathcal{C} & \frac{1}{n}\mathcal{D} \end{bmatrix}$

10.51 $\frac{1}{26}\begin{bmatrix} 12 & -5 \\ -5 & 11 \end{bmatrix}$ ℧

10.52 $\begin{bmatrix} 1.890 - j1.803 & 0.707 + j0.074 \\ 0.707 + j0.074 & 1.266 - j0.972 \end{bmatrix}$ Ω

10.53 $\begin{bmatrix} \frac{668}{183} - j\frac{72}{61} & \frac{487}{915} - j\frac{194}{915} \\ -\frac{487}{915} + j\frac{194}{915} & \frac{373}{3050} + j\frac{146}{4575} \end{bmatrix}$ **10.54** $\begin{bmatrix} 8 & -3 \\ -3 & \frac{153}{16} \end{bmatrix}$ ℧

10.55 $\frac{1}{145}\begin{bmatrix} 74 + j98 & 4 - j77 \\ 4 - j77 & 59 + j205.5 \end{bmatrix}$ ℧

10.56 $\begin{bmatrix} \frac{38}{259} & -\frac{4}{259} \\ -\frac{4}{259} & \frac{69}{518} \end{bmatrix}$ ℧   **10.57** $\begin{bmatrix} \frac{70}{159} & -\frac{2}{53} \\ \frac{308}{2173} & \frac{4680}{2173} \end{bmatrix}$ Ω

10.58 $\frac{2}{675}\begin{bmatrix} 353 & 61 \\ -61 & 43 \end{bmatrix}$ **10.59** $\frac{1}{15}\begin{bmatrix} -5 & -10 \\ 1 & 2 \end{bmatrix}$

10.60 $\frac{4}{257}$ **10.61** $z_a = \frac{5}{7}, z_b = \frac{3}{4}$ Ω

10.62 $\frac{1}{91}\begin{bmatrix} 365 & 27 \\ 27 & 365 \end{bmatrix}$ Ω

10.63 $\begin{bmatrix} 0.12 + j0.16 & -0.02 + j0.04 \\ -0.02 + j0.04 & 0.12 + j0.16 \end{bmatrix}$ ℧

Chapter 11

11.1 $7.5\epsilon^{-2.5t}u(t)$ V   **11.2** $5.547\sin(5t - 146.31°)$ V
11.3 $1.205\cos(2t - 146.62°)$ A   **11.4** 7.609 H
11.5 $\frac{150}{73} + j\frac{476}{73}$ Ω   **11.6** $-4.744 + j22.069$ V
11.7 $E_{oc} = 21.537 + j5.333$ V, $Z_{eq} = 1.546 + j1.449$ Ω
11.8 $j\frac{212}{45}$ Ω   **11.9** $j\frac{5}{191}$ A
11.10 $0.4957 - j2.1590$ A, $20.5707 - j57.0974$ A,
$-1.2392 + j2.8974$ A, $1.7349 - j5.0564$ A
11.11 $-0.0049 - j1.997$ A, $-1.057 - j1.945$ A
11.12 $j\frac{128}{17}$ V, $-j\frac{24}{17}$ V, $-j\frac{104}{17}$ V
11.13 $-0.3633 - j1.2645$ V, $5.4301 + j7.1847$ V, $9.7603 + j4.6847$ V
11.14 $0.3252 - j0.5323$ A, $-0.0174 - j0.2954$ A, $0.1978 - j0.3417$ A
11.15 $0.6386 + j0.0958$ A, $0.5318 + j1.0526$ A, $0.4402 + j0.7043$ A
11.16 $0.2\epsilon^{-0.5t}$ A   **11.17** (a) 25 J, (b) 725 J   **11.18** 0.8216
11.19 $-0.1915 + j0.2101$ A   **11.20** 1.4213 rad/sec, 0.0667
11.21 $\begin{bmatrix} j4 & j2.5 \\ -j4.6 & j6 \end{bmatrix}$ Ω   **11.22** $\begin{bmatrix} 5 - j40 & j\frac{16}{3} \\ -j18 & j\frac{14}{3} \end{bmatrix}$ Ω

11.23 $j\frac{1}{523}\begin{bmatrix} 1448 & 28 \\ 28 & 1113 \end{bmatrix}$ Ω

11.24 $\begin{bmatrix} 1.074 + j0.635 & -0.0333 - j0.464 \\ -0.0333 - j0.464 & 1.040 + j0.4244 \end{bmatrix}$ ℧

11.25 $\frac{44}{7}$ H **11.26** 3 H **11.27** $j\begin{bmatrix} 2 & -\frac{11}{3} \\ -\frac{11}{3} & \frac{37}{6} \end{bmatrix}$ ℧

11.28 (a) $0.0602\underline{/-72.48°}$, (b) $0.3846\underline{/-67.38°}$, (c) $1.814\underline{/-17.33°}$, (d) $1.976\underline{/-3.64°}$, (e) 2

11.29 (a) 0, $63.25\cos(2t + 90°)$ V;
(b) $5.657\cos(2t + 26.57°)$ A, $28.28\cos(2t + 26.57°)$ V;
(c) $6.324\cos 2t$ A, 0

11.30 $\frac{1}{123}(125 + j100)$ A, $\frac{1}{41}(800 - j1000)$ V

11.31 $13.161 + j3.527$ V

Chapter 12

12.1 $\begin{bmatrix} 7.5 & -3.5 & -4 \\ -2.5 & 8.5 & -6 \\ -5 & -5 & 10 \end{bmatrix}$ ℧   **12.2** $\frac{1}{9}\begin{bmatrix} 14 & -12 & -2 \\ -6 & 45 & -39 \\ -8 & -33 & 41 \end{bmatrix}$ ℧

12.3 $\frac{1}{13}\begin{bmatrix} 100 & -31 & -69 \\ -46 & 66 & -20 \\ -54 & -35 & 89 \end{bmatrix}$ ℧

12.4 $\begin{bmatrix} 0.75 + j1.25 & -0.25 + j0.75 & -0.5 - j2 \\ -0.25 + j0.75 & 0.45 + j0.25 & -0.2 - j \\ -0.5 - j2 & -0.2 - j & 0.7 + j3 \end{bmatrix}$ ℧

12.5 $\frac{1}{4}\begin{bmatrix} 11 & -5 & -5 & -1 \\ -5 & 11 & -1 & -5 \\ -5 & -1 & 11 & -5 \\ -1 & -5 & -5 & 11 \end{bmatrix}$ ℧   **12.6** $\begin{bmatrix} \frac{1}{3} & -\frac{5}{6} & \frac{1}{2} \\ 0 & \frac{1}{2} & -\frac{1}{2} \\ -\frac{1}{3} & \frac{1}{3} & 0 \end{bmatrix}$ ℧

12.7 $\frac{1}{1345}\begin{bmatrix} 67 - j114 & 33 - j16 & -100 + j130 \\ 33 - j16 & 217 - j309 & -250 + j325 \\ -100 + j130 & -250 + j325 & 350 - j455 \end{bmatrix}$ ℧

12.8 $\begin{bmatrix} 0.05346 - j0.07859 & -0.03016 + j0.02630 \\ -0.03106 + j0.02630 & 0.03016 - j0.05887 \\ -0.02239 + j0.05229 & 0.00091 + j0.03257 \end{bmatrix}$

$\begin{bmatrix} -0.02239 + j0.05229 \\ 0.00091 + j0.03257 \\ 0.02149 - j0.08486 \end{bmatrix}$ ℧

12.9 $\begin{bmatrix} 0.7 & -0.075 & -0.5 & -0.125 \\ -1.2 & 1.2 & 1 & -1 \\ -0.5 & 0.375 & 1 & -0.875 \\ 1 & -1.5 & -1.5 & 2 \end{bmatrix}$ ℧

12.10 $\begin{bmatrix} 16 & -8 \\ -10 & -5 \end{bmatrix}$ Ω   12.11 $\begin{bmatrix} -3 & 2 \\ -8 & 12 \end{bmatrix}$ Ω

12.12 $\begin{bmatrix} 50 & -35 \\ -22 & 52 \end{bmatrix}$ ℧   12.13 $\begin{bmatrix} -4 & 1 \\ -3 & 15 \end{bmatrix}$ ℧

12.14 $E_{oc} = \frac{90}{7}$ V, $R_{eq} = \frac{1}{7}$ Ω   12.15 $\begin{bmatrix} -3 & 1 \\ 4 & 1 \end{bmatrix}$ ℧   12.16 0

12.17 $\begin{bmatrix} 2 & -1 \\ 3.5 & -4.5 \end{bmatrix}$ ℧   12.18 3.5 ℧   12.19 $\begin{bmatrix} 5 & -3 \\ -5 & 5 \end{bmatrix}$ ℧

12.20 $I = 10$ A, $I_1 = 0$, $I_2 = 22.5$ A, $I_3 = 5$ A, $I_4 = -27.5$ A

12.21 -4 ℧   12.22 $\begin{bmatrix} 4 & -6 \\ -4 & 1 \end{bmatrix}$ ℧

12.23 $\begin{bmatrix} 0 & 6 & -6 \\ 6 & -4 & -2 \\ -6 & -2 & 8 \end{bmatrix}$ ℧   12.24 $\begin{bmatrix} 3 & -2 & 0 & 0 & -1 \\ -1 & 9 & -2 & -2 & -4 \\ 0 & -1 & 6 & -3 & -2 \\ 0 & -3 & -1 & 4 & 0 \\ -2 & -3 & -3 & 1 & 7 \end{bmatrix}$ ℧

12.25 4 A   12.26 5 ℧   12.27 $\begin{bmatrix} -\frac{5}{3} & -10 \\ \frac{44}{3} & 11 \end{bmatrix}$ ℧

12.28 $\frac{1}{7}\begin{bmatrix} 31 & -2 & -29 \\ -2 & 20 & -18 \\ -29 & -18 & 47 \end{bmatrix}$ ℧   12.29 $\frac{1}{3}\begin{bmatrix} 7 & -2 & -5 \\ -2 & 7 & -5 \\ -5 & -5 & 10 \end{bmatrix}$ ℧

12.30 $\frac{1}{38}\begin{bmatrix} 64 & -58 & -6 \\ -22 & 185 & -163 \\ -42 & -127 & 169 \end{bmatrix}$ ℧

12.31 $\begin{bmatrix} \frac{5}{6} & -\frac{3}{8} & -\frac{11}{24} \\ -\frac{3}{8} & \frac{51}{32} & -\frac{39}{32} \\ -\frac{11}{24} & -\frac{39}{32} & \frac{161}{96} \end{bmatrix}$ ℧

12.32 $\frac{1}{1359}\begin{bmatrix} 6316 & -4898 & -1418 \\ -4730 & 5602 & -872 \\ -1586 & -704 & 2290 \end{bmatrix}$ ℧

12.33 $\begin{bmatrix} 2.75 & -0.5 \\ -3.25 & 0.5 \end{bmatrix}$ ℧   12.34 $\frac{1}{45}\begin{bmatrix} 14 & -4 \\ 3 & 12 \end{bmatrix}$ Ω

Index

ABCD matrix, 513
Ac analysis, 369
Ac power, 453-464
Additivity, 131
Adjoint of a matrix, 682
Admittance, 342, 383
  See also Impedance
Admittance locus, 417
Admittance matrix, 506
Alternator, 10
Ampere, 2, 4
Amplification factors, 11
Angular frequency, 371
Apparent power, 469
Available power, 185, 485
Average power, 456

Balanced ladder, 539
Balanced two-port, 539
Bandwidth, 429
Battery, 10
Biasing, 10
Bilinear transformation, 414
Branch, 23

Capacitance, 208
Capacitor, 207-208
  energy in a, 213
  parallel-plate, 208
  with initial energy, 288

Capacitors in parallel, 216
Capacitors in series 215
Cascade connection of two-ports, 564
Chain matrix, 513
Characteristic equation, 264
Charge, 2, 4
Chord, 30
Circle diagram, 414
Circulating current, 81
Closed surface, 23, 25
Coefficient of coupling, 599
Cofactor, 679
Coils, coupled, 587
Column matrix, 677
Complementary function, 264
Complex amplitude, 340
Complex number, 328
Complex power, 471
Conductance, 20, 384
  mutual, 12
Conformability, 680, 688
Conjugate, 334
Conservation of power, 178, 475
Controlled sources, 11, 50-53
Converter
  current-to-voltage, 12, 52
  voltage-to-current, 12, 53
Cotree, 30, 75-76
Coulomb, 2, 9

Coupled coils, 587
   equivalent circuits for, 607
Coupling, coefficient of, 599
Coupling, magnetic, 587
Cps, 371n
Cramer's rule, 85
Critically damped case, 302, 305
Current, 2, 4
   symbol for, 4
Current amplification factor, 11
Current amplifier, 11
Current gain, 11
Current source, 9
Current transformer, 613
Current-division rule, 36, 39, 40, 218, 226, 391
Current-to-voltage converter, 12, 52
Cutset, 23
Cutset analysis, 76

Damping constant, 265
Datum node, 83
Dead network, 150
Default units, 13
Delta function, 240n
Delta-wye transformation, 46-48, 56, 395
Dependent sources, *see* Controlled sources
Determinant, 678
Diagonal matrix, 681
Differentiability, 249
Differentiator, 296
Dot convention, 592
Double-subscript notation, 7
Doublet, 248
Driving-point impedance, 499
Drop, voltage, 6

Duality, 54, 111, 249

E-shift, 169
Effect value, 376, 382
Effective values of sinusoids, 379
Electron, 4
Elastance, 208
Energy, 2, 9
Energy in a capacitor, 213
Energy in an inductor, 223
Energy in coupled coils, 597
Equality of M_{12} and M_{21}, 592
Equivalence of two-terminal network, 139
Euler's formulas, 301, 330-331
Exponential excitation, 327, 336, 349, 361
Exponential form, 330
Exponentially decaying sinusoid, 350
Extended node, 94

F matrix, 513
Farad, 208
Faraday's law, 587
Flux, leakage, 588n
Flux, magnetic, 2
Flux linkage, 221
Follower, voltage, 51
Forced response, 263, 267, 299
Four-terminal network as a two-port, 658
Frequency, 370

g matrix, 509
Gains, 11
Gate function, 236
Gaussian elimination algorithm, 86

INDEX

Generalized node, 94
Ground, 3
Grounded node, 83
Grounded two-port, 500

h matrix, 509
Half-power points, 429
Henry, 220
Hertz, 371
Homogeneous equation, 264
Homogeneous solution, 264
Homogeneity, 130
Hybrid matrices, 509
Hyperbolic function, 301

I-shift, 170
IAM
    definition, 623
    of network with no internal node, 633
    zero-sum property, 629
Ideal op amp, 13-14, 50
Ideal transformer, 543, 614
Impedance, 342, 383
    driving-point, 499
    transfer, 531
    See also Admittance
Impedance locus, 417
Impedance matching, 485
Impedance matrix, 503
Impedances in parallel, 389
Impedances in series, 389
Impulse function, 240
Impulse response, 268, 282
    of parallel RC circuit, 293, 294
    of parallel RL circuit, 293, 294
    of series RC circuit, 282, 294
    of series RL circuit, 267, 294
    of series RLC circuit, 305

In-phase power, 456
Indefinite admittance matrix, *see* IAM
Independent current source, 10
Independent sources, 9
Independent voltage source, 10
Inductance, 220
    mutual, 587
    self, 587
Inductor, 220
    energy in an, 223
    with initial energy, 288
Inductors in parallel, 226
Inductors in series, 225
Initial energy, 288
Integrability, 249
Integrated circuits, 10
Integrator, 296, 297
    inverting, 296
    lossy, 297
Integrity of a port, 497
Inverse of a matrix, 682
Inverting differentiator, 296
Inverting integrator, 296
Inverting terminal, 14
Inverting voltage amplifier, 51

Joules, 2, 9

KCL, 24, 28, 35
KVL, 26, 28, 32
Kirchhoff's laws, 22
    See also KCL and KVL

Ladder analysis, 40
Ladders, 539
Leakage flux, 588n
Linear element, 13, 129
Linearity, 132, 348

Link, 30
Links, number of, 31
Locus of admittance, 417
Locus of impedance, 417
Locus of phasor, 414
Loop, 23
Loop analysis, 76-78
Loop current, 77
Lossy integrator, 297

Magnetic coupling, 587
Magnetic flux, 2
Mapping, 415
Matching, impedance, 485
Matrices of two-ports, 501-524
Matrix multiplication, 677, 680
Matrix theorems, 683
Maximum power transfer, 184, 482
Memory elements, 13, 207
Memoryless circuit, 19
Memoryless elements, 13
Mesh, 23
Mesh analysis, 99-110, 112, 140
Mesh current, 82
Mho, 20
Minor, 678
MKS system, 4
Models for two-ports, 525
Multiplication of matrices, 677, 680
Multiterminal networks, 497, 623
   combination of terminals, 641
   grounding of one terminal, 637
   interconnection of, 646
   internalizing of terminals, 648
   parallel of, 643
   reordering of terminals, 640
Mutual conductance, 12

Mutual inductance, 114, 587
Mutual inductance, dot convention, 592
Mutual resistance, 12

Natural response, 263
Negative resistance, 22
Network function, 357
Network topology, 29
Node, 23, 24
Node analysis, 83-99, 112
Node splitting, 170
Node-pair voltage, 79
Noninverting terminal, 14
Noninverting voltage amplifier, 50
Nonoscillatory case, 301, 305
Nonplanar network, 83
Nonsingular matrix, 682
Norton's branch, 153
Norton's equivalent circuit, 152, 159, 404, 483
Norton's theorem, 157

Ohm, 20
Ohm's law, 19, 343
Ohm's law convention, 20
One-port, 498
Op amp, ideal, 13-14, 50
Op amp circuits, 50, 295
Open circuit, 3, 10, 54
Open-circuit voltage, 148
Operational amplifier, see Op amp
Oscillatory case, 302, 305
Overdamped case, 301, 305

Padding a two-port, 536
Parabola function, 248

INDEX

Parallel, resistances in, 35-38, 55, 56
Parallel connection, 54
   notation for, 37
   of capacitors, 216
   of impedances, 389
   of inductors, 226
   of resistors, 35
Parallel resonance, 431
Parallel-parallel combination of two-ports, 552
Parallel-plate capacitor, 208
Parallel-series combination of two-ports, 555
Particular integral, 263, 276
Particular solution, 263, 276
Partitioning of matrices, 686
Passive elements, 476n
Period, 370
Phase, 371
Phase-shifting network, 424
Phasor, 381, 383, 385
Phasor, reference, 388
Phasor diagram, 385
Phasor locus, 414
Pi two-port, 535
Pi-tee transformation, 49, 56, 395, 535
Pivotal condensation, 650
Planar network, 83, 99
Polar form, 332
Port, 497
Potential difference, see Voltage
Potentiometer rule, 33n
Power, 2, 7, 9, 22
   ac, 453-464
   apparent, 469
   available, 185, 485
   average, 456
   complex, 471
   conservation of, 179, 475
   in-phase, 456
   quadrature, see reactive
   reactive, 458, 460, 462, 464
   real, 456
   triangle, 470
Power factor, 469
Power factor correction, 479
Power triangle, 470
Proportionality, 130
Pseudo-tree, 79n

Q factor, 429
Quadrature power, see Reactive power

Radian frequency, 371
Ramp function, 236
Ramp response
   of series RC circuit, 287
   of series RL circuit, 278
Reactance, 384
Reactive factor, 469
Reactive power, 458, 460, 462, 464
Real power, 456
Reciprocal network, 591
Reciprocal two-port, 533
Reciprocity theorem, 182
Rectangular form, 328
Reference node, 83
Reference, phasor, 388
Relationship of two-port matrices, 516, 522
Resistance, 19, 22, 384
   linear, 20
   mutual, 12
   negative, 22

Resistor, 19
Resonance, 426-432
Rise, voltage, 6
Rms value, 376
Row matrix, 677

Scaling property, 130
Self inductance, 587
Series, resistances in, 32, 55, 56
Series RC circuit, 279
Series RL circuit, 261
Series RLC circuit, 299
Series connection, 54
   of capacitors, 215
   of impedances, 389
   of inductors, 225
   of resistances, 32
Series resonance, 427
Series-parallel combination of
   two-ports, 554
Series-series combination of
   two-ports, 549
Shifting, source, 169-175
Short circuit, 3, 10, 54
Short-circuit current, 147
Siemen, 20n
Signal generator, 10
Simultaneous equations, 685
Singular matrix, 682
Singularity functions, 232-249
Sinusoid, 370
   effective value of, 379
   exponentially decaying, 350
Sinusoidal excitation, 327, 349,
   353, 361
Skew-symmetric matrix, 681
Solution of simultaneous
   equations, 685
Source

controlled, 11, 50-53
current, 9
dependent, 11, 50-53
idle, 10
inactive, 10
independent current, 10
independent voltage, 10
independent, 9
shifting, 169-175
transformation, 153, 162-168
voltage, 9
Source-free response, 268, 284
   of series RC circuit, 284, 294
   of series RL circuit, 269, 294
Source-free response, 268, 284
Special two-ports, 533
Standard Ohm's law convention,
   20
Steady-state analysis, 327, 340,
   353
Step function, 232-236
Step response, 265, 280
   of parallel RC circuit, 291, 294
   of parallel RL circuit, 293, 294
   of parallel RLC circuit, 307
   of series RC circuit, 280, 294
   of series RL circuit, 261, 294
   of series RLC circuit, 299
Submatrices, 687
Supermesh, 106
Supernode, 94
Superposition, 133, 402
Susceptance, 384
Switch, 3
Symmetric lattice, 540
Symmetric matrix, 681
Symmetric two-port, 534

Tee two-port, 535

INDEX

Tee-pi transformation, 49, 56, 395, 535
Tellegen's corollary, 179-182
Tellegen's theorem, 176-178
Terminal pair, *see* Port
Terminal, 3
Terminated two-ports, 529
Thévenin's branch, 153
Thévenin's equivalent circuit, 152, 159, 404, 483
Thévenin's theorem, 157
Theorem,
    maximum power, 184, 482
    superposition, 133, 402
    Norton's, 157
    reciprocity, 182
    Tellegen's, 176-178
    Thévenin's, 157
Theorems on matrices, 683
Three-terminal two-port, 500
Time constant, 265, 281
Topology, 29
Transconductance, 12
Transducer, *see* Converter
Transformer, current, 613
Transformer, ideal, 614
Transformer, voltage, 612
Transient analysis, 261
Transmission matrix, 513
Transpose of a matrix, 682
Transresistance, *see* Mutual resistance
Tree branch, *see* Twig
Tree, 29, 75-76
Triplet, 249
Twig, 29
Twigs, number of, 30
Two-port, 497
    balanced, 539
    convention, 499
    formed from four-terminal network, 658
    grounded, 500
    matrices, 501, 516, 522
    notation, 499
    padding of, 536
    parameters, 501
    pi, 535
    reciprocal, 533
    symmetric, 534
    tee, 535
    three-terminal, 500
Two-ports,
    cascade, 564
    circuit models for, 525
    parallel-parallel, 552
    parallel-series, 555
    series-parallel, 554
    series-series, 549
    special, 533
    terminated, 529
Two-terminal network, 139

Unbalanced ladder, 539
Unbalancing a lattice, 541, 563
Underdamped case, 301, 305
Unit impulse function, 240
Unit ramp function, 236
Unit step function, 232
Units of controlled sources, 13
Units, default, 13

VAR, 458, 460
Vector, 383
Voltage, 2, 6
    node-pair, 79
    symbols for, 6
Voltage amplification factor, 11

Voltage amplifier, 11, 50, 51
Voltage double-subscript
 notation, 7
Voltage drop, 6
Voltage follower, 51
Voltage gain, 11
Voltage rise, 6
Voltage source, 9
Voltage transformer, 612
Voltage-division rule, 32, 216,
 226, 391
Voltage-to-current converter, 12,
 53

Watt, 2, 7, 9
Wattage, 456
Wattless power, *see* Reactive
 power
Wattmeter, 467
Weber, 2, 6
Window function, 236
Wye-delta transformation, 46-48,
 56, 395

y matrix, 506

z matrix, 503